Biorhythms and Human Reproduction

SEMINARS IN HUMAN REPRODUCTION

Series Editors: Raymond L. Vande Wiele (M.D.) and
Ralph M. Richart (M.D.)
International Institute for the Study of
Human Reproduction
Columbia University, New York

BIORHYTHMS AND HUMAN REPRODUCTION

Edited by Michel Ferin, Franz Halberg,
Ralph M. Richart, and Raymond L. Vande Wiele

Biorhythms and Human Reproduction

A Conference Sponsored by the
International Institute for the
Study of Human Reproduction

EDITED BY

MICHEL FERIN, M.D.
FRANZ HALBERG, M.D.
RALPH M. RICHART, M.D.
RAYMOND L. VANDE WIELE, M.D.

ASSISTANT EDITORS
JUDITH A. ANDERSON
KATHERINE F. DARABI

A WILEY BIOMEDICAL-HEALTH PUBLICATION
JOHN WILEY & SONS, New York • London • Sydney • Toronto

Library of Congress Cataloging in Publication Data:

Main entry under title:
Biorhythms and human reproduction.

  "A Wiley Biomedical-Health Publication."
  Includes bibliographical references.
  1. Human reproduction—Congresses.   2. Biology—
Periodicity—Congresses.   I. Ferin, Michel, ed.
Reproduction. [DNLM: 1. Biological clocks—Congresses.
2. Reproduction—Congresses. QT167 B616 1973]

QP251.B568   612.6   73-14700
ISBN 0-471-25761-3

Printed in the United States of America

10 9 8 7 6 5 4 3 2 1

Conference Contributors and Participants

Paul K. Alkon, Ph.D., Professor, Department of English, University of Minnesota, Minneapolis, Minnesota

John A. Anderson, M.D., Professor, Department of Pediatrics, University of Minnesota, Minneapolis, Minnesota

Richard V. Andrews, Ph.D., Professor of Physiology, School of Medicine, Creighton University, Omaha, Nebraska

Edward Batschelet, Ph.D., Biomathematics Division, Mathematics Institute, University of Zürich, Zürich, Switzerland

R. J. Bogumil, International Institute for the Study of Human Reproduction, College of Physicians and Surgeons, Columbia University, New York, New York

Jorge A. Colombo, M.D., Fellow, the Ford Foundation, Department of Anatomy and Brain Research Institute, University of California School of Medicine, Los Angeles, California

Philip A. Corfman, M.D., Director, Center for Population Research, National Institute of Child Health and Human Development, National Institutes of Health, Bethesda, Maryland

George C. Curtis, M.D., Department of Psychiatry, University of Michigan Medical Center, Ann Arbor, Michigan

Thomas D. Downs, Ph.D., School of Public Health, The University of Texas at Houston, Houston, Texas

Inge Dyrenfurth, Ph.D., Research Associate, Department of Obstetrics and Gynecology and the International Institute for the Study of Human Reproduction, College of Physicians and Surgeons, Columbia University, New York, New York

Peter Engel, Prof. Dr. med., Institut für Arbeitsphysiologie und Rehabilitationsforschung de Universitat Marburg/Lahn, West Germany

Michel Ferin, M.D., Assistant Professor in the Departments of Physiology and Obstetrics and Gynecology, Research Associate, International Institute for the Study of Human Reproduction, College of Physicians and Surgeons, Columbia University, New York, New York

John D. Fernstrom, Ph.D., Assistant Professor of Nutritional Biochemistry and Metabolism, Laboratory of Neuroendocrine Regulation, Department of Nutrition and Food Science, Massachusetts Institute of Technology, Cambridge, Massachusetts

Shanna H. Freedman, Ph.D., Oral Contraceptive Drug Study, Kaiser-Permanente Medical Center, Walnut Creek, California

Arthur Giese, Ph.D., Professor Emeritus of Biology, Stanford University, Palo Alto, California

Erna E. Halberg, Research Associate, Chronobiology Laboratories, Department of Pathology, University of Minnesota, Minneapolis, Minnesota

Franz Halberg, M.D., Professor, Chronobiology Laboratories, Department of Pathology, University of Minnesota, Minneapolis, Minnesota

Allen R. Hanson, Ph.D., Assistant Professor, Department of Computer and Information Sciences, University of Minnesota, Minneapolis, Minnesota

R. A. Harkness, M.B., Ph.D., F.R.C.P.E., Senior Lecturer, Department of Paediatric Biochemistry, Royal Hospital for Sick Children and Department of Clinical Chemistry, University of Edinburgh, Edinburgh, Scotland, Consultant in Paediatric Biochemistry, South East Scotland

Erhard Haus, M.D., Ph.D., St. Paul-Ramsey Hospital, St. Paul, Minnesota

Robert I. Henkin, M.D., Ph.D., Chief, Section on Neuroendocrinology, National Heart and Lung Institute, National Institutes of Health, Bethesda, Maryland

Russell K. Hobbie, Ph.D., Professor, School of Physics and Astronomy, University of Minnesota, Minneapolis, Minnesota

Robert B. Jaffe, M.D., Professor and Chairman, Department of Obstetrics and Gynecology, Women's Hospital, University of Michigan Medical Center, Ann Arbor, Michigan

Irwin H. Kaiser, M.D., Ph.D., Professor, Department of Obstetrics and Gynecology, Albert Einstein College of Medicine, Bronx, New York

Edward L. Klaiber, M.D., Worcester Foundation for Experimental Biology, Shrewsbury, Massachusetts

Dorothy T. Krieger, M.D., Professor, Department of Medicine, Division of Endocrinology, Mt. Sinai School of Medicine of the City University of New York, New York, New York

Maciej Krzanowski, Dr. n. med., Institute of Zootechny, Experimental Section at Grodziec Slaski Laboratory of Applied Biochemistry at Gumna/Cieszyn, Poland

Jürgen F. W. Kühl, Chronobiology Laboratories, Department of Pathology, University of Minnesota, Minneapolis, Minnesota

Carolyn Leach, Ph.D., Head, Endocrine Laboratory Section, Johnson Space Center, National Aeronautics and Space Administration, Houston, Texas

Jay A. Leavitt, Ph.D., Associate Professor, Department of Computer and Information Sciences, University of Minnesota, Minneapolis, Minnesota

Jung-Keun Lee, Chronobiology Laboratories, Department of Pathology, University of Minnesota, Minneapolis, Minnesota

Rufus E. Lee, M.D., McGovern Allergy Clinic, Houston, Texas

Howard Levine, M.D., Director of Medical Education, Chief of Medicine, New Britain General Hospital, New Britain, Connecticut, Associate Professor of Medicine, University of Connecticut Health Center, Farmington, Connecticut

Charles W. Lloyd, M.D., Professor, Department of Obstetrics and Gynecology, M. S. Hershey Medical Center, Hershey, Pennsylvania

Ian McIndoe, Department of Psychiatry, University of Minnesota, Minneapolis, Minnesota

Michael Menaker, Ph.D., Professor, Department of Zoology, The University of Texas at Austin, Austin, Texas

Norberto Montalbetti, M.D., Director, Centro Malattie Endocrine E. Metaboliche, Ospedale di Magenta, Milan, Italy

Howard R. Nankin, M.D., Assistant Professor, Department of Medicine, Montefiore Hospital, Pittsburgh, Pennsylvania

John F. O'Connor, M.D., Associate Clinical Professor of Psychiatry, Department of Psychiatry and the International Institute for the Study of Human Reproduction, College of Physicians and Surgeons, Columbia University, New York, New York

John E. Pauly, Ph.D., Professor and Chairman, Department of Anatomy, Medical School, University of Arkansas, Little Rock, Arkansas

Olga Petre-Quadens, M.D., Ph.D., Born-Bunge Research Foundation, Department of Developmental Neurology, Antwerp, Belgium

John C. Porter, Ph.D., Professor of Physiology, Department of Physiology, The University of Texas Southwestern Medical School, Dallas, Texas

Harriet B. Presser, Ph.D., Associate Professor of Public Health (Sociomedical Sciences), International Institute for the Study of Human Reproduction, College of Physicians and Surgeons, Columbia University, New York, New York

F. S. Preston, M.D., Principal Medical Officer (Air), British Overseas Air Corporation, London, England

William M. Rand, Ph.D., Assistant Professor of Biostatistics, Department of Nutrition and Food Science, Massachusetts Institute of Technology, Cambridge, Massachusetts

Alain Reinberg, M.D., Ph.D., Equipe de Recherche de Chronobiologie Humaine (CNRS N° 105) Foundation A. de Rothschild, Laboratoire de Physiologie, Paris, France

Ralph M. Richart, M.D., Director, Biomedical Division, International Institute for the Study of Human Reproduction, Professor of Pathology, College of Physicians and Surgeons, Columbia University, Director Pathology and Cytology Laboratories Obstetrical and Gynecological Division (The Sloane Hospital for Women), Presbyterian Hospital, New York, New York

John A. Rummel, Ph.D., Chief, Environmental Physiology Branch, National Aeronautics and Space Administration, Manned Space Craft Center, Houston, Texas

Kenneth Savard, Ph.D., National Institute of Child Health and Human Development, National Institutes of Health, Bethesda, Maryland

Lawrence E. Scheving, Ph.D., Professor of Anatomy, University of Arkansas, Little Rock, Arkansas

Otto Schmitt, Ph.D., Chief, Biophysics Group, Professor of Bioengineering, University of Minnesota, Minneapolis, Minnesota

Ronald Shiotsuka, Chronobiology Laboratories, Department of Pathology, University of Minnesota, Minneapolis, Minnesota

Hugh Simpson, M.D., Ph.D., University Department of Pathology, Royal Infirmary, Glasgow, Scotland

Michael H. Smolensky, Ph.D., Assistant Professor of Environmental Health, School of Public Health, The University of Texas at Houston, Houston, Texas

Robert B. Sothern, Chronobiology Laboratories, Department of Pathology, University of Minnesota, Minneapolis, Minnesota

Raymond L. Vande Wiele, M.D., Director, International Institute for the Study of Human Reproduction, Willard C. Rappleye Professor and Chairman of the Department of Obstetrics and Gynecology, College of Physicians and Surgeons, Columbia University, New York, New York

Lieve Verdonck, Ph.D., Laboratory for Steroid Hormones, Department of Endocrinology and Metabolism, Medical Clinic, Akedemisch Ziekenhuis, Ghent, Belgium

A. Vermeulen, M.D., Department of Endocrinology and Metabolism, Medical Clinic, Akademisch Ziekenhuis, Ghent, Belgium

Rudolf F. Vollman, M.D., Hiltbrunnerweg, Uerikon, Switzerland

Elliot D. Weitzman, M.D., Professor and Chairman, Department of Neurology, Albert Einstein College of Medicine, Montefiore Hospital and Medical Center, Bronx, New York

Charles M. Winget, Ph.D., Human Studies Branch, Biomedical Research Divisions, Ames Research Center, National Aeronautics and Space Administration, Moffett Field, California

Richard J. Wurtman, M.D., Professor of Endocrinology and Metabolism, Massachusetts Institute of Technology, Cambridge, Massachusetts

F. Eugene Yates, M.D., Department of Biomedical Engineering, University of Southern California, Los Angeles, California

Samuel S. C. Yen, M.D., Professor and Chairman, Department of Obstetrics and Gynecology, University of California at San Diego, La Jolla, California

Preface

The conference on Biorhythms in Human Reproduction is the first in a series organized by the International Institute for the Study of Human Reproduction. The conferences will examine carefully selected subjects in human reproduction, emphasizing those that in recent years have undergone rapid growth and development. They will deal especially with areas where there is controversy, and where there appears to be a timely need for a meeting of the minds that, we hope, will lead to a sorting out of the chaff from the wheat, and could generate a document summarizing the state of the art relevant to this subject. In selecting topics for these conferences we plan also to give preference to areas where the new information has come from disparate disciplines, between which channels of communication are few and dialogues infrequent.

The subject of the present conference admirably suited these goals. In the last few years the scientific study of biorhythms, more appropriately called chronobiology, has made enormous strides. Progress has resulted mainly from the development of new analytical methods, predominantly computer techniques, that have made it feasible to scan rapidly and precisely very large numbers of data, and to uncover frequencies or rhythms that, had the data been analyzed by conventional techniques, would have escaped detection because they were obscured by the existence of multiple rhythms or simply by "noise." Few workers in reproductive biology are aware of the many advances made in the study of biological rhythms and in fact, many look with some suspicion at a discipline that in earlier years had been disserved by the extravagant claims of a small fraction of its devotees. The chronobiologists, too often seem to work in a vacuum and not to take full advantage of recent advances in reproductive biology. There were additional reasons that made a meeting between chronobiologists and reproductive biologists most timely. The rapid advances in chronobiology had coincided with a somewhat similar revolution in reproductive biology. This revolution also had been generated by a methodological breakthrough—in this case the development of radioimmunoassays for gonadotropic and steroid hormones. At last it became possible to determine, simultaneously and at hourly or even shorter intervals, the plasma levels of FSH, LH, estradiol, estrone androstenedione, and testosterone and thus to "dissect" in detail the hormonal events of the menstrual cycle. As this was done, it became evident that in addition to the well known, long-term, cyclic hormonal changes (measured in days) there were rhythmic changes of higher frequency (measured in minutes) and, in fact, that all gonadal hormone secretions are pulsatile in nature. The complexities of these rapid changes in so many interrelated hormones raised many unfamiliar problems to the reproductive biologist for which to some degree the chronobiologist seemed to have answers. Consquently we felt that much might be gained from a discussion between interested chronobiologists and reproductive biologists.

The conference was held at the Sterling Forest Conference Center in New York State, from October 4 to October 8, 1972. More than 70 scientists participated. The meeting was exciting; a great deal of heat was generated and also some new light. There was plenty of dialogue but it occasionally seemed that both parties were not talking in the same language. Perhaps it was too much to hope that complete understanding could be achieved at a first meeting. Even so the meeting was successful in attaining its major goals. All participants left wanting to know more about the work of the "opposite" group and since the meeting, several have gone beyond the discussion and have started cooperative studies.

In concluding I would like to pay tribute to the many individuals, not least the contributors to the meeting, whose untiring efforts made the conference possible. I especially thank my colleagues on the organizing committee, who for many months were saddled with a heavy burden on top of their already heavy schedules. Miss J. Anderson did perhaps more than anybody else for the success of the conference. Only those who have organized such a conference are aware of the myriad chores that have to be accomplished to get a hundred investigators from all over the world to a conference center in time, lodge them, feed them, and keep them happy. That eventually nothing went wrong and everybody had a good time is testimony to her organizational talents.

The conference could not have been held without the generous support of the National Institutes of Health. Funds for travel expenses and accommodations were obtained through a contract with the National Institute for Child Health and Development. We are grateful to Dr. P. Corfman, Director of the Center for Population Research, with whom the contract was negotiated. As always, he went beyond mere arrangements for financial support by also being a wise advisor during the many months of preparation. Finally, I would like to acknowledge the expert help from the publisher who made it possible to publish this book within record time.

RALPH M. RICHART

New York, New York
August 1973

Contents

6 SECONDARY RHYTHMS RELATED TO THE MENSTRUAL CYCLE: BEHAVIORAL CHANGES
John F. O'Connor, *Moderator*

7 PRIMARY AND SECONDARY RHYTHMS OF THE HYPOTHALAMO-PITUITARY-ADRENAL AXIS
Elliot D. Weitzman, *Moderator*

8 RHYTHMS OF THE MALE HYPOTHALAMO-PITUITARY-TESTICULAR AXIS
A. Vermeulen, *Moderator*

9 SECONDARY RHYTHMS RELATED TO TESTICULAR FUNCTION
Robert B. Sothern, *Moderator*

10 SYNCHRONIZERS OF REPRODUCTIVE FUNCTION
Michael Menaker, *Moderator*

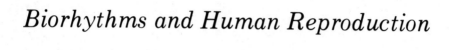

Biorhythms and Human Reproduction

1   Introductory Address

CHAPTER 1

Historical Development of the Concept of Time

PAUL K. ALKON

Department of English, University of Minnesota, Minneapolis, Minnesota

St. Augustine discussed the role of measurement in the study of time without any mention of the clepsydra, the hour glass, or the sun dial. A concern with Christian values underlying the subjective experience of time led him away from the instruments of natural science. Augustine's enduring contribution to knowledge of how time is involved in the central issues of religion, philosophy, and psychology was made in ways that sharply distinguish the inward turn of his *Confessions* from the consideration of external time in Aristotle's *Physics* (1). But after 15 centuries such distinctions have blurred. Human time is now divorced from its religious meaning and most widely studied in relationship to those physical periodicities measured by clocks. Yet even in this spiritually impoverished situation, recent advances are a continuation of the Augustinian approach to time.

The classical questions raised by Augustine are again being formulated, but in new ways that admit valid answers which supplement without contradicting his crucial assertion that there are not three times, past, present, and future, but instead only a present of things past, a present of things present, and a present of things future, existing in the mind and nowhere else (2). Eliot's *Four Quartets* starts with a poetic elaboration of this view (3). Einstein wrote that the separation between past, present, and future has only the meaning of an illusion, although a tenacious one (4). To appreciate how even the most disparate strands of our intellectual life converge in their return to the classical questions posed by Augustine, the best starting place is a figure uniquely conspicuous in modern fiction: the man who tries to lecture about time.

TIME IN MODERN FICTION

Vonnegut's Billy Pilgrim finds an all-night talk show where he can try to discuss his adventures among Tralfamadorian time-travelers. But an alarmed daughter manages to prevent further attempts that year at what Billy regards as the helpful task of comforting people with the truth about time. Finally, in 1976, just after a public lecture in Chicago on the true nature of time, Billy is shot (5).

Van Veen, the hero-narrator of Nabokov's *Ada*, fights insomnia by working on his book *The Texture of Time*. He also delivers three public lectures at a university. Although a little late in arriving for the first, dealing with the past, it goes without incident. At the second lecture, on the present, 5 seconds of silence are requested from the audience to illustrate a point which never does get made effectively because the 5 seconds are filled with snores from a white-bearded sleeper, presumably some symbolic elderly professor, who causes an inappropriate wave of laughter by his unconscious antics. At the third lecture, on the future, Van's tape recorder breaks down. Nervous about speaking before a learned audience, he had taped all three lectures for playing from a microphone concealed in his vest pocket while he simultaneously mouths the words in time with the tape. Suddenly faced with the need to reconstruct from crumpled notes a lecture on the future spoken into a recorder at some point in the past when no audience was present, Van prefers to simulate a heart attack, comically emblematic of his topic, and is "carried out into the night forever (insofar as lecturing was concerned)" (6).

The episode is described in the novel's fourth part, which provides Van's extensive meditations on time as a framework within which is interspersed an account of his reunion in Switzerland with Ada after a 17-year separation. He is 52. She is 50. The time is 1922. At first there is some awkwardness:

The utilitarian trivialities of their table talk—or, rather, of his gloomy monologue—seemed to him positively degrading. He explained at length—

fighting her attentive silence, sloshing across the puddles of pauses, abhorring himself—that he had a long and hard journey; that he slept badly; that he was working on an investigation of the nature of Time, a theme that meant struggling with the octopus of one's own brain. She looked at her wrist watch. "What I'm telling you," he said harshly, "has nothing to do with timepieces." The waiter brought them to their coffee.

That night's insomnia allows him to complete *The Texture of Time*. A sleeping pill then provides some relief along with a dream that he is speaking on the subject amid hecklers in the lecturing hall of a transatlantic liner. Explaining his work next day to Ada, Van implores her not to laugh at his philosophic prose and says that his "aim was to compose a kind of novella in the form of a philosophic treatise on the Texture of Time, an investigation of its veily substance, with illustrative metaphors gradually increasing, very gradually building up a logical love story, and just as gradually reversing analogies and disintegrating again into bland abstraction." Ada's response is to wonder whether the subject is really worth investigating: "We can never know time. Our senses are simply not meant to perceive it. It is like—."

With this incomplete analogy serving as a challenge to readers, part 4 ends. As an alternative to the impossible task of adequately completing Ada's unfinished sentence, it is tempting to apply Van's nearby description of his treatise to Nabokov's book. The reader is thus forced to halt momentarily his forward progress through the novel while deciding whether to complete the analogy if he can, whether to think back over earlier events to see if they conform to Van's description, or whether to turn the page and start part five. Opting for either or both of the alternatives to plunging straight into Part 5 involves the reader in a temporal digression of indeterminate length. Trying to complete the analogy could take seconds, hours, or days. Dismissing the idea that Van's description of his treatise may be Nabokov's clue to the structure of his novel is at least the quicker, certainly the easier, of the available choices because the description can only apply if reversed, and even then, as a few moments thought suffice to show, it doesn't quite fit.

A philosophical treatise on time in the form of a novel is an impossible contradiction in genres (7). Nabokov's *Ada* is certainly a charming love story, his most brilliant. But like all delightful love stories, it is resolutely illogical. Far from disintegrating into abstractions in part four, the narrative is so clearly pursued that all the discussion of time there becomes one facet of the complex characterization of Van. Nor are the abstractions in his meditation on time at all bland. Major writers on the topic from Augustine through Guyeau, Bergson, and Whitrow are alluded to in ways that compel readers to pass the history of time concepts in mental review, more or less rapidly depending on the extent of their familiarity with the sugject. But this review is accompanied by an increasing sense of exasperation because the tone remains sufficiently satirical to dramatize the inadequacy of all abstractly articulated temporal concepts when applied by an individual trying to make sense of his own chronological experience. Here, for example, is Van's comment on one important area of study:

> Maybe the only thing that hints at a sense of Time is rhythm; not the recurrent beats of the rhythm but the gap between two such beats, the grey gap between black beats: the Tender Interval. The regular throb itself merely brings back the miserable idea of measurement, but in between, something like true Time lurks. How can I extract it from its soft hollow? The rhythm should be neither too slow nor too fast. One beat per minute is already far beyond my sense of succession and five oscillations per second make a hopeless blur. The ample rhythm causes Time to dissolve, the rapid one crowds it out. Give me, say, three seconds, then I can do both: perceive the rhythm and probe the interval. A hollow, did I say? A dim pit? But that is only Space, the comedy villain, returning by the back door with the pendulum he peddles, while I grope for the meaning of Time.

This passage, like the extended meditation in which it occurs, recalls important aspects of time-perception only to deflect attention from such factual matters to the narrative issue of their relevance in an unfolding story, which at that point in the reader's initial experience of the novel does at least seem mainly to be going from past to present.

Part 5, however, is described as the true introduction. In it is revealed what has frequently been hinted: the narrator's exact biological location in time as a 97-year old still living happily with 95 year old Ada and a retinue of doctors, nurses, and helpers. Revelation that he completes the narrative in 1967 allows various present-tense interpolations scattered throughout by Van and Ada to be understood, as is the whole story on its necessary second reading, as commentary on their past perceived backwards from their situation in the sixties. The novel's temporal structure is inescapably circular, going from present to past, back to the present, and again into the past. "One can even surmise that if our time-racked, flat-lying couple ever intended to die they would die, as it were, *into* the finished book, into Eden or Hades, into the prose of the book or the poetry of its blurb" (6). The last pages shade off into a parody of a publisher's blurb describing the book, so the reader is directed back to its beginning to start over.

On again arriving at the novel's conclusion, however, the reader still finds no exit, only the now-anticipated invitation to go through its episodes once more by rereading or remembering. Attention is directed to the similarity of these two activities by making recall an attractive but quicker alternative to yet a third reading. *Ada*'s temporal structure thereby dramatizes how easily memories shade off from and are substituted for experience. The stage of this drama is the reader's mind.

During the second reading, moreover, memories of the first encounter with each episode as well as memories of subsequent episodes crowd back to alter the quality of the reader's involvement with each narrative moment by placing every event in the forward-moving story in relationship to events that are in the characters' future but the reader's past. Each event, however, must *also* be considered as taking place in the character's past when it is recalled that the narrative is supposed to be written retrospectively by 97 year old Van with the parenthetical comments of 95 year old Ada, to whose 1967 self the narrator addresses the story, which thus becomes on one level a dialogue between them as they are at the end when the book is written.

The reader's circular experience becomes a metaphor illuminating in miniature the nature of time as experienced by all individuals. Distinctions between past, present, and future become increasingly irrelevant, or at least increasingly hard to maintain as the wall between anticipation and memory crumbles. It is as though Nabokov has illustrated the human meaning of Augustine's denial that past, present, and future exist as three separate times. Absent, however, is the Christian affirmation of a timeless eternity toward which the individual soul must strive. What remains is a predicament that often leads to the tragedy of existentialist nausea, but in Nabokov's novel results in a comic affirmation of the pleasures possible or at least imaginable in an earthly paradise (8, 9).

It is an inaccessible Eden, however, because the temporal location of *Ada*'s world is obscured in ways that prevent readers from imaginatively meshing with the calendar time within which its fictional events unfold. From scattered allusions and partial explanations the reader gradually infers that Van and Ada live on a twin to our earth in a plausible if not possible parallel universe where the years almost correspond to our own but where events have taken a happier turn. The cultures of England, France, and prerevolutionary Russia have harmoniously blended on the North American continent. England had annexed France in 1815. The development of electricity as an instrument of technology has been forbidden although cumbersome and somewhat amusing substitutes have been permitted. What results is a world vaguely reminiscent of the peaceful pre-1914 era with large, beautiful estates sprinkled over prosperous countrysides. The wars, fanaticisms, atrocities, and potential disasters that characterize our twentieth century are known only through the garbled reports of a few individuals claiming visions of a twin world known to them as Terra. The reports are so filled with implausible horrors that the visionaries are dismissed as lunatics. Finally a shrewd film director garbles the accounts still further in turning them into a science-fiction movie.

In sharp contrast to Van Veen's attractive planet is the world of A. J. Wiener's vivid short story, *Capture the Flag*. Its central event is concisely narrated:

A small figure appeared in a second story schoolhouse window. He had been hiding inside! Then he leaped, bayonet first, toward the nest. As he fell three of the riflemen defending the nest turned to meet him, and he was spitted on their bayonets. At the same time the grenades in his belt all went off, and the nest flew apart, amid screams. The charging infantry was under the barbed-wire and over the sandbags, and with a few final bayonet thrusts the flag was captured. (10)

This vigorous attack impresses Mr. Hawkins, District Superintendent of Schools, who is paying a surprise visit to Miss Alison's seventh-grade class and stays to watch the game. After the wounded are taken away, he has a serious talk with Johnny Matthews, the obviously promising captain of the seventh-grade team. The setting is an unspecified future. The story appears without prior introduction after undramatized scenarios for Business Cycle Worlds, A New Pan-European Movement, New Mass or Elitist Movements and Social Controls, in Chapter 8 of Kahn and Wiener's *The Year 2000: A Framework for Speculation on the Next Thirty-Three Years.*

Nabokov's *Ada* portrays an implausible alternative past that might have occurred if some crucial events of our world's history had worked out differently. *Capture the Flag* shows an impossible but equally conceivable future: impossible because it could not occur exactly as described, conceivable because we can imagine it. In both works the reader experiences a dislocation in time accompanied by a temporal confusion that is unique to the twentieth century. He is cut adrift, but not wholly adrift, from familiar time and projected imaginatively into an era that did not, could not, and will not ever exist except in the mind—where, according to Augustine, all times exist for us in any case.

Similar experiences have always been available in literature. Mythological pasts like those found in the Homeric epics, Virgil's *Aeneid*, and the Arthurian legends are still part of the mental landscape of Western civilization. So, to a lesser degree, are the largely fictive pasts that serve as settings for plays like Shakespeare's *Lear, Macbeth* and *Hamlet*. Also familiar, though less appealing to modern tastes, are the dream-like temporal environments encountered in such allegories as *Piers Plowman, The Faerie Queene, Pilgrim's Progress*, or any fable beginning with the traditional "once upon a time." That opening usually serves as a clear signal that the ensuing narrative is to be understood as taking place apart from the time occupied by real people. Other conventional devices, especially the presence of gods, goddesses, magicians, supernatural spirits, or super-heroic individuals, serve to distance epics and romances from historical time.

But in such works there is no confusion. The reader intuitively knows exactly where he stands with respect to the depicted time: it may be either totally unrelated to that of his own world, as in fairy tales, fables, allegories, and dream visions; or it may be explanatively connected to the time of his world as Troy's destruction is linked in the *Aeneid* to the historical world occupied by the Roman empire of Augustus. Even mythologized history of a more explicit kind, as in Shakespeare's *Henry V* or *Richard III*, creates no uncertainty: readers may accept, reject, or debate the versions of real events so presented, but there is no problem deciding which segment of historical time is to be compared with that depicted by the writer. So too most eighteenth and nineteenth century novels clearly locate the depicted time with respect to the reader's sense of historical time. Fielding's Tom Jones encounters soldiers going to fight Prince Charles, thus locating events around 1745. Despite the prophetic distortions of narrative chronology in Sterne's *Tristram Shandy*, Tobey's participation in the siege of Namur puts him in unequivocal relationship to the reader's historical time. Stendhal's Fabrice experiences his famous confusion as he wanders over the field of Waterloo.

DISORIENTATION IN TIME

What sets fiction like *Ada* and *Capture the Flag* decisively apart from earlier traditions of temporal representation are the extreme difficulties created for any reader who tries to relate the depicted time to his own clock, calendar, or historical time. These works avoid conventional means of establishing that relationship, yet they also induce insistent attempts to locate their action with respect to the sequence of events in the reader's world. Actions are thus placed in a kind of temporal limbo which is neither affiliated to historical time in any precisely definable way nor altogether in the fabulous regions of once upon a time which never induce anyone to wonder about relationships between the depicted time and that occupied by themselves.

To the major categories of past, present, and future rendered suspect by Augustine, recent developments in sociology as well as fiction have added another which is neither the Christian eternity outside time altogether nor the long-recognized literary times showing (in Sir Phillip Sidney's phrase) what may be or should be (11). Works like *Ada* do not primarily attempt to show us what may happen if, what might have happened if, or what ethically should or should not happen, although to some extent these traditional intentions may be achieved as subordinate goals. Instead, by nagging at us to locate them temporally without allowing us to do so, such works reflect and create heightened awareness of our situation in time by wrenching us uncomfortably away from the usual possibilities of getting our temporal bearings. Disorientation in time has supplanted displacement in space as a major technique for achieving important narrative as well as satiric effects.

Swift's invitation to consider the ethical direction of modern society took the form of keeping Lemuel Gulliver in the eighteenth century but sending him to Lilliput, Brobdingnag, Laputa, and Houyhnhnmland. George Orwell locates Winston Smith in 1984. This kind of shift from use of space to use of future time as a distancing device grew appealing as the world's geography and past history were so thoroughly explored that few blank spots were left to be filled imaginatively. The rising popularity during the nineteenth century of time-travel and future-located stories is reflected in the works of such transitional figures as Bellamy, Wells, Huxley, and Orwell. But the very specificity of fiction like *Nineteen Eighty-Four* places it in the same category as *Gulliver's Travels*. Readers can easily follow the chronological relationship of events within the story, and they can precisely measure their temporal as well as ethical distance from Orwell's Oceania. The first sentence zeroes in on Winston Smith during a bright cold day in April as the clocks are striking 13. As calendar years go by for the book and its readers, moreover, 1984 will approach, arrive, and recede into the past. Clock time, with its fixed categories of past, present, and future, is still the most conspicuous dimension. However terrifying and effective Orwell's satiric vision, *Nineteen Eighty-Four* neither seeks nor achieves the extraordinary dislocations of subjective time that are a distinctive feature of more recent works.

The disturbing novels of Alain Robbe-Grillet, like his script for *Last Year at Marienbad*, illustrate what can be achieved by sustained resort to various techniques that disorder the subjective time-sense. Consider, for example, the first three paragraphs of *In The Labyrinth*:

I am alone here now, under cover. Outside it is raining, outside you walk through the rain with your head down, shielding your eyes with one hand while you stare ahead nevertheless, a few yards ahead, at a few yards of wet asphalt;

outside it is cold, the wind blows between the bare black branches; the wind blows through the leaves, rocking whole boughs, rocking them, rocking, their shadows swaying across the white roughcast walls. Outside the sun is shining, there is no tree, no bush to cast a shadow, and you walk under the sun shielding your eyes with one hand while you stare ahead, only a few yards in front of you, at a few yards of dusty asphalt where the wind makes patterns of parallel lines, forks, and spirals.

The sun does not get in here, nor the wind, nor the rain, nor the dust. The fine dust which dulls the gloss of the horizontal surfaces, the varnished wood of the table, the waxed floor, the marble shelf over the fireplace, the marble top of the chest, the cracked marble on top of the chest, the only dust comes from the room itself: from the cracks in the floor, maybe, or else from the bed, or from the curtains or from the ashes in the fireplace.

On the polished wood of the table, the dust has marked the places occupied for a while—for a few hours, several days, minutes, weeks—by small objects subsequently removed whose outlines are still distinct for some time, a circle, a square, a rectangle, other less simple shapes, some partly overlapping, already blurred or half obliterated as though by a rag. (12)

The spatial aspect of these paragraphs is unambiguous. Their narrator immediately specifies that he is without companions: a single figure is to be envisioned. The meaning of "here . . . under cover" is never in doubt because the reader's eye takes in and allows him to register the first word of the second sentence, "outside," in the same glance that suffices to absorb the very short first sentence. The period signaling a pause between the two sentences, but noticeably enforcing one only if the passage is read aloud, does not even in that case allow or create any confusion about the sense of "under cover." Quite the contrary, "outside"clarifies the meaning of "under cover" by equating it with the opposite of outside: inside. The entire familiarity of the distinction between inside and outside dispels any worrisome spatial uncertainty.

As the paragraphs proceed, the physical dimension comes into focus for the reader with increasing clarity. Rain, cold, wind, sunlight, and dust are elsewhere, as the first paragraph emphasizes by its triple repetition of the word "outside." The second paragraph summarizes what has been emphasized by explicitly reiterating that "the sun does not get in here, nor the wind, nor the rain, nor the dust." Throughout the rest of that paragraph the meaning of "here" as "inside" is made even more specific by mention of ordinary concrete objects: a varnished wooden table, a waxed floor, a fireplace surmounted by a marble shelf, a marble-topped chest, window-curtains and, finally, a bed. The narrator is thus located in that most familiar of all spatial environments, one that we occupy for a third of our lives, a bedroom. The familiarity of this setting dispels any impulse to wonder much about the exact shapes and colors of the objects, the appearance of the building in which the bedroom is situated, or the name of the city in which the building is found. One *may*, and later in the novel readers *do* wonder about these matters, but nothing in the first three paragraphs encourages such speculation, nor does it ever become the major concern. By the end of the third paragraph, which focuses attention on the comfortably solid table-top, spatial confusion has been minimized.

What gets increasingly problematical as the meaning of the word "here" becomes more specific is the sense of the word "now." "Here now" translates itself into the unaswerable but ever more bothersome question "there *when*?" All verbs are in present tense throughout the first paragraph, in harmony with the usual meaning of "now." The grammatical signals are appropriate for the description of a state that persists through a definable present moment, that is to say, a situation that exists in a distinct relationship to an earlier "then" and a subsequent "afterwards." But contradicting the grammatical time-clues are several closely juxtaposed images of situations that cannot easily be regarded as coexisting in the same "now": rain falling, the sun shining; wet asphalt, dusty asphalt; bare branches, leaves; shadows swaying, no tree, no bush to cast a shadow. Customary distinctions between then and now, between past, present and, by implication, future, are abolished.

The meaning of "now" starts to expand in both directions, as of course in theory but not ordinary practice it can: "now" *may* refer to the specious present, to this day, this month, this decade, or this century. But we seldom remember its infinite expansibility. Customary usage and the time-concepts implied by that usage are suddenly thrown into doubt as the paragraph brings to consciousness implications that we accept but do not like to think about.

The reader's temporal bewilderment is rendered even more acute in the third paragraph. Verb tenses switch from past ("dust has marked. . . objects. . . removed") to present ("whose outlines are still distinct") within its single sentence. Words usually serving to enhance precision in specifying time-relationships ("hours. . . days, minutes, weeks") are employed to render unmistakably ambiguous the vague "a while" which is employed to describe time spent on the table-top by "small objects subsequently removed." The reader is by the words "hours. . . days, minutes, weeks" reminded of the possibility of temporal precision and thus challenged to locate events temporally while he is also denied the information essential for doing so. Subsequently to *what* moment is a question that is made to intrude in the reader's consciousness along with realization that no satisfactory answer is possible.

i

This predicament of unavailing struggle to establish temporal relationships where none exist is sustained with varying degrees of intensity throughout Robbe-Grillet's novel. Recurring conversations between a small boy and a lost soldier trying to deliver a package seem very much alike because whenever these characters meet similar dialogue takes place. The temporal settings for their conversations are so vague—during the day, at night, when it is snowing, when it isn't—that, just as when minutes, days, hours, and weeks are mentioned in the third paragraph, the only purpose served by the text's time-signals is to arouse insatiable temporal curiosity. It is never possible to work out a satisfactory ordering of episodes with respect to one another within the fictional chronology of a coherent plot. Nor is it even possible to keep clearly in mind the sequence of narrated episodes as they have been experienced by the reader in his own clock time from the moment he starts the first sentence to the moment when he finishes the book.

Memory is short-circuited by the piling up of similar incidents and the suppression of temporal clues. It is the reader who gets lost in the labyrinth of his own mind as he loses the ability to remember where he has been within the novel and consequently loses his ability to say, even during a second reading, where, at any particular moment, while reading any particular sentence, he is going next. For him time is abolished insofar as it no longer seems a context within which events—in his case the reading of narrated episodes—flow measurably from one to another. Robbe-Grillet's novel affects its readers' time-perception by impeding the formation of memories.

FUTUROLOGY

A complex alteration of the subjective time-sense must occur throughout the clock-time of a reading experience skillfully designed to collapse perceived distinctions between past and present. But the impulse to make such distinctions is not diminished. The habits of a life in time impel readers to keep searching for some way out of the labyrinth within which they have ventured. From this search stem both the psychological interest of their situation and a variety of aesthetic pleasures peculiarly appealing to the twentieth century mentality. Spatial mazes like the one planted almost three centuries ago in the Hampton Court Palace gardens diminished in popularity after the great age of geographical exploration had ended. But an ample variety of temporal labyrinths are available in a civilization characterized by an unprecedented amount of attention devoted to serious interdisciplinary and governmental exploration of the future.

The American Academy of Arts and Sciences' Commission on the Year 2000, the Hudson Institute, and the Club of Rome are only the most famous of the futurist groups. America, like tne rest of Western civilization since the industrial revolution, has been increasingly oriented forward rather than toward the past (13). But the sheer intensity and novel methods of the newly organized concern with the future are themselves striking developments in the concept of time. And so far as method is concerned, a key development is the diminished role of prediction and hence a turn away from regarding as an end in itself extrapolation based on time-series studies.

This development is most clearly seen at the borderline where futurology borrows from literary practice and criticism the term "scenario" to describe any hypothetical sequence of events. Kahn and Wiener, like David Bell and other serious students of the future, stress equally their need for scenarios and the fact that no scenario is to be regarded as a prediction. This distinction is still widely misunderstood. Futurist scenarios are devised primarily to focus attention on the range of available choices and their implications by identifying possible relationships between events called branch points, whose occurrence would have important consequences. A scenario may be a reconstruction of past events as well as what is now more common, a hypothetical sequence of future events (14). A scenario can be thought of as a non-mathematical model describing the mechanism of interrelationships among events within any particular past or future time-interval. Scenarios are a device for enlarging freedom of present action by enhancing awareness of the temporal dimension of any present state of affairs. Because causal links between events are postulated, it is desirable to envision as many alternative futures—end results of different causal chains—as possible for the purpose of generating the widest conceivable range of scenarios. The resulting models of alternative futures are not primarily employed predictively to say what *will* happen, but as an aid to altering what might without their use have been the future. Extrapolation from time-series studies merely for the purpose of identifying inevitabilities so they can be accommodated to plays as slight a role in futurology as it does in the physical and biological sciences, whose concern with predictive natural laws has always been subordinated to the more exhilarating Faustian interest in achieving control over nature.

In a discussion of how attitudes towards the future influence the present, Pierre Massé, director of the Fourth French Economic Plan, writes that projects like that plan start from an exploration of the future, although not from any single deduced future but rather from a plurality of imagined futures. He adds that the final year of the Fifth French Economic Plan, 1970, was built as much by feedback from 1985 as by projection starting from 1965 (15).

The futurist canon of credibility illustrates their refusal to consider even accurate extrapolation as a sufficient goal, however necessary. Kahn and Wiener argue that in constructing scenarios one should not confine speculation to the most likely contingencies. Freedom to specu-late more widely is necessary because "history is likely to write scenarios that most observers would find implausible not only prospec-tively but sometimes, even, in retrospect. Many sequences of events seem plausible now only because they have actually occurred; a man who knew no history might not believe any. Future events may not be drawn from the restricted list of those we have learned are possible; we should expect to go on being surprised" (10). Hence the expendability of any given scenario. Since their aim is not predictive, futurist scenarios may be expected to date quickly. But this does not diminish their utility. There is great heuristic value in a method that exemplifies and encourages what Kahn and Wiener describe as the virtue of rea-soning by analog or at least by metaphor.

MEMORY AND MEASUREMENT

Not since the Renaissance has analogical thinking been in such high repute. But to describe the renewed prominence of metaphorical rea-soning as an aspect of the historical development of the concept of time may seem as strange as the inclusion under that rubric of novels like *Ada* and *In the Labyrinth* until we recall what is meant by "time."

It was Augustine who decisively clarified this issue for Western thought by remarking early in the eleventh chapter of his *Confessions* that he knows what time is so long as nobody asks him to tell what it is. This often misunderstood remark is used in the full context of that fa-mous chapter not only to draw a distinction between intuitive and ar-ticulated knowledge, but for the more subtle purpose of distinguishing between definition in terms of essence (what time or anything else *is*) and definition that is operational. As Augustine proceeds, he brilliantly avoids the reifying trap of fastening on the operations *of* time and at-tempting to say what *it* does. His discussion is thus not limited to the concept of time as change, although Augustine amply acknowledges that changes are inseparable from existence in time. Nor does his ar-gument lead in the direction of the largely hostile iconography evolved in Western art and literature to embody the concept of time the destroyer (16, 17). Instead, Augustine proceeds to consider the operations involved in measuring time.

For this purpose he firsu distinguishes between time and motions like those of the sun, moon, and stars, which are in his physics primarily spatial. Then he isolates for consideration as a purely temporal phenomenon, occupying no space, a speaking voice reading a verse from St. Ambrose's evening hymm: *Deus creator omnium*. The voice may be external or, when silently reading a book to oneself, internal. In either case, the voice's duration, or that of any part of what is said, can only be measured within the mind by comparing the length of the syllables with one another. The measure of one duration—any given syllable—is thus some other duration, at first a puzzling impasse. Augustine escapes from the circularity by more precisely identifying the location and nature of such measurement:

> What, then, is it that I measure? Where is that short syllable by which I measure? Where is that long syllable which I measure? Both have sounded, have fled away, have gone into the past, and no longer exist. . . . Therefore what I am measuring is not the syllables themselves. . . but something in my memory which remains there fixed.
>
> It is in you, my mind, that I measure time. . . . As things pass by they leave an impression in you; this impression remains after the things have gone into the past, and it is this impression which I measure in the present, not the things which, in their passage, caused the impression. It is this impression which I measure when I measure time. Therefore, either this itself is time or else I do not measure time at all. (18)

Augustine's focus here is equally on the measurement of time and on time as a phenomenon best studied within the mind by attending to the mechanism of memory. It follows from this view that there are as many concepts of time as there are ways of measuring the duration of events and ways of describing or awakening human awareness of duration.

Augustine nowhere denies the reality of time or the physical universe. Quite the contrary. His entire eleventh chapter is addressed to the problem of discovering what is meant by the opening of the *Book of Genesis*: "I want to understand how 'In the beginning [God] created heaven and earth.'" Time is considered by Augustine to explicate *beginning* in the difficult first sentence of the Bible. He takes for granted the reality of everything that God created, including that time whose beginning is mentioned but which is so hard to comprehend. Augustine's position in his *Confessions* builds upon the assumed objective existence of the universe. He could hardly have proceeded otherwise while retaining faith in God. Augustine's theory is subjectivist only insofar as he correctly implies that all temporal measurement is ultimately a mental act: time, like space, is an attribute of the universe, but past, present, and future are delineated within the mind. Nothing is either affirmed *or denied* by Augustine about any biological basis for such measurements. The issue was neither relevant for his purposes nor one that admitted fruitful speculation when he wrote. But a remarkable strength of the Augustinian position is that it points beyond Kant and Bergson to the dominant modes of twentieth-century empiricism (19).

The Augustinian emphasis on memory and measurement now prevails. Novelists may describe the individual experience of time, as in *Ada*, by creating a rich context of memories around particular fictive events, or else go toward the opposite extreme, as throughout *In The Labyrinth*, by destroying memories to create a perpetual present. Current psychological study of time-perception fastens most promisingly on relationships between experienced duration and memory (20). In the physical and biological sciences powerful new techniques are employed to measure the impress of events over time. Literary critics, sociologists, futurologists, and historians have worked out elaborate ways of describing temporal relationships among the events they study (21). All such methods of description and measurement are operational definitions of time.

TEMPORAL EXPERIENCE AND LITERARY FORM

The very proliferation of modern attempts to understand time echoes the questions, though seldom the religious answers, voiced by Augustine as part of his longing to discover the meaning of human lives apparently so confined in time. Yet it is only in the past decade that the twentieth century has systematically started to take stock of the answers on hand. The historians themselves present a striking example of this suddenly accelerated interest in time-concepts. As recently as 1966, C. G. Starr, writing in an issue of *History and Theory* devoted to essays on history and the concept of time, could point out that while "philosophers and metaphysicians have always talked much about time, practicing historians have never done so" (22). Never, that is, until 1966. In English literary criticism the pioneering work appeared in 1952: Mendilow's valuably study *Time and the Novel* (23). The French, as usual somewhat ahead of us above ground as well as under ground (24), had in 1946 Jean Pouillon's *Temps et roman* and in 1950 Georges Poulet's *Etudes sur le temps humain* (25, 26).

Until the seventies, however, the majority of even those critics most closely concerned with aspects of time in literature tended to accept in practice the assumption voiced for the Modern Language Association by Northrup Frye's unequivocal assertion that "the process of academic criticism" begins only *after* a work has been read to the end so that it is possible to see it as a unity comprising "a simultaneous pattern radiating out from a center, not a narrative moving in time" (27). The significant critical departure of the seventies has been articulated by Stanley Fish, who argues persuasively that readers do and critics should respond to the "temporal flow of the reading experience . . . and not to the whole utterance." He advocates in opposition to "the atomism of much stylistic criticism" a method that "involves an analysis of the developing response of the reader in relation to the words as they succeed one another in time." In his view the trouble with criticism that treats novels, poems—or for that matter plays or films—as *objects* of analysis is that "it transforms a temporal experience into a spatial one; it steps back and in a single glance takes in a whole (sentence, page, work) which the reader knows (if at all) only bit by bit, moment by moment" (28). Here is yet another modern application of the Augustinian emphasis on the inward or outward sounding voice as a purely temporal phenomenon.

Fish does not attain the full insight of Augustine's emphasis on the role of memory. But this omission can be remedied by literary criticism that explains the formal unity of particular works while also adequately accounting for the reader's temporal experience of them. Still to be discovered, however, are the limits of accuracy when describing the subjective experiences induced by literature. It is not clear whether, even for such an extreme case as Robbe-Grillet's *In The Labyrinth*, the reader's temporal experiences can be quantified in any meaningful way for precise measurement. This difficult problem is a challenge to literary criticism, the psychology of perception and, perhaps, chronobiology. A combination of time-concepts from all three areas might advance our understanding of how temporal experience is related to literary form, or at least show more clearly the outermost limit of each field.

Methods borrowed from or suggested by one area may advance learning elsewhere. The helpful analogy between microscopy and the resolving power of electronic calculators seems a case in point (29). So perhaps, even if negatively, are the attempts to apply cybernetic feedback models to biological periodicity (30). There is little need to dwell on the pleasant situations where a technique may be directly imported from one field to another. More common, and in the long run probably more helpful, are those concepts which may function as heuristic metaphors. Whether significant advances always depend on an imaginative substructure of analogical thinking fed by cross-currents of contemporary ideas is one of those questions of historical causation that defy any neat answer (31). There is, however, enough evidence of increasing convergence between different disciplines in the area of time study so that the process might well be rendered fully conscious not only for the purpose of comparing ideas, but to accelerate the invention of new time-concepts.

THE CONCEPT OF TIME AS A FUTURE-ORIENTED ACTIVITY

As in other areas of modern thought, attention should be on where we are going, not on where we have been. The past is more accessible than ever (32), but should not attract us by virtue of its pastness. Again in the Augustinian tradition, we should think of it as a present of things past, but considered for the modern purpose of enlarging the scope of future options. The most important aspect of the historical development of the concept of time is that it is just getting under way as a future-oriented activity. We are at the outset of an expansion of time-concepts that has started to require not introductory addresses but interdisciplinary conferences devoted exclusively to the subject for anything like even the beginnings of a full description (33–36). There would be little point in trying to compress such meetings into a few abstractions here. The novelty of the present situation resides in the abundance of available time-concepts that allow new answers to classical questions. The task ahead is that of encouraging the future development of time-concepts that will apply across disciplinary boundaries between the sciences and humanities.

Nabokov has set an example: there is at one point in Van Veen's meditation on time an amusing reference to a real poet, John Shade, and an invented philosopher, Martin Gardner, supposed author of an imaginary book, *The Ambidextrous Universe*. In fact, as Nabokov's admirers will know, it is the other way around. John Shade is the imaginary hero of Nabokov's *Pale Fire*. Martin Gardner is the real author of an excellent book, *The Ambidextrous Universe*, as well as of an article on a related theme published in *Scientific American:* "Can Time Go Backward?" (37).

The issue involved in this question is, to borrow the title of a 1972 article in *Science*, the crisis about the origin of irreversibility and time anisotropy (38). Whether time can go backward is part of the larger question of the extent to which our universe is symmetrical. In a 1969 *Scientific American* article, Oliver Overseth described "Experiments in Time Reversal" (39). And Robert Sachs, director of the Enrico Fermi Institute, in a 1972 article on "Time Reversal," reports recent experiments indicating the breaking of time-reversal symmetry and hence the experimental establishment of the direction of time's arrow (40). The purpose of these articles is to convey some intuitive and especially visual—that is, metaphorical—grasp of the issues at stake. Pictures of mirrors, references to Alice's Adventures through the looking-glass, and discussions of "galaxies of anti-matter that are also mirror-reflected matter" abound (36).

Nabokov has incorporated such metaphoric explanations of symmetry into the structure of *Ada* by calling its planet Antiterra, thus turning it into a mirror image of our world. The setting is therefore both an alternate past *and* a puzzling time-reversed reflection where imaginary poets are real and real authors imaginary. Some books even get turned around: *Ada*'s first sentence reverses the opening of *Anna Karenina* by asserting that "All happy families are more or less dissimilar; all unhappy ones are more or less alike." But the statement is nevertheless attributed to Tolstoy. Within the narration of events leading to a suicide toward the end of the novel is the curious observation that "Lucinda Veen was only five hours old if one reversed the human 'time current.' " By such teasing invitations to consider his Antiterra as a place where time may be running backwards, or at least differently from that experienced in our world, Nabokov induces yet another baffling level of temporal uncertainty. The more any reader understands about the crisis over time anisotropy, the more fully *Ada*'s temporal structure is realized. Here certainly is a beautiful artistic paradigm symbolizing what may be gained by the metaphoric application of borrowed time-concepts. In this elegant context of heightened temporal awareness it is possible to understand why a conspicuous figure in modern fiction is the man who lectures on time, and why that figure recalls St. Augustine.

REFERENCES

1. John F. Callahan, *Four Views of Time in Ancient Philosophy*, Harvard Univ. Press, Cambridge, Mass., 1948.
2. St. Augustine, *Confessions*, Book XI, Chapter 20.

3. T. S. Eliot, *The Complete Poems and Plays 1909–1950*, Harcourt, Brace, New York, 1952, p. 117.

4. Roman Jakobson, "Verbal Communication," *Sci. Amer.*, **227**, 3, (1972), 72.

5. Kurt Vonnegut, Jr., *Slaughterhouse-Five or the Children's Crusade: A Duty-Dance With Death*, Dell, New York, 1970.

6. Vladimir Nabokov, *Ada or Ardor: A Family Chronicle*, McGraw-Hill, New York, 1969.

7. Sheldon Sacks, *Fiction and the Shape of Belief*, Univ. California Press, Berkeley and Los Angeles, 1964.

8. Robert Jordan, "Time and Contingency in St. Augustine," *Rev. Metaphys.*, **8**, 394 (1955).

9. Jean-Paul Sartre, *Nausea* (L. Alexander, trans.), New Directions, New York, 1969.

10. Herman Kahn and Anthony J. Wiener, *The Year 2000: A Framework for Speculation on the Next Thirty-Three Years*, Macmillan, New York, 1967, pp. 354, 264, 32.

11. Sir Philip Sidney, *An Apologie for Poetrie*, 1595.

12. Alain Robbe-Grillet, *In The Labyrinth*, (Richard Howard, trans.), Grove Press, New York, 1960.

13. J. H. Plumb, *The Death of the Past*, Houghton Mifflin, Boston, 1970.

14. Herman Kahn and B. Bruce-Briggs, *Things To Come: Thinking About The Seventies and Eighties*, Macmillan, New York, 1972, p. 175.

15. Pierre Massé, "Attitudes towards the Future and Their Influence on the Present," *Futures*, **4**, 1 (1972), 24.

16. Erwin Panofsky, *Studies In Iconology*, Oxford Univ. Press, Oxford, 1939.

17. William Shakespeare, *Sonnets*, Nos. 12, 13, 15, 16, 55, 60, 63, 64, 65, 116.

18. *The Confessions of St. Augustine* (Rex Warner, trans.), New American Library, 1963, p. 280 (Book XI, Chap. 27).

19. Hugh M. Lacey, "Empiricism and Augustine's Problems about Time," *Rev. Metaphys.*, **22**, (1968), 219.

20. Robert E. Ornstein, *On The Experience of Time*, Penguin, Harmondsworth, 1969.

21. George Kubler, *The Shape of Time*, Yale Univ. Press, New Haven, Conn., 1962.

22. Chester G. Starr, "Historical and Philosophical Time," *History and Theory*, **6**, (1966), 24.

23. A. A. Mendilow, *Time and the Novel*, Peter Nevill, London, 1952.

24. M. Siffre, *Beyond Time*, McGraw-Hill, New York, 1964.

25. Jean Pouillon, *Temps et Roman*, Gallimard, Paris, 1946.

26. Georges Poulet, *Etudes sur le temps humain*, Plon, Paris, 1950.

27. Northrop Frye, "Literary Criticism," in *The Aims and Methods of Scholarship in Modern Languages and Literatures*, 2nd ed. (James Thorpe, ed.), Modern Language Assoc. America, New York, 1970, p. 69.

28. Stanley E. Fish, "Literature in the Reader: Affective Stylistics," *New Literary History*, **2**, 1, (1970), 123.

29. Franz Halberg, "Resolving Power of Electronic Computers in Chronopathology—An Analogy to Microscopy," *Scientia*, **101**, 412 (1966).

30. Gay Gaer Luce, *Biological Rhythms in Psychiatry and Medicine*, U.S. Department of Health, Education and Welfare Public Health Service Publication No. 2088, Washington, D.C., 1970, p. 87.

31. Thomas S. Kuhn, *The Structure of Scientific Revolutions*, 2nd ed., Univ. Chicago Press, Chicago, 1970.

32. Elizabeth L. Eisenstein, "Clio and Chronos: Some Aspects of History-Book Time," *History and Theory,* **6,** (1966), 36.

33. J. T. Fraser, ed., *The Voices of Time: A Cooperative Survey of Man's Views of Time as Expressed by the Sciences and by the Humanities*, Braziller, New York, 1966.

34. Roland Fischer, ed. "Interdisciplinary Perspectives of Time," *Ann. N.Y.Acad. Sci.,* **138,** 2 (1967), 367.

35. Jiri Zeman, ed., *Time in Science and Philosophy: An International Study of Some Current Problems*, Elsevier, Amsterdam, London, New York, 1971.

36. J. T. Fraser, F. C. Haber, G. H. Muller, eds., *The Study of Time: Proceedings of the First Conference of the International Society For the Study of Time*, Springer-Verlag, New York, Heidelberg, Berlin, 1972.

37. Martin Gardner, "Can Time Go Backward?" *Sci. Amer.,* **216,** (1967), 98.

38. Benjamin Gal-Or, "The Crisis about the Origin of Irreversibility and Time Anisotropy," *Science,* **176,** 4030 (1972), 11.

39. Oliver E. Overseth, "Experiments in Time Reversal," *Sci. Amer.,* **221,** 4 (1969), 89.

40. Robert G. Sachs, "Time Reversal," *Science,* **176,** 4035 (1972), 587.

2 Statistical Rhythms Evaluation

EDWARD BATSCHELET, *Moderator*

CHAPTER 2

Statistical Rhythm Evaluation

EDWARD BATSCHELET

Biomathematics Division, Mathematics Institute, University of Zürich, Zürich, Switzerland

In the study of biological rhythms all quantities of concern are approximately periodic functions of time t. If long-term trends occur, they are eliminated mathematically before undertaking analysis of the periodicities. To fit nearly periodic functions, various mathematical models are used. The simplest is the cosine function

$$z = C_0 + C \cos (\omega t - \phi) \tag{1}$$

with C_o denoting the mean level, C the amplitude, ω the angular frequency, and ϕ the acrophase, that is, the phase at which the peak of the quantity z occurs (1). Such a function can be fitted successfully only if the observed graph resembles a sine wave. In all other occasions fitting the function, Eq. 1 would be misleading and could end with spurious results.

An alternative procedure consists in using a Fourier polynomial such as

$$z = C_0 + C_1 \cos (\omega t - \phi_1) + C_2 \cos (2\omega t - \phi_2) + \cdots \tag{2}$$

Formula 2 offers the advantage that an arbitrary periodic function can be fitted with any degree of accuracy. The main disadvantage is that

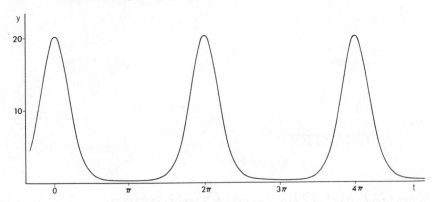

Figure 1   Graph of a nonlinear osciwation. The equation is y = exp (3 cos t).

the frequencies 2ω, 3ω, ... usually have no biological meaning whatsoever.

Model 2 is *linear* in the coefficients C_k as well as in the terms cos ϕ_k and sin ϕ_k. For statistical purposes this is a considerable advantage. However, if a satisfactory fit is of major importance, one has often to turn to *nonlinear* models. As an example we mention

$$y = e^z \tag{3}$$

where z is the quantity defined in Formula 1. Figure 1 depicts the graph of Formula 3 for specific values of the parameters C_o, C, ω, and ϕ.

Formula 3 is strongly related to the von Mises distribution in the analysis of directions (2–4).

A detailed discussion of various models for biological rhythms was given by Sollberger (5).

There exists no unified statistical theory for rhythmic processes. Rather there is a variety of techniques which are applicable only under certain circumstances. There are essentially two groups of data that have to be carefully distinguished:

1. Rhythmic behavior with exactly known period (or frequency).

2. Rhythmic behavior with unknown period or even an unknown multiplicity of periods.

We will deal with both cases in the following sections.

RHYTHMS WITH KNOWN PERIODS

The period T will be used to represent the length of the smallest time interval in which the periodic event produces a full cycle.

The angular frequency ω is related to T by the formula

$$\omega = 2\pi/T \qquad (4)$$

We assume that T and ω are known. We first consider the model of Formula (1) and apply it to suitably chosen time instants t_i ($i = 1, 2, \ldots, n$). Then the observed values z_i can be represented by

$$z_i = C_0 + C \cos(\omega t_i - \phi) + e_i \qquad (5)$$

where e_i denotes an error term. Equation 5 can be rewritten in the form

$$z_i = C_0 + x \cos \omega t_i + y \sin \omega t_i + e_i \qquad (6)$$

where x and y are called regression coefficients and are given by

$$x = C \cos \phi, \qquad y = C \sin \phi \qquad (7)$$

We may plot x and y as rectangular coordinates of a point or a vector. Then C and ϕ are its polar coordinates (Figure 2). This presentation takes into account that the acrophase is only determined to the extent that multiples of 2π are disregarded.

Formula 6 is linear in the unknown parameters C_o, x, and y. If we assume that the errors e_i are independently distributed with mean zero and common variance σ^2, we can apply the standard techniques of analysis of variance, including the Gauss principle of least squares. Let $\hat{C}_o$, $\hat{x}$, and $\hat{y}$ denote the least square estimates of C_o, x, and y. Then the famous Gauss-Markov theorem states: *Among all possible linear estimates the least square estimates $\hat{C}_o$, $\hat{x}$, and $\hat{y}$ have the smallest variance* (6). In this sense the procedure is optimal. If we assume in addition that the errors e_i are normally distributed, then analysis of variance furnishes a test for the null hypothesis $C = 0$. If the null

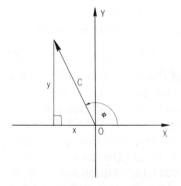

Figure 2   Acrophase ϕ and amplitude C as polar coordinates of a point. Regression coefficients x and y are rectangular coordinates.

hypothesis can be rejected, then we know that the true amplitude is different from zero and that the observed rhythm is not spurious.

A clear presentation of the mathematical details illustrated by instructive examples was given by Bliss (7). Bliss also treats the case of fitting a Fourier polynomial, Formula 2, and of the related analysis of variance. Moreover, the nonlinear model given by Formula 3 can be reduced to the above linear model by applying the transformation $z = \ln y$.

In practical applications it is not always easy to find suitable time instants t_i for which measurements z_i can be taken. What does "suitable" mean in this connection? First of all, the t_i's have to be dense enough to guarantee sufficient information about the process. The t_i's need not be equally spaced even though equal spaces lead to considerable savings in the amount of computation (8).

When we fail to provide for adequate data for a single individual or a single biological unit, then we cannot draw the conclusions we actually want. However, we may *replicate* the measurements for different individuals (or for different biological units). For each individual we get rough least square estimates $\hat{C}$ and $\hat{\phi}$ for the amplitude C and the acrophase ϕ, respectively. Such a $\hat{C}$ need not be significantly different from zero, nor have all conditions required for an analysis of variance to be satisfied. For such a case Halberg and co-workers (9) have elaborated a method known under the term "cosinor." For each out of $j = 1, 2, \ldots, N$ individuals we get estimates $\hat{x}_j, \hat{y}_j$ of the regression coefficients x and y of Eq. 6 by applying the principle of least squares. From now on we drop the hat sign to simplify notation. Thus we get a set of values

$$x_1, x_2, \ldots, x_N$$

$$y_1, y_2, \ldots, y_N$$

which we will treat in the following like measurements.

By applying Formulas 7 for each pair (x_j, y_j), we obtain equations

$$x_j = C_j \cos \phi_j, \qquad y_j = C_j \sin \phi_j \qquad (8)$$

which we can easily solve for estimate C_j of C and estimate ϕ_j of ϕ. We may plot the vectors (x_j, y_j) in a rectangular coordinate system with equal units on the two axes. Instead of rectangular coordinates we may as well employ the polar coordinates C_j, ϕ_j (Figure 3). All N points together form a scatter diagram. Now we assume that this scatter diagram is "well behaved" in the sense that it has the properties of a random sample taken from a bivariate normal distribution with certain mean values μ_x and μ_y, certain standard deviations σ_x and σ_y, and a certain correlation coefficient ρ. In the normal way we estimate μ_x and μ_y by

$$\bar{x} = \frac{1}{N} \sum x_j, \qquad \bar{y} = \frac{1}{N} \sum y_j \qquad (9)$$

Using Formulas 7 we obtain from Eqs. 8 estimates $\bar{C}$ for the amplitude C and $\bar{\phi}$ for the acrophase ϕ:

$$\bar{x} = \bar{C} \cos \bar{\phi}, \qquad \bar{y} = \bar{C} \sin \bar{\phi} \qquad (10)$$

$\bar{C}$ and $\bar{\phi}$ are simply the polar coordinates of the point $(\bar{x}, \bar{y})$.

We also estimate the standard deviations of x_j and y_j by

$$s_x = \frac{\sum (x_j - \bar{x})^2}{N-1}, \qquad s_y = \frac{\sum (y_j - \bar{y})^2}{N-1} \qquad (11)$$

the standard errors by

$$s_{\bar{x}} = s_x/\sqrt{N}, \qquad s_{\bar{y}} = s_y/\sqrt{N} \qquad (12)$$

and finally the correlation coefficient by

$$r = \frac{\sum (x_j - \bar{x})(y_j - \bar{y})/(N-1)}{s_x s_y} \qquad (13)$$

Then it is well-known (10, 11) that the *confidence ellipse* of (μ_x, μ_y) is given

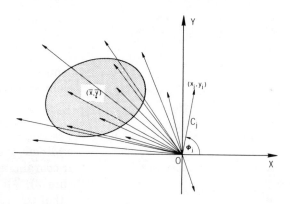

Figure 3   A sample of acrophases and their related amplitudes. The mean of the two-dimensional scatter diagram is the point $(\bar{x}, \bar{y})$. A confidence ellipse for the theoretical midpoint (μ_x, μ_y) is also shown.

by the inequality

$$\left(\frac{\bar{x}-\mu_x}{s_{\bar{x}}}\right)^2 - 2r\left(\frac{\bar{x}-\mu_x}{s_{\bar{x}}}\right)\left(\frac{\bar{y}-\mu_y}{s_{\bar{y}}}\right) + \left(\frac{\bar{y}-\mu_y}{s_{\bar{y}}}\right)^2 \leq 2(1-r^2)\frac{N-1}{N-2}F_{2,N-2,\alpha}$$

(14)

where $F_{2,N-2,\alpha}$ denotes Fisher's F with 2 and $N-2$ degrees of freedom and level of significance α.

Figure 3 depicts the 95% confidence ellipse for the unknown point (μ_x,μ_y). The midpoint of the ellipse is $(\bar{x},\bar{y})$. Computer programs for performing all calculations as well as for plotting the ellipse are available.

We may use the confidence ellipse for testing the null hypothesis H_0 which states

$$H_0: \text{There is no preferred acrophase}$$

When the confidence ellipse contains the origin O. then $\mu_x = 0$ and $\mu_y = 0$ is a possibility that could not be excluded statistically. This happens when the acrophases $\hat{\phi}_j$ are scattered into all directions. In this case we have no reason to reject the null hypothesis H_0.

On the other hand, when the confidence ellipse excludes the origin O, $\bar{C}$ is significantly different from zero. We also see that the acrophases $\hat{\phi}_j$ are concentrated around a mean acrophase $\bar{\phi}$. Hence we reject the null hypothesis H_o.

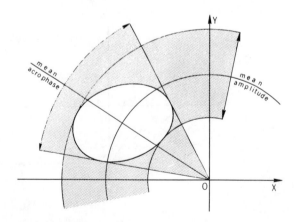

Figure 4 Approximate confidence intervals for the mean acrophase and the mean amplitude.

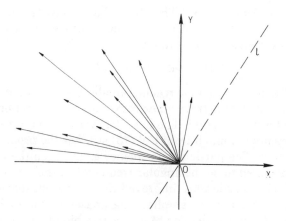

Figure 5   Hodges' bivariate sign test applied to the sample of Figure 3.

In the case of significance the confidence ellipse may also be used to determine approximate confidence intervals for the amplitude C and the acrophase ϕ which are defined implicitly by $\mu_x = C \cos \phi$ and $\mu_y = C \sin \phi$ (see Figure 4).

For applications see, for example, Refs. 12–14.

An alternative test for statistical significance is based on Hodges' *bivariate sign test* (15). We only have to plot the scatter diagram of Figure 3 with vectors (x_j, y_j). Then we draw a straight line l through the origin in such a way that a minimal number of vectors is left on one side of line l (Figure 5). We denote the minimal number of vectors by K and use it as a test statistic. If K is sufficiently small, we reject the null hypothesis. Tables of critical K values may be found in Refs. 4 and 15.

RHYTHMS WITH UNKNOWN PERIOD

Let us begin with the model given by Eq. 5. Whereas we had to deal with only three unknown parameters in the previous section, all four parameters C_o, C, ω, and ϕ are now considered as unknown quantities. The angular frequency ω causes a nasty problem since the model is not linear in ω.

One of the early methods of estimating ω was proposed by Schuster in a paper with the remarkable title "On the Investigation of Hidden Periodicities with Application to a Supposed 26 Day Period of

Meteorological Data" (16). The basic idea consists   in defining a certain functional $I[z(t)]$ with the following property: When $I[z(t)]$ is applied to a known sinusoidal function

$$z(t) \;=\; C_0 + C \cos (\omega_0 t + \phi) \tag{15}$$

then $I[z(t)]$ as a function of ω reaches an absolute maximum at $\omega = \omega_o$, which is quite different from other maxima of this function. $I[z(t)] = f(\omega)$ is called a *periodogram*. The periodogram of an empirical function which differs not too much from Eq. (15) still has the property of a distinct maximum. The particular frequency ω_1 at which the maximum is taken is considered to be the angular frequency of the process.

Stumpff (17) considerably improved Schuster's approach. A further development especially by emphasizing phases is due to Blume (18). He also offered a rich collection of instructive biological examples.

Related methods were proposed by Whittaker and Robinson (8) and by Lanczos (19). For an expository presentation of all sorts of periodogram methods, see Ref. 20.

Despite its merits, periodogram analysis has drawn considerable criticism from statisticians who rightly feared that frequencies suggested by periodograms might sometimes be spurious. They also complained that standard statistical methods are not applicable. To be sure, attempts have been made to introduce statistical tests of significance, but such tests cover only simple cases and can seldom be used in down-to-earth problems (see Refs. 20–22).

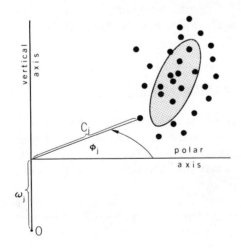

Figure 6   Application of the extended Cosinor method.

For the analysis of biological rhythms another approach seems to be more promising. Let us assume that the period is wobbling to a certain extent. Then the procedure is a straight-forward extension of the cosinor method described in the section on Rhythms with Known Periods.

Consider N individuals (or N biological units) for which we measure a rhythmic quantity $z(t)$ individually. Although we do not know the angular frequency ω, we may still apply least square technique to fit a function of Type Eq. 5 to every individual empirical function. Due to the nonlinearity of the problem, the computer program is somewhat more complicated than in the linear case of the section on Rhythms with Known Periods. Nevertheless we get estimates $\hat{\omega}_j$, $\hat{C}_j$, and $\hat{\phi}_j$ ($j = 1, 2, \ldots, N$) for the hypothetical parameters ω, C, and ϕ. In what follows we drop the hat sign for the ease of writing. Thus we arrive at a set of values

$$\omega_1\ \omega_2\ \cdots\ \omega_N$$

$$C_1\ C_2\ \cdots\ C_N$$

$$\phi_1\ \phi_2\ \cdots\ \phi_N$$

which we may treat like observed quantities.

We may plot each point (ω, C_j, ϕ_j) in the three-dimensional space observing, however, the circular nature of ϕ_j. More specifically, let ω_j be plotted along a vertical axis from an origin O on Figure 6. Then, in a horizontal plane at the "level" ω_j, we plot C_j and ϕ_j as polar coordinates of a point. The pole is on the vertical axis, and the polar axis points into a prefixed direction. The N points (ω_j, C_j, ϕ_j) form a scatter diagram in space. Assume that this scatter diagram qualifies to be considered as a random sample from a three-dimensional normal distribution. In this case we can apply standard techniques of multivariate statistical analysis to determine a confidence ellipsoid for the unknown parameters ω, C, and ϕ (Figure 6). Midpoint of the ellipsoid is the sample center of gravity, $\bar{\omega}$, $\bar{C}$, $\bar{\phi}$, which we determine by

$$\bar{\omega} = \frac{1}{N} \sum \omega_j \tag{16}$$

and by Formula 10.

Testing significance and defining confidence intervals for ω, C, and ϕ are performed along the same lines as in the section on Rhythms with Known Periods.

When a *multitude of frequencies* is involved, periodogram methods as well as an extension of Cosinor may fail. The past two decades have featured a major development of Fourier analysis. The basic model is a "stationary" process with a "spectrum" of finitely or infinitely many frequencies (22–24). The method requires large series of data which biologists can supply only under favorable conditions. An instructive example for a successful application of this type of spectral analysis is the analysis of brain waves (25, 26).

REFERENCES

1. E. Batschelet, *Introduction to Mathematics for Life Scientists*, Springer, Berlin, Heidelberg, New York, 1971, p. 111.

2. K. V. Mardia, *Statistics of Directional Data*, Academic Press, London, New York, 1972, p. 57.

3. E. Batschelet, *Statistical Methods for the Analysis of Problems in Animal Orientation and Certain Biological Rhythms*, Amer. Inst. Biol. Sciences, Washington, D.C., 1965, p. 10.

4. E. Batschelet, in *Animal Orientation and Navigation* S. R. Galler, K. Schmidt-Koenig, G. J. Jacobs, and R. E. Belleville, (Eds.) NASA Symposium at Wallops Station, Va., U.S. Gov. Printing Office, Washington, D.C., 1972, p. 75.

5. A. Sollberger, *Biological Rhythm Research*, Elsevier, Amsterdam, London, New York, 1965, pp. 14, 53.

6. H. Scheffé, *The Analysis of Variance*, Wiley, New York, 1959, p. 14.

7. C. I. Bliss, *Statistics in Biology*, Vol. 2, McGraw-Hill, New York, 1970, Chap. 17.

8. E. Whittaker, and G. Robinson, *The Calculus of Observations*, Blackie, London, Glasgow, 1946, Chap. 10.

9. F. Halberg, Y. L. Tong, and E. A. Johnson, in *Cellular Aspects of Biorhythms*, Springer, 1967, pp. 20–48.

10. A. Hald, *Statistical Theory with Engineering Applications*, Wiley, New York, 1952, p. 607.

11. T. W. Anderson, *An Introduction to Multivariate Statistical Analysis*, Wiley, New York, 1958, pp. 107, 108.

12. F. Halberg, "Chronobiology," *Ann. Rev. Physiol.* **31** 1969, 675.

13. F. Halberg, et al., *Experientia*, **25** (1969), 107.

14. F. Halberg, E. A. Johnson, W. Nelson, W. Runge, and R. Sothern, *Physiol. Teacher* **1**, No. 4 (1972), 1.

15. J. L. Hodges, *Ann. Math. Stat.* **26** (1955), 523.

16. A. Schuster, *Terrestrial Magnetism*, **3** (1898), 13.

17. K. Stumpff, *Grundlagen und Methoden der Periodenforschung*, Springer, Berlin, 1937.

18. J. Blume, *Nachweis von Perioden durch Phasen- und Amplitudendiagramm mit Anwendungen aus der Biologie, Medizin und Psychologie*, Westdeutscher Verlag, Cologne, Opladen, 1965.

19. C. Lanczos, *Applied Analysis*, Prentice-Hall, Englewood Cliffs, N.J., 1956.

20. H. L. Wiener, "A Test of Significance for Periodogram Analysis," Ph.D. Thesis, Catholic University, Washington, D.C., 1971.

21. R. A. Fisher, *Proc. Roy. Soc., Ser. A,* **125** (1929), 54.

22. T. W. Anderson, *The Statistical Analysis of Time Series*, Wiley, New York, 1971.

23. R. B. Blackman, and J. W. Tukey, *The Measurement of Power Spectra*, Dover, New York, 1958.

24. E. Parzen, *Time Series Analysis Papers*, Holden Day, San Francisco, 1967.

25. G. Dumermuth, P. J. Huber, B. Kleiner, and T. Gasser, *IEEE Trans. Audio Electroacoustics*, **AU-18** (1970), 404.

26. D. O. Walter, *Exp. Neurol.*, **8** (1963), 155.

CHAPTER 3

Rhythmometry Made Easy

RUSSELL K. HOBBIE

School of Physics and Astronomy, University of Minnesota,
Minneapolis, Minnesota

FRANZ HALBERG

Departments of Pathology and Physiology, University of
Minnesota, Minneapolis, Minnesota

Although least squares techniques have been used to fit curves to data
for many years, the conceptual simplicity of the technique is often obscured by the algebra necessary to apply it. This paper describes an 8-
minute motion picture, *Rhythmometry in Biology and Medicine*, which
was produced by us in 1971 for visual presentation of the least squares
concept. The data are approximated by a cosine wave, shape of which
is adjusted to give the best fit to the data using the least squares
criterion.

The biological example used is spontaneous human births (1), which
show a pronounced circadian rhythm. This example was chosen because a large amount of data is available so that statistical fluctuations do not obscure the procedure. Nonetheless, the technique can
be used when there are such fluctuations, though the values of the
parameters that are determined will be less certain because of the
noise.

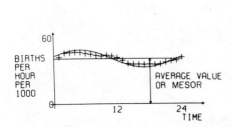

Figure 1 *Birth rate per hour is plotted versus time of day. The mesor is the steady value around which the birth rate fluctuates.*

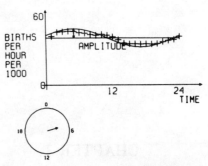

Figure 2 *The amplitude is the maximum excursion of the sine curve above and below the steady value or level. It is represented by the length of the arrow in the clock face in the lower left-hand corner.*

PARAMETERS

Before showing the least squares technique, the film defines the three parameters that will be determined: mesor, amplitude, and acrophase. The *mesor* is the steady, time-independent value around which the data vary (Figure 1). If the data points are equally spaced throughout the day, it is best approximated by the average of all the data. However,

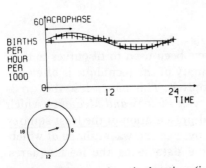

Figure 3 *The acrophase is that time (in hours) or phase (in degrees) at which the sine curve has its peak value. It is represented by the time to which the arrow points on the 24-hour clock in the lower left-hand corner.*

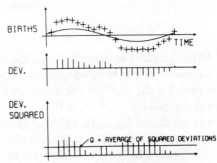

Figure 4 *The deviation is the difference between the data point and the value calculated from the approximation. The parameters in the approximate expression are varied to minimize Q, which is the average value of the squares of the deviations. The level has been subtracted in this and subsequent figures better to display the small differences between the data points and the approximation.*

since the data may not be so conveniently spaced, it is better to think of it as the average value of the curve that we are trying to pass through the data.

The amplitude and the acrophase are the two parameters that describe the shape of the cosign wave that approximated the data. A third parameter, the period of repetition of the sine wave, has been taken here as exactly 24 hours, because data from many days and places have all been lumped together using local time as the independent variable; this procedure averages out any rhythms with periods that are not 24 hours or submultiples of it. The *amplitude* is shown in Figure 2. It is the maximum excursion of the sine wave above and below the level value. The *acrophase* is the time of day at which the sine wave has its peak value (Figure 3). It may be given as a clock time (in this case 5 A.M. or 05⁰⁰) or as the number of degrees around a 24-hour clock face (in this case 75°). Both the acrophase and the amplitude may be conveniently represented by an arrow drawn on a 24-hour clock face, as shown in the lower left-hand corner. The length of the arrow represents the amplitude, and the time to which the arrow points is the acrophase. Such a plot is called a cosinor. The accuracy with which the amplitude and acrophase can be specified.

LEAST SQUARES

The film shows qualitatively how the mesor, amplitude, and acrophase can be adjusted to fit the data. If we wish to have a quantitative measure of how good the fit is, we must adopt some convention for making this measurement. The least squares technique is one such

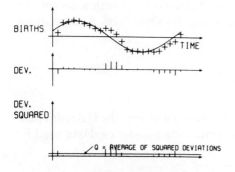

Figure 5   The amplitude and acrophase have been adjusted to give the smallest possible value of Q. *Compare the value of* Q *with that in Figure 4.*

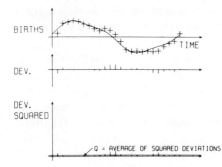

Figure 6   An additional term with a period of 8 hours (the third harmonic) has been added to the approximate expression used in Figure 5. Note the further decrease in Q.

convention. We will call the difference between the experimentally observed points and our "theoretical" curve the deviation (Figure 4). This deviation may be positive or negative. If we square this, we have a quantity which is always positive. The value of the squared deviation averaged over all the data points is the mean square error, which we call Q. We wish to adjust the three parameters—mesor, amplitude, and acrophase—to minimize the mean square error, Q. That is why this technique is called the method of least squares. The film shows how adjustments of the amplitude and acrophase will reduce Q; this may also be seen by comparing Figure 4 and 5.

In many cases an even better fit to the data points can be obtained by adding higher harmonics of the original cosine wave. That is, if the fundamental frequency is 1 cycle/day (a period of 24 hours), a better fit may involve additional cosine waves which go through 2 cycles/day (a period of 12 hours), 3 cycles/day (8 hours), 4 cycles/day (6 hours), and so on. In this case an additional term with a period of 8 hours improved the fit considerably can be seen from a comparison of Figure 5 with Figure 6.

ACKNOWLEDGMENTS

We are grateful to Dennis Villella for helping to run the Calcomp plotter and to Kevin McMahon for programming assistance. Data used for additional examples in the film were supplied by R. Harner, L. Scheving, Jr., R. Sothern, and D. Villella. This work was supported in part by NASA.

REFERENCE

1. I. H. Kaiser and F. Halberg, "Circadian Periodic Aspects of Birth," *Ann. N.Y. Acad. Sci.,* **98**, (1962), 1056–1068.

CHAPTER 4

Temperature Rhythms Analyzed by Multivariate Template

WILLIAM M. RAND

Department of Nutrition & Food Science, Massachusetts Institute of Technology, Cambridge, Massachusetts

A standard approach to the study of a biological system is the gathering of data and the searching for patterns. The next step is the interpretation of these patterns, and here, as in the search for them, mathematical and statistical analyses can be important. The investigator of a biological system has the problem of identifying, and separating if possible, four fundamental types of patterns. The first are those due to the components of the system, the mechanisms by which the system operates; second are those which embody the purpose of the system, what it is for or does; third are those random patterns which arise in any system as complex and variable as a biological one; and fourth are those introduced by the methods of data gathering and analysis.

This fourth type is especially important in the consideration of time-repetitive patterns, biological rhythms, since these are standardly analyzed by a single type of method, some form of periodic regression. [See Bliss (1) for a good exposition of periodic regression.] These methods assume that a biological rhythm is defined by a sequence of measurements taken at specific times or intervals of time and proceed using both the levels of the variable and their associated times of occurrence to estimate the coefficients of some specific continuous model. These coefficients are then assumed to represent the important aspects of the biological system under study. This procedure may introduce not only its own patterns but also may obscure the patterns of mechanism and purpose when they are not directly related to both time and level.

An extended definition of a biorhythm is that it is a repeating sequence of events which may be represented as either levels at fixed times, levels at the occurrence of specific events, or the times of these events. An alternative to periodic regression is proposed to deal with this extended situation.

THE TEMPLATE METHOD

Time-repetitive patterns can be represented by a fixed number, say p, of observations for each repetition. These observations can be the levels of a variable at predetermined times of the pattern, the levels of a variable or different variables at predetermined events of the pattern, or the times of the events defining the pattern. Thus the time series that represents the rhythm splits naturally into segments of length p, where the starting point of the segments is arbitrary. The data can alternatively be conceptualized as a sequence of points (vectors) in p-dimensional space, each time or event being an axis. Due to the inherent variability in any biological system, the points will not coincide and the natural model is then the p-dimensional normal distribution.

Given measurements of a system, a template can be constructed consisting of the average vector and the associated covariance matrix, estimates of the parameters of the proposed multivariate normal distribution. This template is proposed as containing the essentials of the measured biological system and can be used, for example, as a model of the system with which to compare either other systems or the given system under other situations. The natural measure of difference in this p-dimensional space is Hotelling's T^2 which is basically Euclidean distance normalized by the variability along each axis and the correlations between measurements. [See Anderson (2) for a good introduction to multivariate statistics, and Rand (3) for development and discussion of this template method.]

AN EXAMPLE

The biological system to be considered is the temperature cycle of a single white rat which lived for a number of days under one temporal pattern of lighting and then for a number of days under another. The analysis presented considers the data, the temperature of the rat measured every 4 hours, as if each day were a point in six-dimensional space. These points, for the period when the rat has a stable rhythm, can be considered as a sample from a six-dimensional multivariate normal distribution.

The data are shown in Table 1. They represent 21 consecutive daily temperature cycles of a rat whose cage was lighted from noon to midnight for the first 10 days; on the 11th day the light remained on until 8:00 A.M. and thereafter the cage was lighted from 8:00 P.M. until 8:00 A.M. the next day. The data points used are those for hours 0, 4, 8, 12, 16, and 20. The data show a 24-hour rhythm which initially has its maximum at 4 hours and its minimum at 16 hours. Between days 11 and 12 the rhythm changes so that ultimately its maximum is at 12 hours and its minimum at 0 hours.

Table 1   Temperature Data (°C) of a White Rat under Strictly Controlled Lighting Patterns

	Hour					
Day	0	4	8	12	16	20
1	36.16	36.33	35.84	35.52	35.16	35.66
2	36.10	36.60	36.14	35.62	35.29	35.56
3	35.81	36.08	35.82	35.52	34.98	35.33
4	35.80	36.13	35.76	35.30	35.12	35.20
5	35.51	35.89	35.89	35.74	35.21	35.43
6	35.70	36.25	35.93	35.57	35.37	35.79
7	35.95	36.08	35.65	35.45	35.07	35.52
8	35.85	36.08	35.60	35.33	35.14	35.39
9	35.82	36.16	36.06	35.60	35.25	35.54
10	36.08	36.38	35.85	35.34	35.05	35.29
11	35.60	36.21	35.70	35.42	35.14	35.15
12	35.19	35.52	35.89	35.92	35.71	35.60
13	35.60	35.85	35.86	36.11	35.82	35.66
14	35.42	35.71	35.90	36.03	35.62	35.42
15	35.42	35.86	36.09	36.03	35.75	35.52
16	35.35	35.55	36.04	36.51	36.16	35.59
17	35.16	35.70	36.02	36.33	35.72	35.42
18	35.14	35.57	35.93	36.28	35.95	35.68
19	35.44	35.86	36.22	36.46	36.05	35.74
20	35.35	35.88	36.00	36.33	35.64	35.44
21	35.25	35.73	36.04	36.25	35.91	35.58

During the first nine days the system is defined as having a stable rhythm—being in equilibrium. The average values at each time can be written as the vector

$$\mathbf{x}_1 = (35.86, 36.18, 35.85, 35.52, 35.18, 35.49)$$

with the corresponding vector of standard deviations

$$\mathbf{s} = (0.20, 0.20, 0.18, 0.14, 0.12, 0.18)$$

Fundamental to the basic assumption of the existence of a rhythm is that the observations are correlated; here the sample correlation matrix is

$$R = \begin{pmatrix} 1.00 & 0.75 & 0.08 & -0.25 & -0.10 & 0.24 \\ 0.75 & 1.00 & 0.58 & 0.08 & 0.45 & 0.46 \\ 0.08 & 0.58 & 1.00 & 0.71 & 0.64 & 0.40 \\ -0.25 & 0.08 & 0.71 & 1.00 & 0.46 & 0.47 \\ -0.10 & 0.45 & 0.64 & 0.46 & 1.00 & 0.68 \\ 0.24 & 0.46 & 0.40 & 0.47 & 0.68 & 1.00 \end{pmatrix}$$

These data can be combined with an assumption of normality of error and the multivariate normal distribution used. We assume

$$\mathbf{x} \sim N(\mathbf{\mu}, \textstyle\sum)$$

where μ is the mean vector and $\sum$ the covariance matrix. The sample estimate of these parameters will be called the template of the system during this period:

$$T_1 = (\bar{\mathbf{x}}_1, S_1)$$

where S_1 is the sample covariance matrix:

$$S_1 = \begin{pmatrix} 0.039 & 0.030 & 0.003 & -0.007 & -0.002 & 0.008 \\ 0.030 & 0.040 & 0.020 & 0.002 & 0.011 & 0.016 \\ 0.003 & 0.020 & 0.031 & 0.018 & 0.013 & 0.013 \\ -0.007 & 0.002 & 0.018 & 0.020 & 0.008 & 0.012 \\ -0.002 & 0.011 & 0.013 & 0.008 & 0.014 & 0.014 \\ 0.008 & 0.016 & 0.013 & 0.012 & 0.014 & 0.031 \end{pmatrix}$$

Four questions about this rhythm are considered in turn. (1) Is there a rhythm during this initial period? (2) Does it change later? (3) If it changes, how does it change? (4) When does it change?

Is There a Rhythm?

Of first interest is the question of whether or not a rhythm exists, as opposed to there being too much noise in the data to make a conclusive statement. Subject to certain assumptions regarding the periodicity and the sample (3), this can be considered as asking whether the differences in the individual components of the mean vector **x** are statistically significant. An easier, statistically equivalent, approach is to ask whether the successive differences of the individual components are significantly different from zero. Thus, instead of examining the six-dimensional vectors, $\mathbf{x} = (x_1, x_2, x_3, x_4, x_5, x_6)$, we look at the vectors $\mathbf{y} = (y_1, \cdots, y_5) = (x_1 - x_2, \cdots, x_5 - x_6)$. This gives rise to the new template;

$$T = (\bar{\mathbf{y}}, S_y)$$

where

$$\bar{\mathbf{y}} = (-0.322, \quad 0.323, \quad 0.338, \quad 0.340, \quad -0.314)$$

and

$$S_y = \begin{pmatrix} 0.020 & 0.007 & -0.008 & 0.004 & -0.005 \\ 0.007 & 0.030 & 0.005 & -0.013 & -0.006 \\ -0.008 & 0.005 & 0.016 & -0.008 & 0.005 \\ 0.004 & -0.013 & -0.008 & 0.018 & -0.004 \\ -0.005 & -0.006 & 0.005 & -0.004 & 0.017 \end{pmatrix}$$

The problem is then to compare this template with the vector $\mathbf{0} = (0, 0, 0, 0, 0)$. The natural measure is Hotelling's T^2, the multivariate analogue of the Student's t. Given a sample of size N with mean vector $\bar{\mathbf{x}}$ and sample covariance matrix S, the distance to some vector $\boldsymbol{\mu}$ is defined as

$$T^2 = N(\bar{\mathbf{x}} - \boldsymbol{\mu}) S^{-1} (\bar{\mathbf{x}} - \boldsymbol{\mu})'$$

This is the standard statistic used for testing hypotheses and constructing confidence intervals about the mean vector of a population in multivariate normal situations. It essentially is the sum of the squared differences of the individual components of the vectors involved, normalized by the estimated variances and correlations. Its distribution, under the hypothesis that the sample is from a population with μ as its mean vector, is that of $(N - 1)(p/N - p)$ times an F distribution with p and $N - p$ degrees of freedom, where p is the dimension of the vectors involved. For convenience is this presentation, the distance between two mean vectors, given a covariance structure, will be written as $D(\bar{\mathbf{x}},\bar{\mathbf{y}},S)$ and defined as that function of the appropriate T^2 which itself has an F distribution. In the present situation, comparison of $\bar{\mathbf{y}}$ with the zero vector gives

$$D(\bar{\mathbf{y}},\mathbf{0},S_y) = 48.92$$

which is well beyond the 0.005 level of $F(5,4)$, establishing that a rhythm does exist.

Does the Rhythm Change?

The next question is whether the rhythm permanently changes. From inspection of the data it appears that a new rhythm is established by at least day 14. Defining the data from this day on as a second equilibrium, a second template can be calculated:

$$T_2 = (\bar{\mathbf{x}}_2, S_2)$$

$$\bar{\mathbf{x}}_2 = (\ 35.32, \quad 35.73, \quad 36.03, \quad 36.28, \quad 35.85, \quad 35.55)$$

$$S_2 = \begin{pmatrix} 0.014 & 0.008 & 0.005 & -0.004 & -0.001 & 0.000 \\ 0.008 & 0.017 & 0.006 & -0.006 & -0.012 & -0.002 \\ 0.005 & 0.006 & 0.010 & 0.007 & 0.009 & 0.006 \\ -0.004 & -0.006 & 0.007 & 0.031 & 0.024 & 0.010 \\ -0.001 & -0.012 & 0.009 & 0.024 & 0.039 & 0.020 \\ 0.000 & -0.002 & 0.006 & 0.010 & 0.020 & 0.015 \end{pmatrix}$$

Transformation of this template to successive differences results in T_y

$$= (\bar{y}, S_y):$$

$$\bar{y} = (-0.416, \quad -0.298, \quad -0.248, \quad 0.428, \quad 0.301)$$

$$S_y = \begin{pmatrix} 0.014 & -0.006 & -0.003 & -0.009 & 0.008 \\ -0.006 & 0.013 & 0.010 & 0.008 & -0.012 \\ -0.003 & 0.010 & 0.026 & -0.008 & -0.012 \\ -0.009 & 0.008 & -0.008 & 0.021 & -0.005 \\ 0.008 & -0.012 & -0.012 & -0.005 & 0.014 \end{pmatrix}$$

Comparison of this with the zero vector gives

$$D(\bar{y}, 0, S_y) = 45.93$$

significant at the 0.005 level of $F(5,3)$, supporting the establishment of a new rhythm.

Given the existence of both an initial and a final rhythm, the two may be compared:

$$D(\bar{x}_1, \bar{x}_2, S) = 20.38$$

where S is the pooled covariance matrix. Clearly this is significant at the 0.001 level of $F(6,10)$. Thus a new rhythm is established and differs from the initial one.

How Does the Rhythm Change?

Before the experiment it was predicted that the rhythm would follow the lighting scheme and advance 8 hours. This can be tested by comparison of the initial equilibrium template with the proper cyclic permutation of the final equilibrium template. Here we want to compare the initial vectors with the final vectors reordered as $(x_5, x_6, x_1, x_2, x_3, x_4)$. We define this permuted template as

$$T_2^{(2)} = (\bar{x}_2^{(2)}, S_2^{(2)})$$

and calculate

$$D(\bar{x}_1, \bar{x}_2^{(2)}, S) = 1.59$$

Table 2 *"Distance" Between the Initial Equilibrium Rhythm and the Rhythm while the System is Changing*

Day	Hour	x	D_t
10	0	36.08	2.10
	4	36.38	.38
	8	35.85	.98
	12	35.34	.35
	16	35.05	1.05
	20	35.29	.41
11	0	35.60	4.51
	4	36.21	2.41
	8	35.70	2.48
	12	35.42	2.55
	16	35.14	1.85
	20	35.15	2.18
12	0	35.19	7.63
	4	35.52	1.41
	8	35.89	3.08
	12	35.92	2.33
	16	35.71	16.42
	20	35.60	9.26
13	0	35.60	18.06
	4	35.85	11.64
	8	35.86	12.01
	12	36.11	16.06
	16	35.82	20.82
	20	35.66	19.51
14	0	35.42	16.17
	4	35.71	17.45
	8	35.90	16.77
	12	36.03	15.15
	16	35.62	8.23
	20	35.42	11.73

where S is the pooled covariance matrix. This is obviously not significant ($p > .25$) for F distributed with 6 and 10 degrees of freedom, and the hypothesis is accepted that the later rhythm is indistinguishable from the initial equilibrium advanced by 8 hours.

If we were merely examining the data for an indication of trend, the full sequence of D's for each permutation would be of interest (i is the number of components shifted):

	$D(\bar{x}_1,\bar{x}_2{}^{(i)},S)$	Shift (hours)
$i = 0$	20.38	0
1	24.20	4
2	1.59	8
3	37.58	12
4	115.05	16
5	62.64	20

From this it is obvious that only the hypothesis of an 8-hour advance would not have been rejected.

When Does the Rhythm Change?

Finally, just when the rhythm changes is of interest. Examination of this involves the calculation of the sequence of $D_t = D(\bar{x}_1,\mathbf{x}_t,S)$, for each t, where $\mathbf{x}_t$ represents the proper cyclic permutation of the six sampled times ending at time t. This vector $\mathbf{x}_t$ is that permutation of $(x_{t-5}, x_{t-4}, x_{t-3}, x_{t-2}, x_{t-1}, x_t)$ which synchronizes it with the initial equilibrium template. Table 2 lists the data for days 10 through 14 with the associated values of D_t. This data shows that the rhythm is changing on day 11 and 12, after which D_t is consistantly higher. Any number of techniques can be used to define a time of change; the easiest is perhaps the nonparametric test which would choose the first of five consecutive points (since a one-sided test is indicated) which were above the expected median ($F_{(6,3)}(0.5) = 1.13$). This sets the start of the change at time zero on day 11. To define a time of end of change, the sequence can be reversed and the same procedure applied. This defines the end of change at 4:00 P.M. on day 13.

The investigation of this situation can be summarized as follows. Initially there is a definite rhythm which starts to change at time 0 of day 11 to a new rhythm which becomes established at 4:00 P.M. of day 13. This new rhythm is indistinguishable from the old shifted forward by 8 hours.

REFERENCES

1. C. I. Bliss, *Statistics in Biology*, Vol. 2, McGraw-Hill, New York, 1970.
2. T. W. Anderson, *An Introduction to Multivariate Statistical Analysis*, Wiley, New York, 1958.
3. W. M. Rand, Submitted for Publication.

CHAPTER 5

Combined Linear-Nonlinear Chronobiologic Windows by Least Squares Resolve Neighboring Components in a Physiologic Rhythm Spectrum

JOHN RUMMEL

National Aeronautics and Space Administration, Manned Spacecraft Center, Houston, Texas

JUNG-KEUN LEE AND FRANZ HALBERG

Chronobiology Laboratories, University of Minnesota, Minneapolis, Minnesota

The analysis of biologic rhythms necessarily involves physiologic and mathematical techniques for purposes of data collection and interpretation, respectively. The mathematical procedures in use consist largely of techniques introduced earlier for the many nonbiologic purposes of time series analysis (e. g., in astronomy, meteorology, or economics) involving noisy series and hence statistical considerations. The

Noise

Input rhythms (period in hours): $\tau_1 = 50$, $\tau_2 = 24$, $\tau_3 = 17$, and $\tau_4 = 7$

Absent (Series 1)

Present (Series 2)

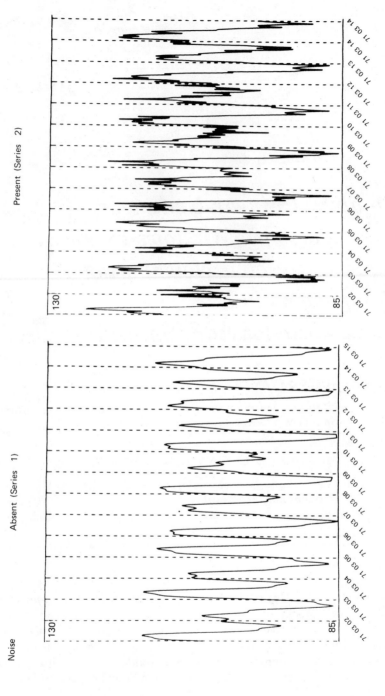

Figure 1 Chronograms of artificial series. (Caption on page 56.)

54

Input rhythms (period in hours): $\tau_1 = 24.8$ and $\tau_2 = 24$

Noise

Absent (Series 3)

Present (Series 4)

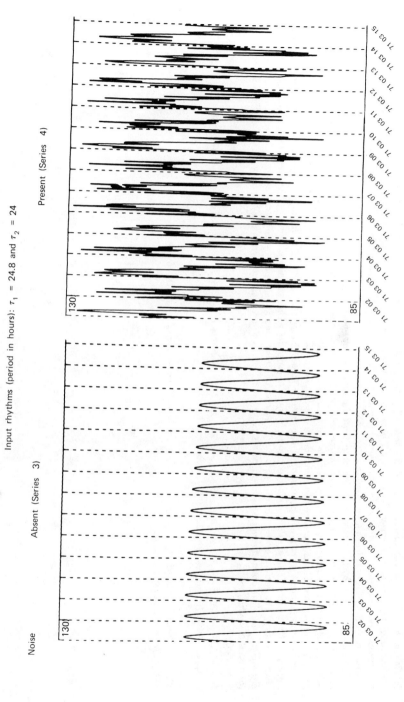

Figure 1 (continued).

55

Figure 1 (continued). Data in arbitrary units are displayed along a horizontal time scale. Vertical dashed lines separate (on this scale) days, labeled at the bottom as year, month, day—in this order. The two numbers displayed at the left provide the upper and lower limits of a convenient range used for all four series. Series 1 and 2 have the same frequency structure, except that Series 2 has noise added to the original four components of Series 1. Series 3 represents the combination of two components with neighboring frequencies, without noise, and Series 4 represents the addition of noise to the two components of Series 3.

Scrutiny of Series 1 by the naked eye leads us to conjecture that several periods occur, yet it is difficult from such macroscopic study to establish any period lengths or other rhythm characteristics. By the same token, Series 3 shows a gradually and steadily decreasing amplitude, suggesting the superposition of two periods and leading us to suspect a beat if the series were longer. No such impression is possible from Series 4 which contains the same two components overlayered by noise.

The point to be made from the inspection of chronograms is that although they can lead us to suspect rhythmicity, it is difficult in the presence of noise and without added (microscopic) techniques to ascertain either the period of such phenomena or other objective rhythm characteristics. The series numbers in this figure refer to the functions used to generate the data, discussed in the section on Illustrative Examples. Thus, for these chronograms, data points $(y_i, e_1 = 1, \cdots, 336)$ are generated from the following trigonometric functions used for all following figures.

We define

$$f(t_i) = 100 + 2 \cos (-\pi/2 + 2\pi/7t_i) + 3 \cos (-\pi/2 + 2\pi/17t_i) + 10 \cos (-\pi/2 + 2\pi/24t_i) + 5 \cos (-\pi/2 + 2\pi/50t_i)$$

$$f^*(t_i) = 100 + 10 \cos (-\pi/2 + 2\pi/24t_i) + 1 \cos (-\pi/2 + 2\pi/24.8t_i).$$

1, $y = f(t_i)$; for Series 2, $y_i = f(t_i) + 10 R(t_i)$; for Series 3, $y_i = f^*(t_i)$; and for Series 4, $y_1 = f^*(t_i) + 20R(t_i)$, respectively, where $\pi = 90°$, t_i are hourly spaced times, and $R(t_i)$ (considered to be noise) are the random numbers chosen from a uniform distribution on the interval [0,1].

wedding of such time series analyses with biologic considerations as well as sampling techniques is most useful to the researcher in physiology if he thus arrives at new displays that clearly reveal to him the biologically important features of a time structure along with an inferential statistical description.

It is the purpose of this paper to introduce some of the displays used in the Chronobiology Laboratories at the University of Minnesota. These involve methods for time series analysis that are based upon contributions by Fourier (1) and Gauss (2) and have more recently been elaborated by Stumpff (3), Blume (4), Chapman (5), Engeli (6), Tong (7), Jurkevich (8), Rummel (9), and others (10–14).

Emphasis will be placed on the results of applying a combination of least squares methods involving a linear model (6) as well as the nonlinear model originally introduced by one of us (J.R.) (9) and presented in the context of displays developed earlier: the chronogram and plexogram (15) (Figures 1 and 2) as classical descriptive statistical summaries (for a first "macroscopic" inspection) and the chronobiologic window (6) (Figure 3), serial section (6, 13) (Figure 4) and cosinor (7) (Figure 5)—"microscopic" or inferential statistical summaries based on the least squares fit of a single cosine function. In these displays the biomedical researcher will wish to pay attention primarily to the given system of coordinates and phase references which furnish our particular "system picture." It was Lotka (16) who wrote that ". . . not a little of the success of scientific investigation depends on the judgment exerted in choosing a suitable system of coordinates, such as will furnish the simplest and most convenient picture." Thus, although in a sense a particular convenient reference frame adds nothing to our knowledge of concrete facts, it nevertheless may constitute a step in the progress of human knowledge.

The term "nonlinear least squares analysis" (8, 17) will be used to describe the simultaneous investigation of rhythm parameters by a suitable approximation and iteration method. Alternatively, the term "linear least squares method" will be reserved for a procedure in which some parameter is fixed in order to linearize a nonlinear model. This latter approach has been most helpful in the past when the probable average period of a rhythm was predictable on biologic grounds; indeed, for this reason the periodogram technique had been successful in relatively early applications to biologic time series analysis (18, 19). By the same token, the linear least squares window has served to quantify as well as detect rhythms, some but not all of them long known to occur since they were prominent from a gross inspection of time plots.

Input rhythms (period in hours): $\tau_1 = 50$, $\tau_2 = 24$, $\tau_3 = 17$, and $\tau_4 = 7$

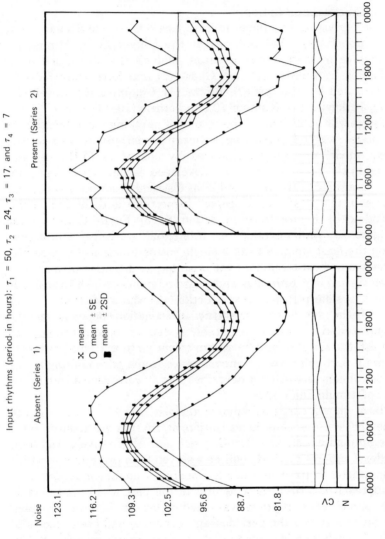

Figure 2 (caption below).

58

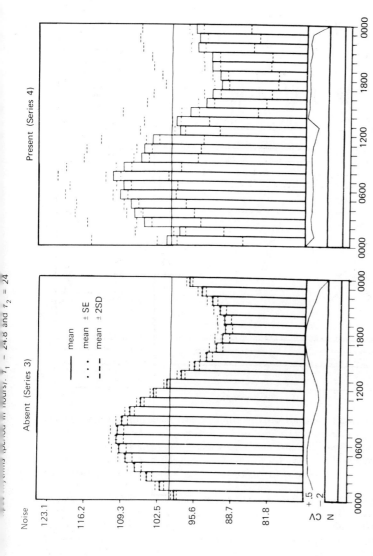

Figure 2 Plexogram of artificial time series covering span of 336 hours folded modulo 24 hours (i.e., as though collected in a single 24-hour span). The mean values of data in each hourly interval are shown as points in Series 1 and 2 and as bars in Series 3 and 4. Confidence limits of (1) standard error and (2) standard deviations about the mean are also indicated and are connected by thin lines in Series 1 and 2 or appear as dotted or dashed lines in Series 3 and 4. The coefficient of variation (CV) and the number of sample points in each hourly interval (N) are shown at the bottom of this figure. By such folding, one can gain in impression as to whether a given periodicity indeed characterizes the data. The periodicity tested may be of a precise 1-day length or it may be some other period found in the same data by separate analysis or anticipated on the basis of earlier evidence.

59

Input rhythms (period in hours):
$\tau_1 = 50$, $\tau_2 = 24$, $\tau_3 = 17$, and $\tau_4 = 7$
Noise absent (Series 1)

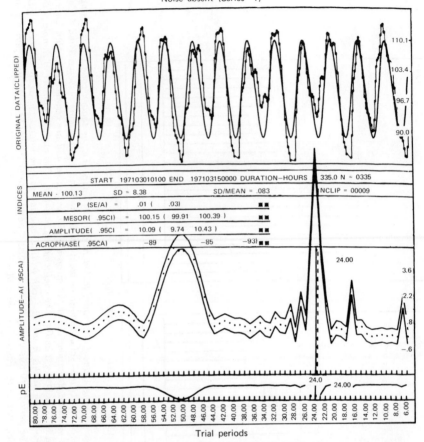

Trial periods

Figure 3a   Chronobiologic windows display captions indicating (as subheading) (1) the period in hours of four input rhythms used to generate the time series and (2) the presence and absence of noise. At the top in each display of artificial time series, the input data are plotted as dots connected by thin lines, and the best-fitting single cosine function determined by the linear least squares analysis is shown as a superimposed heavier curve.

Next to the "Indices" (ordinate), a description of the time series and of some analytical results is provided. Thus start- and end-times are given as numbers in a sequence indicating year, month, day, hour, and minute (the year in four places and the others each in two places). Duration in hours indicates the actual number of hours from start to finish (here 335 hours) and N = total number of data points (here 336 at 1-hour intervals). In the second row the mean, standard deviation (SD) and coefficient of variation (SD/mean) are given along with the number of clipped data in this chronobiologic window—clipping means that data exceeding three standard deviations from the mean in either direction are equalized to the nearest limit. Below the second row are shown results from the best cosine fit (corresponding to the minimum percent error) of original data (not clipped for analysis): p-value, mesor, amplitude and acrophase, as well as the ratio of standard error to the amplitude and the 95% confidence intervals (CI) in parentheses, or the confidence arc (CA) in degrees, the minus sign

60

Noise present (Series 2)

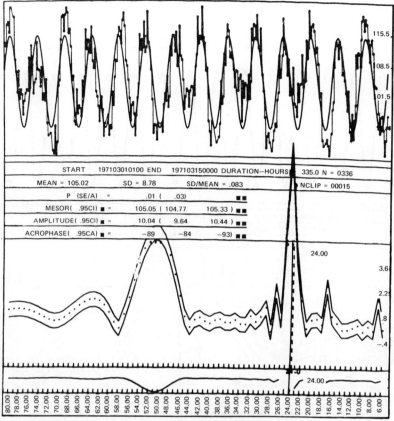

Trial periods

implying a delay from zero φ. Below the indices, 95% CI's envelop the amplitudes plotted as dots as a function of trial periods [the percent error (PE) is shown below]. Vertical solid lines are used to indicate the trial periods attaining the maximal amplitude and minimal percent error, respectively, while the dashed lines indicate the precise 24-hour trial period.

At the bottom the range of trial periods, from 80 to 6 hours, is shown with 1-hour decrements between consecutive trial periods. The linear least squares method here applied precisely detects in both series the prominent 24-hour period and the other true input period of 7, 17, and 50 hours. Only for Series 2 does a print-out reveal (Table 2) that the 50-hour period has been changed to a 51-hour period. Such excellent agreement between input and output constitutes justification for considering all periods detected by the linear least squares method as candidate periods to be checked for validity by other means. Nevertheless, we must not immediately accept all "periods" yielded by the linear method, since it has also produced an artifactual "statistically significant" period (of 27 hours) in Series 1 and two such pseudoperiods (of 27 and 63 hours) in Series 2.

Such statistically significant pseudoperiods will be eliminated by closer scrutiny involving physiologic considerations and/or (as in the case of these artificial series) the concomitant fit in the NLS program of the several periods obtained initially in the LLS program. For this NLS step the initial values consist of one of the mesors and the amplitude and acrophase of each of the several periods yielded by the LLS program.

61

Input rhythms (period in hours): τ_1 = 24.8 and τ_2 = 24

Noise absent (Series 3)

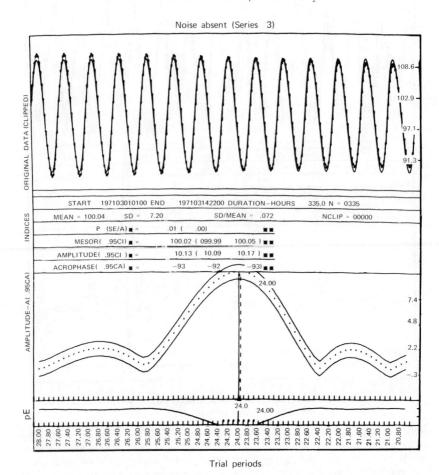

Figure 3b   *Chronobiologic window. Display captions indicate (as subheading) (1) the two input rhythms (of 24 and 24.8 hours) used to generate the time series and (2) the presence and absence of noise. For other details see legend to Figure 3a.*

At the bottom, the range of trial periods (from 28 to 20.8 hours) is indicated with 6-minute intervals between consecutive trial periods. The solid vertical lines used to indicate the trial period attaining the maximal amplitude and minimal percent error, respectively, and the dashed line for the 24-hour period coincide for Series 3. For Series 4 the solid line is separated from the dashed line by only 6 minutes. In this series the

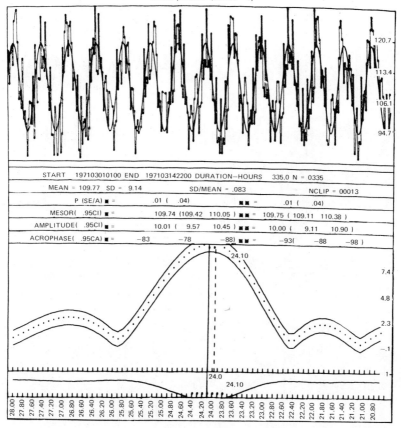

Noise present (Series 4)

Trial periods

numerical outputs obtained from the single cosine fit with a period corresponding to the dashed line (here 24 hours) are also shown—after the *—following the test results from the best-fitting period. The 24- and 24.1-hour periods in Series 3 and 4, respectively, are detected as the prominent periods and show up as peaks in the linear least squares spectrum at those periods. However, instead of detecting the second true input period of 24.8 hours, this linear method yields two statistically significant pseudoperiods for both Series 3 and 4 (Tables 3 and 4).

Such statistically significant pseudoperiods will be eliminated by closer scrutiny involving physiologic considerations and/or (as in the case of these artificial series) the concomitant fit in the NLS program of the several periods obtained initially in the LLS program.

Input rhythms (period in hours): $\tau_1 = 50$, $\tau_2 = 24$, $\tau_3 = 17$, and $\tau_4 = 7$

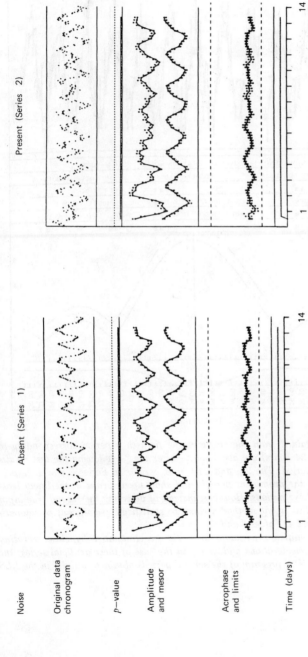

Figure 4 (caption below).

Input rhythms (period in hours): $\tau_1 = 24.8$ and $\tau_2 = 24$

Absent (Series 3)   Present (Series 4)

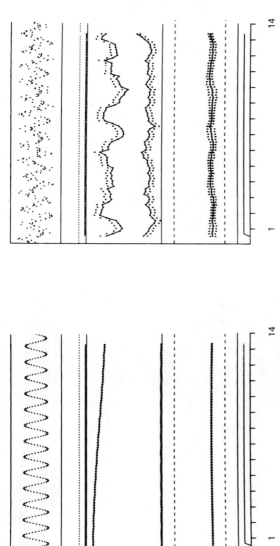

Figure 4   Values shown at the top as original data points are analyzed for consecutive overlapping intervals of 24 hours, displaced in increments of 4 hours. In the second row, p-values, represented on the ordinate by a logarithmic scale, are given to indicate the significance of rhythm description. In a third row, both the amplitude and mesor are given, each with their standard errors, and in a fourth row the acrophase with its limits is shown. At the bottom the time (in days) is indicated (from 1 to 14 days) at 1-day intervals. The 24-hour component analyzed by such serial sections stands out with statistical significance, not only to the naked eye but also in the (second row) p-values computed for these four artificial series.

Changes in mesor, amplitude, and acrophase of Series 1 and 2 reflect the 50-hour period as well as the higher-freuqncy input rhythms. Ther serial sections of Series 3 can be scrutinized for a systematic change in amplitude reflecting the 24.8-hour period superimposed upon the 24-hour period. However, in the presence of noise in Series 4, the naked eye can no longer detect a consistent amplitude change.

65

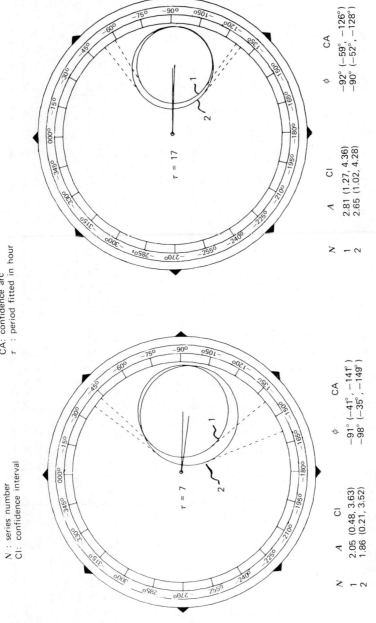

N : series number
CI: confidence interval

CA: confidence arc
τ : period fitted in hour

N	A	CI	φ	CA
1	2.05	(0.48, 3.63)	−91°	(−41°, −141°)
2	1.86	(0.21, 3.52)	−98°	(−35°, −149°)

N	A	CI	φ	CA
1	2.81	(1.27, 4.36)	−92°	(−59°, −126°)
2	2.65	(1.02, 4.28)	−90°	(−52°, −128°)

τ = 7

τ = 17

Figure 5a (caption on page 68).

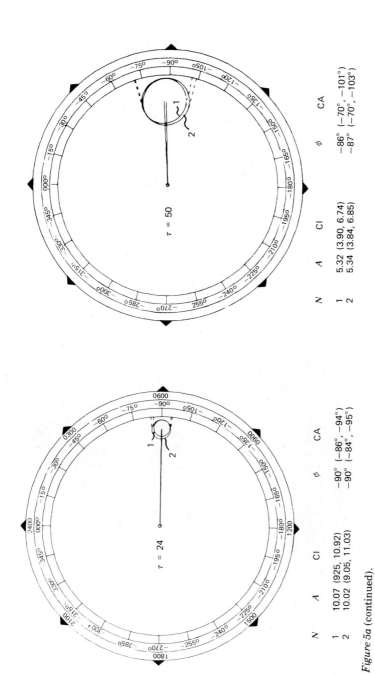

N	A	CI	ϕ	CA
1	5.32	(3.90, 6.74)	−86°	(−70°, −101°)
2	5.34	(3.84, 6.85)	−87°	(−70°, −103°)

$\tau = 50$

N	A	CI	ϕ	CA
1	10.07	(925, 10.92)	−90°	(−86°, −94°)
2	10.02	(9.05, 11.03)	−90°	(−84°, −95°)

$\tau = 24$

Figure 5a (continued).

67

Figure 5a Polar representations of parameters (A, φ) determined by a single cosine fit at the trial period (τ) indicated near each pole are shown with corresponding confidence region and are identified as Series 1 or 2. In Series 1 and 2, the input periods were of 50, 24, 17, and 7 hours and the corresponding amplitudes were 5, 10, 3, and 2, in arbitrary units. All four cosine components had acrophases at −90°. Series 2 contained superimposed noise.

The circle rim shows a scale in neative degrees with the trial period, whatever it may be, equated to 360°. Degrees are given clockwise, hence the negative sign. Corresponding hours are indicated only for the 24-hour period. Under each polar graph the numerical results obtained by fitting a cosine function with the given τ are presented.

When the confidence region overlaps the pole, the trial period chosen for analysis of the series has not been detected as being statistically significant—at least not by fitting a single cosine function. Such overlap does not occur in these illustrations. However, with noise present in Series 2 we find an enlargement of the statistical confidence region as compared to that of Series 1 without noise. Furthermore, the tabulations under each polar presentation show that for each series the prominent period has an amplitude corresponding roughly to the input A. The relative area of the confidence regions seems to be proportional to the percent error (shown at the bottom of Figure 3a).

CI: confidence interval τ : period fitted in hour

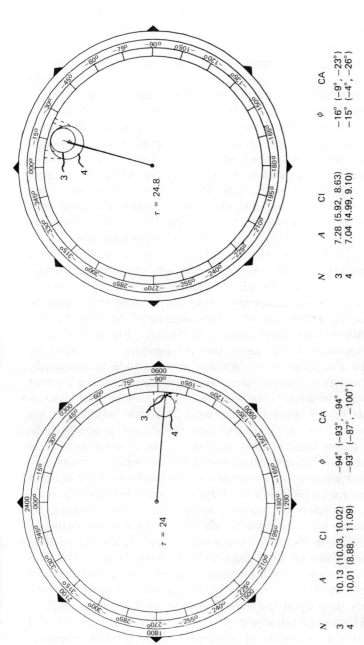

N	A	CI	ϕ	CA
3	10.13	(10.03, 10.02)	−94°	(−93°, −94°)
4	10.01	(8.88, 11.09)	−93°	(−87°, −100°)

$\tau = 24$

N	A	CI	ϕ	CA
3	7.28	(5.92, 8.63)	−16°	(−9°, −23°)
4	7.04	(4.99, 9.10)	−15°	(−4°, −26°)

$\tau = 24.8$

Figure 5b Polar representation of parameters (A, ϕ) obtained from the fit of single cosine functions with either a 24- or a 24.8-hour trial period. Analyses are shown with 95% confidence regions for Series 3 and 4 respectively (see text).

69

Thus Figure 3a shows the resolving power of the linear least squares method in the presence of noise (right) as well as in its absence (left). In this case, four known periods were used to generate the time series. The least squares spectrum resolved these time series clearly, with a peak in the spectrum responding to each known period. Figure 3a shows further the difficulties encountered in the least squares analysis. Whereas in the presence of noise the four prominent (input) periods are all statistically significant, there are two added secondary "statistically significant" pseudoperiods as well, and even in the absence of noise there is one "pseudoperiod" in the linear least squares spectrum—a period that heretofore had to be ignored on the basis of "experience" as a probable side lobe. However, as will be seen, such pseudoperiods can be eliminated by the complementary nonlinear least squares analysis. The functions from which these time series are generated are referred to in the section on Illustrative Examples.

Figure 3b, in turn, shows the limitations of this same linear least squares window. In the chronogram on the left, the presence of two components is macroscopically visible from a progressive decrease in amplitude of a prominent circadian rhythm, a decrease that would be interpreted as the expression of a beat generated possibly by two components with rather similar periods. At the right, Figure 3b shows a second chronogram with the same two components overlayered by noise, and now the progressive decrease in amplitude is completely masked. In these two artificial series each containing both a 24-hour and a 24.8-hour component, the analysis by the linear least squares window reveals the prominent period near 24 hours—whether or not noise is present—and, in addition, two significant pseudoperiods. However, the linear least squares method in the absence as well as presence of noise has failed to detect one of the two input components.

The nonlinear window, as a complement to the linear least squares model, combined with biologic considerations, represents an important step forward: it renders an analysis more reliable (for instance, by eliminating undue focus on certain side lobes) and directly estimates the period along with its confidence interval; that is, the period does not have to be known or fixed separately for each fit as in the LLS model. Most important, as will be demonstrated herein, the combined linear-nonlinear window can resolve in the presence of noise—in time series of hourly data involving even less than 14 days—two rhythms with widely differing amplitudes and with periods as close as 24.0 and 24.8 hours. Such a capability is of great biologic interest whenever lunar effects (generating a 24.8-hour period) might interact with a 24.0-hour serial routine in health or disease.

LINEAR LEAST SQUARES METHOD (LLS)

To assess a biologic rhythm in a given time series we prepare at the outset a chronogram (Figure 1) and, assuming a fixed period, a plexogram (Figure 2) to provide a "macroscopic" view of the data. We then postulate the model of a linear trend (Eq. 1) and estimate its parameters by the least squares method,

$$y_i = M + Bt_i + e_i \qquad i = 1, 2, \ldots, n \qquad (1)$$

where y_i represents a datum obtained at time t_i, n is the number of data points, M and B are parameters, and e_i are independently and identically distributed with normal mean zero and unknown variance. We test the statistical significance of the linear trend (e.g., test the null hypothesis, $H_0:B = 0$) versus the alternative hypothesis, $H_1:B \neq 0$. We next proceed to examine the data for the presence of rhythms, using one of the following two models depending on whether the trend (B) was not significant (Eg. 2) or significant (Eq. 3):

$$y_i = M + A \cos\left[\varphi + (2\pi/\tau)t_i\right] + e_i \qquad (2)$$

$$y_i = M + Bt_i + A \cos\left[\varphi + (2\pi/\tau)t_i\right] + e_i \qquad (3)$$

where M is the mesor (rhythm-adjusted mean), A is the amplitude of cosine, B is the trend detected by Eq. 1 if statistically significant, ϕ is the acrophase (timing of high point of cosine function), and τ is the period of the cosine function. To fit either of these functions to the data, we again use the least squares method. In fact, Models 2 and 3 are not linear in parameters, but one can linearize them by fixing the period, τ, and using the least squares method for estimating parameters in the transformed linear model (6). This method is hereafter called the linear least squares method (LLS)—because of the linearization of the model—distinguishing it from the nonlinear least squares method to be described below.

Although the period fitted at a time is fixed for this procedure, we can select a "best fitting period" by the LLS method provided a time series of appropriate length is available. Given such a time series, we estimate parameters by repeatedly fitting the linearized model to the data, incrementing and/or decrementing the period of the cosine function over an appropriate range, and performing statistical tests for the significance of each period (i.e., $H_0:A = 0$ versus $H_1:A \neq 0$). The results are then plotted as a "spectral window" (amplitude versus period) in order to inspect peaks for which the fitted cosine function is statistically significant. We also determine a "PE" (percent error*) for each fitted cosine function (Figure 3).

Quite often in the case of prominent periods, the LLS method produces satisfactory results, as may be seen from the results of analyses on the artificial data in the table. Sometimes, however, the LLS method may yield incorrect or "significant" pseudoperiods, away from the true periods (see later).

We may also examine a time series by means of a "serial section" (Figure 4), notably when we presume that changes may occur with time in the parameters of (*1*) a period found to be the best-fitting one by the analysis of a series as a whole, or of (*2*) some other period anticipated to characterize the data. For this purpose we choose a fixed length or section of the total series for separate analyses; these sections are called intervals. The analyst also chooses a fixed length for the displacement of this interval, called increment. Accordingly, successive data intervals are either overlapping (increment < interval; progressive serial section) (20) or nonoverlapping (increment = interval; fractionated serial section). At the outset the interval may be equated to the period fitted and the increment (displacement) to the longest Δt (time between successive observations). The interval and increment may have to be lengthened until statistical stability ($p \leq .05$ for zero amplitude) is reached for some part(s) of the series, otherwise the procedure is not applicable. Estimates of acrophase and amplitude are plotted as a function of time.

For each numerical analysis on a new biologic time series, as in chemical determinations, one can use the equivalent of blanks and standards in order to verify the method. For the blanks we generate random series with the same sample mean and standard deviation as the original data and with values at the same times at which original data were obtained. Having generated a number of random series from the standard normal distribution, we fit these series with the same period fitted to the true data and test the significance of amplitude for each random series. One thus can inspect the frequency with which the analyses of such random series can yield "statistically significant " pseudoperiods resembling rhythms.

In the case of standards we generate an artificial series using a cosine or other "regular" model (function), again with values at the same

---

* $PE = \dfrac{\sum_i (y_i - \hat{y}_i)^2}{\sum_i (y_i - \bar{y})^2} \times 100$, where $\hat{y}_i$ presents a predicted value at time t_i and $\bar{y}$ is the sample mean.

times at which the original data were obtained. This standard is fitted and the resulting parameter estimates compared with the known parameters (21) to determine the amount of distortion caused by the sampling schedule.

NONLINEAR LEAST SQUARES METHOD (NLS)

As we mentioned previously, the LLS method may occasionally yield significant pseudoperiods and fail to detect the true period. For instance, it may fail to detect true periods close to each other (say at 24 and 24.8 hours), and instead yield 21-, 24-, and 26-hour periods. This should be kept in mind in studying rhythms in biologic data by LLS.

However, if we are confident from experience that there is one dominant period and that other components are integral divisors of this period, we can use the following harmonic model with the linear least squares analysis:

$$y_i = M + \sum_{k=1}^{s} A_k \cos (\varphi_k + 2k\pi/\tau t_i) + e_i \tag{4}$$

where s is the number of the fitted cosine terms, and we make the same assumption for e_i as in the Model 2. For Model 4 we first fit the presumed dominant cosine component with period τ and test the resulting parameter (i.e., amplitude). If the statistical test is significant, we add the next term (i.e., take $k = 2$), fit, and again test for the significance of the new term. We repeat this procedure until the added term is not significant. This is called the harmonic or waveform analysis.

However, if we suspect that there is more than one period, and in particular when there seems to be no integral relation among periods, we can use the so-called nonlinear least squares method (NLS) to estimate parameters concomitantly according to Model 5 (i.e., Fourier analysis, more general than harmonic analysis):

$$y_i = M + \sum_{k=1}^{s} A_k \cos [\varphi_k + (2\pi/\tau_k)t_i] + e_i \quad i = 1, 2, \ldots, n \tag{5}$$

where n is the number of data points and s is the number of fitted cosine terms. In contrast with the LLS method we now estimate all the parameters in Model 5 simultaneously, varying parameters in the parameter space to minimize the residual sum of squares. This method is called the nonlinear least squares method (NLS). Even though there are many methods, we have used Marquardt's theory and program, described as a maximum neighborhood method. This technique performs an optimal compromise between the gradient method and the fast linearization procedure by the Taylor series. When the latter method is used alone, divergence is frequent, that is, no results will be obtained by the exclusive use of the Taylor series. The gradient method usually converges from an initial guess which can be outside the convergence-region of other methods (17), that is, it tends to yield a numerical result where other methods may not. However, the gradient method may be costly in computer time when used alone, since it often slows down after the first few iterations. Marquardt's compromise combines the merits of the two approaches and avoids some of their shortcomings. For a full description of this technique the reader is referred to original publications (17, 22).

In order to run Marquardt's program, as well as any other program for nonlinear equations, it is necessary to provide a set of initial values for all parameters in the Model 5. The results will depend strongly on these initial values, and so it is useful to utilize for this purpose the statistically significant-appearing periods, amplitudes, and acrophases determined by the LLS method, recalling that this procedure often turns up "significant" pseudoperiods. On the other hand, because we estimate parameters simultaneously in the model for NLS, a period may be indicated by the NLS method that was not statistically significant in the LLS method. For example, in the absence of noise (Table 1) the LLS method shows a statistically significant pseudoperiod ($\tau = 27$) (in addition to the true input periods used to generate the series).

In the first step of the NLS analysis this $\tau = 27$ hours is changed to $\tau = 26.32$ hours. Since the latter period is not statisticaly significant in the LLS output, it is then eliminated from further consideration by the NLS method; the NLS method then is reapplied once again without this particular τ.

On the other hand, the NLS method can identify true periods near major components that obscure them in the LLS. In another example here documented [with a $\tau = 24.0$ and 24.8 hours (used in generating the series) and in absence of noise (Table 3)] the LLS method does not show a secondary dip in percent of error at $\tau = 24.8$ hours in addition to the major minimal percent error at $\tau = 24$ hours, but the LLS provides two pseudo-periods at $\tau = 21.8$ and 26.8 hours.

In the NLS analysis, $\tau = 21.8$ $\tau = 26.8$ hours are merged into the true periods of 24 and 24.8 hours, while the 24-hour period in the NLS remained the same as that in the LLS. In the second step of analysis of the NLS method is reapplied with only τ 24.0 and 24.8 hours as the initial values.

Table 1 Resolution of Function by Combined Linear–Nonlinear Least Squares Spectral Analysis[a]

		Mesor	Periods	Amplitudes	Acrophases
Series 1	Set of initial values from the linear least squares method	$M = 100$	$\tau_1 = 7$	$A_1 = 2.0541$	$\Phi_1 = -91°$
			$\tau_2 = 17$	$A_2 = 2.818$	$\Phi_2 = -92°$
			$\tau_3 = 24$	$A_3 = 10.075$	$\Phi_3 = -90°$
			$\tau_4 = 27$	$A_4 = 2.181$	$\Phi_4 = -346°$
			$\tau_5 = 50$	$A_5 = 5.3232$	$\Phi_5 = -86°$
	Final estimates from the nonlinear method	$M = 100\ (0.000)$	$\tau_1 = 7\ (0.000)$	$A_1 = 2$	$\Phi_1 = -90°$
			$\tau_2 = 17\ (0.000)$	$A_2 = 3$	$\Phi_2 = -90°$
			$\tau_3 = 24\ (0.000)$	$A_3 = 10$	$\Phi_3 = -90°$
			$\tau_4 = $ el[b]	$A_4 = $ el[b]	$\Phi_4 = $ el[b]
			$\tau_5 = 50\ (0.000)$	$A_5 = 5$	$\Phi_5 = -90°$

[a] Input: computer-generated time series (at 1-hour intervals) from combination of cosine functions without noise; 336 data points.
$Y_i = 100 + 2\cos\left[-\pi/2 + (2\pi/7)t_i\right] + 3\cos\left[-\pi/2 + (2\pi/17)t_i\right] + 10\cos\left[-\pi/2 + (2\pi/24)t_i\right] + 5\cos\left[-\pi/2 + (2\pi/50)t_i\right]$
[b] Elimated by the procedure.

In biological data, which presumably are subject to many sources of variation, the initial periods tested by NLS should be only those of biologic interest and with the higher amplitudes; too many periods and small amplitudes may interfere with correct solutions.

We may conjecture the following: Suppose τ_1, τ_2, τ_s are found to be the statistically significant periods detected by the LLS method and also happen to be the true periods in the data. If these periods are used as initial values in the NLS method and are confirmed by that method, statistical tests for zero amplitude will yield significant p-values in the NLS method.

Once we have estimated parameters in any model by either the LLS or NLS method, we can display the results on polar coordinates (Figure 5). Furthermore, we can study group characteristics (7) and compare several rhythms (23).

ILLUSTRATIVE EXAMPLES FOR COMBINED LINEAR-NONLINEAR LEAST SQUARES METHOD (LNLS)

Four artificial (computer-generated) time series will be used to illustrate the combined LNLS method. Each series has 336 data points and the data points are generated from the following trigonometric functions.

Series 1

$$y_i = 100 + 2 \cos\left[-\pi/2 + (2\pi/7)t_i\right] + 3 \cos\left[-\pi/2 + (2\pi/17)t_i\right]$$
$$+ 10 \cos\left[-\pi/2 + (2\pi/24)t_i\right] + 5 \cos\left[-\pi/2 + (2\pi/50)t_i\right]$$
$$\equiv f(t_i)$$

Series 2

$$y_i = f(t_i) + 10R(t_i)$$

Series 3

$$y_i = 100 + 10 \cos\left[-\pi/2 + (2\pi/24)t_i\right] + 1 \cos\left[-\pi/2 + (2\pi/24.8)t_i\right]$$
$$\equiv f^*(t_i)$$

Series 4

$$y_i = f^*(t_i) + 20R(t_i)$$

where $R(t_i)$, i = 1, 2, $\cdots$ 336 are the random numbers selected from a uniform distribution on the interval [0,1]. As compared to a set of real data, the random numbers can be interpreted as noise superimposed on functions.

Now, given the time series (y_i, t_i) $i = 1, 2, \cdots 336$, our first task is to estimate the unknown parameters in the Model 2 by using the LLS method and to decide (if possible from prior biologic evidence coupled with consideration of results from the zero-amplitude test) the number of cosine terms in the NLS Model 5. Thus, as explained above, we can obtain from the LLS method the initial values for the NLS method. In Series 1 and 2 above, periods are not too close to each other, but in Series 3 and 4 they are very close neighbors.

In Series 1 and 2 the LLS method detects all true periods very well, along with some "significant" pseudoperiods (Tables 1 and 2); but in Series 3 and 4, where the periods are very close together, the LLS method fails to detect one of the true periods (Tables 3 and 4). However, the combined linear-nonlinear least squares method, using the initial values from the LLS method, has detected all parameters in close agreement with the known values. The results are found in Tables 1-4.

More specifically, Table 1 shows at the top the function from which artificial time Series 1 was generated by an electronic computer. This model is a combination of mesor = 100 and the four cosine periods of 7, 17, 24, and 50 hours with corresponding amplitudes of 2, 3, 10, and 5 in arbitrary units. The acrophases are all the same $-\pi/2$ in radians $(= -90°)$. The table is divided into two parts under the headings mesor, periods, amplitudes, and acrophases. In its upper part, results obtained from the LLS method are presented as a set of initial values for the NLS analysis shown below. The LLS properly assesses the mesor, and the four true periods (7, 17, 24, and 50 hours) are detected exactly with amplitudes and acrophases in close agreement with the inputs. However, a statistically significant pseudoperiod ($\tau = 27$ hours) also has appeared. The lower part of Table 1 shows the final resolution of rhythm components by the NLS method, utilizing the set of initial values from the upper part. In the final estimates from the NLS method, the statistically significant "pseudoperiod" (27 hours) from the LLS analysis is eliminated and all the time parameters are precisely obtained from the NLS method, all with zero standard errors (in parentheses) of estimates for mesor and period.

Time Series 2, analyzed in Table 2, was generated from the same function as that in Table 1, but with superimposed noise content, 10 R (t_i), in arbitrary units where $R(t_i)$ is as defined previously. Even with noise added, the LLS method has precisely detected the time periods (except for the one at 51 hoursinstead of 50 hours) and, in addition, it

Table 2[a,b]

	Mesor	Periods	Amplitudes	Acrophases
Series 2 Set of initial values from the linear least squares method	$M = 105$	$\tau_1 = 7$	$A_1 = 1.8699$	$\Phi_1 = -90°$
		$\tau_2 = 17$	$A_2 = 2.6536$	$\Phi_2 = -90°$
		$\tau_3 = 24$	$A_3 = 10.023$	$\Phi_3 = -90°$
		$\tau_4 = 27$	$A_4 = 2.3833$	$\Phi_4 = -346°$
		$\tau_5 = 51$	$A_5 = 5.3917$	$\Phi_5 = -63°$
		$\tau_6 = 63$	$A_6 = 1.8244$	$\Phi_6 = -18°$
Final estimates from the nonlinear least squares method	$M = 104$ (0.31)	$\tau_1 = 7.01\ (0.009)$	$A_1 = 1.8212$	$\Phi_1 = -82.09°$
		$\tau_2 = 17.02\ (0.038)$	$A_2 = 2.8089$	$\Phi_2 = -84.1°$
		$\tau_3 = 24.02\ (0.021)$	$A_3 = 9.9413$	$\Phi_3 = -86.71°$
		$\tau_4 = \text{el}^c$	$A_4 = \text{el}^c$	$\Phi_4 = \text{el}^c$
		$\tau_5 = 50.29\ (0.180)$	$A_5 = 5.0617$	$\Phi_5 = -84.42°$
		$\tau_6 = \text{el}^c$	$A_6 = \text{el}^c$	$\Phi_6 = \text{el}^c$

[a] Input: computer-generated time series (at 1-hour intervals) from combination of cosine functions with noise; 336 data points.
$Y_i = 100 + 2\cos\left[-\pi/2 + (2\pi/7)t_i\right] + 3\cos\left[-\pi/2 + (2\pi/17)t_i\right] + 10\cos\left[-\pi/2 + (2\pi/24)t_i\right] + 5\cos\left[-\pi/2 + (2\pi/50)t_i\right] + 10R(t_i)$ (*): standard error of estimate
[b] $R(t_i)$ = random numbers selected from a uniform distribution on the interval [0,1].
[c] Eliminated by the procedure.

Table 3[a]

	Mesor	Period	Amplitudes	Acrophases
Series 3 Set of initial values from the linear least squares method	$M = 100$	$\tau_1 = 21.8$	$A_1 = 2.0896$	$\Phi_1 = -166°$
		$\tau_2 = 24$	$A_2 = 10.13$	$\Phi_2 = -94°$
		$\tau_3 = 26.8$	$A_3 = 2.2746$	$\Phi_3 = -7°$
Final estimates from the nonlinear least squares method	$M = 100 \,(0.000)$	$\tau_1 = \mathrm{el}^b$	$A_1 = \mathrm{el}^b$	$\Phi_1 = \mathrm{el}^b$
		$\tau_2 = 24 \,(0.001)$	$A_2 = 10$	$\Phi_2 = -90°$
		$\tau_3 = 24.8 \,(0.001)$	$A_3 = 1$	$\Phi_3 = -89.99°$

[a] Input: computer-generated time series (at 1-hour intervals) from combination of cosine functions without noise; 336 data points.

$Y_i = 100 + 10 \cos\left[-\pi/2 + (2\pi/24)t_i\right] + 1 \cos\left[-\pi/2 + (2\pi/24.8)t_i\right]$

[b] Eliminated by the procedure.

Table 4[aab]

	Mesor	Periods	Amplitude	Acrophases
Series 4 Set of initial values from the linear least squares method	$M = 110$	$\tau_1 = 21.8$ $\tau_2 = 24$ $\tau_3 = 26.9$	$A_1 = 2.0923$ $A_2 = 10.0170$ $A_3 = 2.5919$	$\Phi_1 = -170°$ $\Phi_2 = -93°$ $\Phi_3 = -358°$
Final estimates from the nonlinear least squares method	$M = 109.8$ (0.63)	$\tau_1 = $ el[c] $\tau_2 = 24.02\ (0.216)$ $\tau_3 = 24.5\ (1.26)$	$A_1 = $ el[c] $A_2 = 9.3169$ $A_3 = 1.7004$	$\Phi_1 = $ el[c] $\Phi_2 = -82.87°$ $\Phi_3 = -98.38°$

[a] Input: computer-generated time series (at 1-hour intervals) from combination of cosine functions with noise; 336 data points.
$Y_i = 100 + 10 \cos\left[-\pi/2 + (2\pi/24)t_i\right] + 1 \cos\left[-\pi/2 + (2\pi/24.8)t_i\right] + 20R(t_i)$
[b] $R(t_i) =$ random numbers selected from a uniform distribution on the interval [0,1].
[c] Eliminated by the procedure.

has produced two statistically significant "pseudoperiods" (at 27 and 63 hours) and a slightly different mesor. With those inputs the NLS analysis has detected all the time parameters in close agreement with the true parameters and has eliminated the two pseudoperiods.

In Table 3 the model consists of mesor = 100, periods of 24 and 24.8 hours, and amplitudes of 10 and 1 in arbitrary units; the acrophases are all the same, that is, $-\pi/2$ in radians ($= -90°$). As noted earlier, the LLS analysis produced two pseudoperiods along with the true 24-hour period but failed to detect the true 24.8-hour period; the NLS analysis revealed all the true parameters precisely. In Table 4 the same functions as in Table 3 are used and much noise is added—$20R(t_i)$, which can be maximally 20% of the mesor. In this case, where the true periods are very close together, and in the presence of a large amount ofnoise, the NLS method has lost its accuracy in the final estimates. These are obtained with higher standard errors (in parentheses) and a notable change in mesor.

ACKNOWLEDGMENTS

The advice of Walter Nelson, Administrative Scientist, Chronobiology Laboratories, and of Arne Sollberger, Associate Professor, Southern Illinois University School of Medicine, Life Science 1, Carbondale, Illinois 62901, is greatly appreciated. This work was supported by the United States Public Health Service (5-K6-Gm-13,191) and by NASA.

REFERENCES

1. J. B. J. Fourier, *La theorie analytique de la chaleur*, Paris, 1822.

2. K. F. Gauss, *Theoria motus corporum coelestium in sectionibus conicus solem ambientium*, Perthes & Besser, Hamburg, 1809.

3. K. Stumpff, *Grundlagen und Methoden der Periodenforschung*, Springer, Berlin, 1937, pp. 332.

4. J. Blume, *Nachweis von Perioden durch Phasen- und Amplitudendiagramm mit Anwendungen aus der Biologie, Medizin, und Psychologie*, Westdeutscher, Colonge, 1965.

5. S. Chapman, and J. Bartels, *Geomagnetism*, 2 vols Clarendon Press, Oxford, 1940, pp. 1049.

6. F. Halberg, M. Engeli, C. Hamburger, and D. Hillman, "Spectral Resolution of Low-Frequency, Small-Amplitude Rhythms in Excreted 17 Ketosteroid; Probable Androgen-Induced Circaseptan Desynchronization, *Acta Endocr.*, Supplement 103 (1965), 54.

7. F. Halberg, Y. L. Tong, and E. A. Johnson, "Circadian System Phase—An Aspect of Temporal Morphology; Procedures and Illustrative Examples, in *The Cellular Aspects of Biorhythms (H. von Mayersbach, Ed.), Springer, Berlin, 1967, pp. 20–48.*

8. I. *Jurkevich, Non-Linear Regression Analysis of Variance of Period Defined by Irregular Observation*, NASA CR-465, 1966.

9. J. A. Rummel, *Rhythmic Variation in Heart Rate and Respiration Rate During Space Flight—Apollo 15*, Environmental Physiology Laboratory, NASA, Manned Spacecraft Center, Houston, Texas. *Proceedings of the International Symposium on Chronobiology*, held at Little Rock, Arkansas, October, 1971, In Press.

10. R. B. Blackman, and J. W. Tukey, *The Measurement of Power Spectra*, Dover, New York, 1958.

11. C. I. Bliss, "Periodic Regression in Biology and Climatology," *Conn. Agric Exp. Station Bull.* **615** (1958) 55.

12. G. M. Jenkins, and D. G. Watts, *Spectral Analysis and Its Applications*, Holden-Day, San Francisco, 1969.

13. F. Halberg, E. A. Johnson, W. Nelson, W. Runge, and R. Sothern "Autorhythmometry—Procedures for Physiologic Self-Measurements and Their Analysis," *Physiol. Teacher,* **1** (1972) 1–11.

14. C. Nute, and P. Naitoh, *A Generalization of Harmonic Analysis for Detection of Long-Period Biorhythmicities from Short Records*, Navy Medical Neuropsychiatric Research Unit, San Diego, Report No. 71-55, 1971.

15. F. Halberg, G. Katinas, Y. Chiba, M. Garcia Sainz, T. Kovats, H. Kükel, N. Montalbetti, A. Reinberg, and R. Scharf. "Chronobiologic Glossary of the International Society for Chronobiology, *Int. Chronobiol.*, In Press.

16. A. J. Lotka, *Elements of Mathematical Biology*. Dover, New York, 1956, p. 465.

17. D. W. Marquardt, "An Algorithm for Least Squares Estimation of Nonlinear Parameters," *J. SIAM,* **2** (1963).

18. F. Koehler, F. K. Okano, L. R. Elveback, F. Halberg, and J. J. Bittner, "Periodograms for the Study of Physiologic Daily Periodicity in Mice and in Man," *Exp Med Surg.*, (1956), **14** 3–30.

19. F. Halberg, E. Halberg, C. P. Barnum, and J. J. Bittner, "Physiologic 24-Hour Periodicity in Human Beings and Mice, the Lighting Regimen and Daily Routine" *Photoperiodism and Related Phenomena in Plants and Animals* (B. Withrow, Ed.) American Association for the Advancement of Science, Washington, D.C., 1959, pp. 803–878.

20. H. Levine and F. Halberg, *Circadian Rhythms of the Circulatory System. Literature Review; Computerized Case Study of Transmeridian Flight and Medication Effects on a Mildly Hypertensive Subject*, U.S. Air Force Report SAM-TR-72-3, April 1972, 64 pp.

21. F. Halberg, "Physiologic Considerations Underlying Rhythmometry, with Special Reference to Emotional Illness," *Symposium Bel-Air III*, Masson, Geneva, 1968, pp. 73–126.

22. N. Draper, and H. Smith, *Applied Regression Analysis*, Wiley, New York, 1966.

23. W. Nelson, F. Halberg, and D.-S. Hwang, "An Evaluation of Time-Dependent Changes in Susceptibility of Mice to Pentobarbital Injection," *Neuropharmacology* **12** (1973) 509–524.

CHAPTER 6

The Applications of Moments and Cluster Analysis to Reproductive Cycles

ALLEN R. HANSON and JAY A. LEAVITT

Department of Computer and Information Sciences,
University of Minnesota, Minneapolis, Minnesota

In this article we consider the application of moments and cluster analysis for the purpose of obtaining insight into and classification of mechanisms in chronobiology, in general, and those involving reproductive cycles, in particular. The procedures that we present should be interpreted as being empirical and not statistical in nature.

GENERALIZED MOMENTS

Generalized moments have been employed in a wide variety of problems because of their ability to reflect geometric features. Applications cover diverse fields of study ranging from ship identification (1) to alphanumeric character recognition (2, 3). Loosely, we define a generalized moment as a parameter we obtain when we form the convolution of a function $f(t)$ we wish to analyze with one of a set of trial functions $\{g_i(t)\}$. Ordinarily, if the function is defined by a set of

83

data points, $f(t_j)$, the ith moment is

$$M_i = \sum_j f(t_j)g_i(t_j)$$

Typical choices of $g_i(t)$ include powers, t^i, trigonometric functions, piecewise constant functions, and pyramid functions. If we are using $\{t^i\}$ as the set of trial functions, M_0 is related to the arithmetic mean of $f(t)$; M_1 is related to its center of mass. A moment may be thought of as a weighted average.

The purpose of introducing moments is to reduce the dimensionality of the space we are investigating. For example, if we were testing subjects so that we had 100 data points, we would not want to have to consider how each subject at each point behaved. Instead properties describing the mean, symmetry, and so forth of the data constitute a smaller set of attributes which could be easier to analyze.

Moments appear in a natural way in linear least squares problems for they are the inhomogeneous terms that appear in the resulting equations (see Appendix). However, unlike the parameters which are obtained in a least squares fit, the moments are independent of the number of terms used and of the other functions in the set. Furthermore, this implies that they are independent of the quality of fit. Thus they form an ideal set of parameters which may be used for subsequent analysis of $f(t)$. The relation between standard experimental procedures and the proposed method is summarized in Figure 1. In the classical experimental procedure (Figure 1a) before one can analyze the data, a fit is obtained and the analysis is based on properties of the fit. This implies that the poorer the quality of the fit, the poorer our ability to analyze. The object of the proposed procedure (Figure 1b) is to avoid this difficulty, that is, to avoid the necessity of obtaining a fit. Instead, we suggest that the analysis should be based on the parameters that lead to a fit. This, of course, alters our technique of analysis and may cost us the loss of some statistical information. In the problems that we have considered we have arbitrarily chosen $\{g_i(t)\} = \{t^i\}$. It is not clear that this is the best choice. It is possible that either pyramid or trigonometric functions would form a better set.

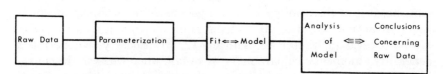

*Figure 1*a   *Classical experimental procedure.*

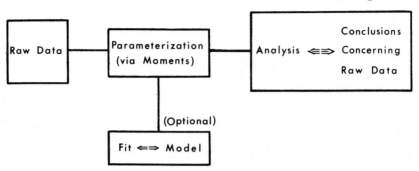

*Figure 1*b   *Proposed experimental procedure.*

CLUSTERING

The problem of detecting subsets of measurements has been considered under the general title of cluster analysis (4–11), subset generation (12, 13), unsupervised learning, learning without a teacher (14–16), decision directed learning (17, 18, 10), learning with a probabilistic teacher (19), decomposition of density functions (14, 20, 21), classificatory strategies (22, 23), and probably many more. We will use the term "clustering" to refer to this general category of procedures. They are all concerned with the problem of grouping together in clusters data points which are 'similar' in some sense.

The general clustering problem may be stated rather briefly; we follow the formulation due to Bonner (5). Suppose we are given a set of objects, each of which is defined by the values of a set of attributes associated with it. Each object is characterized by the same attribute set. It is easy to demonstrate that all the attributes should be reduced to a scale: [0,1]. We want to find "clusters" of objects such that members of a cluster are in some way *similar* to or *look like* each other but are *dissimilar* to objects outside the cluster.

We are being purposely vague in the problem statement, particularly in the definition of the terms *similar, cluster, dissimilar*, and so forth. Indeed, no generally accepted definitions of these terms exist in the context of clustering. They are particularly problem dependent. Consider Figure 2 as an example. Suppose the objects mentioned above were the results of a test performed on a patient. Normally, the test consists of several parts (e.g., blood pressure: systolic and diastolic) which are the attributes of the test. Denote the values of these attributes by $test_1$ and $test_2$. Suppose we test a group of people and plot the results as indicated in Figure 2. Intuitively we would like to say

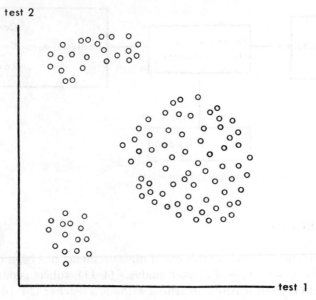

Figure 2   An example of clustering.

that the individuals comprising each cluster are "different" from individuals in the other clusters; furthermore, this difference should be a medically verifiable difference (e.g., hyper/hypo/normo tensive). The data shown in Figure 2 are for illustrative purposes only; actual data rarely show such clear separation.

In general the number of clusters and their characteristics are unknown. How many different types of clouds can be observed in satellite photographs? How many different atomic fingerprints are detectable in cloud chamber photographs? How many different handprinted A's are there? How many varieties of heart disease are there? Can these varieties be detected and separated on the basis of electrocardiograph records? Vectorcardiograph records? Does the accumulated records of thousands of electroencephalograms contain a clue as to the possible varieties of brain disease? Obviously, the list is endless. According to Nagy (24), "... aside from their common quest for some manner of grouping, the outstanding feature shared by the above questions is vagueness. ..." A perusal of the literature reveals the extent of the vagueness; most results are quoted for special problems or classes of problems sharing common characteristics and/or underlying assumptions.

As previously mentioned, the clusters shown in Figure 2 are ideal clusters. Figure 3 illustrates clusters more likely to be encountered in practical applications (11).

We have used a linear clustering procedure (25) to extract information from the moments. Note that in Figures 3c, d, and h a linear procedure would not separate the clusters. The moments correspond to

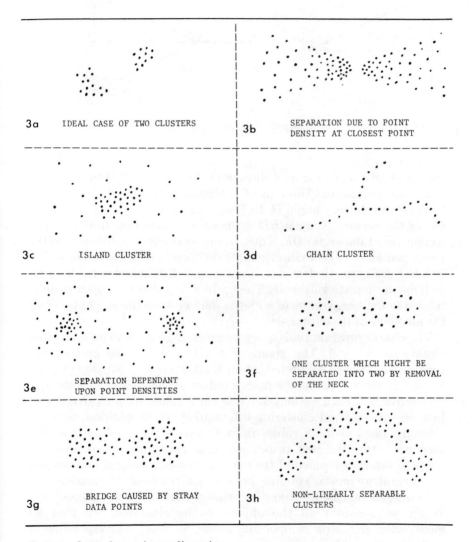

3a	IDEAL CASE OF TWO CLUSTERS	3b	SEPARATION DUE TO POINT DENSITY AT CLOSEST POINT
3c	ISLAND CLUSTER	3d	CHAIN CLUSTER
3e	SEPARATION DEPENDANT UPON POINT DENSITIES	3f	ONE CLUSTER WHICH MIGHT BE SEPARATED INTO TWO BY REMOVAL OF THE NECK
3g	BRIDGE CAUSED BY STRAY DATA POINTS	3h	NON-LINEARLY SEPARABLE CLUSTERS

Figure 3   Some clusters in two dimensions.

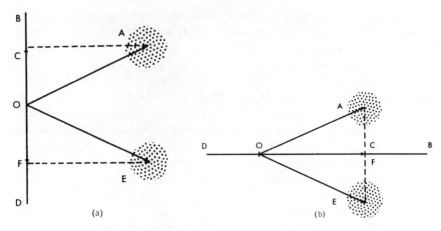

Figure 4   Contrasting projection lines. (a) Ideal choice of BD. *(b) Poor choice of* BD.

the set of attributes. Figure 4 illustrates the strategy used in the linear clustering procedure. The aim of the method is to find the line BD through an arbitrary origin O. In Figure 4a we note that the projection OC of the vector OA upon BD is opposite in direction from the projection OF of the vector OE. Thus, in this example, the direction of the projection on BD acts to discriminate the members of the two sets. The line BD in Figure 4b does not function as a discriminator for the projections are indistinguishable. The optimal line BD to project upon is related to the eigenvectors of a matrix derived from the attributes of all the members of the cluster sets.

The cluster program finds n eigenvectors, E_j, each with an associated eigenvalue, $\lambda_j > 0$. The greater the value for λ_j, the greater the separation obtained by projecting on E_j. Assume $\lambda_1 \geq \lambda_2 \geq \lambda_3 \cdots \geq \lambda_n$. We define the s value as the inner product of the attribute vector and an eigenvector. Then we may say that the s values from E_1 give us the best one-dimensional clustering information. If, in addition, we consider the plots of the s values from E_1 versus the s values from E_2, then we obtain the best two-dimensional clustering information. In order to determine which of the one-and two-dimensional clusters are meaningful, we must reexamine the biological/medical information.

In the general case an object in a cluster has attributes $c_1, c_2, \cdots c_N$. When we consider all the objects to be clustered, we find an eigenvector, BD, with components $w_1, w_2, \cdots w_N$. Corresponding to each object we form $s = c_1 w_1 + c_2 w_2 + \cdots + c_N w_N$. The w's are weights, scaling the relative importance of each associated attribute c, so that the resulting s values represent clusters which are optimally separated. The *s* values may then be used to form a histogram which allows us to classify the object according to its position on the graph. Ordinarily one encounters two types of histograms. The first type has the appearance of a normal curve. Those subjects with *s* values in the middle may be considered normal, while those with *s* values toward the outside may be considered abnormal. The second type has the appearance of being composed of the superposition of two or more normal curves. We would assume then that the members corresponding to distinct normal curves possessing different properties. The individuals comprising those normal curves may then be considered separately.

The *s* values are a function of a set of measurements made on an individual which may vary with time. As a result the value of *s* may vary with time. A plot of the position (i.e., *s*-value) of an individual in the histogram versus time will readily reveal the cyclic properties (if any) of the total measurement. This may occur even though the individual components of the measurement do not readily reveal any cyclic characteristics.

APPLICATIONS

The first test of this technique involved an analysis of circadian rhythms of systolic and diastolic blood pressure of 37 human subjects.

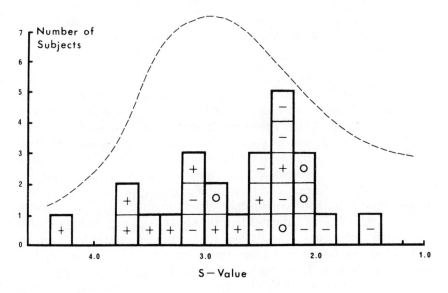

Figure 5   Extract from blood pressure histogram.

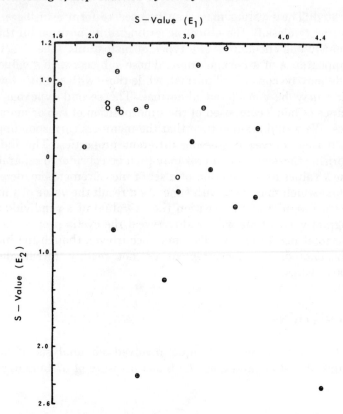

Figure 6   Two-dimensional blood pressure data (see text).

Our measurements were restricted to a 12-hour interval, 8 A.M. to 8 P.M. Eight attribute values were input to the clustering program for each subject. These values corresponded to M_0 through M_3 of the systolic and diastolic blood pressures. Twenty-four of the subjects were clinically tested. Figure 5 is an extract from the histogram. Symbols identify the subject; a + indicates a clinical classification of hypertensive, − hypotensive, and 0 normotensive. A good approximation to the histogram can be obtained by adding the mean systolic blood pressure to the mean diastolic blood pressure; the other moments may be omitted. This indicates how the clustering program may lead to simplified techniques. The dashed line indicates the envelope of the full histogram. Figure 6 locates these same 24 subjects on the best plane of projection. This graph was obtained by plotting the s values of E_1 against the s values of E_2. Observe that a straight line may be drawn on this graph which separates all the hypertensives from the hypotensives. We note that $\lambda_1 = 5.2$, $\lambda_2 = 0.34$, and $\lambda_3 = 0.03$. Little information was contained in graphs utilizing E_3 to E_8.

The second test of the procedure was a study of two experiments yielding temperature series of 46 rats which were implanted with a temperature monitoring device. Each rat, each day, was considered as a different subject in the clustering program. Four parameters, M_0 through M_3 of the circadian temperatures, were input to the clustering program. The envelope of the histogram for the experiment is shown in Figure 7. After the s values were obtained, a plot (see Figure 8) was made of s versus day for each rat. The resulting curve is a representation of the estrus cycle for that rat. The peak-valley sequences appear at 4-day intervals.

Figure 9 is a partial representation of the data concerning rats #6 and #40 from Experiment 12. The curves, reading from bottom to top, are of M_0, M_1, M_2, M_3, and s versus date. It is clear that M_3 forms a good approximation to s. These curves indicate that the higher moments and s offer a better indication of the reproductive cycle of the rat than M_0, the mean temperature. This, however, may not be the case with all reproductive cycles. For example, some biological systems may possess a cycle that is exhibited by a stronger interaction of several moments. This information is revealed only after clustering procedures. In the

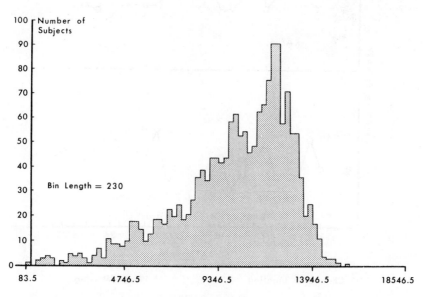

Figure 7   Histogram envelope computed from moments from rat temperature series.

current example, however, all moments serve to position the animal on the histogram, that is, rate it high, normal, or low. On April 13 both rats have a peak:

$$\text{for rat } \#1 \qquad s = 13,690$$
$$\text{for rat } \#6 \qquad s = 7,000 \qquad \text{(see Figure 7)}$$

A feature of this procedure is that the contribution of the circadian rhythm is separated from that of the estrus cycle. Furthermore, the quality of the information concerning the circadian rhythm does not seem important. It is evident that these procedures can be used in studying human reproductive cycles. It is felt that a more clearly identifiable reproductive cycle can be obtained if other data, such as grip strength or gross body movement, were included in the studies.

These procedures seem useful for identifying an individual within a group. These same procedures, when applied to the individual over a period of time, can be used to see if the subject changes with regard to his own norms. Furthermore, the method is tolerant to sparse and noisy data. In one case the subject measured his blood pressure over a long

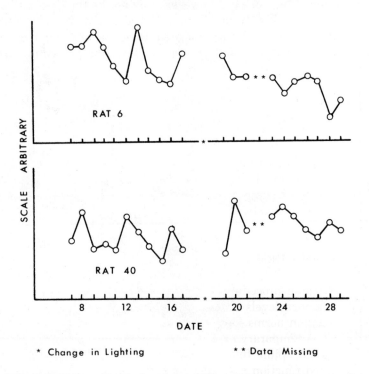

* Change in Lighting   * * Data Missing

Figure 8   Estrus cycle obtained from a daily position in histogram.

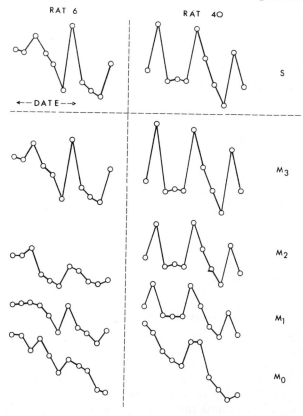

Figure 9   Excerpts of rat data versus date.

period of time, repeating each measurement after 5 minutes. The two readings varied by $\pm 20\%$. The clustering program could not distinguish between the subject identified by the first set of data with that of the subject identified by the second set of data.

CONCLUDING REMARKS

Currently, these procedures cannot be described in statistical terms. Thus confidence levels cannot be attached to the results. However, after population norms have been established, statistical methods can be applied. A temporary expedient may be found in the use of splines. By finding the spline function which possesses the same moments as the measured function $f(t)$, we obtain an approximation to $f(t)$. Statistical means may then be applied to the spline approximation to give an indication of quality of fit. However, the statistical quality varies according to the number of terms used in the fit. The moments are independent of the number of terms. Thus it is possible to have a poor fit and a good clustering analysis.

ACKNOWLEDGMENTS

We wish to express our gratitude to Franz Halberg and to his associates of the Chronobiology Laboratory, University of Minnesota, for bringing our attention to and aiding us in these problems and for providing us with the data sets.

REFERENCES

1. F. W. Smith and M. H. Wright, *IEEE Trans. Comp*, **C20**, (1971), 1089.

2. M. K. Hu, *IRE Trans. Inform. Theory*, **IT 12**, (1962), 179.

3. F. L. Alt, in *Optical Character Recognition*, (G. L. Fischer, Ed.) Spartan, Washington, D.C., 1962, p. 153.

4. G. H. Ball, *1965 Fall Joint Comp. Conf. AFIPS Proc.*, **27**,1 (1965), 533.

5. R. E. Bonner, *IBM J. Res. Develop.*, **8**, (1964), 22.

6. O. Firschen and M. Fischler, *IEEE Trans. Electronic Comp.*, **EC 12**, (1963), 137.

7. G. H. Ball and D. J. Hall, *ISODATA, A Novel Method of Data Analysis and Pattern Classification*, Stanford Research Institute, Menlo Park, Tech Rep., 1965.

8. A. W. F. Edwards and L. T. Cavalli-Sforza, *Biometrics*, **21**, (1965), 362.

9. E. H. Ruspini, *Inform. Contr.*, **15**, (1969), 22.

10. A. J. Lerner, *Comp. J.*, **12**, (1969), 29.

11. C. T. Zahn, *IEEE Trans. Comp.*, **C20**, (1971), 68.

12. R. L. Mattson and J. E. Dammann, *IBM J. Res. Develop.*, **9**, (1965),294.

13. H. P. Friedman and J. Rubin, *Amer. Stat. Assoc. J.*, **62**, (1967), 1159.

14. A. A. Dorofeyuk, *Automation Remote Control*, **27**, (1966), 1728.

15. S. C. Fralick, *IEEE Trans. Inform. Theory*, **IT13**, (1967), 57.

16. J. Spragins, *IEEE Trans. Inform. Theory*, **IT12**, (1966), 223.

17. E. R. Ide, C. E. Kiessling, and C. J. Tunis, *1968 Fall Joint Comp. Conf. AFIPS Conf. Proc.*, **33**, (1968), 1117.

18. W. L. G. Koontz and K. Fukunaga, *IEEE Trans. Comp.*, **C21**, (1972), 171.

19. R. M. Haralick and I. Dinstein, *IEEE Trans. Systems, Man, Cyber.*, **SMC1**, (1971), 275.

20. G. Sebestyen and J. Edie, *IEEE Trans. Electron. Comp.*, **EC15**, (1966), 908.

21. E. M. Braverman, *Automat. Remote Control (USSR)*, **27**, (1966), 1748.

22. G. N. Lance and W. T. Williams, *Comput. J.,* **8,** (1967), 373.
23. G. N. Lance and W. T. Williams, *Aust. Comput. J.,* **1,** (1968), 82.
24. G. Nagy, *Proc. IEEE,* **56,** (1968), 836.
25. R. L. Mattson and J. E. Dammann, *IBM J.,* **9,** (1965), 294.

APPENDIX

The general form of the linear least squares problem for a function described by data is:

Minimize with respect to the coefficients a_i

$$\sum_j (f(t_j) - [a_0 g_0(t_j) + a_1 g_1(t_j) + \cdots + a_N g_N(t_j)])^2$$

where the functions $g_k(t)$ are given a priori and may be nonlinear. The solution to this problem is given by a linear system of equations:

$$\sum_j g_0(t_j)[a_0 g_0(t_j) + a_1 g_1(t_j) + \cdots + a_N g_N(t_j)] = \sum_j g_0(t_j) f(t_j)$$

$$\sum_j g_1(t_j)[a_0 g_0(t_j) + a_1 g_1(t_j) + \cdots + a_N g_N(t_j)] = \sum_j g_1(t_j) f(t_j)$$

$$\vdots \qquad\qquad \vdots$$

$$\sum_j g_k(t_j)[a_0 g_0(t_j) + a_1 g_1(t_j) + \cdots + a_N g_N(t_j)] = \sum_j g_k(t_j) f(t_j)$$

$$\vdots \qquad\qquad \vdots$$

$$\sum_j g_N(t_j)[a_0 g_0(t_j) + a_1 g_1(t_j) + \cdots + a_N g_N(t_j)] = \sum_j g_N(t_j) f(t_j)$$

The expressions on the right-hand side of the equations, $\sum g_k(t_j) f(t_j)$, are called the *inhomogeneous terms.* If $g_k(t) = t^k$, we are seeking the least squares po'ynomial of degree N that approximates $f(t)$. The inhomogeneous terms then have the appearance

$$\sum_j t_j^k f(t_j)$$

If the $g_k(t)$ are the trigonometric functions

$$\cos (kt/\tau)$$

then the solution is called the truncated Fourier series of $f(t)$ [$f(t)$ is assumed to be an even function in this example]. The numbers a_k are called the Fourier coefficients and the period τ is assumed known. If τ is not known then the resulting problem is nonlinear and the above set of equations do not provide the solution. Instead a set of nonlinear equations result which are usually difficult to resolve.

CHAPTER 7

Rotational Angular Correlations

THOMAS D. DOWNS

The University of Texas, School of Public Health at Houston, Houston, Texas

Important phenomena are frequently expressed as time of day, or time of year, or a certain time of the menstrual cycle in the study of biological rhythms. For instance, diastolic blood pressure varies according to a circadian cycle, and normally attains a maximum value around midnight. This cycle repeats itself daily. The time at which the maximum pressure occurs is properly considered to be an angle, say a, which conceivably could vary from 0 to 360°. This time, or angle a, could be plotted as a point on a circle. One day, or 24 hours, thus corresponds to 360° on the circle, and 1 hour to 1/24 this, or 15°.

When considering the peak time (time of day of maximum pressure) for a single individual it is simple and natural to plot the angle a, corresponding to the peak time, on a straight line whose end points are 0 and 360°. However, two or more peak times should be plotted on a circle. The straight line plot can be very misleading. For example, taking 0° to be midnight, 90° to be 0600, etc., the straight line plot suggests that 0100 and 2300 are 22 hours apart; the circular plot shows clearly that they are only 2 hours apart. The application of standard statistical techniques to the straight line plot can be disastrous: the mean of the above two observations is taken to be noon, and the variation is seen to be large. Yet it is self-evident from the circular plot that the mean ought to be midnight, and the variation ought to be small. These two sets of conclusions are, speaking literally, 180° apart!

A SUITABLE GEOMETRIC FRAMEWORK

Given n pairs of angles (a, b), each member of each pair potentially ranging from 0 to 360°, we wish to devise some appropriate measure of correlation between the a's and the b's.

For a simple example consider the case where n is 4 and the observations are

$$a: \quad 45° \quad 135° \quad 225° \quad 315°$$

$$b: \quad 225° \quad 315° \quad 45° \quad 135°$$

A graph of these four points appears in the planar plot of Figure 1. If we proceed with the usual linear methods we obtain $\bar{a} = \bar{b} = 180°$ for the means, $S_a^2 = S_b^2 = 5(45°)^2$ for the variances, $S_{ab} = -3(45°)^2$ for the covariance, and $r = S_{ab}/S_a S_b = -3/5$ for the correlation coefficient.

But, we must be exceedingly cautious in accepting these results. It has been shown that straight line plots can lead to erroneous conclusions, and planar plots are precisely the extension of straight line plots to an additional dimension. The planar plot is *not* an appropriate geometric framework. To obtain an appropriate one consider, again, the straight line ranging from 0 to 360°. By looping the line around and joining its end points we obtain the appropriate circle. An analogous procedure can be performed on the square of Figure 1: the points of the top line are supposed to be the same as the corresponding points of the

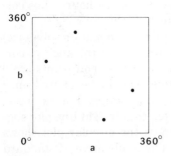

Figure 1

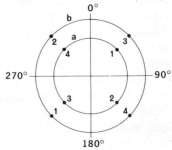

Figure 2

bottom line. We make them be the same by looping the square into a cylinder with the top and bottom lines touching and the circular ends open. Also, the circular ends are supposed to be the same. We make them coincide by looping the cylinder around until the circular ends meet. The resulting surface is a torus (like a donut), and is the appropriate geometric surface for studying angular correlation.

The torus is difficult to visualize and requires three dimensions to exhibit. Instead of plotting the angle pairs (*a, b*) on a torus, a compromise can be achieved by drawing two concentric circles and plotting the *a*'s, say, on the inner circle, and the *b*'s on the outer one. This is done in Figure 2 for the four data points above. It is at once evident from this figure that the results of the usual linear methods are erroneous in their entirety. Indeed, no suitable choice of a mean for either the *a*'s or the *b*'s seems possible, so we may say that these means *do not exist*. This suggests that the variances should be infinite, or at least attain some fixed upper bound. Finally, it will be seen that a rotation of the *a*-circle through 180° brings the *a*'s into a one-to-one correspondence with the *b*'s. Therefore the correlation ought to be positive and complete—not negative and partial. Appropriate techniques for quantifying these statements are given in the next section.

QUANTITATIVE TECHNIQUES

From a sample of n angle pairs (*a, b*) calculate the four arithmetic means:

$$\bar{s}_a = \sum \sin a/n, \qquad \bar{c}_a = \sum \cos a/n$$
$$\bar{s}_b = \sum \sin b/n, \qquad \bar{c}_b = \sum \cos b/n \qquad (1)$$

and the 2 × 2 matrix

$$T = \frac{1}{n} \sum \begin{pmatrix} \cos b - \bar{c}_b \\ \sin b - \bar{s}_b \end{pmatrix} \begin{pmatrix} \cos a - \bar{c}_a \\ \sin a - \bar{s}_a \end{pmatrix}' \qquad (2)$$

All required future calculations depend only on these four arithmetic means and the 2 × 2 matrix T. Next, we define the *trigonometric angle means* $\bar{a}_t$ and $\bar{b}_t$ by

$$\bar{a}_t = \arctan (\bar{s}_a/\bar{c}_a), \qquad \bar{b}_t = \arctan (\bar{s}_b/\bar{c}_b) \qquad (3)$$

where due attention must be paid to the algebraic signs of the arithmetic means so that $\bar{a}_t$ and $\bar{b}_t$ will be in the proper quadrants. If, say, $\bar{s}_a = \bar{c}_a = 0$, we say that $\bar{a}_t$ *does not exist*, since in such cases the arctan function is not well-defined. For the next step, we define the *trigonometric variances* t_a^2

and $t_b{}^2$ by

$$t_a{}^2 = \frac{1}{n} \sum \begin{pmatrix} \cos a - \bar{c}_a \\ \sin a - \bar{s}_a \end{pmatrix}' \begin{pmatrix} \cos a - \bar{c}_a \\ \sin a - \bar{s}_a \end{pmatrix}$$

$$t_b{}^2 = \frac{1}{n} \sum \begin{pmatrix} \cos b - \bar{c}_b \\ \sin b - \bar{s}_b \end{pmatrix}' \begin{pmatrix} \cos b - \bar{c}_b \\ \sin b - \bar{s}_\flat \end{pmatrix} \tag{4}$$

It is easy to show from these definitions that

$$t_a{}^2 = 1 - (\bar{c}_a{}^2 + \bar{s}_a{}^2), \qquad t_b{}^2 = 1 - (\bar{c}_b{}^2 + \bar{s}_a{}^2) \tag{5}$$

from which it follows, for instance, that $t_a{}^2$ lies between 0 and 1, and is 0 when all the a's coincide, and attains its fixed upper bound, 1, when the a's are so widely scattered that the trigonometric angle mean $\bar{a}_t$ does not exist. Next, for reasons to be given later, we define the *trigonometric covariance* t_{ab} by

$$t_{ab} = |R| \text{ trace } (T'T)^{1/2} \tag{6}$$

where $|R|$ is the determinant of the orthogonal matrix R given by

$$R = T[(T'T)^{1/2}]^{-1} \tag{7}$$

and $(T'T)^{1/2}$ is the unique symmetric positive definite matrix whose square is equal to $T'T$. (For details on this, see Ref. 1). Finally, the *trigonometric correlation coefficient* r_t is defined by

$$r_t = t_{ab}/t_a t_b \tag{8}$$

Equations 1, 2, 5, 6, 7, and 8 allow us to compute the trigonometric correlation coefficient r_t.

The above procedures are based on the following rationale: The a's, say, are plotted as points on a unit circle. The direction from the origin to the center of gravity of these points is the trigonometric angle mean $\bar{a}_t$. The trigonometric variance $t_a{}^2$ is one minus the squared distance from the center of gravity to the center of the circle. In fact, $t_a{}^2$ is equal to the sum of the variances of the $\sin a$ and $\cos a$. Similar remarks apply to the b's plotted on a concentric circle whose radius is tacitly taken as unity.

Now the natural analogue of linear correlation here is rotational correlation. Thus we estimate the extent to which the direction of b from the center of gravity of the b's is a constant rotation (the orthogonal matrix R) of the corresponding direction for a. That is, we choose an orthogonal

2×2 R to maximize the expression

$$Q = \frac{1}{n} \sum \begin{pmatrix} \cos a - \bar{c}_a \\ \sin a - \bar{s}_a \end{pmatrix}' R' \begin{pmatrix} \cos b - \bar{c}_b \\ \sin b - \bar{c}_b \end{pmatrix} \tag{9}$$

and the R so obtained is given by Eq. 7. When this value of R is substituted in Eq. 9 it is not difficult to show that the resulting Q is related to the trigonometric covariance t_{ab} by $t_{ab} = |R| Q$. The analogy between the trigonometric variances and covariances in Eqs. 4 and 9 with the corresponding linear quantities is clear.

The factor $|R|$ requires explanation. Since R is an orthogonal matrix, its determinant is either $+1$ or -1. If the determinant of R is $+1$, then R must have the form

$$R = \begin{pmatrix} \cos \theta & -\sin \theta \\ \sin \theta & \cos \theta \end{pmatrix} \tag{10}$$

in which case R represents a clockwise pure rotation through $\theta°$, where $0 \leq \theta < 360$. If the determinant of R is -1, then R must have the form

$$R = \begin{pmatrix} \cos \theta & \sin \theta \\ \sin \theta & -\cos \theta \end{pmatrix} \tag{11}$$

in which case R represents a reflection in the line through the origin whose slope is $\tan \theta/2$, where $0 \leq \theta < 360$.

When $|R| = +1$ the correspondence between the a's and b's is positive, in the sense that both tend to progress clockwise simultaneously. When $|R| = -1$ the correspondence is negative, in the sense that the b's tend to progress clockwise as the a's progress counterclockwise. Since R is chosen to maximize Q, then Q is always nonnegative, and so the factor $|R|$ appears: $t_{ab} = |R| Q$, to allow the trigonometric covariance and correlation to be negative.

It can be shown that r_t lies between -1 and $+1$, inclusive, and attains one of these extreme values only when the b deviation is a constant multiple of an orthogonal transformation of the a deviation for every pair (a, b) in the sample. The statistics defined above possess essential invariance properties when the origins $(0°)$ are changed on the a and b circles. In such cases

1. $\bar{a}_t$ and $\bar{b}_t$ are changed accordingly.
2. t_a^2, t_b^2, and t_{ab} are unchanged.
3. r_t is unchanged.
4. R is changed accordingly.

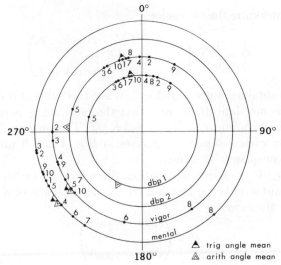

None of Statements (1), (2), (3) hold, in general, when linear methods are used directly on the angles.

Applying the above techniques to the data of Figure 2, we obtain $\bar{s}_a = \bar{c}_a = \bar{s}_b = \bar{c}_b = 0$, so that $\bar{a}_t$ and $\bar{b}_t$ do not exist, and $t_a^2 = t_b^2 = 1$. Letting $e = 2^{-1/2} = \sin 45° = \cos 45°$, we obtain

$$T = \begin{pmatrix} -e^2 & 0 \\ 0 & -e^2 \end{pmatrix} = \frac{1}{2} \begin{pmatrix} -1 & 0 \\ 0 & -1 \end{pmatrix}$$

$$T'T = \frac{1}{4} \begin{pmatrix} 1 & 0 \\ 0 & 1 \end{pmatrix}, \qquad (T'T)^{1/2} = \frac{1}{2} \begin{pmatrix} 1 & 0 \\ 0 & 1 \end{pmatrix}$$

and

$$R = T[(T'T)^{1/2}]^{-1} = \begin{pmatrix} -1 & 0 \\ 0 & -1 \end{pmatrix}, \qquad |R| = +1$$

Thus $t_{ab} = |R|$ trace $(T'T)^{1/2} = 1$, and the trigonometric correlation coefficient is $r_t = 1/\sqrt{1 \cdot 1} = 1$. R has the form of Eq. 10, so $\cos \theta = -1$ and $\theta = 180°$. These results are in exact agreement with our intuitive conclusions based on examination of Figure 2.

BIOLOGICAL EXAMPLES

Ten medical students in Austria measured their blood pressures, body weight, oral temperatures, and so forth several times daily for a period

Table 1 Correlation Coefficients for Selected Variables

Analysis	Variable	dbp1	dbp2	Vigor	Mental	Mean	Variance
Trig	dbp1	1.00	.96	−.19	−.23	350°	.21
	dbp2		1.00	−.24	−.12	345°	.16
	Vigor			1.00	.88	233°	.30
	Mental				1.00	233°	.28
Linear	dbp1	1.00	.53	.02	.27	205°	—
	dbp2		.100	−.43	−.39	273°	—
	Vigor			1.00	.90	230°	—
	Mental				1.00	230°	—

of several weeks. This was done under the direction of Doctors Robert Gunther and Edwin Knapp of the University of Innsbruck. This raw data was then run through the microscopic analysis, developed under the direction of Professor Franz Halberg of the University of Minnesota, to determine the peak times for the above variables. The peak times of four selected variables for each of the ten medical students are analyzed later. I am indebted to the above investigators for making this data available to me. The four selected variables are two successive measurements of diastolic blood pressure (designated as dbp1 and dbp2), a measure of physical vigor, and a measure of mental exuberance.

The data analyzed are the 40 angles representing the estimated peak times of these four variables for each of the 10 students. These data are graphed on the four concentric circles of Figure 3. The students are labeled 1 through 10, and their corresponding estimated peak times so designated in Figure 3. It will be evident from this figure that dbp1 and dbp2 are highly correlated, as are physical vigor and mental exuberance. The degrees of correlation between members of these two sets are not so apparent.

Table 2 Angles θ for R

Variable	dbp1	dbp2	Vigor	Mental
dbp1	0°	2°	100°	153°
dbp2		0°	60°	110°
Vigor			0°	357°
Mental				0°

Table 1 contrasts the correlation results obtained by the trigonometric method with those obtained linearly. The disparity between the two methods for the correlation of dbp1 with dbp2 is striking. This disparity is brought about by the clustering of the angles near the point, $0° = 360°$, of discontinuity. The two methods agree well for the other two variables. *This is because both trigonometric variances are small and neither set of angles clusters near $0° = 360°$.*

Table 2 gives the angles θ for the orthogonal matrices R. For pure rotations ($r_t > 0$), θ represents the phase shift between the peak time angles. Note that the phase shifts of 2 and 357° in the table are both virtually nil.

DISCUSSION

Statistical methods for treating a sample of points on a circle, sphere, and so forth have been developed and improved upon by Fisher (2), Watson and Williams (3), Batschelet (4), and Downs and Liebman (5), as well as others.

To the author's knowledge, however, this paper is the first of its kind in treating pairs of angles on concentric circles. All the techniques given herein are readily generalized to concentric spheres and so forth. Numerous unsolved problems remain, though. The sampling distribution of r_t is unknown, but it is suspected that under certain conditions it will be approximately the same as that of the linear correlation coefficient. Another problem is that the angles may be correlated, but not strictly rotationally. This could conceivably be accounted for by transforming the angles, analogous to a logarithmic transformation, for instance, for linear variables. The author knows of no work in this area.

REFERENCES

1. T. Downs, "Orientation Statistics," *Biometrika,* **59,** (1972), 665–676.
2. R. A. Fisher, "Dispersion on a Sphere," *Proc. Roy. Soc., Ser. A,* **217,** (1953), 295–305.
3. G. Watson and E. Williams, "On the Construction of Significance Tests on the Circle and the Sphere," *Biometrika,* **43,** (1956), 344–352.
4. E. Batschelet, "Statistical Methods for the Analysis of Problems in Animal Orientation and certain Biological Rhythms," *AIBS Monograph,* 1965.
5. T. Downs and J. Liebman, "Statistical Methods for Vectorcardiographic Directions", *Trans. Bio-Med. Eng.,* **16,** (1969), 87–94.

3   *Mathematical Models and Spectral Structures*

RAYMOND L. VANDE WIELE, *Moderator*

CHAPTER 8

Pulsatile Variations in Hormone Levels*

R. J. BOGUMIL

International Institute for the study of Human Reproduction, College of Physicians and Surgeons, Columbia University, New York, New York

The objectives of this article are first to review briefly recent literature pertaining to the discovery of pulse-like fluctuations in plasma hormone levels and second to outline an original theoretical treatment of the subject of endocrine communication and control which provides a possible rationale for the existence (or at least the utility) of the rapid level changes. The emphasis, by example and reference, is on the human female reproductive system. The more general aspects of endocrine regulation are covered in standard texts and have been reviewed by others in this volume and elsewhere (1).

* This work was supported by private contribution. Abbreviations used in this article: ACTH, adrenocorticotropic hormone; CRF, corticotropin releasing factor; EEG, electroencephalogram; FSH, follicle stimulating hormone; FRF, follicle stimulating hormone releasing factor; LH, luteinizing hormone; LRF, luteinizing hormone releasing factor; REM, rapid eye movement (sleep); 11-OHCS, 11-hydroxycorticosteroids; and 17-OHCS, 17-hydroxycorticosteroids.

REVIEW

The existence of gradual level changes of gonadotropic and ovarian steroid hormones accompanying the menstral cycle has been recognized since the early 1930's (2, 3). A pattern for the normal cycle was established and the interactions of the hormones and target organs was postulated as a mechanism of control. Development of improved hormone assays, synthetic steroids, and purified gonadotropin preparations led to experiments which further substantiated the nature of the cyclic changes and gave direct evidence of the hormonal control and feedback relationships.

Evidence that the adrenal system is characterized by a diurnal fluctuation in steroids was reported in 1943 (4). Although the interrelationships among biological systems and rhythms was a topic of considerable interest (5), endocrine research on the female reproductive system continued to focus on the cycle (approximately 28 days in the human) revealed by once daily samples.

Studies of the menstrual cycle by serial hormone determinations are plagued by a number of methodologic problems. One fundamental difficulty is that the ovarian morphological changes—the objective of control—are not experimentally measurable on a repetitive basis. A second limitation is that even with the recently developed radioimmunologic

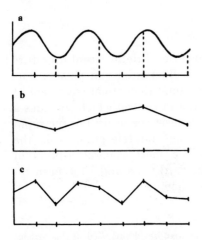

Figure 1   Sequential equally spaced samples of a time varying quantity (a) will not properly detect harmonic components more rapid than one-half the sampling frequency (b). The high frequency signal power will be aliased and appear at specific lower frequencies (b) unless some method of filtering is used to average the rapid variations before the samples are taken or the sampling interval is reduced (c).

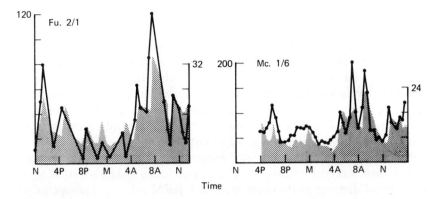

Figure 2   ACTH (line) and 17-OHCS (shading) in peripheral plasma of normal subjects during a 24-hour period [S. A. Berson and R. S. Yalow (is), by permission].

and competitive protein binding assays, the volume of plasma required for simultaneous determination of gonadotropins and ovarian steroids (approximately 3 ml) limits the number of samples which may be drawn from a subject in a specific interval. To avoid errors and possible misinterpretation of data due to *aliasing* (see Figure 1 and Ref. 6), the samples must either be drawn at intervals shorter than the hormone half-life or else the samples must be taken in a fashion which will average out any variations during the interval between samples. The latter approach involves the tacit assumption that any variations occurring more rapidly than one-half the sampling frequency are of no interest. Hormone half-lives range from a few minutes for some steroids to between 1/4 and several hours for protein hormones. Thus, even with present techniques, it is not possible to sample as frequently as desirable over the course of a complete menstrual cycle. Assay sensitivity, specificity, and reproducibility present additional methodologic problems.

Efforts to more precisely characterize the diurnal adrenal cycle faced similar, though less extreme, difficulties. Clinical and physiological evidence that adrenal cortical and medullary secretions adjust rapidly (within minutes to an hour) to changes in activity (and metabolic demand), stress, and experimental stimuli, in addition to the general diurnal rhythm, was established by 1950 (7). However, direct measurement of the complex waveforms likely to result from the superposition of such effects was not feasible with the available techniques. Thus, although the importance of the interaction of the diverse stimuli which affect endocrine secretions was recognized (7, 8), experimental efforts were limited either to acute study of the response to a specific controlled stimulus or to documentation of the more general features of the cyclic hormone level changes in normal and pathological states.

Measurement of 17-OHCS in adrenal vein blood of surgically prepared dogs led Hume (9) to suggest, in 1956, that physiologic corticosteroid levels are maintained by occasional secretory episodes, even in the resting state. A similar experimental preparation was used by Galicich and co-workers (10) who reported the possible interpretation of a 3-hour secretory rhythm in addition to diurnal fluctuations.

Weitzman et al. (11) sampled peripheral plasma in human subjects at 1/2 hour intervals during sleep and concluded that 17-OHCS levels had several distinct peaks correlated with REM activity during this period. Orth, Island, and Liddle (12) made hourly determinations for 48 hour periods under a variety of sleep-wake schedules and demonstrated an irregular pattern in 17-OHCS level changes.

Competitive protein binding assays and radioimmunoassays developed during the early 1960's permitted more percise measurement of hormone concentrations in a smaller sample volume than had been possible by bioassay or chemical methods. Immunochemical methods and their advantages and limitations are the subject of extensive literature (13, 14).

Berson and Yallow analyzed plasma samples taken at 1/2 hour intervals for 24 hours and reported distinct peaks in ACTH levels occurring at 2 to 3 hour intervals (with apparent parallelism in cortisol level changes, as assayed by Krieger and co-workers) (Figure 2) (15). Hellman et al. (16, 17) and Krieger et al. (18) (see also Chapters 23 and 41) have studied this matter with particular care and demonstrated the pulsatile character of the plasma levels of ACTH, 17-OHCS, and 11-OHCS. Comparison of changes in cortisol plasma levels with cortisol specific activity after rapid intravenous administration of trace amounts of ^{14}C-cortisol during the observation period revealed that the cortisol fluctuations were in large measure due to variations in secretion rate rather than clearance (16).

Interest in the possibility of rapid fluctuations in gonadotropins and ovarian steroids did not develop as quickly. By 1961 a highly effective contraceptive steroid had been marketed and early reports of treatment of female infertility by exogenous gonadotropins appeared (19). The success of these therapies in the absence of any specific knowledge of diurnal or other rapid hormone variations tended to minimize concern with such possibilities. Once daily samples seemed to yield reasonably smooth time series within the assay error. That more interest was not expressed in occasional marked deviations can be attributed to the caution used in interpretation of results obtained with the new and sometimes troublesome immunoassays and lack of a specific physiologic reason to expect significant diurnal variation in this system. Thus, with the exception of the characteristic midcycle peak, there seemed to be no need for more frequent measurement. The relatively long primate cycle and its uncertain timing complicates experimental work on these species. Rodents have a short (4 or 5 day for the rat) predictable estrus cycle but also less blood. Other species used for experimental purposes (e.g., livestock) have estrus cycles of moderate length and ample blood; however, the need to develop immunoassay systems for these animals delayed the investigations.

By 1968 several laboratories had become interested in detection of possible diurnal variations in gonadotropins. Gonadotropins in the human male were sampled hourly (20) and every 2 hours (21); however, no distinct pattern of level changes was detected. Other studies (21–23) with less frequent sampling indicated a possible diurnal rhythm.

Midgley and Jaffe (25) reported a significant diurnal variation in LH (four samples per day) during the follicular phase of the cycle in women and the occasional appearance of marked elevations in the normally depressed LH levels during the luteal phase (daily samples). This latter observation was at variance with the hormone pattern established by bioassay and suggested the need for closer study. FSH variations were not as pronounced. Thomas et al. (26) measured plasma LH and FSH in women every 6 hours during 7-day periods estimated to coincide with the midcycle peak. They reported the frequent appearance of a "biphasic" LH peak. The elevated mean gonadotropin levels and the absence of a gonadal cycle in castrate and menopausal subjects somewhat simplified the study of possible diurnal secretion patterns under such conditions. Approximately hourly fluctuations of LH in castrate female monkeys were reported by Atkinson et al. (27) and described in detail by the same laboratory [Dierschke et al. (28)]. Similar though less conclusive observations were reported in cycling ewes (29), the castrate rat (30, 31), in women with normal cycles (32) and menstrual disorders (33), and for plasma testosterone levels in man (34). These phenomena have also been observed in the ram and bull (35, 36), in intact man (37–39) and menopausal women (40). Midgley and Jaffe (41) and Yen et al. (42) (see also Chapter 13) have undertaken the task of demonstrating the existence of pulsatile LH patterns during limited periods in all phases of the human menstrual cycle (Figures 3, 4). Corresponding variations in ovarian steroids have been reported informally by several investigators.

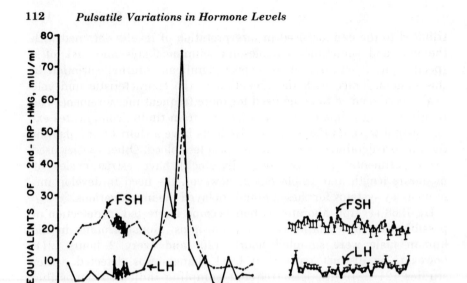

Figure 3 *FSH and LH in peripheral plasma sampled daily during a normal menstrual cycle and hourly for 2 days during the follicular phase [A. R. Midgley Jr. and R. B. Jaffe (41), by permission].*

There has been considerable speculation as to the origin of the pulse-like fluctuations. It is generally presumed that hormone release by the anterior pituitary is largely under hypothalamic control. The possibility that neural components are important determinants of the observed pattern has led to studies of the secretory episodes in relation to EEG measurements (11, 16, 17, 32, 34, 38, 39). A variety of hormone infusion experiments using gonadotropins, steroids, and releasing factors have been designed in an attempt to correlate the pulse release with specific hormonal triggers. These efforts have begun only recently and few conclusions can be drawn from the experimental evidence in the literature to date.

Complementing the efforts to measure naturally occurring hormonal time series by direct *in vivo* techniques are the studies, both *in vivo* and *in vitro*, designed to measure the response of the systems or system components to controlled stimuli. Under this broad heading come the studies of systemic and target organ response to drugs and synthetic and natural hormones administered either locally or peripherally; the measurement of rates of synthesis, release, metabolism, interconversion, protein binding, and clearance of hormones (either *in vivo* or by organ perfusion, tissue superfusion, incubation, or in cell free systems); and morphological studies in both treated and untreated subjects.

The objective of the various techniques of mathematical systems analysis is to provide a link between the data obtained by the experimental approach and the observation time series obtained by sampling the naturally occurring hormone level changes. If the knowledge of isolated subcomponent performance is adequate and if a satisfactory mathematical (or at least programmable) representation of each of the important subcomponents can be developed, then it may be hoped that solution (simulation) of the complete system will yield results similar to available data and have some predictive value. However, failing at this task can be instructive as well, particularly if specific problems can be identified.

Another broad class of mathematical techniques, referred to as *time series analysis*, provides the reciprocal relationship. These methods are used to represent directly measured time series as spectral or other components which define a general system structure (as coefficients of orthogonal functions). These methods may be used for data smoothing,

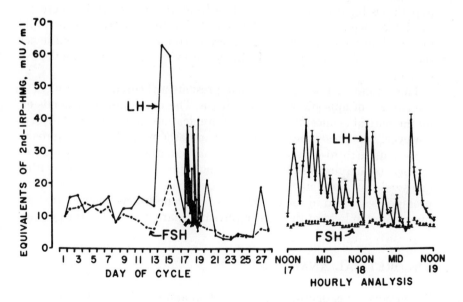

Figure 4   FSH and LH in peripheral plasma sampled daily during a normal menstrual cycle and hourly for 2 days during the luteal phase [A. R. Midgley Jr. and R. B. Jaffe (41), by permission].

prediction, and spectral (harmonic) analysis or, in conjunction with other techniques, to develop simulation models. The literature on both times series analysis and simulation is extensive.

The work of Yates and Brennan (43) on modeling adrenal cortical function is one example of the application of these mathematical tools in endocrinology. A mathematical representation was developed for this system from well-established physiological concepts, the sophisticated dynamic studies of canine adrenal response of Urquhart (44), and a patchwork of other evidence. The weaknesses and the value of cross species comparisons are well known. The point is merely that the mathematical tools are sufficiently flexible to permit the formulation of a complex model from quantitative as well as qualitative data without losing the ability to relate specific features of the simulation results to critical assumptions.

Numerous conceptual models of the nonpregnant female reproductive cycle have been proposed (1, 2, 29, 45–58). Some of these have been accompanied by published computer or analytic solutions (45, 47–49, 54–58) while others represent an English language or block diagram construct designed to present a schematized description of the process. Using a model based on the extensive literature in reproductive physiology and evaluation of multiple alternate physiological hypotheses by simulation where data from direct observation were not available, Bogumil, Ferin, and Vande Wiele (59) suggested that rapid low level fluctuations in hormone levels could considerably influence the menstrual cycle and might in fact provide an effective method of control.

The second section of this article presents a theoretical treatment of the subject of episodic hormone release. The development depends on mathematical concepts from communication and control theory; however, the presentation itself is nonmathematical. This approach represents a third method whereby mathematical techniques may be applied to acquire insight into biological phenomena. Data is neither analyzed nor simulated; rather the endocrine communication-control problem is studied in the abstract and the conclusions reached are then interpreted in light of the known physical characteristics of the system.

THEORETICAL ANALYSIS

The process of endocrine communication and control is conducted over a communication channel composed of the circulatory system plus those tissues, other than target organs, involved in the peripheral degradation of hormones (Fig. 5). The simultaneous use of this channel by many endocrine and metabolic systems has two effects: first to accomplish purposeful cross-links among the various systems and second to complicate the regulatory task of an individual system due to the presence of similar signals of different origin (e.g., both adrenal and ovarian steroids and plasma hormone level variations due to changes in release, reflux from extravascular compartments, and the effects of intermittent activity-rest cycles contributing to changes in blood flow, hepatic clearance and peripheral metabolism of hormones).

Signals sent by the hypopthalamo-pituitary controller are encoded first in releasing factors and then in variations in the secretion/release rate (and resultant plasma concentration) of the trophic hormone(s) for each of the target organs. These signals (represented for convenience by simple rectangular waveforms in Figures 5, 7, and 8) are distorted by the channel (for reasons just cited) and appear at the target organ reduced in amplitude, shifted in phase, and contaminated by noise. In general the pituitary hormone stimulates both slow morphological and rapid secretory response in the target organ. When released, the target organ secretions are distributed through the channel to the entire organism and may in turn affect many physiological processes, but in

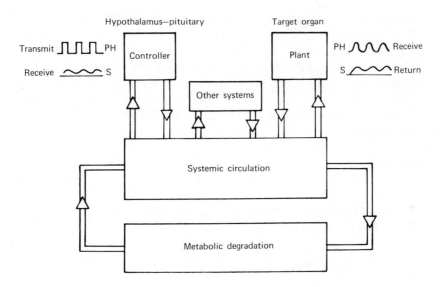

Figure 5 Pituitary hormones (PH) are distributed to target organs by the systemic circulation. Target organs respond with characteristic morphological changes and secretions (S). The response is monitored by the controller. Hormones are removed from the system principally by peripheral metabolism and hepatic clearance.

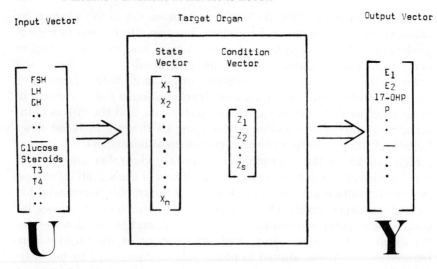

Figure 6   Vector-matrix representation of the ovary: the vector U of control inputs (organ specific trophic hormones, principally FSH and LH) and other inputs (all blood borne substances) affects the system state vector X and condition vector Z. Outputs (vector Y) are a function of the inputs and the state. There are organ specific outputs (e.g., ovarian steroids) and others.

particular the hypothalamo-pituitary feedback sites. The return signal is also distorted in some fashion. The task of the controller is to achieve or maintain a desirable metabolic condition and/or target organ state in the face of constantly changing and unpredictable circumstances.

It should be noted that feedback is not a prerequisite of control. For example, a control system could, under certain conditions, accomplish its task by periodically transmitting a signal guaranteed to achieve some desired effect (open loop control). This could be facilitated by a controller-accessible model for the plant. Open loop control is, however, unsatisfactory or extremely inefficient when unpredictable variations in plant performance may occur due either to "external disturbances" or complex plant structure.

Figure 6 illustrates a vector-matrix representation of an endocrine organ. The equivalent equations are:

$$\dot{X} = AX + [B_1 \mid B_2]\begin{bmatrix} U_1 \\ U_2 \end{bmatrix} \tag{1}$$

$$Y = CX + DU \tag{2}$$

$$Z = EX + [F_1 \mid F_2]\begin{bmatrix} U_1 \\ U_2 \end{bmatrix} \tag{3}$$

where:

X = the n dimensional state vector
$\dot{X}$ = the n dimensional vector of first derivatives of the components of X
Y = the m dimensional vector of outputs

$$Y = \left[\frac{Y_1}{Y_2}\right]$$

with Y_1 = the r dimensional vector of target organ secretions (directly observable outputs)
  Y_2 = the $m - r$ dimensional vector of other outputs (nonspecific metabolites and other factors)
  U = the k dimensional vector of inputs

$$U = \left[\frac{U_1}{U_2}\right]$$

with U_1 = the g dimensional vector of target organ trophic hormones (control inputs)
  U_2 = the $k - g$ dimensional vector of other factors in the peripheral circulation
  Z = the s dimensional condition vector
  A, B = $[B_1 \mid B_2]$, $C, D, E, F = [F_1 \mid F_2]$ are matricies of appropriate dimensions, which indicate the system couplings between inputs, states, outputs, and conditions.

This formulation will be used to develop some communication and control principles of direct relevance to the subject of pulsatile hormone waveforms. While the mathematical structure is quite general in certain respects, the subsequent development requires the assumption that the matrix coefficients are constant or, at most, slowly varying functions of the independent parameter *time* (i.e., the postulate of target organ linearity). This is never a valid assumption for the global performance of a biological system, and only rarely reasonable within the physiologic domain. In the present application the assumption of linearity is used as a working hypothesis. This is because the analysis can be limited to the study of physical perturbations of small amplitude, short duration, and zero mean value (the variational problem). The representation of a nonlinear system by a linear approximation for small perturbations about a known trajectory is often justified. Finally, the results of the linear treatment will be evaluated in light of the known characteristics of the system. The postulate of linearity makes it possible to utilize results obtained by mathematicians in the fields of communication theory and optimal control.

Optimal control is the term applied to a highly mathematical subtopic in control theory. However, it is characterized by a very simple premise which has great relevance to biological control problems. Namely, the *optimal control policy* is the input signal which regulates the plant to achieve the best possible result (according to some specified criterion) subject to system constraints. Typical criteria for optimization are: reaching a specified state in minimum or fixed time or at minimum cost (effort). Common constraints are: limitations on controller peak power or average power (i.e., limitations on the maximum secretion rate of a trophic hormone).

Two fundamental concepts in modern control theory are *controllability* and *observability* (60–63). A brief description of these terms and some mathematical results are contained in the Appendix.

Quite simply, those aspects of target organ performance that require hypothalamo-pituitary regulation must be *controllable* by the appropriate trophic hormones. Furthermore, because the optimal control policy is a function of the state of the plant, it is advantageous to determine the new state at each point as rapidly as possible (given the system constraints). Determination of the state from observation of the outputs (target organ hormones) requires *observability*.

It is clear that the precise state of the target organ need not be determined by hypothalamo-pituitary control. The human ovary contains several hundred thousand follicles of which several hundred undergo development and atresia in each cycle, while only one ovulates. This selection need not be accomplished entirely by gonadotropic regulation. Rather, the gonadotropins may only regulate certain intraovarian conditions which facilitate a single ovulation. Ovarian mechanisms complement this action so that, in combination, proper performance is obtained. A simple mathematical example of complete (state) *controllability* and *observability* versus condition *controllability* and *observability* is presented in the Appendix.

A final concept essential to the development of this article is that of *channel capacity* (64). The *channel capacity* is the maximum rate at which information can be transmitted over a particular channel. Any attempt to send information more rapidly merely results in a greater number of errors. For a linear continuous channel the capacity is limited by the bandwidth and the signal-to-noise ratio. An upper limit on the bandpass of an endocrine communication channel is determined by the hormone half life. Communication signals—encoded as variations in the hormone release rate and propagated as fluctuations in the plasma level—can convey an amount of information (to other organs) which is limited by the bandpass of the communication system. The communication system includes the mechanisms of hormone release, transport, and detection as well as clearance. The dynamic response of each subsystem has been studied experimentally by numerous investigators. The characteristics of the adrenal and ovarian systems are such that signal frequencies above 1 cycle/hour are greatly attenuated.

It is only necessary to answer the following question: Given the known physical attributes of hormonal control systems, what techniques might be used to optimize their performance relative to the characteristic trajectories (diurnal adrenal rhythm, menstrual cycle, or nonperiodic regulation)? From the preceeding material it is clear that both the gradual cyclic changes and the concept of a constant level represent signal frequencies far below the high frequency cutoff and thus do not take advantage of the full system bandwidth.* Our question may be rephrased: How might the control signal be applied to accomplish the same slow morphological transformations that would be achieved by the slowly varying signal and yet utilize the full channel capacity to obtain information on the condition of the target organ as rapidly as possible? Figure 7 illustrates several modulation techniques used in communication systems. For a given signal envelope (top of Figure 7) the same amount of hormone is released over the time interval of several pulses by each method. Thus, if the timing of the pulses is such that they occur rapidly relative to the morphological response times and yet below the communication system cutoff frequency, it would be possible to monitor the target organ hormonal response to individual pulses of the trophic hormone (see Figure 8) as an index of the morphological condition of the target organ and, at the same time, accomplish the appropriate control. To be of value the sequential feedback samples must depend principally on the condition of the target organ and not on the state of the channel itself. The channel bandpass and the minimum reliably detectable level change limit the rate at which channel independent measurements can be made. Critical to the physiological importance of such a scheme is the assumption that hypothalamic steroid receptor mechanisms can in fact detect plasma hormone level changes of the magnitude and frequency observed and perform the required signal analysis. The complexity of

---

* It can be advantageous to concentrate signal power at low frequencies in systems where the high end of the band is extremely noisy. It may be argued that the observed pulses constitute such noise. However, the available data indicate that ACTH and cortisol pulses are correlated. Observations, to date, of LH and ovarian steroids have been less detailed.

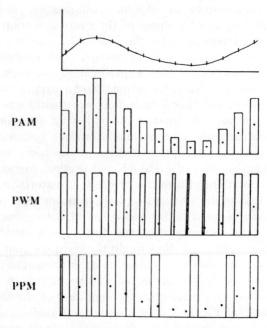

PAM

PWM

PPM

Figure 7 A time varying signal (top) may be encoded by many methods, such as by a series of pulses of varying height (amplitude) (PAM), width (PWM), or frequency (position) (PPM). Dots indicate the average level over a time interval centered at that point.

neuro-endocrine structures suggests that sophisticated signal processing may occur; however, very difficult experimental work will be required to resolve this issue (see Chapter 40).

Additional information provided by the high-frequency feedback would facilitate *subtle* correction of the control signal to provide more nearly optimal control. The adrenal steroids are themselves essential to the maintenance of many physiologic processes. In this system, high-frequency modulation of CRF → ACTH could facilitate hierarchical control of corticosteroid dependent processes. The trophic hormone pulse repetition rate for both the adrenal (CRF → ACTH → steroid) and ovarian (LRF → LH → steroid) systems has an upper limit of approximately 1 pulse/hour. The secretory episode must, of course, occupy only a fraction of the interval. Of the many possible modulation schemes, a combination of pulse width and pulse position modulation has attractive features and is consistent with the observations reviewed in the first section of this article. In any case, it is not necessary that the biological waveforms be rectangular or the temporal relationships as simple as those used here for illustration.

Mathematical analysis of physical systems invariably requires assumptions or approximations. This is, of course, particularly true in the study of biological systems. The most that can be hoped for is that the assumptions are sufficiently correct for the conclusion to provide new and useful insight into the problem. It is then necessary to reassess our result in these terms. Urquhart (44) has shown (under reasonably physiologic, experimental conditions) that ACTH stimulation of adrenal cortisol release has nonlinear properties. Nevertheless, the character of these nonlinearities does not necessitate modification of the development given above. Similar though dynamically less detailed observations have been made of ovarian steroidogenesis. While the postulate of target organ and channel linearity is necessary to the derivation, qualitative aspects of the conclusions are of more general application. The conclusions complement previously reported performance of a nonlinear model of the human menstrual cycle (59). Certain aspects of the analytic treatment will facilitate subsequent modeling efforts.

The success of substitutional therapy in the treatment of hypophysectomized subjects, anovulation, Addison's disease, diabetes, and other endocrine disorders offers simple evidence that high-frequency modulation of hormones is not essential to the control of target organs. However, this does not weaken the arguments for the advantages of a physiologic control mechanism based on pulsatile release of trophic

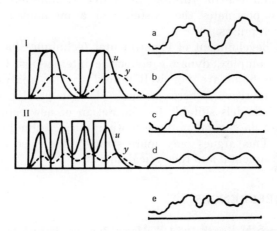

Figure 8 For a pulse repetition rate (I) within the system bandpass, the response (y, b) to individual pulses of the received train (u) will be distinguishable (a) from the response to previous pulses and noise (c). At excessively high rates (II) this is no longer true (d, e), yielding, in the limit, smoothly varying (i.e., nonpulsatile) plasma levels (at least with respect to hormone production).

hormones. On the other hand, it must be emphasized that in no sense does the work outlined in this article constitute a mathematical proof for such a mechanism. The observed oscillations may, at least in part, be attributed to lags and delays in the system (see Appendix) and may, in fact, have no important physiologic consequence.

It has been suggested that the mutual coupling among nonlinear oscillators formed by neural nets in the cerebral cortex could have the effect of producing approximately synchronous neuron firing and thus create rhythms associated with large populations of cells not characteristic of the isolated cell (65). A large number of neurons connected in some indeterminate fashion, each with a response time measured in milliseconds, exhibits a group oscillatory behavior with strong components at distinct and characteristic frequencies in the range of 2 to 20 Hz. It is reasonable to suggest that such performance could also characterize the neuro-secretory behavior of the hypothalamic regulatory sites. The time scale of events is greatly lengthened (relative to the EEG) due to the inertial elements of the secretory mechanism. The basic prerequisites of nonlinear response and local or "short" feedback have been suggested by numerous workers for both the hypothalamus and anterior pituitary. The possibility that the specific physiological origin of the observed pulsatile waveforms may lie with such "self-organizing" behavior is compatible with the material presented in this article. The theoretical principle of inter-organ communication and control through pulse-like modulation of hormone levels simply postulates the existence of a mechanism capable of producing such pulses.

The rapid development of experimental techniques has led to the discovery of complex, dynamic endocrine interactions. Conventional statistical methods are not well suited to the analysis of such phenomena; moreover, these methods do not provide interpretive mathematical models suitable for the design of new experiments. In fact, there are no general purpose mathematical techniques adequate to the task. This argues only more strongly for close collaboration between experimental and theoretical analysts.

REFERENCES

1. F. E. Yates, S. M. Russell, and J. W. Maran, *Ann. Rev. Physiol.* **33**, (1971), 393.
2. C. Moore, and D. Price, *Proc. Soc. Exp. Biol. Med.* **28**, (1930), 38.
3. R. T. Frank, *J. Amer. Med Assoc.* **97**, (1931), 1852.
4. G. Pincus, *J. Clin. Endocr.* **3**, (1943), 195.
5. N. Kleitman, *Physiol. Rev.* **29**, (1949), 1.

6. R. W. Hamming, *Numerical Methods for Scientists and Engineers*, McGraw-Hill, New York, 1962. p. 276.

7. E. J. Kepler, and W. Locke, in *Textbook of Endocrinology*, (R. H. Williams, Ed.), Saunders, Philadelphia, 1950, p. 180.

8. F. Halberg, *J. Lancet*, **73**, (1953), 20.

9. D. M. Hume, in *Pathophysiologia Diencephalica* (S. B. Curri and L. Martini, (Eds.), Springer Vienna, 1958, p. 217.

10. J. H. Galicich, E. Haus, F. Halberg, and L. A. French, *Ann. N.Y. Acad. Sci.*, **117**, (1964), 281.

11. E. D. Weitzman, H. Schaumburg, and W. Fishbein, *J. Clin. Endocr. Metab.*, **26**, (1966), 121.

12. D. N. Orth, D. P. Island, and G. W. Liddle, *J. Clin. Endocr. Metab.*, **27**, (1967), 549.

13. S. A. Berson, and R. S. Yalow, in *The Harvey Lectures (1966-1967)*, Vol. 62, Academic Press, New York, 1968, p. 107.

14. F. Peron, and B. Caldwell (Eds.), *Immunological Methods in Steroid Determination*, Appleton-Century-Crofts, New York, 1971.

15. S. A. Berson, and R. S. Yalow, *J. Clin. Invest.*, **47**, (1968), 2725.

16. L. Hellman, F. Nakada, J. Curti, E. D. Weitzman, J. Kream, H. Roffwarg, S. Ellman, D. K. Fukushima, and T. F. Gallagher, *J. Clin. Endocr.*, **30**, (1970), 411.

17. L. Hellman, E. D. Weitzman, H. Roffwarg, D. K. Fukushima, K. Yoshida, and T. Gallagher, *J. Clin. Endocr.*, **30**, (1970), 686.

18. D. T. Krieger, W. Allen, F. Rizzo, and H. Krieger, *J. Clin. Endocr.*, **32**, (1971), 266.

19. C. A. Gemzell, *Fertility Sterility.*, **13**, (1962), 153.

20. D. T. Peterson, A. R. Midgley, Jr., and R. B. Jaffe, *J. Clin. Endocr.*, **28**, (1968), 1473.

21. R. S. Swerdloff, and W. P. Odell, in *Gonadotropins, Proceedings of a Workshop Conference, Vista Hermosa Mor, Mexico*, (E. Rosemberg, (Ed.), Geron-X,Los Altos, 1968, p. 155.

22. C. Faiman, and R. J. Ryan, *Nature*, **215**, (1967), 857.

23. B. B. Saxena, G. Leyendecker, W. Chen. H. M. Gandy, and R. E. Peterson, *Acta Endocr.*, (Copenhagen), **63**, Suppl. 142 (1969), 185.

24. C. Faiman, and J. S. D. Winter, *J. Clin. Endocr.*, **33**, (1971), 186.

25. A. R. Midgley, Jr., and R. B. Jaffe, *J. Clin. Endocr.*, **28**, (1968), 1699.

26. K. Thomas, R. Walckiers, and J. Ferin, *J. Clin. Endocr.*, **30**, (1970), 269.

27. L. E. Atkinson, A. N. Bhattacharya, S. E. Monroe, D. J. Dierschke, and E. Knobil, *Endocrinology*, **87**, (1970), 847.

28. D. J. Dierschke, A. N. Bhattacharya, L. E. Atkinson, and E. Knobil, *Endocrinology*, **87**, (1970), 850.

29. I. A. Cumming, J. M. Brown, M. A. de B. Blockey, and J. R. Goding, *J. Reprod. Fert.*, **24**, (1971), 148, (Abstract).

30. J. E. Lawton, and N. B. Schwartz, *Amer. J. Physiol.*, **214**, (1968), 213.

31. V. L. Gay, and A. R. Midgley, Jr., *Endocrinology*, **84**, (1969), 1359.

32. S. Kapen, R. Boyar, L. Hellman, and E. D. Weitzman, *Psychophysiology*, **7**, (1970), 337 (Abstract).

33. W. J. Dignam, A. F. Parlow and T. A. Deane, *Am. J. Obstet. Gynecol.,* **105,** (1969), 679.

34. J. I. Evans, A. W. MacLean, A. A. A. Ismail and D. Love, *Nature,* **229,** (1971), 261.

35. D. J. Bolt, *J. Reprod. Fert.,* **24,** (1971), 435.

36. C. B. Katongole, F. Naftolin, and R. V. Short, *J. Endocr.,* **50,** (1971), 457.

37. H. R. Nankin, and P. Troen, *J. Clin. Endocr.,* **33,** (1971) 508.

38. R. T. Rubin, A. Kales, R. Adler, T. Fagan, and W. Odell, *Science,* **175,** (1972), 196.

39. R. Boyar, M. Perlow, L. Hellman, S. Kapen, and E. Weitzman, *J. Clin. Endocr.,* **35,** (1972), 73.

40. E. Silverman, M. P. Warren, U. Bruce, and R. L. Vande Wiele, Unpublished Observations.

41. A. R. Midgley, Jr., and R. B. Jaffe, *J. Clin. Endocr.,* **33,** (1971), 962.

42. S. S. C. Yen, C. C. Tsai, F. Naftolin, G. Vandenberg, and L. Ajabor, *J. Clin. Endocr.,* **34,** (1972) 671.

43. F. E. Yates, and R. D. Brennan, *Math. Biosci.,* Supp **1** (1968), 20.

44. J. Urquhart, *Physiologist* **13,** (1970) 7.

45. H. Lamport, *Endocrinology,* **27,** (1940), 673.

46. O. W. Smith, and G. V. S. Smith, *J. Clin. Endocr.,* **6,** (1946), 483.

47. A. Rapoport, *Bull. Math. Biophys.,* **14,** (1952), 171.

48. L. Danziger, and G. L. Elmergreen, *Bull. Math. Biophys.* **19,** (1957), 9.

49. H. E. Thompson, J. D. Horgan,and E. Delfs, *Biophys. J.,* **9,** (1969), 278.

50. J. W. Goldzieher, in *The Ovary* (H. C. Mack, Ed.), Thomas, Springfield, Ill., 1968, p. 106.

51. N. B. Schwartz and J. C. Hoffman, *Excerpta Med. Int. Congr.,* **132,** (1967), 997.

52. S. M. McCann, D. B. Crighton, S. Watanabe, A. P. S. Dhariwal, and J. T. Watson, *Math. Biosci.,* Suppl **1** (1968), 193.

53. D. F. Horrobin, *J. Theor. Biol.,* **22,** (1969), 80.

54. R. L. Vande Wiele, J. Bogumil, I. Dyrenfurth, M. Ferin, R. Jewelewicz, M. P. Warren, T. Rizkallah, and G. Mikhail, *Rec. Progr. Horm. Res.,* **26,** (1970), 63.

55. N. B. Schwartz, and P. Waltz, *Fed. Proc.,* **29,** (1970), 1907.

56. S. Inoüe, T. Nakamura, and T. Sekiguchi, *Endocr. Japan,* **17,** (1970), 567.

57. W. J. Shack, P. Y. Tam, and T. J. Lardner, *Biophys. J.,* **11,** (1971), 835.

58. R. J. Bogumil, M. Ferin, J. Rootenberg, L. Speroff, and R. L. Vande Wiele, *J. Clin. Endocr.,* **35,** (1972), 126.

59. R. J. Bogumil, M. Ferin, and R. L. Vande Wiele, *J. Clin. Endocr.,* **35,** (1972), 144.

60. R. E. Kalman, *Proceedings of the First International Congress on Automatic Control, Moscow, 1960,* Butterworths, London, 1961, p. 481.

61. R. E. Kalman, *Proc. Nat. Acad. Sci., U.S.,* **48,** (1962), 596.

62. R. E. Kalman, Y. C. Ho, and K. S. Narendra, in *Contributions to Differential Equations,* Vol. 1, Wiley-Interscience, New York, 1963, p. 189.

63. E. Kreindler and P. E. Sarachik, *IEEE Trans. Auto. Contr.,* AC-9 (1964), 129.

64. C. E. Shannon and Weaver, *The Mathematical Theory of Communication,* Univ. Illinois Press, Urbana, 1949.

65. N. Wiener, Cybernetics, 2nd ed., M.I.T. Press, Cambridge, Mass., 1961, p. 199.

APPENDIX

Kalman (60–62) and others have applied the methods of linear algebra to analyze the relationships that must hold for it to be possible to control a linear dynamical plant (by manipulation of the inputs) or determine its state (by observation of the outputs). He has called these concepts *controllability* and *observability*. *Controllability* requires that it be possible to attain any desired state from any initial state by the application of an input for finite time. *Observability* requires that it be possible to determine any initial state by observation of the output for finite time.

For the linear fixed system described by Eqs. 1, and 2

$$\dot{X} = AX + BU \tag{1}$$

$$Y = CX + DU \tag{2}$$

where $\quad X$ = the n-dimensional state vector
$\quad\quad\quad\ Y$ = the r-dimensional output vector
$\quad\quad\quad\ U$ = the g-dimensional control input vector
$\quad\quad\quad\ A$ = a $n \times n$ constant coefficient matrix
$\quad\quad\quad\ B$ = a $n \times g$ constant coefficient matrix
$\quad\quad\quad\ C$ = a $r \times n$ constant coefficient matrix
$\quad\quad\quad\ D$ = a $r \times g$ constant coefficient matrix

Kalman (60) has shown that state controllability requires rank $P = n$ where P is the composite $n \times ng$ matrix

$$P = [B \,|\, AB \,|\, \cdots \,|\, A^{n-1}B]$$

Note that A^h, $h \geqq n$, is expressible in terms of $A, \ldots, A^{n-1}$ (Cayley-Hamilton theorem). The result indicates that the system must be connected in such a fashion that P contains a basis for the n-dimensional state space. Since the range space of P ($R(P)$) is the subspace of $R(A)$ attainable from the input, it follows that the system is controllable. Observability requires rank $Q = n$ where Q is the composite $n \times nr$ matrix.

$$Q = [C^* \,|\, A^*C^* \,|\, \cdots \,|\, (A^*)^{n-1}C^*]$$

where $\quad\quad\quad A^*$ = conjugate transpose of A.

This result establishes that Q must contain a basis for the n-dimensional state space. Since $R(Q)$ is the subspace spanned by the state-dependent output sequence, it follows that the system is completely observable.

Kalman (61) has also shown that any system of the above form can be decomposed into subspaces which are controllable (but not observable), observable (but not controllable), both observable and controllable, and neither observable nor controllable. The controllable subspace, $R(P)$, and the unobservable subspace, $R(Q)^{\perp}$, are invariant under A. The requirement of constant parameters is equivalent to the assumption of fixed rate constants in isotope tracer experiments. However, these results have also been extended to linear systems with time varying parameters (62). While this latter material [and the interesting problem of partitioning U into *(1)* control inputs and *(2)* other inputs] is more general than the preceeding, the complexity of the notation precludes description here. Instead, the constant coefficient results will be used to illustrate principles also applicable to the time variant case.

A slight modification permits description of the case where it is not necessary to control or observe the system state but only some system condition expressible as a linear transformation of the state vector. In addition to Eqs. 1 and 2 we have

$$Z = EX + FU \tag{3}$$

where: $\quad\quad\quad Z =$ the s dimensional condition vector
$\quad\quad\quad\quad\quad\quad E =$ a $s \times n$ constant coefficient matrix
$\quad\quad\quad\quad\quad\quad F =$ a $s \times g$ constant coefficient matrix

The condition (Z vector) controllability requirement is

$$V = [EP \mid F] \text{ of rank } s$$

Condition (Z vector) observability requires that E be an orthogonal projection of the state space on a subspace of $R(Q)$. In the special case where elements of Z represent simple, inclusive sums of identical components, C has the form:

$$C = CE^{\mathrm{T}}E$$

with $c_{ij} = c_{ij}/l_j$, and $l_j =$ the number of components identical to x_j. The condition observability requirement is that Q have rank s (see example below).

These results are obtained by use of Kalman's state space decomposition theorems. The condition vector is controllable if $R(P)$ under linear transformation E is of dimension s. Any direct effect of the control input U on Z (represented by F) augments the controllable space [this result is analogous to output controllability (63)].

Condition controllability and observability can be much less stringent requirements than state controllability and observability when $s \ll$

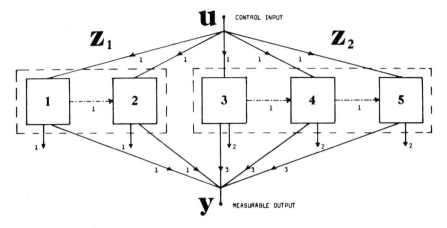

Figure 9 System diagram corresponding to the numerical example (Appendix). Note that the column sum constraint of isotope tracer compartmental analysis does not apply to multisubstrate systems (i.e., compound y may not have a simple molecular relationship to compound u; moreover, factors in addition to control inputs and observable outputs must be considered in applying mass balance).

n. This is demonstrated by the following simple example of a system which is neither state controllable nor observable but is condition controllable and observable. The example, though highly contrived, does illustrate difficulties associated with control and observation of a plant composed of classes of virtually identical components (see Figure 9).

$$X = \begin{bmatrix} -1 & & & & \\ & -1 & & 0 & \\ & & -2 & & \\ & 0 & & -2 & \\ & & & & -2 \end{bmatrix} X + \begin{bmatrix} 1 \\ 1 \\ 1 \\ 1 \\ 1 \end{bmatrix} u$$

$$y = \begin{bmatrix} 1 & 1 & 3 & 3 & 3 \end{bmatrix} X$$

$$Z = \begin{bmatrix} 1 & 1 & 0 & 0 & 0 \\ 0 & 0 & 1 & 1 & 1 \end{bmatrix} X$$

$$P = \begin{bmatrix}1\\1\\1\\1\\1\end{bmatrix}\begin{bmatrix}-1 & & & & \\ & -1 & & 0 & \\ & & -2 & & \\ & 0 & & -2 & \\ & & & & -2\end{bmatrix}\begin{bmatrix}1\\1\\1\\1\\1\end{bmatrix}\Bigg|\begin{bmatrix}1\\1\\1\\1\\1\end{bmatrix}\begin{bmatrix}1 & & & & \\ & 1 & & 0 & \\ & & 4 & & \\ & 0 & & 4 & \\ & & & & 4\end{bmatrix}\begin{bmatrix}1\\1\\1\\1\\1\end{bmatrix}$$

$$\begin{bmatrix}-1 & & & & \\ & -1 & & 0 & \\ & & -8 & & \\ & 0 & & -8 & \\ & & & & -8\end{bmatrix}\begin{bmatrix}1\\1\\1\\1\\1\end{bmatrix}\Bigg|\begin{bmatrix}1\\1\\1\\1\\1\end{bmatrix}\begin{bmatrix}1 & & & & \\ & 1 & & 0 & \\ & & 16 & & \\ & 0 & & 16 & \\ & & & & 16\end{bmatrix}\begin{bmatrix}1\\1\\1\\1\\1\end{bmatrix}$$

$$P = \begin{bmatrix} 1 & -1 & 1 & -1 & 1 \\ 1 & -1 & 1 & -1 & 1 \\ 1 & -2 & 4 & -8 & 16 \\ 1 & -2 & 4 & -8 & 16 \\ 1 & -2 & 4 & -8 & 16 \end{bmatrix} \qquad \text{rank} = 2 \neq n$$

$$V = [EP \mid F] = \begin{bmatrix} 1 & 1 & 0 & 0 & 0 \\ 0 & 0 & 1 & 1 & 1 \end{bmatrix}\begin{bmatrix} 1 & -1 & 1 & -1 & 1 \\ 1 & -1 & 1 & -1 & 1 \\ 1 & -2 & 4 & -8 & 16 \\ 1 & -2 & 4 & -8 & 16 \\ 1 & -2 & 4 & -8 & 16 \end{bmatrix}$$

$$V = \begin{bmatrix} 2 & -2 & 2 & -2 & 2 \\ 3 & -6 & 12 & -24 & 48 \end{bmatrix} \qquad \text{rank} = 2 = s$$

$$Q = \begin{bmatrix}1\\1\\3\\3\\3\end{bmatrix} \left|\; \begin{bmatrix}-1 & & \\ & -1 & 0 \\ & & -2 & \\ & 0 & -2 \\ & & & -2\end{bmatrix}\begin{bmatrix}1\\1\\3\\3\\3\end{bmatrix} \;\right|\; \begin{bmatrix}1 & & 0 \\ 1 & & \\ & 4 & \\ 0 & & 4 \\ & & & 4\end{bmatrix}\begin{bmatrix}1\\1\\3\\3\\3\end{bmatrix}$$

$$\left|\; \begin{bmatrix}-1 & & \\ & -1 & 0 \\ & & -8 & \\ & 0 & -8 \\ & & & -8\end{bmatrix}\begin{bmatrix}1\\1\\3\\3\\3\end{bmatrix} \;\right|\; \begin{bmatrix}1 & & 0 \\ 1 & & \\ & 16 & \\ 0 & & 16 \\ & & & 16\end{bmatrix}\begin{bmatrix}1\\1\\3\\3\\3\end{bmatrix}\right]$$

$$Q = \begin{bmatrix}1 & -1 & 1 & -1 & 1 \\ 1 & -1 & 1 & -1 & 1 \\ 3 & -6 & 12 & -24 & 48 \\ 3 & -6 & 12 & -24 & 48 \\ 3 & -6 & 12 & -24 & 48\end{bmatrix} \qquad \text{rank} = 2 = s \neq n$$

The addition of asymmetric coupling terms will yield a system which is state controllable and observable. For example, if

$$A = \begin{bmatrix} -1 & 0 & 0 & 0 & 0 \\ 1 & -1 & 0 & 0 & 0 \\ 0 & 0 & -2 & 0 & 0 \\ 0 & 0 & 1 & -2 & 0 \\ 0 & 0 & 0 & 1 & -2 \end{bmatrix}$$

then

$$P = \begin{bmatrix} 1 & -1 & 1 & -1 & 1 \\ 1 & 0 & -1 & 2 & -3 \\ 1 & -2 & 4 & -8 & 16 \\ 1 & -1 & 0 & 4 & -16 \\ 1 & -1 & 1 & -2 & 8 \end{bmatrix} \qquad \text{rank} = 5 = n$$

and

$$Q = \begin{bmatrix} 1 & -1 & 1 & -1 & 1 \\ 1 & 0 & -1 & 2 & -3 \\ 3 & -6 & 12 & -24 & 48 \\ 3 & -3 & 0 & 12 & -48 \\ 3 & -3 & 3 & -6 & 24 \end{bmatrix} \qquad \text{rank} = 5 = n$$

The example illustrates several interesting points relevant to the endocrine control problem. It should be noted that the input (a single component in this instance) affects all state variables directly, all states affect the output, and there are two classes of identical dynamic components (two members in the first class and three in the second).

Despite the direct effect of the input on each state variable, the system is not state controllable because of the inseparable behavior of the members of each class. The system is not state observable for the same reason. However, the class members are controllable and observable as a group. It may be noted that the addition of certain couplings among components would yield a system that is state controllable and observable. In endocrine systems asymmetric coupling would result from, for example, anisotropic organ structure (e.g., organ vasculature) or as an artifact of linearization of nonlinear component interaction. The added terms provide simple illustration of the importance of local mechanisms in the neural-endocrine-cellular control hierarchy. Finally, in the present example, it is easily recognized that it is necessary to make at least two distinct observations of the output, on an interval, in order to determine the initial Z condition; moreover, it requires more than the application of an impulse to reach any desired Z condition from some initial Z condition. These latter two points illustrate that even when the control inputs directly affect each state variable it may require finite time to reach a specific condition, due to linear dependence of the Z conditions with respect to the control inputs. Only when rank $[EB|F] = s$ is it possible to reach any condition with an s dimension impulse. This would, of course, require a number of control inputs equal to the number of component classes. Similar arguments apply to the observation problem. In the presence of external disturbances, the control and observation time delays will result in unavoidable deviations from the desired condition. Noise and/or constraints on the controller caused, for example, by deterioration of the communication channel, may render the plant uncontrollable. Commonly, controller identification of unknown or time varying system parameters is also required. To be *identifiable* the value of a parameter must be uniquely calculable from estimates of the observable states (conditions). Moreover, errors in initial estimates (of both parameters and states) must not preclude convergent estimation. Of course, it may not be necessary to accomplish totally error free identification and control in order to obtain satisfactory system function. These matters are the subject of extensive literature in communication and control theory.

The classical physiological concept of neuro-endocrine "negative feedback," though useful as a guide to experimental and clinical procedures since Cannon's time, is a vague and inadequate description for the probable mechanisms of endocrine regulation. Negative feedback does not, of itself, deal with the requirements and implications of *controllability* and *observability*. It should further be noted that the treatment above provides an interpretation for the time delays (1, 29, 58) known to characterize endocrine feedback relationships. The delays may arise from the need to use finite length endocrine signals (i.e., long codes) to span the controllable, observable space. This is illustrated by the preceeding example. As is known from the work of Kolmogorov, Wiener, and others, random or pseudo-random sequences can, under certain conditions, provide an effective alternative to the use of wholly deterministic inputs. These concepts may prove valuable to the interpretation and understanding of neuro-endocrine control.

CHAPTER 9

Modeling Periodicities in Reproductive, Adrenocortical, and Metabolic Systems

F. EUGENE YATES

Department of Biomedical Engineering, University of Southern California, Los Angeles, California

The human female reproductive system has one periodicity, approximately that of the lunar cycle, so prominent that no doubts concerning its existence have ever been seriously put forth. The human adrenal glucocorticoid system was postulated by Pincus in 1943 to possess a rhythm with an approximately 24-hour period (1), and later the existence of this rhythm was demonstrated by Migeon et al. (2), and subsequently confirmed by many other investigators [see Conroy and Mills (3) or Yates and Urquhart (4) for further references]. In 1964 Galicich et al. noted that an ultradian rhythm might be present in the adrenal glucocorticoid system of anesthetized dogs, with a period of about 3.5 hours (5). This second rhythm was later found in humans also by Hellman et al. (6), Krieger et al. (7), Tourniaire et al. (8), and Weitzman et al. (9).

The three rhythms described above are all so conspicuous that no elaborate mathematical methods are required to describe them empirically. In contrast, periodicities in the levels of important metabolites such as free fatty acids, or glucose, in blood are not so well established, and mathematical techniques for extracting periodicities from noisy backgrounds may be required for a convincing demonstration of their existence. Hansen (10), Anderson (11, 12) and Iberall et al. (13) have reported spontaneous oscillations in blood glucose levels at relatively high frequencies in humans, dogs, and rats. In studies of unanesthetized dogs we have failed to confirm the presence of very high-frequency oscillations but have found others to be present with periods of about 2.5 hours. Recently, spontaneous oscillations in free fatty acid levels in fasting humans have also been reported (14), and we have found similar oscillations in unanesthetized, fasted dogs (unpublished data).

The above oscillations, which involve both information fluxes (hormones) and power fluxes (fuels), are representative of the large set of periodicities found within living systems. The set of periodicities represents more than a scientific curiosity: it is instead a manifestation of a general design principle (15).

STABILITY AND PERIODICITY IN NONLINEAR SYSTEMS

A physical basis for the presence of oscillations in both the power flux and information flux domains within autonomous, open thermodynamic systems of all kinds has recently been proposed (15). Since all real, thermodynamic systems operating in the scale between molecules and the solar system are dissipative, any oscillations observed will not be of the linear, harmonic type, but rather of the nonlinear, limit-cycle type. Although a linear, harmonic form may be fitted empirically to the oscillation for descriptive purposes, physical insights into their origins emerge only from recognition of the importance of dissipations and nonlinearities. As Poincaré has shown, one set of nonlinear differential equations containing dissipative terms that has a stable solution reveals a limit-cycle, asymptotic, orbital type of stability in the phase plane representation. Since biological systems are persistent, we should expect from stability requirements alone the existence of periodic structures in time within living systems. In fact, living systems may be viewed as ensembles of weakly coupled, nonlinear oscillators (16) on thermodynamic grounds (17).

DUALISMS IN MODELING SPECTRAL DATA

Given that we should expect living systems to manifest periodic behavior, and that we find such behavior, we are justified in adopting a spectroscopic approach to biological systems. However, as soon as we do so, we are confronted with three classes of experimental and mathematical problems. These are: (*1*) the problem of designing proper strategies for sampling and collecting data to avoid introduction of spurious frequencies or deletion of true frequencies; (*2*) the problem of identifying periodicities that may be present in data collected by our sampling strategy when some noise is present in the periods as well as in the amplitudes, the time series is nonstationary, and the data trains are short; and (*3*) the problem of constructing a mathematical model of oscillatory systems found by data collection and analysis. The present paper concerns the third problem.

Models of oscillatory processes can take a variety of forms ranging from simple, formal empirical models, through empirical-explanatory models, to unit-process, isomorphic, explanatory models. A model is "explanatory" if it obtains its structure from the topography of components and connections believed to be present in the real system, regardless of how poorly known these may be. Examples of primitive, formal models that are not explanatory but that may be used to describe periodic processes are given in Eq. 1, 2, and 3:

$$y = a \sin \omega t + b \tag{1}$$

(this equation is one of the solutions to the equation below)

$$M \frac{d^2y}{dt^2} + Ky = 0 \tag{2}$$

and the van der Pol equation:

$$M \frac{d^2y}{dt^2} + \mu(1 - y^2) \frac{dy}{dt} + Ky = 0 \qquad \mu < 0 \tag{3}$$

where a, b, ω, M, K, and μ are parameters; t is time, and y is a state variable or displacement. Equations 1 and 2 generate a fixed circle in the phase plane. Equation 3 generates a family of limit cycles whose shapes are influenced by μ.

Obviously, such primitive, formal "models" do not imply any physical structure, but instead represent equations which produce periodic plots or solutions that may be fitted to data.

An outstanding example of the usefulness of empirical-explanatory modeling for industrial and social processes has been provided by Forrester (18–21). The same kind of approach has been used to provide a model of the human female reproductive system (22). Relationships leading to a similar model of the reproductive system have been presented by Yates et al. (23), who also view the system as an intrinsic, nonlinear oscillator. Both of these models of the reproductive system assert a biological structure out of which oscillations arise, and are more than mere equations serving as descriptions. Instead, these models lie between the simplest empirical form and the isomorphic, unit-process ideal.

The strict requirements for the approach to isomorphic, unit-process modeling have been described in detail (24–26). The choice of approach finally depends upon the purposes (and temperament) of the modeler. There is no general guideline. However, unit-process modeling comes closer to the physicochemical or thermodynamic issues than does empirical modeling, however explanatory. Unit-process models are more likely to provide insights into structure and function, but they are also more demanding of data, and therefore are more difficult to construct. They may also lead to overmodeling of a system (i.e., specification of more detail than matters at the level of phenomena of interest).

Regardless of type, simulations of oscillatory (or other) biological subsystems ultimately reach a limit of value, even if they approach the isomorphic, unit-process ideal. They all fail to provide generalizable principles because they are special cases. The only generalizable approach for the formulation of periodic processes lies through nonlinear mechanics and statistical thermodynamics (15–17), but because of its general nature, this approach so far fails to describe specific cases in the life sciences. We therefore have the familiar dualism that isomorphic, unit-process or other computer simulations can describe the behavior of specific systems and account for the origin of oscillations by proposed structures, but that they cannot give us general principles of design. On the other hand, the thermodynamic approach leads to a general principle of design of life, involving periodicities, but it does not describe specific cases. We are left with a desire to proceed at two levels simultaneously. At one level we wish to develop explanatory models of either the empirical or isomorphic types, while at another level we wish to develop a general, thermodynamic principle to explain the origins of life and the tendencies toward spatial and temporal organization which may provide a basis for development of a theoretical biology (15). We ask that this theoretical biology be sufficient to lead us to expect ensembles of nonlinear oscillators throughout living systems, which we can then describe individually by special case simulations.

ORIGINS OF SPECIAL CASE OSCILLATIONS

If, by whatever means, it is established that believable periodicities are present at the output of a biological subsystem of interest, or within it, then several possibilities exist concerning the origin of those oscillations. These possibilities will be exemplified by consideration of the adrenal glucocorticoid system and the human female reproductive system.

ADRENAL GLUCOCORTICOID SYSTEM

In the simplest case, when we find an oscillatory process we may be looking at a system that is not itself oscillatory, but that is forced by or coupled to a periodic system, either by direct or by parametric forcing or coupling. An example of this case is provided by the adrenal glucocorticoid system, with its prominent 3.5 and 24 hours approximate rhythms.

The global structure of the adrenal glucocorticoid system has been presented in detail elsewhere (27) and will not be repeated here. The components of the system consist of neural input pathways converging on the hypothalamic median eminence, certain neurons of the median eminence itself, the pituitary gland, the adrenal gland, the liver, and the blood channels flowing between these regions. The dominant time constants of the chief processes within the system range from less than 1 second to as long as 1 week (27). We want to know whether the available knowledge about the topography of the system, and its range of time constants, accounts for the two strong periodicities seen in its basal behavior.

To answer the above question, it is useful to begin by determining whether or not any of the major components named above is itself an oscillator that drives the overall system containing multiple, negative feedback loops, into oscillation. The placement of negative feedback loops around an oscillator would ordinarily, in principle, decrease the sensitivity of the variables within the system to the influence of the oscillator, but would also serve to transmit oscillations upstream from the source component to other components in the closed loop. The approximate 24-hour periodicity in the adrenal glucocorticoid system has been found at every point within the system at which it has been sought. It has been found in the median eminence of the hypothalamus (28–30) and in the levels of the pituitary output variable (31, 32) as well as in the adrenal output variables, as described in the introduction to this paper. But these data merely state that the overall negative feedback system is sufficiently linear so that frequencies are not changed or lost around the loop. They do not establish the origin of the oscillations.

The only component of the adrenal glucocorticoid system that has been proposed to have inherent oscillatory characteristics itself is the adrenal gland. The isolated, cultured hamster adrenal has been reported to show a 24-hour variation in its responsiveness to its input variable (33). Thus this component could provide a source of oscillations for the whole system through its periodically varying parameters. However, if some of the anterior, neural pathways that serve as inputs to the hypothalamic component of the system are cut, then the 24-hour periodicity is lost (34) even though the cut is not known to damage any of the components. Therefore, it does not appear likely that the adrenal gland itself provides the periodicity. We conclude instead that the 24-hour rhythm in this system is imposed at its anterior, hypothalamic inputs, and that it arises from another, as yet unidentified, endogenous oscillator. It is known that the periodicity is not imposed from the environment directly (3) although environmental cues can entrain and phase lock the system.

At the present time, five groups of investigators have designed various computer simulations of the adrenal glucocorticoid system based upon the topography alluded to above (35–44). All of these models are deterministic, and most are continuous system types. None oscillates spontaneously under its normal parameter values. None supposes that the adrenal glucocorticoid system is intrinsically an oscillator.

The adrenal glucocorticoid system has other periodicities than those described here, and these are reviewed in detail elsewhere (27).

From the little that is so far known about the 3.5-hour periodicity within the adrenal glucocorticoid system, it appears that this oscillation is also imposed at a neural input and does not arise within the system itself. Thus the adrenal glucocorticoid system appears to be responding to some other oscillator and may in turn itself cause oscillations in other systems. Because some of the components of the adrenal system have nonlinear, memory characteristics, if they have been driven in an oscillatory fashion, their parameters will vary periodically, so that removal of a component and study of its characteristics *in vitro* will demonstrate periodicities in the performance of the component. Such data have been obtained (see Figures 17 and 20*a* in Ref. 45). These data, though interesting, demonstrate only that some of the components have memory properties, and not that they themselves are intrinsic oscillators.

FEMALE REPRODUCTIVE SYSTEM IN MAMMALS

The female laboratory rat has a 4- or 5-day estrus cycle that appears to be established by an oscillator in the anterior, preoptic region of the hypothalamus serving as an input to the reproductive system. A model of this system has been developed (46), and the oscillations arise from the oscillator at the input and not from the intrinsic properties of the system itself (thus this system is similar to the adrenal glucocorticoid system in that it manifests oscillations but is not itself an intrinsic oscillator). In contrast, the reproductive system of the human female appears to be timed by properties of the ovary itself, so that the system is an intrinsic oscillator (22, 23). In this interesting case the growth and senesence properties of the ovarian follicle and corpus luteum provide the nonlinear processes that drive the system periodically. Again, the models available describe special cases and there is no general design applicable even to all mammalian, female reproductive systems. Some are intrinsic oscillators and others are not. All, however, show periodic behavior: it is the origin of the periodicities that differs in the different species. The available models of the mammalian, female reproductive system have been reviewed elsewhere (23).

COUPLING BETWEEN THE ADRENAL GLUCOCORTICOID SYSTEM AND THE FEMALE REPRODUCTIVE SYSTEM

Kitay and his co-workers have established that gonadal steroid hormones affect the function of the adrenocortical system at several points (47–49). Furthermore, it has long been known that the level of protein binding of adrenal glucocorticoid hormones is affected by estrogens (4, 50). Conversely, it has recently become apparent that adrenal androgens, whose production rate fluctuates with the production rate of glucocorticoid hormones, may be responsible for maintenance of libido in primate females (51). The multiple, bidirectional couplings between the gonadal and adrenal systems in females suggests that a full understanding of the periodic behavior of each of these systems must take into account the periodic behavior of the other. At the present time, no model has been developed that demonstrates the consequences of these mutual interactions.

COUPLING BETWEEN THE ADRENAL GLUCOCORTICOID SYSTEM AND INTERMEDIARY METABOLISM

Adrenal glucocorticoid hormones alter the sensitivity of fat cells to certain lipolytic stimuli, and the same pituitary hormone that drives the adrenal also can act as a lipolytic stimulus itself (52). Furthermore, adrenal glucocorticoid hormones have been known to have profound effects on carbohydrate metabolism through their ability to stimulate gluconeogenesis. We have examined trains of data on blood glucose and free fatty acid levels in unanesthetized dogs and have found strong periodicities. Others have reported such periodicities also, as mentioned in the introductory remarks. Although it is true that correlations of frequencies, like correlations of amplitudes, suggest causal relationships, they do not prove them. At the present time we can only state that the metabolic system, the reproductive system, and the adrenal glucocorticoid system all have strong periodicities, and strong to weak interactions among them. Out of these features a comprehensive model of important aspects of temporal structuring by mammalian organisms may arise.

ACKNOWLEDGMENT

This work was supported by USPHS Grant AM15145-02.

REFERENCES

1. G. Pincus, *J. Clin. Endocr.*, **3**, (1943), 195.
2. C. J. Migeon, F. H. Tyler, J. P. Mahoney, A. A. Florentin, H. Castle, E. L. Bliss, and L. T. Samuels, *J. Clin. Endocr.*, **17**, (1956), 1051.
3. R. T. W. L. Conroy and J. M. Mills, *Human Circadian Rhythms*, Churchill, London, 1970.
4. F. E. Yates and J. Urquhart, *Physiol. Rev.*, **42**, (1962), 359.
5. J. H. Galicich, E. Haus, F. Halberg, and L. A. French, *Ann. N.Y. Acad. Sci.*, **117**, (1964), 281.
6. L. Hellman, F. Nakada, J. Curti, E. D. Weitzman, J. Kream, H. Roffwarg, S. Ellman, D. K. Fukushima, and T. G. Gallagher, *J. Clin. Endocr.*, **30**, (1970), 411.
7. D. T. Krieger, W. Allen, F. Rizzo, and H. P. Krieger, *J. Clin. Endocr.*, **32**, (1971), 266.
8. J. Tourniaire, J. Urgiazzi, J. F. Riviere, and H. Rousset, *J. Clin. Endocr.*, **32**, (1971), 666.
9. E. D. Weitzman, D. Fukushima, C. Nogeire, H. Roffwarg, T. F. Gallagher, and L. Hellman *J. Clin. Endocr.*, **33**, (1971), 14.
10. K. Hansen, *Acta Med. Scand. Suppl.*, **4**, (1923), 27.
11. G. Anderson, R. Hillman, I. Van Elk, and A. Perfetto, *Amer. J. Clin. Nutr.*, **4**, (1956), 673.
12. G. Anderson, Y. Kologlu, and C. Papadopoulus, *Metabolism*, **16**, (1967), 586.
13. A. Iberall, M. Weinberg, and A. Schindler, Contr Report No. 1806, NASA, June 1971.
14. C. M. Court, M. F. Dunlop, and R. F. Leonard, *J. Appl. Physiol.*, **31**, (1971), 345.

15. F. E. Yates, D. J. Marsh, and A. S. Iberall, "Integration of the whole organism – A foundation for a theoretical biology," In: *Challenging Biological Problems* (J. Behnke, Ed.), Oxford Univ. Press, New York, 1972, p. 110.

16. A. S. Iberall, *Theoria to Theory,* **4,** (1970), 40.

17. A. S. Iberall, *Toward a General Science of Viable Systems,* McGraw-Hill, New York, 1972.

18. J. W. Forrester, *Principles of Systems,* Wright-Allen Press, Cambridge, Mass., 1969.

19. J. W. Forrester, *Industrial Dynamics,* MIT Press, Cambridge, Mass., 1961.

20. J. W. Forrester, *Urban Dynamics,* MIT Press, Cambridge, Mass., 1969.

21. J. W. Forrester, *World Dynamics,* Wright-Allen Press, Cambridge, Mass., 1971.

22. R. L. Vande Wiele, J. Bogumil, I. Dyenfurth, M. Ferin, R. Jewelewicz, M. Warren, T. Rizkallah, and G. Mikhail, *Rec. Progr. Hormone Res.,* **26,** (1970), 63.

23. F. E. Yates, S. M. Russell, and J. W. Maran, *Ann. Rev. Physiol.,* **33,** (1971), 393.

24. F. E. Yates, R. D. Brennan, J. Urquhart, M. F. Dallman, C. C. Li, and W. Halpern, in *Systems Theory and Biology* (M. Mesarovic, Ed.), Springer, New York, 1968, p. 141.

25. F. E. Yates, *Physiologist,* **11,** (1968), 115.

26. F. E. Yates, in *Biomedical Engineering* (J. H. U. Brown, J. E. Jacobs, and L. Stark, Eds.), Davis, Philadelphia, 1971, p. 3.

27. F. E. Yates and J. W. Maran, in *Hypothalamo-Hypophyseal System (The Adenohypophysis), The Handbook of Physiology* (W. Sawyer and E. Knobil, Eds.), American Physiological Society, 1973 (In Press).

28. P. N. Cheifetz, N. T. Gafferd, and J. F. Dingman, *J. Endocr.,* **43,** (1969), 521.

29. M. A. David-Nelson and A. Brodish, *Endocrinology,* **85,** (1969), 861.

30. T. Hiroshige, M. Sakakura, and S. Itoh, *Endocr. Japan,* **16,** (1969), 465.

31. S. A. Berson and R. S. Yalow, *J. Clin. Invest.,* **47,** (1968), 2725.

32. H. Demura, C. D. West, C. A. Nugent, K. Nakagawa, and F. H. Tyler, *J. Clin. Endocr.,* **26,** (1966), 1297.

33. R. V. Andrews and G. E. Fold, Jr., *Comp. Biochem. Physiol.,* **11,** (1964), 393.

34. M. A. Slusher, *Amer. J. Physiol.,* **206,** (1964), 1161.

35. F. E. Yates, in *The Adrenal Cortex* (A. B. Eisenstein, Ed.), Little, Brown, Boston, 1967, p. 133.

36. F. E. Yates, *J. Basic Eng.,* (June 1969), 305.

37. F. E. Yates and R. D. Brennan, *Proceedings of the IBM Symposium on Computers in Chemistry,* IBM Data Processing Division, White Plains, New York, 1969, Form 320-2954-0, p. 213.

38. F. E. Yates and R. D. Brennan, in *Hormonal Control Systems* (E. B. Stear and A. H. Kadish, Eds.), American Elsevier, New York, 1969, p. 20.

39. F. E. Yates, R. D. Brennan, and J. Urquhart, *Fed. Proc.,* **28,** (1969), 71.

40. D. S. Gann, *Amer. J. Surg.,* **114,** (1967), 95.

41. D. S. Gann, L. E. Ostrander, and J. D. Schoeffler, in *Systems Theory and Biology* (M. Masarovic, Ed.), Springer, New York, 1968, p. 185.

42. C. B. Dolkas and M. A. Leon, *IEEE Trans. Biomed. Eng.,* **17,** (1970), 1.

43. S. B. Fand and R. P. Spencer, in *Statistics in Endocrinology* (J. W. McArthur and T. Colton, Eds.), MIT Press, Cambridge, Mass., 1970, p. 287.

44. E. M. Stokely and L. L. Howard, *IEEE Trans. Biomed. Eng.*, **19**, (1972), 13.
45. F. Halberg, in *Walter Reed Army Institute for Research Symposium on Medical Aspects of Stress in the Military Climate*, U.S. Govt. Printing Office, Washington, D.C., 1965, p. 1.
46. N. B. Schwartz, *Rec. Prog. Hormone Res.*, **25**, (1969), 1.
47. J. I. Kitay, M. D. Coyne, and N. H. Swygert, *Endocrinology*, **87**, (1970), 1257.
48. J. I. Kitay, M. D. Coyne, N. H. Swygert, and K. E. Gaines, *Endocrinology*, **89**, (1971), 565.
49. J. H. Gaskin and J. I. Kitay, *Endocrinology*, **89**, (1971), 1047.
50. N. Keller, L. R. Sendelbeck, U. I. Richardson, C. Moore, and F. E. Yates, *Endocrinology*, **79**, (1966), 884.
51. B. Everitt and J. Hebert, *Nature*, **222**, (1969), 1065.
52. T. Braun and O. Hechter, in *Adipose Tissue* (B. Jeanrenaud and B. Hepp, Eds.), Academic Press, New York, 1970, p. 11.

4 Studies on the Female Hypothalamo-Pituitary-Ovarian Axis

RAYMOND L. VANDE WIELE, *Moderator*

CHAPTER 10

Temporal Data Relating to the Human Menstrual Cycle

HARRIET B. PRESSER

International Institute for the Study of Human Reproduction, College of Physicians and Surgeons, Columbia University, New York, New York

What do we know about the length and variability of the human menstrual cycle? What is the average cycle length? What range of cycle lengths is considered to be "normal"? How much variability is to be expected of a "healthy" woman? It is the purpose of this paper to bring together data from numerous studies concerned specifically with such questions. Studies were selected for review if they were written in English and readily accessible. As we shall see, a considerable amount of data is available. The lack of comparability between studies, however, makes it difficult to draw definitive conclusions; there are major differences in the design of the studies and in the investigator's methodological approach. Nevertheless, we can consider the extent to which the findings of these studies support various hypotheses.

This paper is divided into two sections. First, we shall discuss some of the methodological considerations that affect the interpretation of the data. Second, we shall compare the findings of different studies on specific aspects of menstrual timing.

145

METHODOLOGICAL CONSIDERATIONS

As is to be expected in a review of the literature, the samples of different studies differ with respect to nationality, occupation of women, and other social variables. The extent to which social factors produce variation in menstrual timing cannot be determined from these studies; variation may be more a function of differences in study design than in the social characteristics of the sample. For many studies it is questionable whether the samples are representative of the population being studied.

Of the studies to be reviewed, all are based on the use of calendar records rather than relying on retrospective reporting. As Matsumoto et al. (1) have demonstrated, there is considerable error in the reporting of cycle lengths when based on memory as compared to record keeping. The criteria for the inclusion of certain cycles differs, however, between studies. Some studies exclude anovulatory cycles, some exclude women who do not "menstruate normally." Not all studies consider only "healthy" women, and the definition of "health" varies considerably—if defined at all. There is also the methodological problem of sample selection: women who agree to participate in such studies and keep records may menstruate more regularly than women who decline to participate; we do not know. But it is clear that low response rates are characteristic of these types of studies, and there is high attrition during the period of observation.

The analytical approach to the data also varies by study. There are two distinct approaches that are taken; one may be termed aggregate; the other, individual. In the aggregate approach all menstrual cycles experienced by a sample of women are combined into a single distribution, giving each cycle equal weighting in the computation of summary measures for the total sample (such as means and standard deviations). The individual approach analyzes the distribution of menstrual cycles for each woman separately. Means and standard deviations are first computed for each woman and then averaged for the total sample. Accordingly, the experience of each woman, rather than each cycle, is given equal weighting in the summary measures for the total sample. As Burch et al. (2) have indicated, it is only when summary measures relating to the individual are the unit of analysis that we can consider the average length and variability of the menstrual cycle for *the average woman*. The aggregate approach focuses, instead, on *the average cycle* and does not necessarily characterize the typical woman.

DATA ON LENGTH AND VARIABILITY

Most investigators take an aggregate approach in their analysis of data on menstrual timing. Table 1 reports the summary measures of such studies. We see that the studies vary considerably in the size of the sample and the number of cycles included. The mean number of cycles per woman varies from 2.3 [Collett et al. (14)] to 36.3 [Vollman, (13)]. For many of the studies the range in the number of cycles each woman has contributed is substantial; the widest range is from King's study (5) in which women contributed from three to 104 cycles each. As Burch et al. (2) have noted, differences in the number of cycles observed per woman can seriously affect the results of a study. For studies of a fixed time period there is generally a relationship between the respondent's mean cycle length and the number of cycles she has contributed. Accordingly, in such studies short cycles are more numerous than long cycles, tending to shorten the average length of the menstrual cycle.

Bearing in mind this bias, what do the studies reveal about the average length and variability of the menstrual cycle? As may be seen in Table 1, the mean cycle lengths vary by study from 27.3 [Latz and Reiner, (3)] to 33.9 [Engle and Shelesnyak, (19)]. Studies also differ as to the extent of variation observed; the standard deviations range from 2.2 days [Haman, (8)] to 11.8 days [Engle and Shelesnyak, (19)]. In general it appears that the longer the mean cycle length reported for a study, the greater the variability around the mean.

The distribution of cycle lengths only roughly approximates a normal curve. Table 2 shows the distribution for studies that provide such data. In almost all of these studies the majority of cycles are 25 to 30 days in length. There are, however, considerable differences between studies in the proportion of cycles falling within this range. Whereas the proportion is 35.9% in the study by Engle and Shelesnyak (19), it is 78.3% in the study by Issmer (6). With only three exceptions the distributions are skewed to the right; that is, proportionately more cycles are 30 days or more in length than 24 days or less in length.

There are few studies that focus on the mean cycle length and mean variation *per woman*. Kreitner (4), in his sample of 80 women, found the mean cycle length per woman to be 27.6, with a mean standard deviation of 2.14. The mean cycle length per woman in the study by Gunn et al. (24) of over 10,000 women was 28.7; the mean standard deviation, 4.31. Chiazze et al. (11), in their study of 2316 women, calculated a mean cycle length per woman of 29.4 and a mean standard de-

Table 1 Summary Measures Relating to the Length of the Menstrual Cycle from Selected Studies (Listed in Order of Mean Cycle Length)[a,b]

Investigator and Year of Publication	No. of Women	No. of Cycles	Mean Cycles per Woman	No. of Cycles Reported per Woman		Reported Length of Cycle		Mean Length of Cycle	Standard Deviation
				Lowest	Highest	Shortest	Longest		
Latz and Reiner 1937 (3)	100	1,336	13.4	7	49	19	101	27.3	3.65
Kreitner 1970 (4)	80	949	11.9	10	12	16	49	27.6	2.14
King 1926 (5)	17	523	30.8	3	104	18	53	27.7	3.68
Issmer 1889 (6)	12	120	10.0	10	10	20	40	27.8	2.93
Bailey and Marshall 1970 (7)	1,353	12,247	9.1	n.a.	n.a.	11	72	28.2	3.78
Haman 1934 (8)	150	2,460	16.4	7	49	10	55	28.4	2.2
Latz and Reiner 1935 (9)	102	1,113	10.9	6	44	15	51	28.4	3.27
Larsen 1947 (10)	17	130	7.6	n.a.	n.a.	6	61	28.9	7.11

Chiazze et al. 1968 (11)	2,316	30,655	13.2	10	n.a.	n.a.	n.a.	29.1	7.47
King 1933 (12)	37	354	9.6	4	27	16	57	29.1	5.46
Vollman 1956 (13)	592	21,499	36.3	n.a.	n.a.	6	409	29.2	na..
Collett et al. 1954 (14)	146	333	2.3	n.a.	n.a.	16	92	29.6	n.a.
Larsen 1947 (10)	22	147	6.7	n.a.	n.a.	21	47	29.9	4.19
Larsen 1947 (10)	26	189	7.3	n.a.	n.a.	21	45	30.1	4.51
Scipiades 1935 (15)	50	339	6.8	n.a.	n.a.	20	91	30.2	7.17
Moos 1969 (16)	839	n.a.	n.a.	n.a.	n.a.	n.a.	n.a.	30.3	4.7
Fluhmann 1934 (17)	76	747	9.8	3	19	11	144+	30.4	11.53
Matsumoto et al. 1962 (1)	701	18,213	26.0	6+	n.a.	8	147	30.4	6.54
Matsumoto et al. 1962 (1)	n.a.	3,000	n.a.	n.a.	n.a.	n.a.	n.a.	31.0	6.75
Rork and Hellebrandt 1940 (18)	231	1,690	7.3	n.a.	n.a.	8	122	31.9	8.55
Engle and Shelesnyak 1934 (19)	100	3,140	31.4	20	69	7	256	33.9	11.80

[a] From Burch et al. (2), Larsen (10), and original sources.
[b] n.a. = not available

Table 2   Percent Distribution of Menstrual Cycles by Length of Cycle for Selected Studies (Listed in Order of Percent of Cycles 25 to 30 days)[a] [b]

Investigator and Year of Publication	No. of Cycles	Total Percent	Percent Distribution by Length of Cycle		
			24 Days or Less	25–30 Days	31 Days or More
Engle and Shelesnyak 1934 (19)	3,140	100.0	24.0	35.9	40.1
Fluhmann 1934 (17)	747	100.0	18.5	46.9	34.6
Goldzieher et al. 1947 (20)	500	100.0	24.6[c]	53.6[d]	21.8[e]
Collett et al. 1954 (14)	333	100.0	n.a.	54.4	n.a.
Scipiades 1935 (15)	339	100.0	5.5	55.2	39.3
Allen 1933 (21)	1,522	100.0	14.4	61.5	24.1
King 1933 (12)	354	100.0	14.3	64.0	21.7
Pratt et al. 1929 (22)	299	100.0	22.8	66.5	10.7
Chiazze et al. 1968 (11)	2,316	100.0	2.8	67.3	29.9
Bailey and Marshall 1970 (7)	12,247	100.0	10.0	70.3	19.7
Latz and Reiner 1935 (9)	102	100.0	7.2	71.0	21.8
Haman 1934 (8)	2,460	100.0	9.6	71.4	19.0
King 1926 (5)	523	100.0	13.8	73.7	12.5
Latz and Reiner 1937 (9)	1,336	100.0	11.8	75.3	12.9
Issmer 1889 (6)	120	100.0	9.2	78.3	12.5

[a] From Cox (23) and original sources.
[b] n.a. = not available.
[c] 25 days or less
[d] 26 to 29 days.
[e] 30 to 36 days.

viation of 3.27. (This can be compared with the aggregate measures based on all *cycles* reported for the same study—a mean cycle length of 29.1 and a standard deviation of 7.47.)

Age of Woman

The samples differ considerably in their age composition. Since age may be related to the length and variability of the menstrual cycle, it is important to control for this variable when possible.

Table 3 reports the mean length and standard deviation of cycles by age of woman for studies that provide such data. Looking across the rows—particularly at the studies by Chiazze et al. (11), Bailey and

Marshall (7), and Collett et al. (14)—there appears to be some relationship between age of woman and mean cycle length. Specifically, the
mean length of cycle seems to decline with age. Cycles appear to be
least variable for women in their mid-20s to late 30s.

It is important to consider here that these studies are cross-sectional
in nature, examining different women at different ages. This is not the
same as following a single group of women from menarche to menopause. The study by Treloar et al. (25) comes closest to such a design
in that many women are followed for a great many years (although all
not for the same number); to adjust for the lack of data during the im-
mediate postmenarche years, data on the daughters of the initial participants were obtained. In this study the unit of analysis is the individual. For relatively young women, age was considered as the number
of postmenarche years that had passed; for women near the end of the
reproductive span, age was considered as the number of
premenopausal years remaining. Chronologic age was used for women
aged 20 to 40. Table 4 reports the findings of this study with respect to
the mean "person year" length of the menstrual cycle and the mean
"person year" standard deviation. Apparently "person years" refers to
all persons for whom data are available for a particular year.

It may be seen that the mean length of the menstrual cycle per
woman declines with age, increasing again during the premenopausal
years. The variability of the menstrual cycle also appears to be curvi-
linearly associated with age. The mean standard deviation is quite high
during the first few years following menarche, then declines and rises
again in the premenopausal years.

Studies that examine the mean length and mean standard deviation
per woman by age group are reported in Table 5. We see that the data
based on individuals follows a pattern similar to data based on cycles:
the mean cycle length per woman declines with age. The data in this
table do not show a clear pattern, however, with regard to the relationship between the mean standard deviation and age. It is of interest to
note that Krietner's study (4) shows lower mean cycle lengths and
lower mean standard deviations than do the other studies. It may be
highly relevant that the sample in Krietner's study is comprised of ce-
libate nuns. This suggests that sexual activity may increase variation in
menstrual timing. This may be largely a function of spontaneous abor-
tion experience among the sexually active. Unfortunately, there are no
studies which investigate this possibility; indeed, it may be exceedingly
difficult to do so, since abortions may occur without the knowledge of
the woman or physician.

Table 3. The Mean Length ($\overline{X}$) and Standard Deviation (S.D.) of the Menstrual Cycle by Age of Woman for Selected Studies (Listed in Order of the Age Range of the Sample)[a]

Investigator and Year of Publication	No. of Women	Age Range of Women	Mean Age	$\overline{X}$ or S.D.	Age of Woman							
					<15	15–19	20–24	25–29	30–34	35–39	40–44	45–49
Engle and Shelesnyak 1934 (19)	100	11–15	13.1	$\overline{X}$ S.D.	33.9[c] 11.80							
Larsen 1947 (10)	17	16–20	17.5	$\overline{X}$ S.D.		28.9[d] 7.11						
Larsen 1947 (10)	26	18–21	19.7	$\overline{X}$ S.D.			30.1[e] 4.51					
Larsen 1947 (10)	22	18–21	20.2	$\overline{X}$ S.D.			29.9[e] 4.19					
King 1933 (12)	21[b]	17–24	n.a.	$\overline{X}$ S.D.		28.6[f] n.a.	27.2 n.a.					

Study	N	Age range		Stat							
Fluhmann 1934 (17)	76	18–29	21.5	$\bar{X}$	30.4[a]						
				S.D.	11.53						
Moos 1969 (16)	839	20–30	25.2	$\bar{X}$		30.3[h]					
			n.a.	S.D.		4.7					
Chiazze et al. 1968 (11)	2316	15–44	n.a.	$\bar{X}$	30.5	30.3	29.4	28.8	28.3	28.1	
				S.D.	8.95	8.22	6.54	6.53	6.61	7.71	
Bailey and Marshall 1970 (7)	1353	18–49	n.a.	$\bar{X}$	30.2[i]	29.2	28.7	27.9	27.8	27.1	26.1
				S.D.	3.00	3.69	3.57	3.55	3.62	4.24	7.26
Collett et al. 1954 (14)	146	17–50	n.a.	$\bar{X}$	30.6[d]	29.2	29.4	28.7	26.5	30.1[j]	
				S.D.	n.a.	n.a.	n.a.	n.a.	n.a.	n.a.	

[a] n.a. = not available.
[b] Subsample of college women.
[c] Age group 11–15.
[d] Age group 16–20.
[e] Age group 18–21.
[f] Age group 17–19.
[g] Age group 18–29.
[h] Age group 20–30.
[i] Age group 18–29.
[j] Age group 40–50.

153

Table 4   Person-Year Mean Length of Menstrual Cycle and Mean Standard Deviation for a Study of Mothers and Daughters[a]

Age Scale	Year	No. of Person-Years	Mean Length of Cycle	Mean Standard Deviation
Postmenarche	0, 1	361	36.90	11.28
	2	245	34.09	8.65
	3	253	32.40	6.38
	4	334	31.94	6.16
	5	543	31.22	4.37
	6	736	30.65	4.47
	7	898	30.27	4.19
Chronologic	20	452	30.09	3.94
	21	775	29.92	3.73
	22	969	29.77	3.62
	23	1013	30.01	3.64
	24	1002	30.18	3.61
	25	1005	29.84	3.45
	26	977	29.99	3.47
	27	963	29.44	3.21
	28	950	29.56	3.07
	29	940	29.26	3.17
	30	916	29.30	3.16
	31	906	29.12	2.93
	32	897	28.70	2.86
	33	877	28.60	2.81
	34	868	28.43	2.91
	35	850	28.22	2.67
	36	812	27.98	2.51
	37	810	27.92	2.74
	38	787	27.76	2.85
	39	748	27.53	2.76
	40	730	27.26	2.83
Premenopause	−9	113	27.04	3.15
	−8	112	26.52	3.06
	−7	116	26.73	3.49
	−6	113	26.88	4.35
	−5	117	28.38	6.39
	−4	118	30.02	8.23
	−3	119	33.20	14.24
	−2	117	43.51	19.54
	−1, 0	212	57.14	35.49

[a] From Table IV and V of Treloar et al. (25).

Table 5 The Mean Length ($\bar{X}$) and Mean Standard Deviation (S.D.) of the Menstrual Cycle per Woman by Age of Woman for Selected Studies (Listed in Order of Mean Cycle Length)

Investigator and Year of Publication	No. of Women	No. of Cycles per Woman	$\bar{X}$ or S.D.	Age of Women						
				Total	15–19	20–24	25–29	30–34	35–39	40–44
Kreitner 1970 (4)	80	11.9	$\bar{X}$	27.6	—	28.2[c]		27.6	26.4	—
			S.D.	2.14	—	2.40		1.86	2.39	—
Gunn et al. 1937 (24)	208[a]	n.a.	$\bar{X}$	29.0	32.1	30.1	29.9	27.6	27.4	27.0[d]
			S.D.[b]	4.31	3.64	3.84	5.71	2.74	2.40	2.55
Chiazze et al. 1968 (11)	2,316	13.2	$\bar{X}$	29.4	30.8	30.5	29.6	29.0	28.5	28.3
			S.D.	3.27	3.38	3.99	2.68	2.92	2.58	2.77

[a] Subsample of "reliable" cases only.
[b] Derived from data on standard error and sample size.
[c] Age group 20 to 29.
[d] Age group 40 to 53.

155

Table 6   Incidence of Anovulatory Cycles by Age of Woman[a]

Age of Women	No. of Women	No. of Cycles	Anovulatory Cycles	
			No.	%
17–18	59	81	25	30.8
20–24	33	112	11	9.8
25–29	21	45	2	4.4
30–34	11	23	0	0
35–39	13	33	4	12.1
40–50	11	33	5	15.1
Total	146	327	47	14.3

[a] From Table 5 of Collett et al. (14).

Anovulation

Age appears to be highly correlated with the incidence of anovulatory cycles. A curvilinear association is evident in the study by Collett et al. (14); see Table 6. The proportion of cycles anovulatory (determined by basal body temperature graphs) is especially high among women aged 17 and 18, declining with age until the early 30s and rising again in the late 30s and 40s. Could anovulation be producing the variation observed in the length and variability of the menstrual cycle by age?

The answer seems to be negative. The study by Bailey and Marshall (7) reported in Table 3 is based on ovulatory cycles only, and the relationship generally noted between age and cycle length and variability still obtains.

Length of Pre- Versus Postovulatory Phase of Cycle

Several investigators have questioned whether the preovulatory phase produces more variability in the length of the menstrual cycle than the postovulatory phase. The data suggest this is the case.

Collett et al. (14) note that the range of the preovulatory phase in their sample was 4 to 39 days as compared with 4 to 26 days for the postovulatory phase. Goldzieher et al. (20) indicate that 79.6% of the cycles had a preovulatory length of 10 to 16 days, whereas 94.0% had had a postovulatory length within this range. Potter et al. (26) suggest, on the basis of their review of seven studies, that the length of the pre- and postovulatory phases are only weakly correlated, if at all.

Table 7   *Mean Length of the Pre- and Postovulatory Phase of the Menstrual Cycle by Total Length of Cycle*[a]

		Mean Length of Ovulatory Phases	
Length of Cycle	No. of Cycles	Pre-ovulatory (days)	Post-ovulatory (days)
21–26	23	11	13
27–28	27	14	14
29–33	31	18	12
34–35	29	20	14
36–39	43	22	15
40–45	43	27	15
46–60	31	36	15

[a] From Chart 1 of Fluhmann, (27).

Fluhmann (27) presents data on the mean length of each phase by the total length of the cycle (see Table 7). It is evident that the longer cycles have relatively long preovulatory phases; the postovulatory phase is not as variable by length of total cycle.

The study by Matsumoto et al. (1) provides a detailed analysis of the length and variability of each phase of the menstrual cycle by age of the woman (see Table 8). It is apparent that for every age group the mean length and the standard deviation are greater for the preovulatory relative to the postovulatory phase.

Correlation between Consecutive Cycles

While the length of the pre- and postovulatory phase may not be highly correlated, what can be said of the correlation of the length between two consecutive cycles? Gunn et al. (24) conclude, on the basis of their analysis, that the length of each cycle is not related to the length of its predecessor. Matsumoto et al. (1) found a perfect agreement in length in only about 10% of the cases. Very short cycles (less than 24 days) and very long cycles (over 45 days) were generally followed by cycles within the range of 25 to 39 days. In other words, it was relatively infrequent to find a very short or a very long cycle consecutively repeated. Only one-third of the cycles were within 1 day of the next

Table 8 *Comparison of Summary Values Relating to the Length and Variability of the Pre- and Postovulatory Phases of the Menstrual Cycle by Age of Woman*[a]

Age of Woman	No. of Cycles	Preovulatory Phase (in Days)				Postovulatory Phase (in Days)			
		Mean Length	S.D.	Skewness	10-90% Range	Mean Length	S.D.	Skewness	10-90% Range
13-17	280	23.0	9.39	+0.32	12-33	11.7	1.54	+0.19	10-14
18-19	100	21.0	7.07	+0.42	14-31	12.2	1.89	+0.42	10-16
20-24	350	18.4	5.65	+0.25	13-25	12.4	1.56	+0.25	10-15
25-29	1,000	18.6	7.43	+0.34	13-25	12.7	1.63	+0.41	11-15
30-34	750	17.2	5.31	+0.23	12-23	12.8	1.64	+0.15	11-15
35-39	400	16.7	4.54	+0.16	12-22	12.6	1.62	+0.39	11-15
40-52	120	15.9	4.28	+0.21	11-20	12.5	1.76	+0.26	10-15
Total	3,000	18.4	6.75	+0.35	13-25	12.5	1.66	+0.33	10-15

[a] From Table 1 of Matsumoto et al. (1).

cycle, less than one-half were within 2 days, and slightly over three-fourths were within 5 days.

Marshall (28) raises the question of the probability that the length of a future cycle will fall within a particular range. Specifically, how does the knowledge of the past 3, 6, 9, or 12 cycles help to predict the range of the next 3 cycles? The proportion of subsequent cycles that fell within the range of past cycles averaged from 64% when based on the range of the past 3 cycles to 90% when based on the range of the past 12 cycles.

CONCLUSION

The major conclusion to be drawn from this review of the literature is that the length of the menstrual cycle is highly variable. While most cycles range from 25 to 30 days, it is not uncommon for cycles to fall outside this range.

The age of a woman appears to be related to the length and variability of the menstrual cycle. This relationship is evident even when considering ovulatory cycles only.

The preovulatory phase of the menstrual cycle was found to be more variable than the postovulatory phase. The length of each phase is only weakly correlated, if at all. There is only a weak correlation, also, between the length of consecutive cycles.

We are limited to these general conclusions because of the lack of comparability between studies. Moreover, we have had to rely upon reported findings and do not have the raw data available to consider additional questions. There is a clear need for more systematic research, rigorously designed, on the temporal aspects of the menstrual cycle.

ACKNOWLEDGMENT

The author gratefully acknowledges the research assistance of Lenore Klopotoski in compiling the bibliography for this paper.

REFERENCES

1. S. Matsumoto, Y. Nogami, and S. Ohkuri, *Gunma J. Med. Sci.,* **11,** (1962), 294.
2. T. K. Burch, J. J. Macisco, and M. P. Parker, *Int. J. Fertil.,* **12,** (1967), 67.
3. L. J. Latz and E. Reiner, *Illinois Med. J.,* **71,** (1937), 210.

4. P. C. Kreitner, "Temperature Rhythm Abstinence: A Quantitative Analysis," Unpublished Thesis Submitted to the Graduate School of Georgetown University, 1970.

5. J. L. King, *Contrib. Embryol.*, **18**, (1926), 79.

6. E. Issmer, *Arch. Gynaekol.*, **35**, (1889), 310.

7. J. Bailey and J. Marshall, *J. Biosoc. Sci.*, **2**, (1970), 123.

8. J. O. Haman, *Amer. J. Obstet. Gynecol.*, **27**, (1934), 73.

9. L. J. Lantz and E. Reiner, *J. Amer. Med. Assoc.*, **105**, (1935), 1241.

10. E. M. Larsen, *Amer. J. Obstet. Gynecol.*, **54**, (1947), 1069.

11. L. Chiazze, F. T. Brayer, J. J. Macisco, M. P. Parker, and B. J. Duffy, *J. Amer. Med. Assoc.*, **203**, (1968), 377.

12. J. L. King, *Amer. J. Obstet. Gynecol.*, **25**, (1933), 583.

13. R. F. Vollman, *Gynaecologia*, **142**, (1956), 310.

14. M. E. Collett, G. E. Wertenberger, and V. M. Fiske, *Fertility Sterility*, **5**, (1954), 437.

15. E. Scipiades, *Arch. Gynaekol.*, **159**, (1935), 360.

16. R. H. Moos, *Amer. J. Obstet. Gynecol.*, **103**, (1969), 390.

17. C. F. Fluhmann, *Amer. J. Obstet. Gynecol.*, **27**, (1934), 73.

18. R. Rork and F. A. Hellebrandt, *Amer. J. Physiol.*, **129**, (1940), 450.

19. E. T. Engle and M. C. Shelesnyak, *Human Biol.*, **6**, (1934), 431.

20. J. W. Goldzieher, A. E. Henkin, and E. C. Hamblen, *Amer. J. Obstet. Gynecol.*, **54**, (1947), 668.

21. E. Allen, *Amer. J. Obstet. Gynecol.*, **25**, (1933), 705.

22. J. P. Pratt, E. Allen, D. E. Newell, and L. J. Bland, *J. Amer. Med. Assoc.*, **93**, (1929), 834.

23. H. J. E. Cox, *Brit. Med. J.*, **1**, (1968), 252.

24. D. L. Gunn, P. M. Jenkin, and A. L. Gunn, *J. Obstet. Gynecol. Brit. Empire*, **44**, (1937), 839.

25. A. E. Treloar, R. E. Boynton, B. G. Behn, and B. W. Brown, *Int. J. Fertil.*, **12**, (1967), 77.

26. R. G. Potter, T. K. Burch, and S. Matsumoto, *Int. J. Fertil.*, **12**, (1967), 127.

27. D. F. Fluhmann, *West. J. Surg.*, **65**, (1957), 265.

28. J. Marshall, *Lancet* **1**, (1965), 263.

CHAPTER 11

Some Conceptual and Methodological Problems in Longitudinal Studies on Human Reproduction

RUDOLF F. VOLLMAN

Uerikon, Switzerland

The period of human reproduction extends over the definite interval between the menarche and the menopause. This implies that the anatomic and physiologic processes involved must pass through definable stages of development. During the period of adolescence the infantile ovaries and uterus grow to adult size, shape, and function. Through coordination with the hypothalamic-pituitary endocrine system they achieve their optimal functional capacity, culminating in conception, implantation, and uncomplicated pregnancy, and resulting in a term life-born baby. In the phase of the premenopause the functional endocrine coordination deteriorates and the ovaries and uterus atrophy. The life story of the human reproductive function, therefore, is best described as a developmental process, with an ascending, optimal, and descending branch. Quantitative and qualitative measurements of this developmental process may be deduced from the changing length and degree of variability of the menstrual cycles, the length of the post- and premenstrual phases, the incidence of the anovulatory menstrual

161

cycles, the age-specific prevalence of conceptions, and the outcome of pregnancy.

By concept such studies must be longitudinal, optimally covering the whole span from menarche to menopause. They can be successful only with the free cooperation of intelligent volunteers. These studies are, unfortunately, slow data producers and they have constantly to fight the loss of participants. They establish, however, the level, the degree of change, and the range of variability of the study variables in the individual participant. In medicine, and especially in obstetrics, the source of investigation and the point of reference can only be the individual, real case, not the fictitious, collective Norma and Norman of the classical anthropologist. The whole study population, finally, serves to determine the range of variability between the individual women.

DESCRIPTION OF DATA AND METHOD OF ANALYSIS

Since 1935 a group of healthy Swiss women has volunteered to collect data on their reproductive physiology. The participants received a booklet with preprinted columns for the entry of the following information: calendar date, onset and duration of menstruation, date and localization of intermenstrual pain, cervical mucorrhoea, intermenstrual bleeding, and daily basal body temperature. In addition, records were kept on intercourse, conception, course and outcome of pregnancy and lactation, intercurrent disease, medication, growth, and body weight. Each woman received a printed instruction sheet, an initial interview, and repeated physical examinations. The data were originally reviewed monthly; later in 3-month intervals. Women who terminated their cooperation by withdrawal, by reaching the menopause, or by serious disease or death were replaced by new volunteers, preferably by sisters, daughters, or other close relatives of the completed case.

The study population comprises 691 women with 31.676 recorded menstrual cycles. In addition to the daily basal body temperature (BBT), 280 women reported intermenstrual pain in 3442 menstrual cycles, 138 cervical mucorrhoea in 2022 menstrual cycles, and 43 intermenstrual bleeding in 105 menstrual cycles. Information on the day of conception, the course, and the outcome of 502 pregnancies has been collected. Observations were recorded in 56 girls prior and after menarche and in 41 women during the menopausal and postclimacteric phase.

The data are analyzed as follows:

1. By detailed description of typical individual cases.
2. By descriptive statistical summaries of the total study population by means, medians, percentiles, and coefficents of regression in time series of chronologic and gynecologic ages.
3. By graphic representations.

THE MENSTRUAL CYCLE

Menstruation is scientifically defined as "a periodic-cyclic bleeding from the uterus accompanied by desquamation of the endometrium" (1). In a longitudinal study, however, menstruation must be defined as a periodic vaginal bleeding which the woman herself must diagnose as "menstruation," based on her previous experience and on accompanying symptoms. Consequently, not all reported "menstruations" will in fact be physiologic menstrual bleedings, however, some of the "vaginal bleedings" not reported will be true menstruations. Tested against the course of the basal body temperature, the probability of the over- or underreporting of a menstrual date is less than 1 per 1000 reported cycles.

During the first years after menarche the length of the menstrual cycle may oscillate irregularly between 10 and >60 days. The range of variation is so wide that in many adolescent girls no typical pattern of menstrual cycles can be discerned. The mean and median length of the menstrual cycles decreases systematically with the progress of the adolescent development. Parallel runs an increasing concentration to certain lengths of the menstrual cycle which are characteristic for the individual woman, that is, predominant frequencies of cycles of 22, 23, 24; 24, 25, 26, 27; 27, 28, 29, 30; or 29, 30, 31, 32 days; and so on. This indicates that the level of the length of the menstrual cycle is established during the course of the adolescent phase. But the slope of the level measured by the median cycle length and the variability continue to decline during the years of maturity, though in a smaller angle. The lowest mean cycle length and the smallest degree of variability of the length of the menstrual cycles in any woman are reached just prior to the onset of the climacteric phase at the end of the phase of maturity. Approximately 40 to 50 cycles prior to the menopause, the length of the menstrual cycles increases abruptly, as does the degree of variability. This pattern of the menstrual cycles has been noted in all participating women with extended records. The length and variability of the menstrual cycle, therefore, are a function of personal characteristics of the individual woman and her age.

TESTS OF OVULATION

Carl Hartman, by daily rectal palpation, could diagnose with his index finger the growing and freshly ruptured follicle in the living Rhesus monkey, and he pinpointed the occurrence of ovulation with a precision of ± 1 day (2). Veterinarians use the same method successfully in breeding cows and mares. But in women, though tried by Dickinson, this method is impractical (3). The only direct test of ovulation in women presently used is direct visualization of the ovaries through culdoscopy or laparotomy with observation of the freshly ruptured follicle, preferably with documentation by still or time-lapse film. This special technique requires clinical facilities and a definite indication. It is not a practical test for the sequential observation of ovulation in the same woman over scores of menstrual cycles. All other tests of ovulation in use today are indirect tests that use physiologic changes related to the ovulation process: blood levels of pituitary follicle stimulating hormone, luteinizing hormone, estrogens and progesterone, pre- and postovulatory histologic changes in the endometrium, the vaginal epithelium, chemical and physical changes in the cervical mucus, and a host of biochemical variations in the blood, saliva, urine, and cervical and vaginal contents. Some of these tests are of a high degree of specificity, sensitivity, sharpness, and reproducibility, but of a high degree of variability within and between women (Table 1). They all require a special, delicate apparatus and skilled tech-

Table 1   Criteria for Tests of Ovulation

---

1. Specificity
2. Sensitivity
3. Sharpness
4. Noninterference
5. Stability
6. Reproducibility
7. Innocuousness
8. Low range of variability within and between women
9. Applicability over long periods of time, age, and in all socioeducational classes
10. Technical simplicity (easy to teach and to learn)

Table 2   Characteristics of Basal Body Temperature (BBT), Intermenstrual Pain (IP), Cervical Mucorrhoea (FL) and Intermenstrual Bleeding (FL.s) as Tests of Ovulation

Criteria	BBT	IP	FL	FL.s
Specificity	Low	High	Low	Low
Sensitivity	Low	High	Low	Low
Sharpness	High	High	Low	Low
Noninterference	None	None	None	None
Stability	High	Low	Low	Low
Reproducibility	High	High	Low	Low
Innocuousness	+	+	+	+
Range of variability				
Within women	Low	Low	High	High
Between women	Low	Low	High	High
Applicability				
Over long periods of time	Good	Good	Good	Good
Technical simplicity	High	High	High	High

niques. They are therefore only applicable in small series of menstrual cycles, which prohibits the generalization of the results over any period of time in a individual woman or any population of women.

There are, however, four signs and symptoms related to ovulation which the woman herself can observe and report. Their value may therefore be tried in a longitudinal study. These four signs and symptoms are: the course of the BBT, the intermenstrual pain, the cervical mucorrhoea, and the intermenstrual bleeding (Table 2). Their respective quality as tests of ovulation may be rated, based on 11 criteria as excellent, for the intermenstrual pain (10 favorable over 11), good for the BBT (9/11), and satisfactory for the cervical mucorrhoea and the intermenstrual bleeding (4/11 for each).

The Course of the Basal Body Temperature (BBT)

The temperature curve in relation to menstruation ". . . shows that the temperature is lowest at the middle of the intermenstrual period, it gradually rises, attaining its maximum two days before menstruation. There is a sudden drop on the day preceding the flow, with a second slighter drop at the end of the period. It rises slightly for the first week after the cessation of menstruation and there is a third fall at the beginning of the intermenstrual period" (4). Giles gave a precise description of the course of the basal body temperature between two menstrual bleedings but he interpreted the course of the temperature in the sense of the then prevalent "wave theory of bodily functions in women" (blood pressure, pulse rate, respiratory frequency and volume, CO_2 production, quantitative fluctuations in urinary components, changes in reflexes, and dynamometer values, as investigated by Mary Putnam-Jacobi (5). However, in 1905 van de Velde presented stringent evidence that the fluctuation of the BBT must be associated with the function of the ovaries and that the changes of the body temperature can be used to assess the occurrence and the time of ovulation in women (6).

The BBT allows a concurrent estimation of the length of the post- and premenstrual phases of the menstrual cycle over practically the whole reproductive life span. Its technical simplicity and noninterference with the physiology of the genital cycle has made it possible, for the first time, to survey quantitatively the development and course of the reproductive function from menarche to menopause in an individual woman as well as in larger study populations of women.

In individual women the duration of the low phase of the BBT, the postmenstrual phase, correlates strictly with the length of the menstrual cycle. The length of the menstrual cycle is a function of the length of the postmenstrual phase. Therefore, over the years of their reproductive lives of the women in the study population, the length of the postmenstrual phases demonstrate the same individual and age characteristic patterns described for the menstrual cycles.

The length of the phase of the elevated temperature, the premenstrual phase, varies independently from the length of the postmenstrual phase and the menstrual cycles. The length of the premenstrual phase by BBT may range from 0 to 16–17 days, where 0 days probably indicates an anovulatory menstrual cycle and a duration of 18 days and more testifies to an early pregnancy. During the period of adolescence, premenstrual phases (BBT) of short duration (≤ 9 days) prevail; during maturity, in approximately 90% of the cycles, the length of the premenstrual phases ranges from 10–15 days; while in the premenopause the frequency of the occurrence of shorter premenstrual phases again increases.

Intermenstrual Pain

Intermenstrual pain "is an acute pain starting in the lower abdomen, on one side or the other, soon irradiating all over the hypogastrium, with heaviness in the lumbar region and pressure in the perineum. The pain is not aggravated by pressure but it may cause nausea and it may become severe enough to make it impossible for the woman to stand erect. The interval between the onset of the pain and the following menstruation is more constant than the interval between the pain and the onset of the preceding menstrual period" (7). The etiology of the pain is obscure. The clinical symptomatology suggests spastic tubal contractions. It can be induced experimentally by estrone injection (8). In approximately 50% of the instances the intermenstrual pain is located in the right lower quadrant, in 25% in the left lower quadrant, but in 15% it occupies simultaneously the left and the right quadrants.

The intermenstrual pain is a very sharp test that may be used as a caesura between the post- and premenstrual phases. With the intermenstrual pain as an indicator, the same associations which have been detailed for the BBT can be demonstrated in the individual woman and for the whole study population. The length of the postmenstrual phase by intermenstrual pain is correlated in straight regression with the length of the menstrual cycle; the length of the premenstrual phase by intermenstrual pain varies independently of the length of the menstrual cycles. However, independently of the length of the menstrual cycle, the mean length of the postmenstrual phase by intermenstrual pain is 2 days shorter than by BBT; and inversely, the mean length of the premenstrual phase by intermenstrual pain is 2 days longer than by BBT. The curve of the frequency distribution of the intermenstrual pain by different lengths of menstrual cycles is markedly leptokurtic, corresponding to a high degree of sharpness of the test.

Cervical Mucorrhoea

Some time ago F. A. Pouchet (9) published a pertinent description of cervical mucorrhoea or intermenstrual vaginal discharge: "From the 10th to the 15th day another phenomenon can be observed regularly. During these days the utero-vaginal mucus which had become thick and non-glary-white (opaque) becomes more liquid and more abundant than ever. Often there is such a quantity of discharge that it moistens the genital organs and overflows unto the adjacent parts." The intermenstrual cervical mucorrhoea is another sign for the estimation of the phases of the menstrual cycle. However, its frequency distribution is markedly platykurtic due to a low degree of sharpness and a wider range of variation than the BBT or the intermenstrual pain. The mean length of the postmenstrual phase estimated by cervical mucorrhoea is 2 days shorter than by intermenstrual pain and 4 days shorter than by BBT. The mean length of the premenstrual phases follows in the inverse sequence. In association with the age of women, the mean lengths of the phases of the menstrual cycle ran in parallel curves of different levels.

Intermenstrual Bleeding

In the study population 42 women reported intermenstrual bleeding or spotting in a total of 105 menstrual cycles. By careful and repeated examinations any possible vaginal, cervical, endo-, and myometrial pathology has been excluded in these women. The mean length of the postmenstrual phase by intermenstrual bleeding is still 1 day shorter than by cervical mucorrhoea. The curve of the frequency distribution of the episodes of intermenstrual bleeding is again markedly platykurtic. Because of its low prevalence, the intermenstrual bleeding is but of limited value within the present study population.

THE ANOVULATORY CYCLE

During the year after menarche close to 50% of all BBT curves do not demonstrate a premenstrual temperature elevation. They are monophasic curves. The prevalence of monophasic BBT curves declines in the course of adolescent development but continues at a level of 2–3% during the years of maturity. With the approach to the menopause the prevalence of monophasic BBT curves manifestly increases. Based on rather skimpy and circumstantial evidence and on the notion of reduced fertility in women during early adolescence and the premenopauseal years, monophasic BBT curves have been generally if glibly assumed to indicate anovulatory menstrual cycles. The BBT curve, however, is presently the only technical means for the presumptive diagnosis of anovulatory cycles that require further investigation with more specific techniques.

CONCEPTION AND PREGNANCY

During the postmenstrual phase by BBT up to 10 days prior to the rise of the BBT curve, and after the second day following the elevation of the temperatures, no conceptions have been observed in any of the women studied. Conceptions are concentrated on days 1, 2, and 3 before the rise of the BBT with a spread, in a skewed frequency distribution, between day 9 before and the first day after the rise of the temperature. With respect to the intermenstrual pain, conceptions are concentrated on day 1 and 2 before and on the day of the intermenstrual pain. This time relationship between fertile intercourse and conception reflects the degree of specificity of the intermenstrual pain and the low specificity and sensitivity of the BBT for the timing of ovulation. The temperature curve, however, is a convenient means for the early diagnosis of pregnancy and of the distribution of fetal losses in the course of pregnancy.

THE ADOLESCENT PHASE

The adolescent phase is characterized by instability and variability in the length of the menstrual cycles, by a preponderance of monophasic BBT curves, or of short premenstrual temperature rises of 3–9 days' duration. The course of the adolescent development is more adequately studied by years elapsed since the menarche, that is, in gynecologic years, than by comparing women of different chronologic ages (10). The BBT curve allows for the estimatation of the individual duration of the adolescent development, which may last longer than the physical growth development.

THE CLIMACTERIC PHASE

The climacteric phase is indicated by an increasing instability of the length of the menstrual cycles and by a progressive reappearance of BBT curves with premenstrual temperature elevations of less than 9 days or monophasic curves. In the individual woman the onset of the preclimacteric phase can be diagnosed 40 to 50 menstrual cycles before the menopause. The course of the climacteric phase is not marked by a straight, progressive decline of the reproductive functions, but it proceeds as an oscillating process of runs of menstrual cycles of various length and with short premenstrual phases, or monophasic temperature curves, alternating with normal, mature cycles. Frequently the very last menstrual cycle shows the same characteristics as the cylce in the mature woman.

SUMMARY

The whole course of woman's reproductive physiology from menarche to menopause can be adequately studied with available indirect tests: basal body temperature, intermenstrual pain, cervical mucorrhoea, and intermenstrual bleeding.

REFERENCES

1. G. W. Bartelmez, *Amer. J. Obstet. Gynecol.*, **74,** (1957), 931.

2. C. G. Hartman, *Anat. Rec.*, **45,** (1930), 263.

3. R. L. Dickinson, *Amer. J. Obstet. Gynecol.*, **33,** (1937), 1027.

4. A. E. Giles, *Amer. J. Obstet. Dis. Wom. Child.*, **35,** (1897), 713.

5. M. Putnam-Jacobi, "The question of rest for women during Menstruation," *Boylston Prize Essay for 1876*, Putnam, New York, 1877.

6. T. H. van de Velde, *Ueber den Zusammenhang zwischen Ovarialfunction, Wellenbewegung und Menstrualblutung, und ueber die Entstehung des sogenannten Mittelschmerzes*, Fischer, Jena, 1905.

7. Sorel, *Gaz. Méd. Picardie*, **4,** (1886), 167.

8. R. F. Vollman, *Mschr. Geburtsh. Gynaek.*, **110,** (1940), 115, 193.

9. F. A. Pouchet, *Théorie Positive de l'Ovulation Spontanée* et de la Fécondation des Mammifères et de l'Espéce Humaine, etc., Paris, 1847.

10. R. F. Vollman, "Patterns of Menstrual Performance in Adolescent Girls, in *Proceedings of the Second World Congress on Fertility and Sterility, Naples, 1956*, p. 27.

CHAPTER 12

Temporal Relationships of Hormonal Variables in the Menstrual Cycle

I. DYRENFURTH, R. JEWELEWICZ, M. WARREN, M. FERIN, and R. L. VANDE WIELE

Department of Obstetrics and Gynecology and International Institute for the Study of Human Reproduction, College of Physicians and Surgeons, Columbia University, New York, New York

This paper will concentrate on the temporal relationships between the changes of blood levels of the individual hormones as they are observed from one menses to the next. Daily measurements result in very characteristic patterns. The monthly recurrence of these patterns forms a rhythm of a very low frequency. Newer studies indicate, however, that underlying the monthly hormone patterns may be secretory rhythms of the individual hormones of a much higher frequency, measuring possibly even in minutes (1–3).*

The knowledge we have today of the changes in the blood levels of the essential hormones has been obtained by the use of the newer radioligand methods, which have been developed in this field during

---

* Since the time of the Conference a paper appeared by H. L. Judd and S. S. C. Yen: Serum androstenedione and testosterone levels during the menstrual cycle. *J. Clin. Endocr.* **36:**475, 1973.

the last 7 years. Their extreme sensitivity furnishes the possibility of simultaneous daily analysis of a variety of hormones. Such data were needed to arrive at concepts about the delicate interplay of the hormones regulating the various phases of the menstrual cycle. The first successful attempts to integrate such data into detailed physiologic comprehension were made in 1969, when Ross et al. (4) and Vande Wiele et al. (5), in their Laurentian Hormone Conference papers, presented an extensive overview and statistical analysis of the then available data on all pertinent hormones: luteinizing hormone (LH), follicle-stimulating hormone (FSH), progesterone, 17α-hydroxy-progesterone and, in some instances, estrogens. Tentative concepts about the hormonal control of the menstrual cycle, which could be derived from these data, were formulated. These concepts were confirmed and amplified in the following years by further experimental work in animals and in humans, and several reviews of the entire field, as well as parts of it, have recently been published. (6–13).

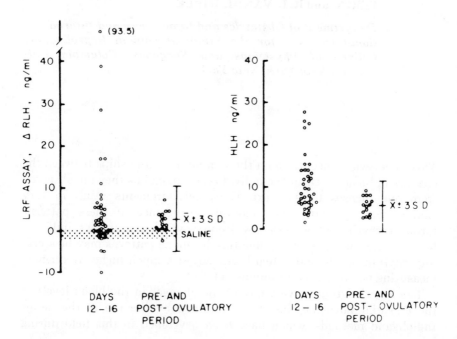

Figure 1   Plasma LRF activity and LH determinations at different stages of the menstrual cycle (14).

Table 1   Total FSH and LH Activity in Human Pituitaries during the Menstrual Cycle (15)

Phase of Cycle	Day of Cycle	FSH IU/pituitary[a]				LH IU/pituitary[a]			
		N	$\bar{X}$	S.D.	p	N	$\bar{X}$	S.D.	p
Early menstrual phase	28–2	5	283	82		4	12368	1929	
					NS				.001
Late menstrual phase	3–5	4	173	73		4	745	687	
					NS				.005
Early follicular phase	6–8	5	292	78		5	5598	2191	
					NS				.020
Late follicular phase	9–13	2	300	70		3	12333	3821	
					.05				.010
Early luteal phase	15–17	2	163	64		3	335	129	
					NS				.005
Mid luteal phase	18–20	1	161	—		2	4755	940	
					NS				.001
Late luteal phase	21–27	4	209	56		3	1097	237	

[a] N = number of cases, $\bar{X}$ = mean, S.D. = standard deviation, p = probability, NS = nonsignificant.

HYPOTHALAMIC AND GONADOTROPIC HORMONES

While articles dealing with the effects of administered Gonadotropin Releasing Factor (LRF) are rapidly accumulating, there is presently still only one publication available that reports the measurement of endogenous LRF in healthy untreated women. Malacara et al. (14) collected blood in 36 women between day 12 and 16 of their cycles and using an elaborate bioassay found significantly elevated LRF values in six of these samples. LH determinations indicated midcycle peak levels in 14 samples (Figure 1).

The suggestion that LH is stored in the pituitary and periodically released while FSH is synthesized and secreted more evenly finds support not only in the largely fluctuating LH plasma levels and more stable levels of FSH which were obtained by measuring blood samples at frequent intervals (2), but also from a study of Bischoff et al. (15) who analyzed the pituitary content of LH and FSH in accident victims. Endometrial histology was used to date the time of the cycle. Table 1

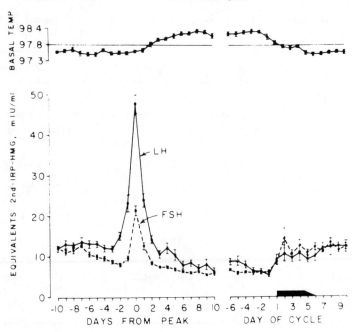

Figure 2   Composite curves of basal body temperature and concentrations of LH and FSH in sera during single menstrual cycles from 37 women. The data have been plotted in relation to the number of days before and after the LH peak and the number of days before and after the onset of menses. Vertical bars represent one standard error of the mean (16).

shows a clear accumulation of LH from day 3 to days 9–13, which is followed by a very low pituitary LH content on days 15 to 17, indicating the midcycle release of this hormone and depletion of pituitary stores. Another buildup (days 18 to 20) and decline (days 21 to 27) follows in the pituitary. Much lower and more constant levels of pituitary FSH were observed throughout the cycle.

In Figure 2, serum levels of LH and FSH obtained daily throughout 37 normal menstrual cycles are shown in the form of composite curves (16). The basic characteristics of these curves have been confirmed by a number of other authors (10, 17–22). Most characteristic for the course of LH is the sharp peak at midcycle, when for the duration of 24 to 48 hours the serum levels rise approximately 3 to 10-fold over baseline levels. This peak coincides with the shift in basal body temperature and is often directly related to its nadir. The course before and after this peak shows low levels but there are occasional secondary elevations, particularly during the luteal phase. The need for small amounts of LH for the maintenance of normal corpus luteum function has been demonstrated by experiments in which ovulation was induced in hypophysectomized patients with HMG and HLH (Human Luteinizing Hormone) (5).

The characteristic features of the normal FSH curve are high levels in the beginning of the cycle, a very typical nadir 1 or 2 days before the LH peak, and a peak either coinciding with or occurring a day after that of LH (19, 20) (Figure 2). In comparing FSH and LH curves the difference in half-life of the two hormones must be kept in mind. During the luteal phase FSH levels are generally lower than in the follicular phase. Shortly before the onset of the next menses they start to rise again.

If blood samples are assayed more often than once a day, the LH pattern changes. The seemingly sharp midcycle peak dissolves into a double peak, as was pointed out by Thomas et al. (1), who sampled every 4 to 8 hours. The more recent studies of Midgley and Jaffe (2) and Yen et al. (3), who sampled as often as every 20 minutes, showed that the LH peak consists of a series of rapid oscillations and suggested that the double or triple peaks of Thomas et al. were incidental to the frequency of their sampling.

The relationship between the hormonal events at the time of midcycle and ovulation was studied by Yussman et al (23), who measured plasma LH at intervals of 8-hours in patients undergoing laparoscopy. Ovulation time was dated from the histology of corpus luteum biopsies. LH was always significantly elevated 24 hours before ovulation. Occasionally the elevation started as early as 40 hours prior to ovulation.

ESTROGENS

Our initial understanding of the course of estrogens during the menstrual cycle was gained in the fifties and early sixties through the extensive studies of Brown and co-workers, who measured urinary glucuronides and sulfates of estrone, estradiol, and estriol (24). Later other metabolites were also assayed. From these studies the picture of a midcycle peak and luteal maximum of the estrogens arose. However, more detailed understanding of the relationship of estrogen secretion to the gonadotropins, and possibly other hormones, could be obtained only from a knowledge of the levels of these hormones in the circulating blood.

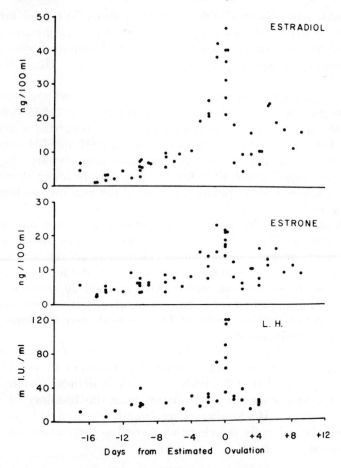

Figure 3   Concentration of LH and unconjugated estrone and estradiol-17β in peripheral plasma of women during the menstrual cycle. (25).

    Baird and Guevara (25) reported the first cycles studied by a double
isotope method for plasma estradiol and estrone, and correlated the
results with LH levels (Figure 3). Other early results were obtained by
utilizing uterine receptor protein or pregnancy plasma in competitive
protein binding procedures (26,27). Since radioimmunoassay became
available (28, 29), the typical estrogen patterns in plasma unraveled (5,
6, 30). Figure 4 presents a composite curve of free (directly ether-ex-
tractable) estradiol for 10 cycles. In the early part of follicular
development, when FSH levels are high, estrogens remain low. This period may last for 7 to 10 days from the onset of menses. Then follows a

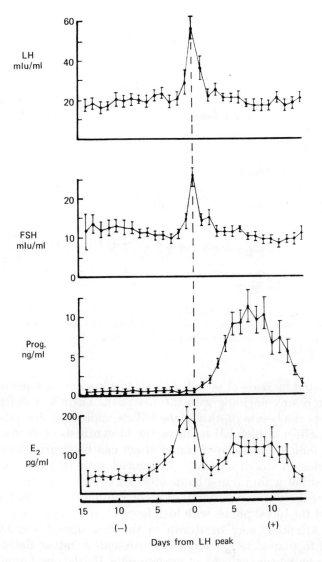

Figure 4 Mean values of LH, FSH, progesterone, and estradiol in the same aliquots of daily serum samples obtained from 10 women during ovulatory cycles. The vertical bar represents one standard error of the mean (6).

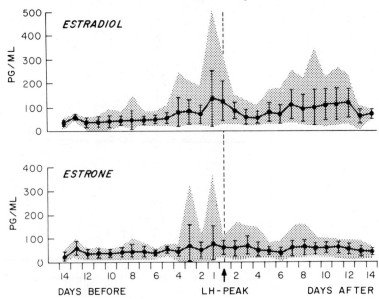

Figure 5   Daily levels of free plasma estradiol and estrone in 22 normal cycles. Mean and standard deviation given in the vertical bars. The shaded area indicates the entire range over which measurements have been obtained.

rather rapid increase of estradiol levels to a maximum generally occurring before or coinciding with the LH peak. There is a sharp drop in estradiol even before rupture of the follicle, indicating that this must be a direct effect of high LH levels on the biosynthesis of estrogens in the mature follicle. This gonadotropin effect can be clearly demonstrated in patients in whom ovulation is induced with HCG. In such cases, estrogens fall within 3 to 4 hours after HCG administration. The third period of the estrogen pattern is a more gradual rise and fall over the 2 weeks of the luteal phase, with high levels during days 5 to 10 postovulatory. Attention may be drawn to the fact that this course is not parallel to that of progesterone, but presents a rather flat plateau in contrast to the convex peak of progesterone. By the onset of menses the estradiol levels have returned to baseline.

Figure 5, presenting both free plasma estradiol and estrone in 22 normal subjects studied by us, shows that the typical pattern residues with estradiol and not so much with estrone, which often shows a rather even course throughout the cycle. This is not surprising in view of the fact that estrone derives partly from Δ^4-androstenedione of adrenal origin.

Preedy, in 1965, isolated estrone sulfate as the main plasma estrogen (31). More recently Loriaux et al. (32) described the measurement of this conjugate in plasma using a radioimmunoassay as end point. Brown and Smyth (33) applied fluorimetry to the systematic quantification of sulfates and glucuronides of the blood estrogens during the cycle, and their main compound seemed, indeed, to be estrone sulfate. We have observed a very diversified distribution of estrone and estradiol between the free and the sulfate fraction. Examples of this are shown in Figures 6 and 7. Subject R.R. presents a distribution presently considered typical: a pronounced pattern in the free

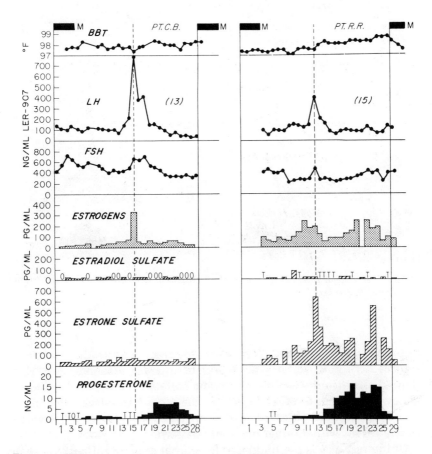

Figure 6 Hormonal patterns of two healthy individuals, C.B. and R.R. Menses are indicated by black horizontal bars. The cycle length is given by the large figure at the end of the time scale; luteal phase length by the bracketed figure in the upper right of the graph. T = traces only were measured. 0 = not detected. For comment see text.

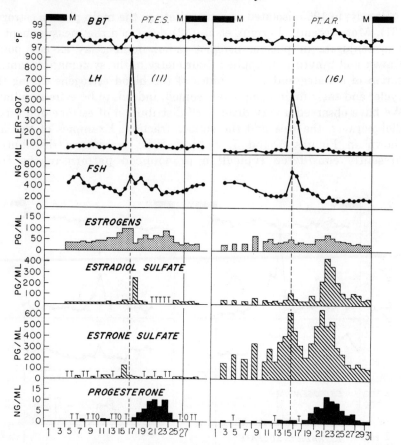

Figure 7   Hormonal patterns of two healthy individuals, E.S. and A.R. Explanations as in Figure 6. See text for comment.

estrogens and in estrone sulfate with only traces of estradiol sulfate. In subject C.B. both sulfates were present in trace amounts only. E. S. showed significant amounts of estrone sulfate and estradiol sulfate at midcycle, but not at other times. On the other hand, large amounts of both sulfates were observed in A.R. together with rather low levels of free estrogens with a rather indistinct pattern.

Of interest also is the finding of Rosenthal et al. (34) that 98% of the estrone sulfate in plasma is bound to albumin in contrast to free estrogens, which are bound to the specific estrogen binding β-globulin of the plasma.

PROGESTINS

The principal features of the progesterone pattern are the low levels during the 2 weeks of follicular development and a gradual elevation and decline connected with corpus luteum growth and regression. The same curve was obtained by earlier studies of urinary pregnanediol, the chief metabolite of progesterone.

The small amounts of progesterone (generally around 1 ng/ml or even less) which are measured in the first half of the cycle have been shown by Zander et al. (35) in isolation studies to represent the hor-

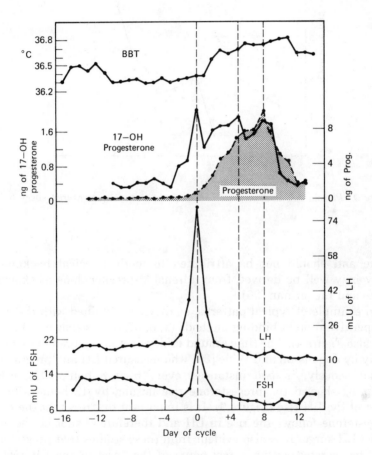

Figure 8   Mean daily basal body temperature, plasma 17α-hydroxyprogesterone and progesterone concentrations, and mean daily LH and FSH concentrations during 16 presumptively ovulatory cycles synchronized around the day of the LH midcycle peak. (4).

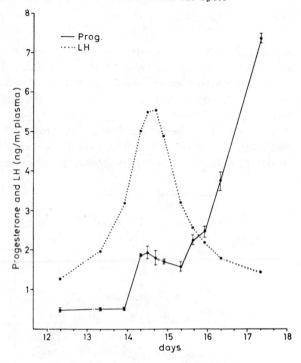

Figure 9   Plasma LH and progesterone levels at midcycle, measured every 6 hours (39).

mone and should not be attributed to methodological background. They may well be derived from adrenal Δ^5-pregnenolone as shown by Arcos and Lieberman (36).

An example of typical patterns, as they are obtained with the newer competitive protein binding methods (4, 6, 37, 38), is shown in Figure 8 (see also Figure 4). The time around midcycle was explored more extensively by Johansson and Wide [39] who measured LH and progesterone simultaneously, in two instances, every 6 or 8 hours. The 6-hour sampling showed the progesterone rise lagging by 12 hours after the onset of the LH rise (Figure 9). It is important to stress that the rise in progesterone follows the rise in LH and therefore cannot be the cause of the LH surge. It is also evident from these studies that progesterone may be secreted within a few hours of the onset of the LH rise but before ovulation, which occurs, as mentioned above, 24 to 40 hours after the onset of the LH surge.

Progesterone has been measured at 2 to 4-hour intervals by Runne-baum et al. (40). Random fluctuations but no systematic diurnal varia-tions were observed during the luteal phase in contrast to late preg-nancy, when a diurnal variation became significant with maxima at 4 P.M. and minima at 8 A.M. Johansson (41), studying four cycles in the same women at different times of the year (February, April, June, and October), found no seasonal changes.

In addition to progesterone, three other progestins were isolated and identified by Runnebaum et al. (42) in human plasma collected on days 22 to 26 of the normal menstrual cycle: 17α-hydroxyprogesterone, 20α-dihydroprogesterone, and 20β-dihydroprogesterone. Measure-ments of 17α-hydroxyprogesterone by Strott et al. (43) during the entire cycle are shown in Figure 10. The curve is parallel to that of progesterone in the luteal phase, but shows in addition a rise and peak at midcycle. The latter coincides approximately with the peak of LH. The authors postulate that the changes in this compound may be used as an additional index of follicular maturation. Abraham et al. (30, 44), using radioimmunoassay for progesterone, 17α-hydroxyprogesterone, and estradiol, were able to confirm the parallelism between the course of 17α-hydroxyprogesterone and estradiol (Figure 11).

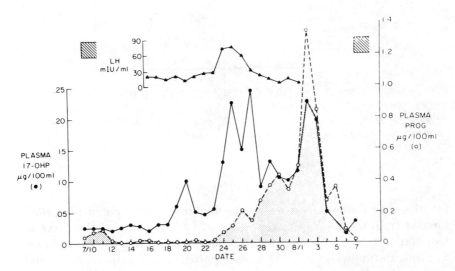

Figure 10   Plasma concentrations of 17α-hydroxyprogesterone, progesterone, and LH in a normal menstrual cycle. The progesterone scale is 4 times that of 17α-hydro-xyprogesterone. (43).

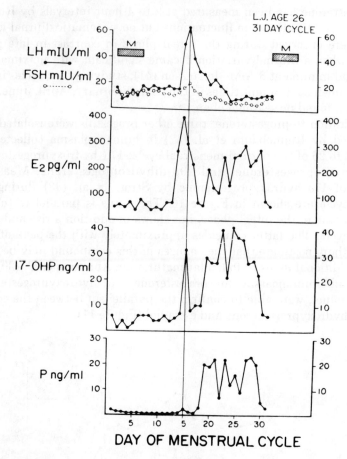

Figure 11   Daily plasma levels of FSH, LH, progesterone, 17α-hydroxyprogesterone, and estradiol-17β during a normal 31-day-long cycle. (44).

Figure 12 shows the course of 20α-dihydroprogesterone in 17 normal cycles from the work of Saxena et al. (21). The shape of the curve is similar to that of progesterone, but the maximum level stays below 4 ng/ml. Bermudez et al. (45) have reported a radioimmunoassay and values for Δ^5-pregnenolone, and Loriaux and Lipsett (46) for 17χ-hydroxy-Δ^5-pregnenolone. Both compounds were found lower in follicular and higher in the luteal phase, but never exceeded 4 ng/ml.

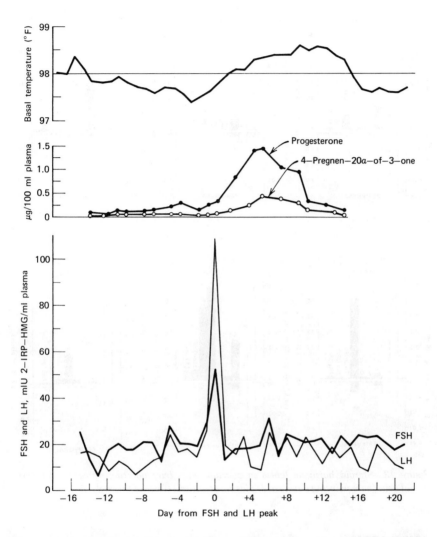

Figure 12 Correlation of mean FSH and LH values for 17 cycles with concentrations of progesterone and Δ⁴-pregnene-20α-ol-3-one in plasma (21).

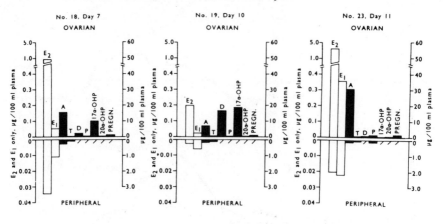

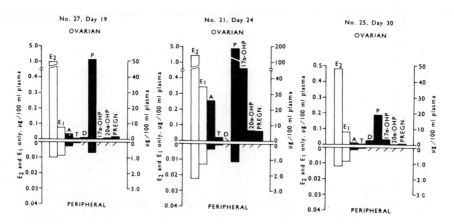

Figure 13   *Steroid hormone levels in ovarian vein (bars pointing upward) and peripheral vein (bars pointing downward) plasma at various times of the cycle (50).*

ANDROGENS

Several androgens have been measured in the blood of women, but few data are available which distinguish between the two phases of the cycle. With competitive protein binding methods Δ_4-androstenedione was analyzed and found to have slightly higher levels in the luteal phase, with a maximum of approximately 2 ng/ml (47, 48). $\Delta^{5\text{-}Androstene}$-$3\beta,17\beta$-diol presented no significant changes during the cycle (46). The levels were around 1 ng/ml. Testosterone was measured by a double isotope assay over the entire cycle by Lobotsky et al. (49). Slight elevations around midcycle, and even lesser ones during the luteal phase, were indicated.

The most informative study comes from the same group of workers (50). Figure 13 describes the entire spectrum of steroidal hormones including the androgens Δ^4-androstenedione, testosterone, and dehydroisoandrosterone in ovarian vein and peripheral plasma. Patients were selected on days 7, 10, 11, 19, 24, and 30 of their respective cycles. The figure gives a clear impression of the small amounts of testosterone secreted by the normal ovary as compared to the substantially larger amounts of Δ^4-androstenedione. In peripheral plasma the levels of both these hormones are more similar, partly due to adrenal secretion, partly to interconversion. The amounts of dehydroisoandrosterone were negligible in the effluents of all but one ovary. Similar observations in ovarian vein blood had been made by Mikhail (51).

ADRENAL CORTICOSTEROIDS

The course of plasma cortisol and corticosterone needs to be restudied, now that competitive protein binding and/or radioimmunoassays have become available for the measurement of these substances.

Aldosterone has always been of particular interest in connection with the ovarian hormones, since estrogens as well as progesterone are known to be intimately connected with the fluid balance. Gray et al. (52) measured aldosterone secretion rates in 15 women on two midfollicular (90 to 190 μg/day) and two midluteal days (170 to 490 μg/day). These results were correlated with the pregnanediol levels. Body water

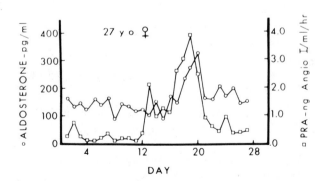

Figure 14   Plasma aldosterone and plasma renin activity in a Gr ii P ii woman during a normal cycle (53).

and electrolytes did not change in these healthy women. Katz and Romfh (53) showed by means of radioimmunoassay the course of plasma aldosterone and plasma renin activity in relation to each other in four cycles. Figure 14 demonstrates their most distinct pattern which indicates for aldosterone a course like that of progesterone, whereas renin has a midcycle peak in addition to the luteal maximum similar to estradiol.

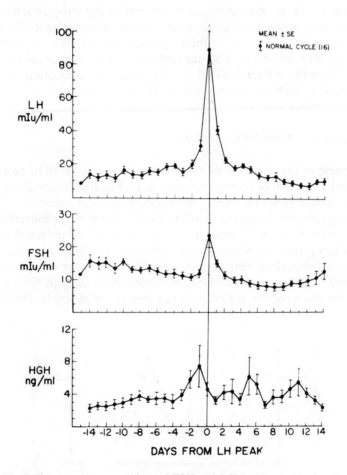

Figure 15   Daily serum concentrations of FSH and LH in 16 menstrual cycles. Mean ± SE and daily growth hormone concentrations taken while patients were fasting in six cycles (56).

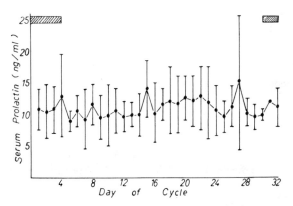

Figure 16   Serum prolactin concentrations during the menstrual cycle. The mean ±SD is shown for nine subjects (62).

GROWTH HORMONE AND PROLACTIN

The relation of growth hormone to the ovarian hormones and its course during the menstrual cycle is difficult to establish, largely because of the extreme sensitivity of this hormone to the most varied influences and the resulting greatly fluctuating levels. From the studies of several groups (54–56) a pattern seems to be emerging, indicating a midcycle peak and several elevations during the luteal phase. An example is shown in Figure 15.

Blood prolactin had been studied in earlier years with the pigeon crop sac bioassay (57, 58). This work described elevated levels during the latter part of the luteal phase, in agreement with even earlier measurements of this hormone in urine (59). The specificity of these methods is open to question, even more so because the findings could not be confirmed by newer methods using radioimmunoassay and extensively purified standards. Levels of 9 to 11 ng/ml prolactin in normal female blood were reported with these newer techniques, with only random variations over the menstrual cycle (60–62). Figure 16 presents data of nine healthy women. This observation—no characteristic pattern—was confirmed by Midgley and Jaffee (63).

DISCUSSION

A number of features are common to all normal cycles. These become evident from the composite curves of cycles from a larger number of

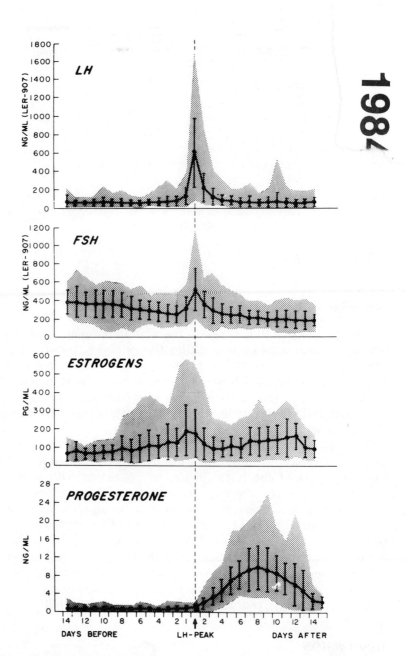

Figure 17 Mean ± SD of daily plasma hormone levels of 40 normal cycles. The shaded area indicates the entire range over which measurements were obtained. Estrogens represent free estradiol and estrone.

healthy women. Figure 17 shows an example from our work. The mean hormonal values with their standard deviation are plotted and set against the background of a shaded area which represents the entire range over which measurements have been obtained. The women were 21 to 39 years of age. The length of the cycles ranged from 24 to 34 days and the interval from LH peak to onset of menses from 11 to 16 days. Inclusion in the group was solely on the basis of reported regular menses and biphasic temperature curves. Methods used throughout the study were uniform. In addition, we followed 16 cycles in which method modifications were tried, different gonadotropin standards were used, or only selected hormones were measured. Although these added to our overall experience, they were not included in the group for the composite curves.

From the onset of bleeding for approximately 1 week a clear elevation of FSH is seen, while estrogens remain low at baseline levels. This corresponds to the early phase of follicular development. During the following week estrogens begin to rise, first slowly then often very rapidly, while FSH declines to its nadir. The midcycle events are characterized by sudden peaks in LH and FSH, the peak of LH being much more pronounced in magnitude. This LH surge follows the final and almost geometrically-progressing increase of the estrogens which appear to be the trigger of the LH surge. Estrogens have been shown in numerous experiments to exert this effect. Immediately following the gonadotropin surge the estrogens fall abruptly. The fall may extend over 2 or 3 days but is steepest prior to follicular rupture. The hormonal changes during the second 2 weeks of the normal cycle are characterized by the endocrine function of the growing and regressing corpus luteum. During this period both progesterone and estrogens are secreted in amounts increasing over the first and decreasing over the second week until they return to baseline levels and menstruation occurs. Gonadotropin levels are low during the luteal phase. While the corpus luteum regresses, FSH begins gradually to rise (this is not seen in the composite curve centered around the LH peak, but can be seen clearly in individual patterns) and initiates the development of a new set of follicles.

Though the composite curves reflect in a very informative way the typical events, the individual cycles show a most surprising degree of diversification and variations of the hormonal patterns.

Cycle U.B. (Figure 18) presents a very typical hormonal pattern for LH and progesterone. FSH shows clearly rising levels at both menses and a nadir preceding the midcycle peak by 2 days. The absence of any significant amounts of estradiol sulfate and a substantial amount of

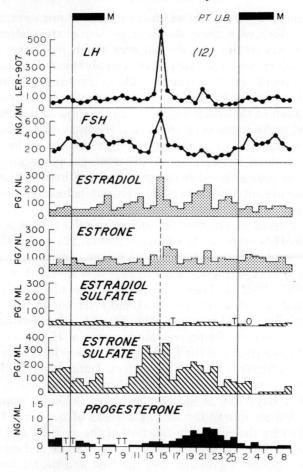

Figure 18   Hormonal pattern of a healthy individual, U.B. Explanation as in Figure 6. For comment see text.

estrone sulfate may also be considered typical. The interesting feature is the three distinct peaks in free estradiol in the follicular phase, the final one associated with the LH surge. The same peaks are indicated, although not with the same clarity, in the patterns of free and sulfate estrone. The physiological mechanisms leading to these events are not known, but may correspond to separate processes of follicular maturation and atresia.

Both the LH and FSH curves of U.B. have a secondary elevation on day 22 which is followed on day 23 by a peak in free estrogens and progesterone. The correlation between secondary peaks in luteal phase LH and a second peak in progesterone is even more clearly demonstrated in cycles EP and CH in Figure 19.

In addition EP shows another much rarer pecularity: Here the estradiol peak precedes the LH surge by 4 days. The same early elevation is seen in estrone, which gives additional importance to the estradiol peak. Similar observations (i.e., estrogen peaks preceding the LH surge by up to 4 or 5 days) have also been reported elsewhere (27). Unfortunately, we have no measurements of estrogen sulfates in this case, which might clarify this observation. Interestingly, the early estrogen peak is accompanied by a moderate rise in LH which, however, does not lead immediately to the ovulatory surge. It is conceivable that later spikes in the free estrogens did so, but were missed by the once-a-day mode of blood sampling. An example from a cooperative study with Drs. J. Ferin and K. Thomas (64) will illustrate this point. In Figure 20 plasma LH and estrogens were measured simultaneously every 2 hours during 5 days at midcycle. The dotted bars represent the 8 to 10 A.M.

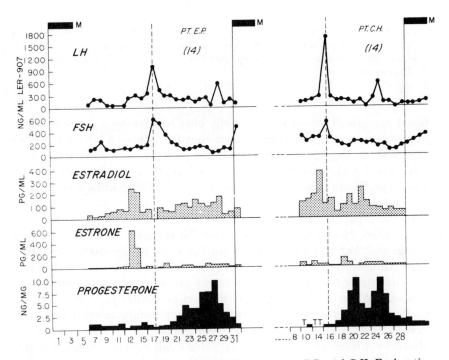

Figure 19   Hormonal patterns in two healthy individuals, E.P. and C.H. Explanations as in Figure 6. For comment see text.

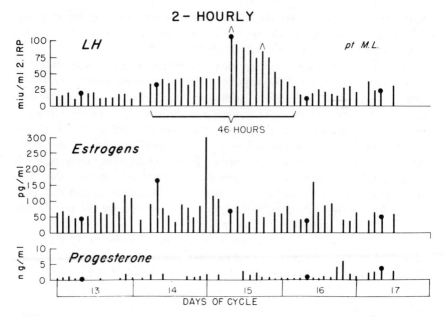

Figure 20   LH, estrogen, and progesterone levels at midcycle in subject M.L. Plasma samples were obtained every 2 hours. Bars with rounded tops indicate the 8 to 10 A.M. samples. The triangular roof over a bar signifies a peak level. (In collaboration with K. Thomas and J. Ferin)

blood samples. If these were the only ones to have been measured, the impression would have been created that the LH peak which occurred at 8 A.M. on day 15 was preceded by an estrogen level of 170 pg/ml on day 14 which appears to be a rather low level for the precipitation of an LH surge. There was, however, another much higher spike (300 pg/ml) in the estrogens at midnight between day 14 and 15 which may have been the trigger for LH release.

In Figure 21 a cycle of 41 days duration is shown (BM). The woman stated that her menses had been regular prior to the study, occurring approximately every 28 days. It is remarkable that in spite of the rise in FSH during the first menses, estrogen secretion is delayed for almost 2 weeks. When it starts there are three rises similar to U.B. (compare to Figure 18) which finally precipitate apparently normal midcycle events and a probably normal luteal phase of 14 days duration. However, progesterone and estrone sulfate fall rather rapidly, 5 and 6 days, respectively, prior to the onset of the next menses. There are also no secondary rises in the LH curve during the luteal phase.

Variations in the distribution of estradiol and estrone between the free and the sulfate fraction have been pointed out in the discussion of Figures 6 and 7.

Variations similar to those seen in the hormonal pattern of the cycles of different women may also be observed between various cycles in the same woman. This is illustrated in Figure 22, which shows two consecutive cycles in M.W., the second one depicting the beginning of a normal pregnancy.

In considering the variations in the individual cycles the possibility that some of the irregularities are due to analytical problems must ob-

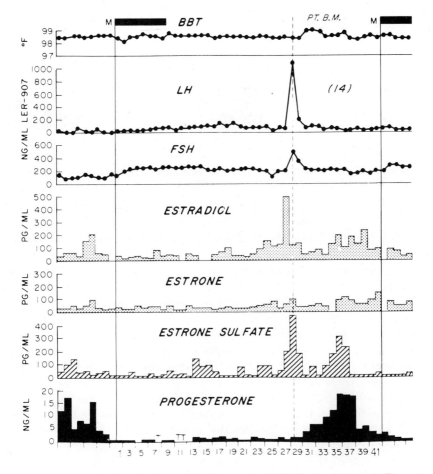

Figure 21   Hormonal pattern in a 41-day-long cycle, B.M. Explanations as in Figure 6. For comment see text.

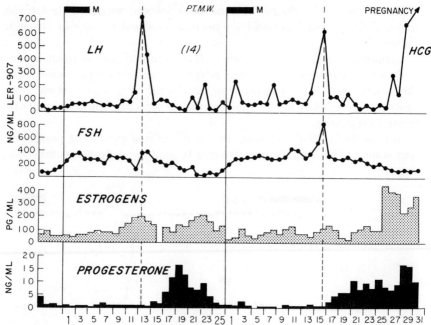

Figure 22   Hormonal patterns of two subsequent cycles and beginning pregnancy in a healthy individual, M.W. Explanations as in Figure 6. For comment see text.

viously be considered. Some of the variations, however, are clearly beyond methodological errors, particularly in these cases when irregularities are observed in more than one variable (see earlier). It appears that it is more likely that the deviations are real and that their study is important and may lead to a further insight into the mechanisms that control the events of the menstrual cycle. The menstrual cycle then appears to be not a sequence of events that proceed ineluctably once it has started, but a sequence of processes undergoing continuous readjustments, perhaps in response to disturbances originating outside the reproductive system. Relevant to this is a recent study by Bogumil et al. (65) who developed a mathematical model of the menstrual cycle incorporating most of what is presently known about morphological and hormonal changes during the cycle. With such a model it is possible to analyze by computer simulations the effect of changes in one or more of the variables upon the remainder of the variables. In one such study, the effect of random noise, superimposed upon the cyclical changes during a "typical" cycle was analyzed. Physiological sources of such random noise could be the effect of changes in adrenal function, sleep and activity cycle, psychological influences, and so forth. These simulations showed clearly that such random, short-term disturbances resulted in long-term changes in the hormonal and morphological events of the cycle. In fact, when successive cycles were simulated, they had the same variations in the length of the proliferative and secretory phases that are actually seen in patients. Furthermore, changes in LH, progesterone, and estrogens gave results similar to those observed in the individual cycles discussed earlier. Finally, in long-term simulations there was an occasional occurrence of transient anovulation, an observation sometimes made in normal women. More work is obviously needed, but the possibility should be seriously considered that deviations from typical cycles are the result of processes such as proposed by Bogumil et al.

NOTE ON METHODOLOGY

Gonadotropins

LH and FSH were measured by the radioimmunoassay technique of Midgley (66, 67). Reference preparation for the standard curves was LER-907. In some of the earlier work 2.IRP-HMG was used.

Progesterone

Progesterone was determined by modification of the method of Neill et al. (37) using chloroform ethylacetate (4:1) as the TLC system and a rigidly timed exposure to florisil of 30 seconds.

Free Estrogens

The solid phase of RIA of Abraham (28) was used. With many plasmas, it was found necessary to add defatting steps. Neutral fats were removed by a pre-extraction with hexane. Phospholipids—by far the worse interfering substances—were precipitated from the crude ether extracts in a 1% $CaCl_2$ solution. The precipitate was filtered and the estrogens back extracted from the filtrate into ether. The procedure is similar to that reported by Ismail et al. (68). Estrone and estradiol were separated on 3 g Sephadex LH-20 columns with benzene methanol (85:15) as eluate. The separation is safe on these large columns. Any displacement which may occur with plasma extracts as compared to standards takes the same direction as in chromatographic columns, that is, small amounts of estradiol (i.e., up to 10%) may be eluted with the estrone fraction, but no estrone will be displaced into the estradiol fraction. This was important to know for the cycles where estradiol sulfate was measured. Mass recovery of estrone and estradiol from buffer, including the defatting and gel filtration steps, ranged from 70 to 80% over the entire range of the standard curve, which was arranged to extend from 25 to 500 or up to 700 pg estrogen. Blanks were below 25 pg. Quantitative recoveries of estrogen added to plasma were harder to achieve and even varied to some extent with the individual plasma in such experiments. They could be considerably improved by the addition of gamma-globulin to the RIA mixture according to the suggestion of Leyendecker et al. (69).

Estrogen Sulfates

When sulfates were measured, the plasma remaining after ether extraction of the free estrogens was acidified and salted and the hydrosulfates extracted into ethyl acetate. Incubation at 37°C overnight yielded the free estrogens. The extracts, after bicarbonate and water washings, were evaporated and then subjected to Sephadex gel filtration and RIA as outlined earlier. This solvolysis procedure yielded between 50 and 55% recovery, so that for comparison with data from Loriaux et al. (32) and Preedy (31) the figures reported here, which are not corrected for losses, should be multiplied roughly by a factor of 2. Because of the instability of the estrogen sulfates during chemical manipulations, no effort was made to purify the conjugates as such on TLC or otherwise as described by Loriaux et al. (32).

ACKNOWLEDGMENT

Supported by Program Project 1 PO1 HD 05077 of the National Institutes of Health.

REFERENCES

1. K. Thomas, R. Walkiers, and J. Ferin, *J. Clin. Endocr.*, **30,** (1970), 269.
2. A. R. Midgley, and R. B. Jaffe, *J. Clin. Endocr.*, **33,** (9171), 962.
3. S. S. C. Yen, C. C. Tsai, F. Naftolin, G. Vandenberg, and L. Ajabor, *J. Clin. Endocr.*, **34,** (1972), 671.
4. G. T. Ross, C. M. Cargille, M. B. Lipsett, P. L. Rayford, J. R. Marshall, C. A. Strott, and D. Rodbard, *Rec. Prog. Hormone, Res.*, **26,** (1970), 1.

5. R. L. Vande Wiele, J. Bogumil, I. Dyrenfurth, M. Ferin, R. Jewelewicz, M. Warren, T. Rizkallah, and G. Mikhail, *Rec. Prog. Hormone Res.*, **26**, (1970), 63.

6. D. R. Mishell, R. M. Nakamura, P. G. Crosignani, S. Stone, K. Kharma, K. Nagata, and I. H. Thorneycroft, *Amer. J. Obstet. Gynecol.*, **111**, (1971), 60.

7. M. R. Henzyl and E. J. Segre, *Contraception*, **1**, (1970), 315.

8. G. D. Niswender, K. M. J. Menon, and R. B. Jaffe, *Fertility Sterility*, **23**, (1972), 432.

9. W. Goebbelsman, A. R. Midgley, and R. B. Jaffe, *J. Clin. Endocr.*, **29**, (1969), 1222.

10. M. L. Taymor and J. Miyata, in*Karolinska Symposia on Research Methods in Reproductive Endocrinology, Vol. 1, Stockholm 1969* (E. Diczfalusy, Ed.), p. 324.

11. B. T. Donovan, *Acta Endocr.*, **66**, (1971), 1.

12. L. Speroff, and R. L. Vande Wiele, *Amer. J. Obstet. Gynecol.*, **109**, (1971), 234.

13. W. D. Odell, and D. L. Meyer, *Physiology of Reproduction*, Mosby, St. Louis, 1971, p. 66.

14. J. M. Malacara, L. E. Seyler, and S. Reichlin, *J. Clin. Endocr.*, **34**, (1972), 271.

15. K. Bischoff, G. Bettendorf, and H. E. Stegner, *Arch. Gynaekol.*, **208**, (1969), 44.

16. A. R. Midgley, and R. B. Jaffe, *J. Clin. Endocr.*, **28**, (1968), 1699.

17. M. Taymor, T. Aono, and C. Pheteplace, *Acta Endocr.*, **59**, (1968), 298.

18. W. D. Odell, A. F. Parlow, C. M. Cargille, and G. T. Ross, *J. Clin. Invest.*, **47**, (1968), 2551.

19. V. C. Stevens, *J. Clin. Endocr.*, **29**, (1969), 904.

20. C. Fayman and R. J. Ryan, *J. Clin. Endocr.*, **27**, (1967), 1711.

21. B. B. Saxena, H. Demura, H. M. Gandy, and R. E. Peterson, *J. Clin. Endocr.*, **28**, (1968), 519.

22. D. S. Schalch, A. F. Parlow, R. C. Boon, and S. Reichlin, *J. Clin. Invest.*, **47**, (1968), 665.

23. M. A. Yussman, M. L. Taymor, J. Miyata, and C. Pheteplace, *Fertility Sterility*, **21**, (1970), 119.

24. J. A. Loraine and E. T. Bell, *Hormone Assays*, Williams & Wilkins, Baltimore, 1966, p. 225.

25. D. T. Baird, and A. Guevara, *J. Clin. Endocr.*, **29**, (1969), 149.

26. S. G. Korenman, L. E. Perrin, and T. P. McCallum, *J. Clin. Endocr.*, **29**, (1969) 879.

27. K. J. Catt, in *Karolinska Symposia on Research Methods in Reproductive Endocrinology, Vol. 1 Stockholm 1969*, (E. Diczfalusy, Ed.), p. 222.

28. G. E. Abraham, *J. Clin. Endocr.*, **29**, (1969), 866.

29. G. Mikhail, C. Wu, M. Ferin, and R. L. Vande Wiele, *Steroids*, **15**, (1970), 333.

30. G. E. Abraham, R. Swerdloff, D. Tulchinsky, and W. D. Odell, *J. Clin. Endocr.*, **32**, (1971), 619.

31. J. R. K. Preedy, in *Estrogen Assay in Clinical Medicine* (C. A. Paulsen, Ed.), Univ. Washington Press, Seattle, 1965, p. 162.

32. D. L. Loriaux, H. J. Ruder, and M. B. Lipsett, *Steroids*, **18**, (1971), 463.

33. J. B. Brown, and E. J. Smyth, *J. Reprod. Fertility*, **24**, (1971) 142, (Abstract).

34. H. E. Rosenthal, E. Pietrzak, W. R. Slaunwhite, and A. A. Sandberg, *J.Clin. Endocr.,* **34,** (1972), 805.

35. J. Zander, T. R. Forbes, A. M. von Münstermann, and R. Neher, *J. Clin. Endocr.,* **18,** (1958), 337.

36. M. Arcos and S. Liebermann, *Biochemistry,* **6,** (1967), 2032.

37. J. D. Neill, E. D. B. Johansson, J. K. Datta, and E. Knobil, *J. Clin. Endocr.,* **27,** (1967), 1167.

38. C. M. Cargille, G. T. Ross, and T. Yoshimi, *J. Clin. Endocr.,* **29** (1969), 12.

39. E. D. B. Johansson, and L. Wide, *Acta Endocr.,* **62,** (1969) 82.

40. B. Runnebaum, W. Rieben, A. M. Bierwirth-. Münstermann, and J. Zander, *Acta Endocr.,* **69,** (1972), 731.

41. E. D. B. Johansson, *Acta Endocr.,* **61,** (1969), 592.

42. B. Runnebaum, H. van der Molen and J. Zander, *Steroids Suppl.* **2,** (1965), 189.

43. C. A. Strott, T. Yoshimi, G. T. Ross, and M. B. Lipsett, *J. Clin. Endocr.,* **29,** (1969), 1157.

44. G. E. Abraham, R.Swerdloff, D. Tulchinsky, K. Hopper, and W. D. Odell, *J. Clin. Endocr.,* **33,** (1971), 42.

45. J. Bermudez, P. Doerr, and M. B. Lipsett, *Steroids,* **16,** (1970), 505.

46. D. L. Loriaux, and M. B. Lipsett, *Steroids,* **19,** (1972), 681.

47. E. Horton, *J. Clin. Endocr.,* **25,** (1965), 1237.

48. L. R. Rosenfield, *Steroids,* **14,** (1969), 251.

49. J. Lobotsky, H. I. Wyss, E. J. Segre, and C. W. Lloyd, *J. Clin. Endocr.,* **24,** (1964), 1261.

50. C. W. Lloyd, J. Lobotsky, D. T. Baird, J. A. McCracken, J. Weisz, M. Pupkin, J. Zanartu, and J. Puga, *J. Clin. Endocr.,* **32,** (1971), 155.

51. G. Mikhail, *Clin. Obstet. Gynecol.,* **10,** (1967), 29.

52. M. J. Gray, K. S. Strausfeld, M. Watanabe, E. A. Sims, and S. Solomon, *J. Clin. Endocr.,* **28,** (1968), 1269.

53. F. H. Katz, and P. Romfh, *J. Clin. Endocr.,* **34,** (1972), 819.

54. A. G. Frantz, and M. T. Rabkin, *J. Clin. Endocr.,* **25,** (1965), 1470.

55. W. N. Spellacy, W. C. Buhi, and R. P. Bendel, *Amer. J. Obstet. Gynecol.,* **104,** (1969), 1138.

56. S. S. C. Yen, P. Vela, J. Rankin, and A. S. Littell, *J. Amer. Med Assoc.* **211,** (1970), 1513.

57. B. Simkin, and R. Arle, *Proc. Soc. Exp. Biol. Med.* **113,** (1963), 485.

58. I. Gati, J. Doszpod, and J. Preisz, *Acta Physiol.* (Budapest), **32,** (1967), 115.

59. R. L. Coppedge, and A. Segaloff, *J. Clin. Endocr.,* **11,** (1951), 465.

60. L. S. Jacobs, I. K. Mariz, and W. H. Doughaday, *J. Clin. Endocr.,* **34,** (1972), 484.

61. H. G. Friesen, *Clin. Obstet. Gynecol.,* **14,** (1971), 669.

62. P. H. Wang, H. Guyda, and H. G. Friesen, *Proc. Nat. Acad. Sci, U.S.,* **68,** (1971), 1902.

63. A. R. Midgley, and R. B. Jaffe, Personal Communication.

64. J. Ferin, K. Thomas, and R. L. Vande Wiele, To Be Published.

65. R. J. Bogumil, M. Ferin, and R. L. Vande Wiele, *J. Clin. Endocr.,* **35,** (1972), 144.
66. A. R. Midgley, *J. Clin. Endocr.,* **27,** (1967), 295.
67. A. R. Midgley, *Endocrinology,* **79,** (1966), 10.
68. H. A. A. Ismail, D. N. Love, and R. W. McKinney, *Steroids,* **19,** (1972), 689.
69. G. Leyendecker, S. Wardlaw, and W. Nocke, *J. Clin. Endocr.,* **34,** (1972), 430.

CHAPTER 13

Ultradian Fluctuations of Gonadotropins

**S. S. C. YEN, G. VANDENBERG, C. C. TSAI, and
D. C. PARKER**

*Departments of Obstetrics and Gynecology, and Medicine,
University of California at San Diego, La Jolla, California*

Inquiry into the biologic rhythm of gonadotropin secretion encounters
the problem of ascertaining whether or not short-term periodicity exists
in the midst of a long trend. Previous attempts to answer the question
of a possible diurnal variation in circulating gonadotropin levels have
produced conflicting results. For men, using 3 to 6 hour sampling frequency, Saxena et al. (1) reported a diurnal variation for both luteinizing hormone (LH) and follicle-stimulating hormone (FSH).
Faimen and Ryan (2) detected a diurnal cycle for FSH but not LH,
whereas Peterson et al. (3) did not detect such variation in either hormone. For women, Midgley and Jaffe (4) found significant diurnal
variation in both FSH and LH during the follicular phase of the cycle
and Saxena et al. (1) reported similar findings in both the follicular
and luteal phases. In postmenopausal women, using similar sampling
frequency, we have observed a much wider variation (with no diurnal
pattern) than those found in subjects with intact hypothalamic-
pituitary-gonadal axis for both LH and FSH (5). These inconsistent
findings have lead us to suspect that the existence of rapid fluctuations

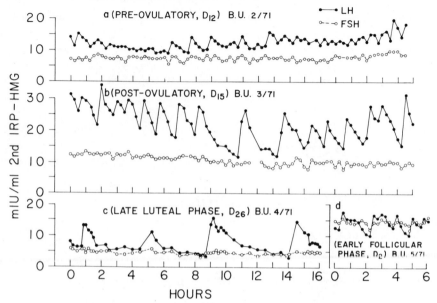

Figure 1   *Variation in the frequency and magnitude of the pulsatile pattern of circulating gonadotropins during different phases of the menstrual cycle.*

is a highly likely possibility and that the observed random fluctuation is probably related to the infrequent sampling. Studies were initiated to determine the presence of short-term rhythm by frequent sampling with a time interval of less than the biologic half-life of the hormone concerned (6, 7). In the meantime the observation of circhoral oscillations of LH in ovariectomized rhesus monkeys (8) have added further impetus for the disclosure of short-term rhythm in man.

It is now realized that within the well-established cyclic rhythm with a monthly periodicity, gonadotropin levels in subjects with intact hypothalamic-pituitary-ovarian axis are not maintained by continuous rates of secretion as previously supposed. Rather, they represent the integration of episodic burst of pituitary discharge (9, 10) with a frequency and magnitude which varies according to different phases of the cycle (10). The acyclic gonadotropin levels found in eugonadal males are similarly composed of episodic pituitary release (11, 12).

When a sampling frequency of 10 to 15 minute intervals was made, a well-defined pattern of episodic fluctuation of LH was disclosed (Figure 1). Because of the rapid onset in the incremental changes of LH (within 5 minutes), we have elected to use the term "pulsatile" rather than episodic to describe the dynamic nature of pituitary LH release. This rapid release was followed by a relatively slow decline. The approximated rate of decay for the downside of LH pulses was linear in most instances with a mean $t_{1/2}$ of 60 minutes, which was approximately 3 times longer than the endogenous $t_{1/2}$ determined following hypophysectomy in humans (6). It is likely that the pulsatile component of LH release was superimposed on a continuous secretion, which may account for the slower decay during the downside of the pulses (13).

A periodicity of about 90 minutes was seen in all phases of the cycle, except during mid and late luteal phases when a diminished frequency

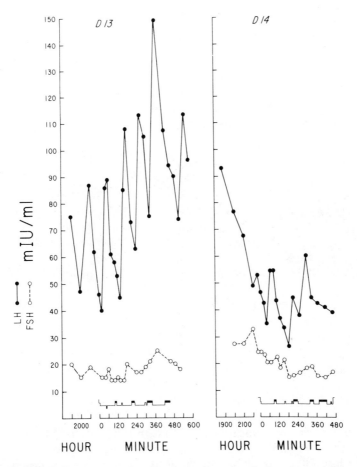

Figure 2   Pulsatile pattern of circulating gonadotropins during the ascending (D13) and descending (D14) limbs of the midcycle surge. (The sleep stages were monitored polygraphically.)

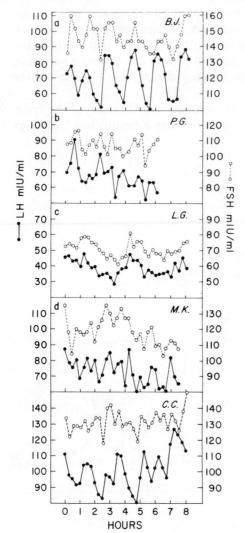

Figure 3 Pulsatile patterns of circulating gonadotropins in patients with gonadal dysgenesis.

of 3 to 4 hours was observed. The magnitude of the LH pulses was found to be greatest during the midcycle surge (Figure 2) and least during late follicular phase. In the latter, LH pulses are less discrete. For descriptive purposes we have referred to this pattern as "oscillations" (Figure 1). Since experimentally infusion of 17β-estradiol induced a rapid conversion of the pulsatile pattern to a pattern of oscillation (13), it is likely the observed change between early and late follicular phase is causally related to the feedback modulation by the increasing levels of estradiol. It has been shown that administered progesterone localizes in the higher neuronal centers, possesses potent general anesthetic properties, and alters EEG and behavioral activity (14–16). The possibility of a dampening effect on the central neuronal signal of progesterone to account for the diminished frequency seen during the mid and late luteal phase should be considered and remains to be elucidated.

Pulsatile fluctuations of circulating FSH are not as well defined and often not discernible (Figures 1 and 2). However, it does appear that there are more oscillations during the early follicular phase (a period of low estradiol levels) and during midcycle surge than during other phases of the cycle.

In the absence of the feedback loop, as in postmenopausal women (10) and in patients with gonadal dysgenesis, the elevated levels of FSH and LH are maintained by an increase in the amplitude of the pulses without a change in frequency (Figures 3 and 4). In marked contrast to the findings observed in premenopausal women, coincidental peaks of FSH and LH of comparable magnitude are discernible but obvious dissimilarities in the patterns are apparent. The latter may be accounted for at least in part by a severalfold longer half-life of FSH

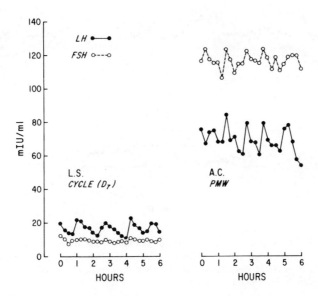

Figure 4   Differences in the amplitude and the disparity of LH and FSH pulses between pre- and postmenopausal women.

than LH (6, 7). These data suggest that without gonadal steroid feed-back regulation, a "programmed" central neuronal mechanism exists which triggers the periodic discharge of both pituitary LH and FSH at a maximal effort via hypothalamic neurohormonal events (13). The addition of nonlinear gonadal steroid output as during the menstrual cycle produces a general dampening effect of the pulsatile signal and pituitary discharge. Two integrated events may be operative: (1) *A frequency modulation* associated with maximal ovarian estradiol output (high frequency and low amplitude) and maximal progesterone output from the corpus luteum (low frequency and high amplitude), and (2) *a preferential dampening* of the pulsatile event of pituitary release of FSH than LH.

Attempts were made to elucidate the central neuronal and steroid feedback regulatory mechanisms of the pulsatile gonadotropin release in a variety of physiological and pharmacological experiments.

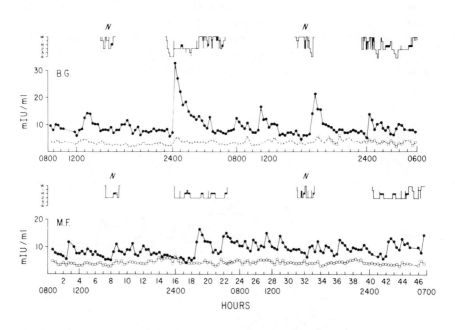

Figure 5a   Pulsatile patterns of circulating gonadotropins obtained during a continuous 48-hour study with polygraphically monitored sleep stages. (W = wake, 1 = stage I sleep, R = REM sleep, 2 = stage II sleep, 3 = stage III sleep, and 4 = stage IV sleep. N = nap during the day hours.)

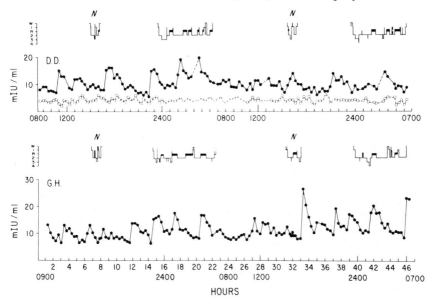

Figure 5b Pulsatile patterns of circulating gonadotropins obtained during a contin-uous 48-hour study with polygraphically monitored sleep stages. (W = wake, 1 = stage I sleep, R = REM sleep, 2 = stage II sleep, 3 = stage III sleep, and 4 = stage IV sleep. N = nap during the day hours.)

EFFECT OF WAKE-SLEEP CYCLE

Major alterations in CNS activity and neuroendocrine functions characteristically occurring in different stages of sleep have been described (17–21). Studies to ascertain the presence or absence of diurnal variation, circadian rhythm, and the temporal relationship between different stages of sleep cycle and pulsatile event of gonadotropin release were made. The acyclic gonadotropin pattern seen in male subjects would be better suited for these studies. As shown in Figures 5a and 5b, during a continuous 48-hour study with poly-graphically monitored sleep, there was no significant differences in the pulsatile pattern between wake-sleep cycles and there appears to be a variation in the pattern from day to day within the same individuals. In addition, correlation between initiations and peaks of pulsatile fluc-tuations and different stages of sleep, particularly the rapid eye movement (REM) periods, was not found. Boyar et al. (22) have

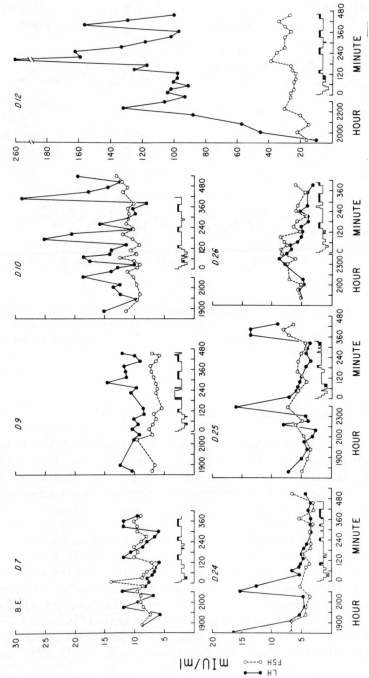

Figure 6 Pulsatile pattern of circulating gonadotropins during wake-sleep cycle at different phases of the menstrual cycle. (Poly-graphically monitored sleep stages.)

recently reported a similar finding in man during a 24-hour wake-sleep cycle study. This is at variance with the finding of LH values 14% greater during REM periods when compared with other sleep stages in man as reported by Rubin et al. (23). A lack of correlation between nocturnal wake-sleep cycle and between stages of sleep was also found in cycling women during different phases of the menstrual cycle (Figures 2 and 6). It is now clear that all pituitary hormones are secreted in episodic fashion (17–21) and that at least four hypothalamic-pituitary systems have temporal patterns of secretion which are closely linked to the 24-hour sleep-wake activity in man: ACTH (17, 18), HGH (19), prolactin (20), and thyroid-stimulating hormone (TSH) (21). The recent demonstration of an intimate association of LH release during early and late puberty, but not in pre-pubertal children, in both boys and girls (24) would indicate that the hypothalamic-pituitary gonadotropin system is also linked with the sleep-wake cycle during a transient period of sexual maturation. From our observations, no circadian rhythm or sleep-linked gonadotropin release in adult men and women are discernible.

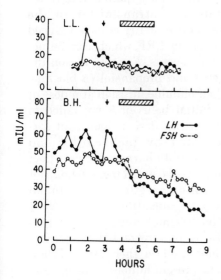

Figure 7   Effect of sodium pentothal anesthesia (with morphine and atropine as pre-medication, as indicated by the arrow) on the pulsatile pattern of circulating gonadotropins in a premenopausal (L.L.) and a postmenopausal woman (B.H.). The hatched area indicates the duration of anesthesia.

EFFECTS OF ANESTHESIA

Nine adult female patients, 21 to 52 years of age and free of intercurrent disease, were selected at random for this study during elective pelvic surgery for benign conditions. Eight patients were subjected to sodium pentothal anesthesia and one patient received spinal anesthesia. Of these nine studies, only two patients (one premenopausal and one postmenopausal) exhibited a clear-cut suppression of pulsatile gonadotropin patterns (Figure 7). Both patients had in common the combination of premedication which consisted of morphine and atropine given at 1 and 1.5 hours prior to sodium pentothal anesthesia. Nonetheless, this finding represents the first demonstration of a dampening effect of general anesthesia on the central mechanism of pulsatile regulation on gonadotropin release in humans.

EFFECTS OF ADRENERGIC BLOCKING AGENTS

Experimental evidence in rats as reported by Schnider and McCann (25) and Kamberi et al. (26) indicates that the hypothalamic α-adrenergic mechanism is an important neurotransmitter in the hypothalamic LRF release. In rats the increased LH release induced by dopamine is occasioned by an increase in LRF which can be negated by α- but not β-adrenergic blocking agents. For this reason the effect of phentolamine (α-blocking agent) and propanolol (β-blocking agent) was evaluated in six patients. Phentolamine infusion at a dose of 30 mg/hour for 2 hours following an initial bolus of 5 mg failed to abolish the pulsatile pattern (Figure 8). At this dose a significant cardiovascular effect occurred as reflected by tachycardia and hypotension, indicating an adequate peripheral pharmacological effect had been achieved. Propanolol infusion at a dose of 0.6 mg/kg/hour for 2 hours, despite a significant cardiovascular effect of profound bradycardia, produced no change in the pulsatile pattern. Very recently Bhattacharya et al. showed that phentolamine, but not propanolol, effectively abolished the pulsatile pattern of LH in ovariectomized rhesus monkeys (27). The dose used in their study is approximately 10 times the dose used in the present human investigation. The negative finding in our study is probably related to a partial blood-brain barrier of phentolamine, and optimal concentration in the hypothalamus may not have been reached. Therefore it is highly probable that the pulsatile discharge of LH and FSH observed is the consequence of an intermediate adrenergic signal which initiates the release of LRF into the portal system and results in episodic pituitary discharge of gonadotropins. Clearly, a demonstration of pulsatile nature of LRF release in the portal vessel would provide a critical step in the delineation of the regulatory mechanism of pulsatile gonadotropin release.

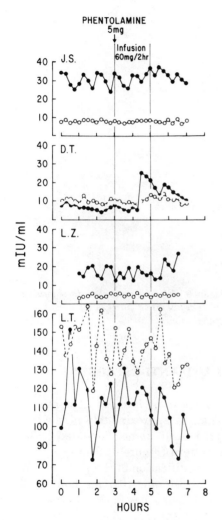

Figure 8 Effect of phentolamine on the pulsatile pattern of circulating gonadotropins in subjects with and without (L.T.) gonadal function.

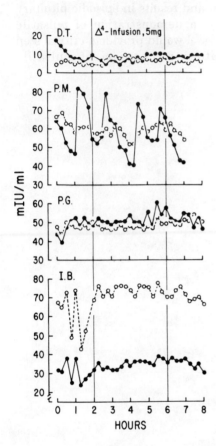

Figure 9   *Effect of androstenedione infusion on the pulsatile pattern of circulating gonadotropins.*

EFFECTS OF GONADAL STEROIDS AND ANTI-ESTROGENS

Administration of androstenedione by constant infusion at a rate of 1.25 mg/hour for a 4-hour period, which is approximately nine times the daily production rate for premenopausal women, resulted in no clear-cut suppression of the pulsatile pattern (Figure 9). However, in a patient with gonadal dysgenesis (IB) a modification of the pattern may have occurred. Similarly infusion of testosterone (0.5 mg/hour × 4 hours) and progesterone (2 mg/hour × 4 hours) has failed to elicit any change in the pulsatile pattern in five subjects. The question of whether or not these steroids may exert a chronic and/or synergistic modulatory effect cannot be answered from these acute experiments.

In contrast, infusion of 17β-estradiol (50 μg/hour × 4 hours) elicited a clear and reproducible response in the abolishment of the pulsatile pattern of LH and promptly converted into a pattern of oscillations (Figure 10). This resulted in a rapid reduction in circulating concentrations which is greater for LH than for FSH. This negative feedback event was sustained for 5 to 6 hours despite the rapid disappearance of circulating E_2 following the infusion. This finding may reflect the kinetics of cellular binding of estradiol in the hypothalamic-pituitary system which has been shown to be rapid (within 5 minutes) and retained for several hours in ovariectomized rats (28, 29). The time course (several hours) and the rebound increase in the magnitude (but not frequency) of the pulsatile discharge observed during the recovery phase following E_2 suppression are interpreted as an overlapping event between the negative and positive feedback effect of E_2 on the pituitary sensitivity to LRF as well as in the LRF release consequent to the disappearance of the biologic action of E_2 at the receptor sites within the hypothalamic-pituitary system. Since clomid has been shown to

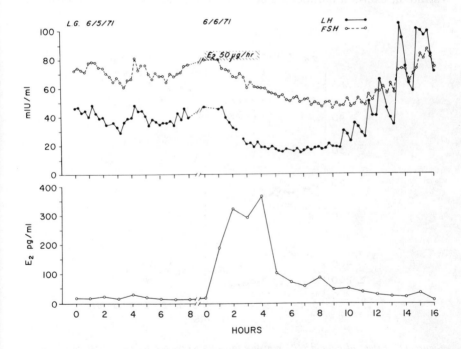

Figure 10 Effect of E_2 infusion on the pulsatile pattern of circulating gonadotropins in a patient with gonadal dysgenesis (circulating E_2 concentration before, during, and after E_2 infusion is also shown.)

promote gonadotropin release by competition of E_2 recptor sites in the hypothalamic-pituitary unit (30, 31), the dampening effect of E_2 on the pulsatile pattern of gonadotropin release may further be evaluated by the responsiveness to clomid. As shown in Figure 11, an increase in the amplitude of the pulsatile release, more for LH than FSH, in response to clomid treatment was observed. This finding provides additional evidence that E_2 modulates the amplitude of pulsatile release with preferential inhibition of the FSH component.

Thus the pulsatile event of gonadotropin release in the absence of gonadal feedback may be formulated at the present time as a consequence of periodic "programmed" central neuronal signal (at an undefined site) which is mediated through α-adrenergic mechanism to initiate episodic release of LRF into the portal system, resulting in the pulsatile pituitary discharges with quantitatively and qualitatively similar release for FSH and LH. This event may be described as an *inherent ultradian rhythm of CNS-pituitary gonadotropin regulation* (Figure 12). Addition of cyclic ovarian steroid feedback, particularly estradiol, exerts a dampening effect on the α-adrenergic mechanism of

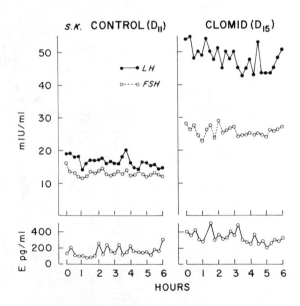

Figure 11   Effect of clomid (100 mg/day for 4 days) on the pulsatile pattern of circulating gonadotropins during the follicular phase of the cycle. The expected midcycle surge of LH was delayed for several days in this case.

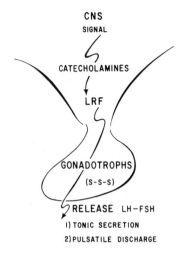

Figure 12   Diagramatic illustration of the postulated sequence of events governing the pulsatile pituitary discharge of gonadotropins. (S-S-S = synthesis, storage, and sensitivity of the gonadotrophs.)

LRF release concomitant with a differential action on the pituitary gonadotropin, producing cells with a preferential inhibition of FSH component.

ACKNOWLEDGMENT

Research supported by Rockefeller Foundation (RF 70029).

REFERENCES

1.   B. B. Saxena, G. Leyendecker, W. Chen, H. M. Gandy, and R. E. Peterson, in *Immunoassay of Gonadotropins* (E. Diczfalusy, Ed.), Stockholm, 1969, p. 185.
2.   C. Faimen and R. J. Ryan, *Nature,* **215,** (1967), 857.
3.   N. T. Peterson, A. R. Midgley, Jr., and R. B. Jaffee, *J. Clin. Endocr.,* **28,** (1968), 1473.
4.   A. R. Midgley, Jr., and R. B. Jaffe, *J. Clin. Endocr.,* **28,** (1968), 1699.
5.   S. S. C. Yen and C. C. Tsai, *J. Clin. Endocr.,* **33,** (1971), 882.
6.   S. S. C. Yen, O. Llerena, B. Little, and O. H. Pearson, *J. Clin. Endocr.,* **28,** (1968), 1763.

7. S. S. C. Yen, L. A. Llerena, O. H. Pearson, and A. S. Littell, *J. Clin. Endocr.,* **30,** (1970), 325.

8. D. J. Dierschke, A. N. Bhattacharya, L. E. Atkinson, and E. Knobil, *Endocrinology,* **87,** (1970), 850.

9. A. R. Midgley, Jr., and R. B. Jaffe, *J. Clin. Endocr.,* **33,** (1971), 962.

10. S. S. C. Yen, C. C. Tsai, F. Naftolin, G. VandenBerg, and L. Ajabor, *J. Clin. Endocr.,* **34,** (1972), 671.

11. H. R. Nankin and P. Troen, *J. Clin. Endocr.,* **33,** (1971), 558.

12. F. Naftolin, S. S. C. Yen, and C. C. Tsai, *Nature New Biol.,* **236,** (1972), 92.

13. S. S. C. Yen, C. C. Tsai, G. VandenBerg, and R. Rebar, *J. Clin. Endocr.,* **35,** (1972), 897.

14. B. J. Meyerson, *Endocrinology,* **81,** (1967), 369.

15. M. Kawakami and C. H. Sawyer, *Endocrinology,* **65,** (1959), 652.

16. B. S. Kopell, *Metabolic Effects of Gonadol Hormones and Contraceptive Steroids* in (H. A. Salhanick, D. M. Kipnis and R. L. Vande Weile, Eds.), Plenum Press, New York, 1969, pp. 668–675.

17. D. T. Krieger, W. Allen, F. Rizzo, and H. P. Krieger, *J. Clin. Endocr.,* **32,** (1971), 266.

18. L. Hellman, N. Fujinori, J. Curt, E. D. Weitzman, J. Kream, H. Roffwarg, S. Ellman, D. K. Fukushima, and T. F. Gallagher, *J. Clin. Endocr.,* **30,** (1970), 411.

19. D. C. Parker, J. F. Sassin, J. W. Mace, R. W. Gotlin, and L. G. Rossman, *J. Clin. Endocr.,* **29,** (1969), 871.

20. J. F. Sassin, A. G. Frantz, E. D. Weitzman, and S. Kapen, *Science,* **177,** (1972), 1205.

21. L. VanHaelst, E. VanCauter, J. P. Degaute, and J. Golstein, *J. Clin. Endocr.,* **35,** (1972), 61.

22. R. Boyar, M. Perlow, L. Hellman, S. Kapen, and E. D. Weitzman, *J. Clin. Endocr.,* **35,** (1972), 61.

23. R. T. Rubin, A. Kales, R. Adler, T. Fagan, and W. Odell, *Science,* **175,** (1972), 196.

24. R. Boyar, J. Finkelstein, H. Roffwarg, S. Kapen, E. D. Weitzman, and L. Hellman, *New Eng. J. Med.,* **287,** (1972), 582.

25. H. P. G. Schneider and S. M. McCann, *Endocrinology,* **85,** (1969), 121.

26. I. A. Kamberi, R. S. Mical, and J. C. Porter, *Endocrinology,* **89,** (1971), 1042.

27. A. N. Bhattacharya, D. J. Dierschke, T. Yamaji, L. E. Atkinson, and E. Knobil, *Endocrinology,* **90,** (1972), 778.

28. T. F. Mowles, B. Ashkanazy, E. Mix, Jr., and H. Sheppard, *Endocrinology,* **89,** (1971), 484.

29. J. L. McGuire and R. D. Lisk, *Proc. Nat. Acad. Sci., U.S.,* **61,** (1968), 497.

30. M. Igarashi, Y. Ibuki, H. Kubo, J. Kamioka, N. Yokota, Y. Ehara, and S. Matsumoto, *Amer. J. Obstet. Gynecol.,* **97,** (1967), 120.

31. J. Kato, T. Kobayashi, and C. A. Villee, *Endocrinology,* **82,** (1968), 1049.

CHAPTER 14

Causal Relationship Between the Hormonal Variables in the Menstrual Cycle

S. S. C. YEN, G. VANDENBERG, C. C. TSAI, and T. SILER

Department of Obstetrics and Gynecology, University of California at San Diego, La Jolla, California

The phenomenon of the cyclic ovarian function is dependent upon the complex interplay between two integrated endocrine components: the ovarian unit and the hypothalamic-pituitary system. A multitude of interactions between, as well as within, the components are possible. The background information has been reviewed and important concepts were advanced by Vande Weile et al. in 1970 (1).

A basic premise in arriving at an understanding of the performance of a system in which an input and an output are causally related is to describe the operational characteristics of that relationship. Unlike the ACTH-cortisol system, the physiologic regulation of hypothalamic-pituitary-ovarian system (H-P-O) concerns additional trophic hormones with different biologic actions [i.e., follicle-stimulating hormone (FSH) and luteinizing hormone (LH)] and a nonlinear feedback function of ovarian steroids (i.e., estradiol). These two systems differ further in the fact that both negative and positive feedback in a closed-loop model are operational in H-P-O system but not in the ACTH-cortisol system. On the other hand the latter system is complicated by the presence of an inherent circadian rhythm which is independent of the negative feedback loop (2).

Evidence has recently been advanced that within the well-established cyclic rhythm with a monthly periodicity, gonadotropins are released in a pulsatile fashion (3, 4, and Chapter 13) and that gonadal steroids, particularly estradiol, exert a modulatory feedback action on the amplitude of the pulsatile release (3, 5). These findings, together with the recent availability of synthetic LRF for human studies (6–9), have afforded a new dimension in the elucidation of H-P-O interaction. A satisfactory account of the control system from a broad operational viewpoint may now be formulated.

THE TWO COMPONENT SYSTEM

Based on anatomical and functional characteristics, the H-P-O system can be divided into the upper (H-P) and the lower (ovarian) units.

1. The ovary, inherently a *cyclic* organ composed of two interrelated endocrine and morphological structures: the follicular complex and the corpus luteum. The optimal life span of both is functionally limited (approximately 14 days) and the development of each is overlapped in their time course.

2. The H-P system as an independent functional unit is inherently *acyclic* and releases maximal amount of gonadotropin in a rhythmic

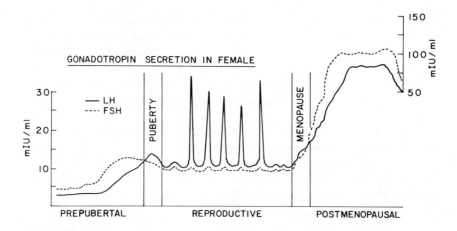

Figure 1 A diagrammatic overview of the life cycle of gonadotropin secretion in the human female.

fashion of high frequency (about 90 minutes) and large amplitude for both FSH and LH. The cyclic gonadotropin release during reproductive life is a secondary phenomenon consequent to the inherent property of the ovarian function. An overview of gonadotropin life cycle may be seen in Figure 1.

3. Operationally, the H-P system may be viewed as a *provider* and the cyclic ovarian steroid output as a *modulator*. The latter acting as signals ideally should provide two built-in biological principles: (1) exert an overlapping and interrelated sequence of negative and positive feedback action, and (2) a differential effect between the release of FSH and LH.

NEGATIVE FEEDBACK CONTROL OF GONADOTROPIN SECRETION

This is a prerequisite for the positive feedback control.

Although it has been recognized for several decades that gonadotropin secretion is under negative feedback control by the gonadal hormones, the dynamics of this control system have become apparent only recently. Quantitative relationships between the negative feedback action of gonadal steroids and gonadotropin release can readily be demonstrated by ovarian ablation. Interruption of the negative feedback loop by ovariectomy results within 2 days in a significant increase in circulating LH and FSH concentrations which continue to rise until a plateau is reached at approximately 10 times the initial concentration at about 3 weeks after the operations (10). There was a significantly greater rise in both FSH and LH during the initial phase (first week) in subjects ovariectomized during the follicular phase of the cycle than in subjects ovariectomized during the luteal phase of the cycle. These data provide indirect evidence that the phases of the ovarian cycle may exert a quantitative influence in the gonadotropin turnover rate within the hypothalamic-pituitary system (synthesis, storage, and sensitivity). As will be shown later, this observation is pertinent to the finding that pituitary gonadotrophs sensitivity to LRF changes during different phases of the menstrual cycle.

It has been shown that the LRF release is significantly increased after castration (11). Thus, in the absence of gonadal function, as a functional unit the hypothalamic-pituitary system synthesizes and releases FSH and LH at a maximum rate, probably under the influence of the increased LRF activity. It has been shown that the increased level of gonadotropins in subjects without gonadal function is a consequence of the increase in the magnitude of the pulsatile discharge for both FSH and LH (5). Indirect evidence suggests that gonadotropin secretion is maintained by two components of pituitary secretion: The pulsatile discharge superimposed on a small continuous secretion (5). The possibility exists that hypothalamic LRF release may similarly be composed of a two-component system.

With the addition of cyclic ovarian steroid production the gonadotropin release is modified and becomes cyclic in pattern. Two independent but *overlapping* mechanisms are involved: the tonic mechanism (through negative feedback) and the surge mechanism (through positive feedback). The pulsatile nature of gonadotropin release exists for both tonic and surge mechanisms, but differs in their magnitude. As discussed elsewhere in this conference, estradiol constitutes the most important ovarian steroid to account for the negative feedback event. The rapid reduction of circulating gonadotropin is greater for LH than for FSH in response to E_2 infusion (Figures 2a and 2b). The mechanism for the acute negative feedback action of 17β-estradiol appears to be related to a dual action of an in-

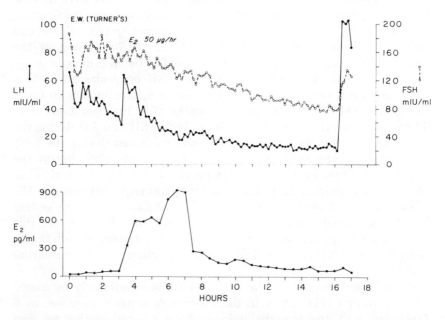

Figure 2a   *The negative feedback event followed by a rebound increase in gonadotropin levels during and after E_2 infusion (50 µg/hour × 4 hours).*

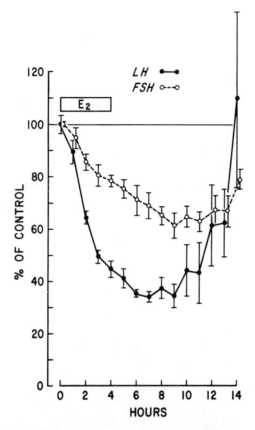

Figure 2b　*The negative feedback event followed by a rebound increase in gonadotropin levels during and after* E_2 *infusion (50 µg/hour* × *4 hours).*

hibitory effect on hypothalamic LRF release concomitant with a diminished sensitivity of pituitary gonadotrophs to LRF. This assumption is supported by the finding that E_2 infusion or single injection consistently attenuates the LRF responsiveness in both pre- and postmenopausal subjects (Figure 3). Since it has been shown that estrogen inhibits the dopaminergic induction of LRF release of the hypothalamus (12), modulation of adernergic mechanism on LRF release may have occurred. The time course and the rebound increase in the magnitude of pulsatile pattern observed during the recovery phase following E_2 suppression (Figure 2a) are interpreted as an expression of an overlapping event of the negative and positive feedback consequent to the disappearance of biologic action of E_2 from hypothalamic-pituitary receptor sites.

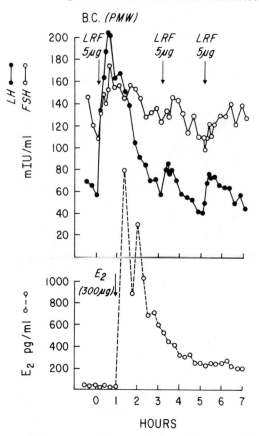

Figure 3   Pituitary responsiveness to LRF before and after intravenous administration of a single dose of E_2 (300 μg) in a postmenopausal woman.

As we have previously shown, continuous administration of estradiol for a duration of 6 to 8 days elicited a biphasic pattern which involves primarily LH release (13). This biphasic phenomena with a differential effect between LH and FSH is a transient phenomena limited to a duration of 2 to 3 weeks, since prolonged estradiol treatment exerts a consistent negative feedback action for both FSH and LH (Figure 4). This time related event of estradiol in the feedback modulation of the hypothalamic-pituitary system can further be elucidated by sequential responsiveness to synthetic LRF at weekly intervals during small doses of estradiol treatment (1 μg/kg). Initially, during the first week, estradiol promotes the responsiveness to LRF which was followed by a gradual diminish in the pituitary responsiveness to LRF stimulation

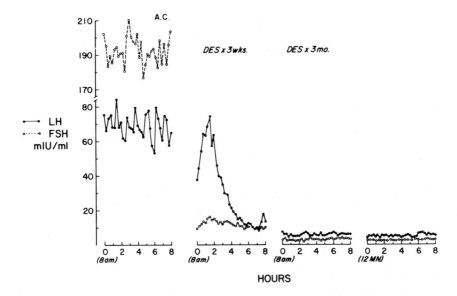

Figure 4 *A transient biphasic effect of estrogen feedback action was seen at 3 weeks of treatment which was followed by a consistent negative feedback for both LH and FSH (day and night) in a postmenopausal woman studied at 3 weeks and 3 months after diethylstilbesterol (1 mg/day).*

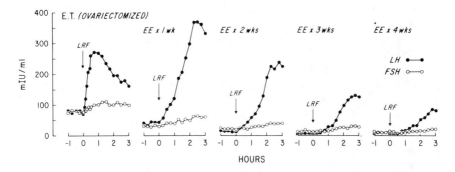

Figure 5 *Pituitary responsiveness to synthetic LRF (150 µg) before and during the administration of ethinyl estradiol (1 µg/kg/day) studied at weekly intervals in an ovariectomized subject.*

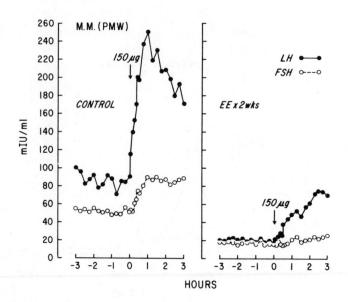

Figure 6 Pituitary responsiveness to synthetic LRF (150 μg) before and after administration of large dose of EE (2 μg/kg/day) in a postmenopausal woman.

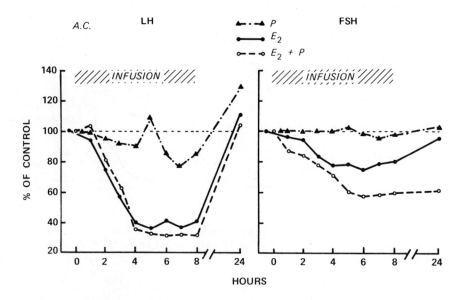

Figure 7 The effect of progesterone infusion (2 mg/hour), E₂ infusion (50 μg/hour) and combined progesterone and E₂ infusion for a duration of 8 hours on gonadotropin levels studied at monthly intervals in a postmenopausal woman.

(Figure 5). When the estradiol dose was increased to 100 µg/day, a prompt impairment of pituitary responsiveness to LRF occurred (Figure 6). These data provide the first clear demonstration of the time and dose related phenomenon in estradiol feedback action, and the pituitary gonadotrophs constitute an important site of the feedback event.

The acute administration of progesterone alone revealed no negative feedback effect on the pulsatile gonadotropin release. However, combination of progesterone and estradiol infusion does appear to exert synergistic negative feedback action primarily involving FSH, which is more than that observed with the infusion of estradiol alone (Figure 7). Other ovarian steroids, such as estrone, 17α-hydroxyprogesterone, 20α-hydroxyprogesterone, androstenedione, and testosterone, are considered to be secondary steroids from the ovary and require conversion to biologically active hormones for their action. Hence, the feedback action of these steroids, if any, should be viewed in a different context.

POSITIVE FEEDBACK CONTROL OF THE SURGE MECHANISM OF GONADOTROPIN RELEASE

This is a consequence to the negative feedback action.

Compelling evidence that increasing levels of circulating estradiol trigger the ovulatory surge of LH is afforded by the ability of exogenous estradiol to induce acute release of LH in rats (14, 15), sheep (16, 17) and monkeys (18). Additionally, administration of antibody to 17β-estradiol blocks spontaneous and PMS-induced LH surges in rats which can be restored by the treatment with a noncross-reacting synthetic

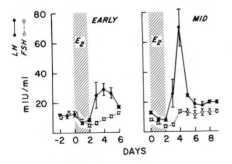

Figure 8   The difference in the positive feedback action of EE (200 µg/day × 3 days) between the early and midfollicular phases of the normal cycle.

estrogen, diethylstilbestrol (19, 20). We have recently demonstrated in humans that a rapid increase in serum estradiol concentrations from exogenous source during the midfollicular phase of the cycle elicited an acute release of gonadotropins resembling that occurring spontaneously at midcycle (Figure 8). This induced surge involved both LH and FSH (21).

In a similar study performed during the early follicular phase of the cycle, we have found a significant increase in LH following estradiol treatment (22), but it was not accompanied by a rise in FSH and the magnitude was not comparable to that observed at midcycle (Figure 8). These findings were interpreted to indicate that the sensitivity of the hypothalamic-pituitary complex to the feedback action of estradiol varied during the menstrual cycle. Thus, during the midfollicular phase when endogenous estradiol begins to rise, an increased sensitivity may be anticipated. More important is our recent finding that the sensitivity of pituitary gonadotrophs to LRF is preferentially increased for LH release during the late follicular phase of the cycle (Figure 9a), and that a "window" of maximal pituitary responsiveness to LRF for both LH and FSH release appears to occur at the midcycle (Figure 9b). It is likely that this event is brought about by a direct feedback action of increasing levels of serum estradiol. This is consistent with the

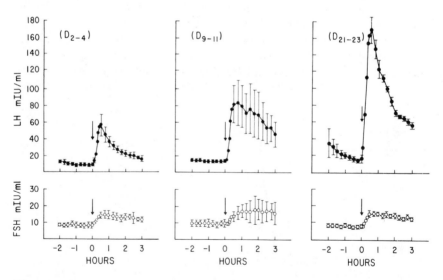

Figure 9a Serum LH and FSH concentrations (mean ±SE) before and after LRF (150 μg single IV dose) during different phases of the cycle studied serially in six subjects.

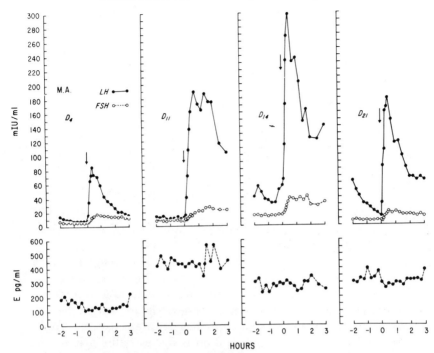

Figure 9b Serial responses in serum LH and FSH levels before and after LRF (150 µg single dose) during different phases of the cycle in one subject. (Spontaneous mid-cycle surge occurred on day 15.)

finding that optimal amount and duration of estradiol treatment enhanced the LRF responsiveness in humans (Figure 5).

A change in the feedback sensitivity during the course of the follicular phase of the cycle was further demonstrated by the responsiveness to clomid which induced a significantly greater rise in serum LH but not FSH during the late follicular phase than during the early follicular phase (Figure 10). This finding indicates that a progressive decrease in the negative feedback action of estradiol occurs toward midcycle.

Unlike postmenopausal women, the hypothalamic-pituitary system in patients with gonadal dysgenesis and primary amenorrhea may never have been exposed to cyclic ovarian feedback interaction. These patients offer a unique opportunity for further characterization of the hypothalamic-pituitary system as an independent functional unit in the positive feedback action of estradiol. As can be seen, the pulsatile

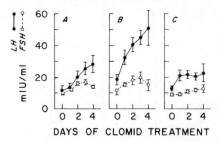

DAYS OF CLOMID TREATMENT

Figure 10   Effect of clomid (100 µg/day × 4 days) on serum LH and FSH levels (mean ±SE) during early follicular phase (A), late follicular phase (B), and during midluteal phase (C) of ovulatory cycles in five subjects.

fluctuation of both FSH and LH are suppressed by daily administration of ethinyl estradiol (Figure 11). During the sustained suppression, a concomitant release of LH, and to a lesser degree FSH, was induced within 4 hours following im injection of estradiol benzoate. It would appear that both negative and positive feedback response is operative in patients with gonadal dysgenesis in whom the cyclic gonadal steroid interaction presumably has not previously occurred.

Although progesterone administration under estradiol-primed situations can trigger an acute release for both LH and FSH in postmenopausal women (Figure 12a) as well as in patients with gonadal dysgenesis (Figure 12b), it is unlikely that it does so in a physiological context, since the preovulatory discharge has not been found to be preceded by an increment in plasma progesterone

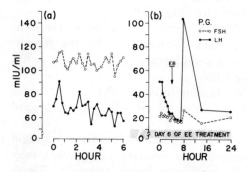

Figure 11   Pulsatile pattern of circulating LH and FSH (a) and the concomitant release of LH and FSH induced by estradiol benzoate injection (EB, 1 mg) during the sustained suppression by ethinyl estradiol (EE 200 µg/day) in a patient with gonadal dysgenesis.

concentrations (23). Additionally, administration of antibody to progesterone during the preovulatory period did not effectively inhibit ovulation in the rat (20). However, since a greater LRF response was found during the midluteal phase than during the follicular phase (Figure 9), the possibility of an additive and synergistic action of progestins other than progesterone (such as 17α-hydroxyprogesterone), particularly at midcycle, in the positive feedback action must be considered and remains to be elucidated.

The qualitative and quantitative relationship between estradiol levels and the dimension of LH surge cannot be clearly established. The dose may not be critical, but an incremental pattern of estrogen levels which are maintained for a period of 24 to 36 hours appears to be a critical factor (Figure 13). In no instance did the acute release occur during estradiol administration which exhibited uniform suppression of

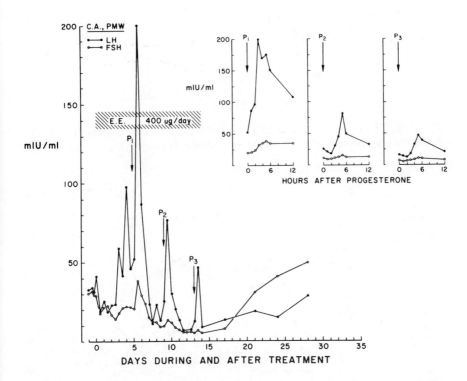

Figure 12a   Progesterone (10 mg im) induced acute release of LH–FSH during EE suppression in a postmenopausal woman. A diminished response was seen during repeated injection at 3-day intervals.

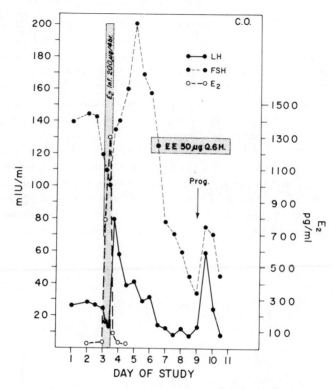

*Figure 12*b   *A concomitant rise of FSH and LH induced by progresterone (10 mg im) during the suppression by EE in a patient with gonadal dysgenesis.*

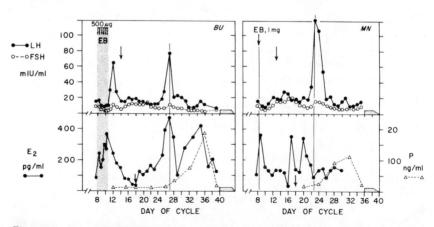

Figure 13   Serum LH, FSH, and E_2 patterns in a subject receiving repeated EB injection (500 µg every 12 hours) and in a subject receiving one single injection (1 mg).

232

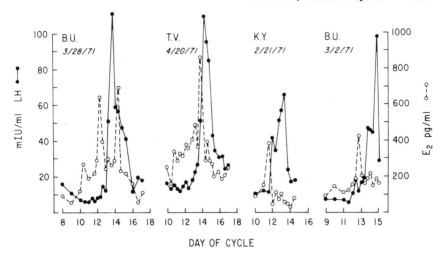

Figure 14   Serum LH and E₂ fluctuation in samples obtained at 4 to 8 hour intervals around midcyle in four subjects.

both gonadotropins. The onset of this acute release occurred about 12 to 24 hours after the termination of estradiol treatment when a consistent drop in serum estradiol concentrations was observed. This finding would suggest that an acute fluctuation in serum estradiol concentrations may be important in the triggering event. This assumption need not be inconsistent with the observed preovulatory rise in serum levels based on daily measurements, since major fluctuations in serum estradiol concentrations have been detected in samples obtained to 4 to 8 hour intervals (Figure 14). Nonetheless, the precise time course between estradiol fluctuations and the initiation of LH surge remains to be determined.

THE ROLE OF GONADOTROPINS IN THE MAINTENANCE OF OPTIMAL OVARIAN FUNCTION

Sufficient evidence is now available which indicates that the circulating gonadotropin pattern observed during different phases of the cycle is finely adjusted via minute-to-minute feedback modulation of ovarian steroids. When the rise of circulating FSH and LH during the early follicular phase was effectively abolished through the negative feedback action of EE (200 to 400 μ/day) for 3 days, a disturbance of the normal sequence of folliculogenesis and follicular maturation oc-

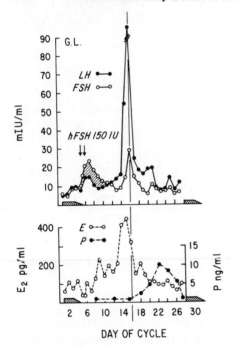

Figure 15   Effect of im administration of hFSH (150 IU/day) for 2 days during mid-follicular phase of the cycle (hatched area represents the serum FSH increments induced by exogenous FSH).

curred (22). This is reflected by the prolongation of the follicular phase (7 to 9 days) and a delayed preovulatory rise of circulating estradiol. Similar results were observed when the normal sequence of gonadotropin secretion was interrupted at the midfollicular phase of the cycle by administration of EE (21). These data provide evidence for the implication that the gonadotropin levels are functionally coordinated with the follicular development for a given cycle.

The question of the effect of additional gonadotropins from exogenous source or endogenous source (induced by clomid) on the follicular development during a course of normal cycle has not been evaluated. We have recently carried out a series of studies dealing with this question. As shown in Figure 15, a large rise of circulating FSH lasting 2 to 3 days induced by administration of two daily injections of 150 IU of human pituitary FSH (courtesy of NPA) on days 6 to 7 of the cycle resulted in no significant changes in either follicular development or steroidogenesis as judged by the duration of follicular phase and estradiol and progesterone concentrations. Administration of human pituitary LH of 2000 IU for two doses at 8 hour intervals induced a rise in circulating LH resembling that seen at the midcycle surge, but was similarly ineffective in changing either follicular development or steroidogenesis (Figure 16). On the other hand, a prolonged exposure to an excessive amount of LH-like activity for 5 to 6 days as seen in response to HCG administration (single 10,000 IU, im) during midfollicular phase resulted in a marked increase in circulating estradiol and a delay in preovulatory surge of LH with a shortened luteal phase (Figure 17). Taken together, these data suggest that the sensitivity of the ovarian follicular complex to gonadotropin stimulation is progressively increased from the mid to late follicular maturation and that the maintenance of optimal amount of FSH and LH stimulation is essential for the functional life span of the follicular complex during preovulatory phase of the cycle.

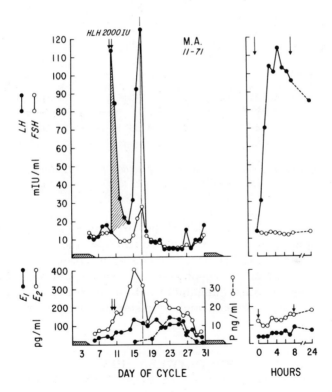

Figure 16 Effect of im administration of hLH at 8-hour intervals for 2 doses (2000 IU) during midfollicular phase of the cycle (hatched area represents the serum LH increments induced by exogenous LH).

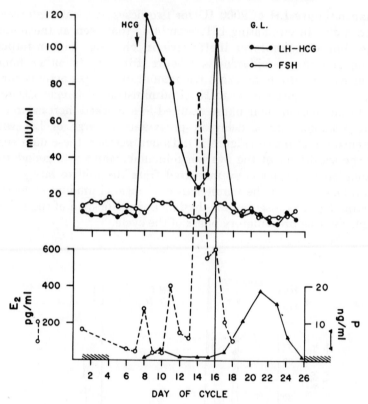

Figure 17   Effect of a single im injection of 10,000 IU HCG during midfollicular phase of the cycle. Note the marked increase in E_2 levels, a prolonged follicular phase, and a short luteal phase (10 days).

The regulation of the life span of the corpus luteum and its secretion of progesterone is also intimately dependent upon the presence of small amounts of circulating LH. This relationship has been demonstrated in the induced ovulation in hypophysectomized women by Vande Wiele et al. (1). Unless an optimal amount of LH (or HCG) is provided, the life span of the corpus luteum and progesterone secretion will be very much limited.

CAUSAL RELATIONSHIP BETWEEN HORMONAL VARIABLES IN THE MENSTRUAL CYCLE (A FUNCTIONAL EXPRESSION OF H-P-O AXIS)

The results of these studies in humans are consistent with the view that the cyclic pattern of gonadotropin release throughout the menstrual cycle can be accounted for by the change in ovarian estrogen output. It is implied that the ovary but not the hypothalamic-pituitary system constitutes the primary regulator for the rhythmic cyclicity. Optimal amount of FSH and LH release is regulated by circulating estradiol via modulation of the magnitude of pulsatile rhythm with a preferential inhibitory effect on FSH pulses. Concomitantly, estradiol induces a progressive diminish in negative feedback action and a progressive increase in the sensitivity of positive feedback action within the hypothalamic-pituitary unit. The triggering signal can be accounted for by the incremental change with episodic fluctuation of circulating estradiol—an expression of the end stage of the maturational event of the Graafian follicle. The site governing the surge mechanism is not clear. However, studies thus far obtained would indicate the pituitary is a major site of feedback action through an incremental increase in the sensitivity of gonadotrophs to estradiol which may lead to a "window" of maximal sensitivity at midcycle and account for the acute discharge of both FSH and LH as previously suggested by Bogdanove (24). The recent finding of an increased circulating bioassayable LRF-like activity at the midcycle in humans (25), if confirmed, would indicate that the event of cyclic gonadotropin release which leads to the acute surge at midcycle is regulated by a dual mechanism of a progressive increase in pituitary sensitivity to LRF and an enhanced LRF release at midcycle.

ACKNOWLEDGMENT

Research supported by Rockefeller Foundation (RF 70029).

REFERENCES

1. R. L. Vande Wiele, J. Bogumil, I. Dyrenfurth, M. Ferin, R. Jewelewicz, M. Warren, T. Rizkallah, and G. Mikhail, *Rec. Progr. Hormone Res.,* **26,**(1970), 63.

2. E. G. Yates, in *The Adrenal Cortex* (A. B. Eisenstein, Ed.), Little, Brown, Boston, 1967.

3. S. S. C. Yen, C. C. Tsai, F. Naftolin. G. VandenBerg, and L. Ajabor, *J. Clin. Endocr.,* **34,** (1972), 671.

4. A. R. Midgley, Jr., and R. B. Jaffe, *J. Clin. Endocr.,* **33,** (1971), 962.

5. S. S. C. Yen, C. C. Tsai, G. VandenBerg, and R. Rebar, *J. Clin. Endocr.,* (1972), 000.

6. S. S. C. Yen, R. Rebar, G. VandenBerg, F. Naftolin, Y. Ehara, S. Engblom, K. J. Ryan, K. Benirschke, J. Rivier, M. Amoss, and R. Guillemin, *J. Clin. Endocr.,* **34,** (1972), 1108.

7. R. Rebar, S. S. C. Yen, G. VandenBerg, F. Naftolin, Y. Ehara, S. Engblom, K. J. Ryan, J. Rivier, M. Amoss, and R. Guillemin, *J. Clin. Endocr.,* **36,** (1973), 10.

8. S. S. C. Yen, G. VandenBerg, R. Rebar, and Y. Ehara, *J. Clin. Endocr.,* **35,** (1972), 931.

9. S. S. C. Yen, R. Rebar, G. VandenBerg, F. Naftolin, Y. Ehara, K. J. Ryan, and R. Guillemin, in *Hypothalamic Hypophysiotropic Hormones: Clinical and Physiological Studies* (C. Gual and E. M. Rosemberg, Eds.), Excerpta Medica, In Press.

10. S. S. C. Yen and C. C. Tsai, *J. Clin. Invest.,* **50,** (1971), 1149.

11. E. E. Piacsek and J. Meites, *Endocrinology,* **79,** (1966), 432.

12. H. P. G. Schneider and S. M. McCann, *Endocrinology,* **87,** (1970), 330.

13. S. S. C. Yen and C. C. Tsai, *J. Clin. Endocr.,* **33,** (1971), 882.

14. N. B. Schwartz, *Rec. Progr. Hormone Res.,* **25,** (1969), 1.

15. L. Caligaris, J. J. Astrada, and S. Taleisnik, *Endocrinology,* **88,** (1971), 810.

16. J. R. Goding, K. J. Catt, J. M. Brown, C. C. Kaltenbach, I. L. Cumming, and B. J. Mole, *Endocrinolgoy,* **85,** (1969), 133.

17. R. J. Scaramuzzi, S. A. Tillson, I. H. Thorneycroft, and B. V. Caldwell, *Endocrinology,* **88,** (1971), 1184.

18. D. J. Dierschke, J. Hotchkiss, T. Yamaji, A. N. Bhattacharya, L. E. Atkinson, A. H. Surve, and E. Knobil, *Endocrinology,* **89,** (1971), 1034.

19. M. Ferin, P. E. Zimmering, and R. Vande Wiele, *Endocrinology,* **84,** (1969), 893.

20. M. Ferin, A. Tempone, P. E. Zimmering, and R. Vande Wiele, *Endocrinology,* **85,** (1969), 1070.

21. S. S. C. Yen and C. C. Tsai, *J. Clin. Endocr.,* **34,** (1972), 298.

22. C. C. Tsai and S. S. C. Yen, *J. Clin. Endocr.,* **33,** (1971), 917.

23. G. E. Abraham, W. D. Odell, R. S. Swerdloff, and K. Hopper, *J. Clin. Endocr.,* **34,** (1972), 312.

24. E. M. Bogdanove, *Vitamins Hormones, 2,* (1964), 205.

25. J. M. Malacara, E. Seyler, Jr., and S. Reichlin, *J. Clin. Endocr.,* **34,** (1972), 271.

5 Secondary Rhythms Related to the Menstrual Cycle: Biochemical and Physiological Changes

MICHAEL H. SMOLENSKY, *Moderator*

CHAPTER 15

Circatrigintan Secondary Rhythms Related to Hormonal Changes in the Menstrual Cycle: General Considerations

ALAIN REINBERG

Equipe de Recherche de Chronobiologie Humaine (CNRS N0. 105) and Fondation A. de Rothschild, Laboratoire de Physiologie, 29 Rue Manin, Paris, France

MICHAEL H. SMOLENSKY

School of Public Health, University of Texas, Houston, Texas

The menstrual cycle is undoubtedly one of the most studied and earliest observed low frequency rhythms known to mankind. Nevertheless, it was not until the end of the nineteenth century that definite scientific interest in rhythms secondary to the reproductive function in females began. With the development of improved experimental procedures during the twentieth century for the documentation of female hypothalamic, pituitary, and gonadal hormone rhythms, and because of concern about population stabilization and alteration of re-

241

productive functions due to contraceptive medication, investigation of rhythmic physiologic and biochemical activities dependent on qualitative and quantitative changes in gonadal hormones commenced with renewed vigor.*

Results reported during the past decade deal with a multitude of biochemical and physiologic variables. Such work has been carried out at various levels of organization ranging from the level of the system to that of the cell. Using classical (nonrhythmometric) methods of analysis, reports were published suggesting rhythms in a large number of biological activities including neurological (4–8), cardiovascular (9–11), and respiratory (12, 13) as well as enzymatic (14–21) and hormonal (11, 22–30).

The purpose of this work is to summarize, by means of appropriate statistical analyses (31, 32), circatrigintan (about 30-day) changes in biological function secondary to alterations of hormonal secretions (mainly but not exclusively gonadal hormones) in healthy regularly menstruating women. Although rhythms of other frequencies such as ultradian (high frequency), circadian (about 24-hour), and infradian (lower frequency non-30-day) are of equal importance for delineating relationships at various levels of organization as well as understanding reproductive regulatory mechanisms, it will not be possible herein to treat these latter aspects adequately. Thus, this endeavor is limited to a review of studies published, for the most part, during the last decade and which deal with those functions thought to be circatrigintan.

DATA ACQUISITION

For exact details of the methodology used in collecting data for this review, the reader is referred to the original articles cited. Deficiencies in experimental techniques as well as unknown factors related to data collection by various investigators are major concerns in interpreting previously reported data and must not be ignored. For the purposes of this review, these known or unknown deficiencies have been classified as "noise." Noise, as used here, represents effects of any and all chance influences—chance being defined as the sum total of sources of variation due to unknown or uncontrolled factors.

One example of difficulties encountered in reviewing the multitude

---

* Let us remember that rhythms of about 30 days have been reported and described not only for adult women and males but also for plants and invertebrates (Palolo worm and sea urchin) in addition to vertebrates other than primates. In general, about 30-day rhythms appear to be rather commonplace; they probably reflect hormonal, neural, and other influences (1–3).

of work on menstrual rhythms is the interpretation of what on the surface appear to be representative findings. In some investigations
measurements obtained from various women, each studied daily, independent of clock hour and on different days of the menstrual cycle, are
combined into a table or a single graph to demonstrate a 28-day group
rhythmicity. In order for the data arising from such studies to be accepted as valid, it must be assumed that all women comprising the
sample can be considered as having an equal menstrual cycle (as well
as menstruation, follicular, and luteal stage) duration. Moreover, in
such studies it is assumed that the base line (circatrigintan rhythm
adjusted time series mean) and amplitude (rhythmic within period
differences) are nonvarying and comparable between women. Individual variation in physiology is the rule rather than the exception,
and, at least moderate variation in such values between women is
expected.

Still another problem in interpreting previously reported data is the
failure in some studies to relate, by means of cytological, hormonal, or
body temperature determinations, menstrual day to menstrual stage.
Lack of such information complicates the interpretation of data from
different subjects each of whom has a different menstrual cycle length.
The same cycle day need not reflect identical circatrigintan-system
phases in different women. Thus, reliance on external time markers,
such as calendar date or menstrual day rather than internal ones,
can complicate and confound rhythm quantification and in some cases
its detection as a group phenomenon as well.

Another important aspect of interpreting findings is the experimental
design used to gather data. In most cases investigators, in researching
circatrigintan rhythms, neglect the possible influence of other periodicities, especially 24-hour (circadian) rhythms, which may be
characterized by larger amplitudes than the 30-day rhythm. Thus, by
sampling once daily at random clock hours—or for that matter even at
the same time of day in subjects adhering to nonstandardized societal
routines (as in shiftwork or transmeridian flight, for example),
collected data may reflect, to a greater degree, circadian rather than
circatrigintan rhythmicity. As an example, one may cite the rhythm of
cutaneous sensitivity of skin to histamine. The circadian amplitude of
this rhythm is greater in magnitude than that of the circatrigintan (33).
Similarly, the circadian amplitude of body temperature is greater than
the circatrigintan amplitude (34).

Moreover, it is necessary to consider the possibility of circatrigintan
(menstrual) modulation of either one or all the circadian (or other)
rhythm variables besides the rhythm adjusted time series mean (M).

Menstrual modulation of amplitude (within period rhythmic variation, A) acrophase (the peak of the rhythm approximated by curve fitting, ϕ) and period (τ) can occur. Evidence for changes in A, ϕ, and/or possibly τ comes from two recently completed studies—one on circatrigintan rhythms in cutaneous reactivity to histamine (33) and the other on pain threshold to radiant heat (35). When menstrual modulation of the circadian φ and/or τ occurs, sampling once daily may not only fail to reveal such events, but may lead to erroneous findings (33). In such cases, sampling once daily may reveal a biphasic rhythm; however, more frequent sampling would reveal a monophasic waveform with a change in circadian φ during the menstrual cycle (33).

Thus, in order to more thoroughly study circatrigintan rhythms, it is strongly suggested, when appropriate, that investigators utilize a rhythmometric approach including the utilization of profiles (36) of 24-hour duration scheduled at least at weekly intervals throughout the menstrual cycle and microscopic techniques of analysis (31, 32).

DATA PROCESSING

Values found in the literature were first coded on punch cards for subsequent analysis by cosinor and related methods (31, 32). Detection of circatrigintan rhythms was explored by fitting cosine curves of appropriate periods to individual time series by the method of least squares (31). If rhythms were demonstrable, an objective quantification and description of their characteristics were attempted by the same curve fitting and, when possible, by added steps (32).

A rhythm detected ($p < .05$) by curve fitting may be described by several endpoints along with 95% confidence intervals. One of these is a rhythm-adjusted level, M. The amplitude A, another characteristic of the rhythm, is a measure of the extent of change predictable by the cosine approximation used. Herein, the A obtained by the fit of a 30-day cosine curve approximates the extent of rhythmic within menstrual change. The computative acrophase ϕ is an index locating in time the peak of the cosine function found to best approximate the data. Detection and quantification of circatrigintan rhythms were carried out by fitting a set of cosine curves with periods in the region of 30.0 ± 7.0 days to time series, covering spans of about 30 days or longer, by the method of least squares. In these analyses the interval between consecutive trial periods was selected linear in period and equated to 24 hours. Upon completion of such spectral windows, the parameters M, A, and ϕ (and corresponding dispersion indices) were examined at the 30-day period.

When only a single time series was available, the ratio between the standard error (SE) of the A and the A itself (the SE/A) was used for tentative rhythm detection. Thus, one tests the rhythm by the statistical significance of its amplitude (37).

When three or more comparable time series were available, A and ϕ values from least squares spectral analysis were tested by cosinor (32). Cosinor analysis detects a rhythm if, at a given test frequency such as 1 cycle/30 days, separate time series constituting a sample exhibit similar acrophases and amplitudes. A rhythm substantiated at an acceptable level of statistical significance ($p < .05$) is described by a sample average level, amplitude, and acrophase with their respective 95% confidence limits.

In this paper acrophases are expressed by convention in degrees as negative values or delays (31) from an arbitrarily chosen phase reference. The period τ of a given single cosine curve, serving to approximate circatrigintan rhythms, equals 360°. Thus when $\tau = 30$ days $= 360°$, each 12° is equivalent to one menstrual (calendar) day.

In order to compare acrophases of different biochemical and physiologic variables and samples, the acrophase has been referenced to local midnight of day 1 (menses) of the menstrual cycle in each case. Ideally it is desirable to reference the acrophase of one rhythmic function to that of another studied concomitantly. However, since in many studies data on only one rhythmic function were given, local midnight of day 1 serves as the phase reference for the acrophases. In this paper, only the circatrigintan acrophases and their 95% confidence arcs are presented so as to provide a mapping of temporal integration of various biochemical and physiologic functions during the menstrual cycle. For this reason amplitude values are not included. However, since amplitude values are of crucial importance in interpreting the magnitude of circatrigintan (menstrual) variation for individual variables, these will be included in a subsequent paper.

RESULTS

Results of least squares spectral and cosinor analyses are presented in Figures 1 to 4. For this review, variables have been categorized according to the most pertinent physiologic system. In addition to physiologic system and variable, included in each figure is the name of the first author of each work and the timing of the circatrigintan acrophase (the temporal occurrence of the rhythm's crest approximated by the fit of a 30-day cosine curve) as well as its 95% confidence interval. Additional information pertaining to the number of women and cycles

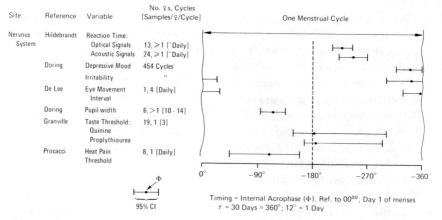

Figure 1 Summary of the acrophase (the temporal peak of the 30-day cosine curve approximating time series data) and the 95% confidence interval for various menstrual rhythms of the nervous system.

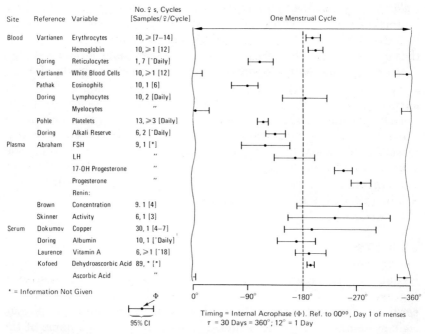

Figure 2 Summary of the acrophase (the temporal peak of the 30-day cosine curve approximating time series data) and the 95% confidence interval for various menstrual rhythms for aspects of blood, plasma, and serum.

246

studied as well as the average number of determinations per woman per cycle is also indicated for each variable, when available. In this review only statistically significant ($p < .05$) rhythms are listed in the figures. The values for the time series mean (M) or amplitude (A) are not included for each variable.

Figure 1 summarizes circatrigintan variations in neural function. As shown, statistically significant rhythms are documented for reaction time (4), mood (5), irritability (5), threshold to taste and pain (6, 35, 38), as well as eye movement interval (7) and pupil width (8). As might be expected, the highest ratings for depressive mood state as well as irritability occurs premenstrually. While the greatest taste threshold to

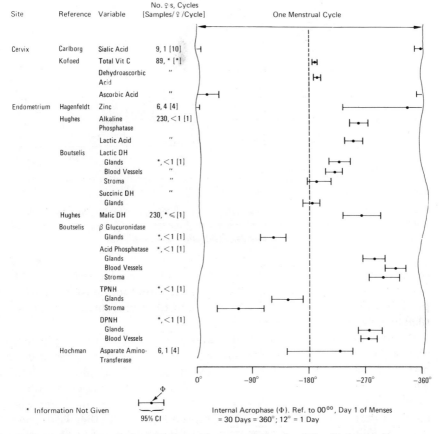

Figure 3 Summary of the acrophase (the temporal peak of the 30-day cosine curve approximating time series data) for several menstrual rhythms for different aspects of cervical and endometrial physiology.

FEMALE CIRCATRIGINTAN (MENSTRUAL) SYSTEM

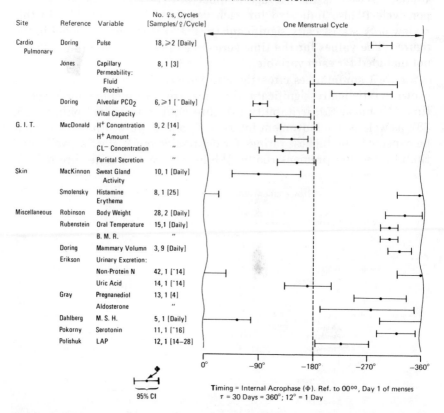

Site	Reference	Variable	No. ♀s, Cycles [Samples/♀/Cycle]	One Menstrual Cycle
Cardio Pulmonary	Doring	Pulse	18, ⩾2 [Daily]	
	Jones	Capillary Permeability: Fluid Protein	8, 1 [3]	
	Doring	Alveolar PCO_2	6, ⩾1 [˜Daily]	
		Vital Capacity	"	
G. I. T.	MacDonald	H^+ Concentration	9, 2 [14]	
		H^+ Amount	"	
		CL^- Concentration	"	
		Parietal Secretion	"	
Skin	MacKinnon	Sweat Gland Activity	10, 1 [Daily]	
	Smolensky	Histamine Erythema	8, 1 [25]	
Miscellaneous	Robinson	Body Weight	28, 2 [Daily]	
	Rubenstein	Oral Temperature	15,1 [Daily]	
		B. M. R.	"	
	Doring	Mammary Volumn	3, 9 [Daily]	
	Erikson	Urinary Excretion:		
		Non-Protein N	42, 1 [˜14]	
		Uric Acid	14, 1 [˜14]	
	Gray	Pregnanediol	13, 1 [4]	
		Aldosterone	"	
	Dahlberg	M. S. H.	5, 1 [Daily]	
	Pokorny	Serotonin	11, 1 [˜16]	
	Polishuk	LAP	12, 1 [14–28]	

0°   −90°   −180°   −270°   −360°

95% CI

Timing = Internal Acrophase (Φ). Ref. to 00⁰⁰, Day 1 of menses
τ = 30 Days = 360°; 12° = 1 Day

Figure 4   Summary of the acrophase (the temporal peak of the 30-day cosine curve approximating time series data) for a number of menstrual rhythms including those of the cardiovascular, respiratory, and gastrointestinal systems.

quinine and proplythiourea occurs around midcycle, highest threshold for thermally induced pain occurs earlier during the menstrual cycle, prior to midcycle. Increased reaction time to either optical or acoustical signals occurs after midcycle.

Figure 2 summarizes menstrual studies carried out on blood, plasma, and serum. Circatrigintan rhythms are verified for most formed elements of the blood, including erthrocytes (39, 40), reticulocytes (39), white blood cells (40), eosinophils (41, 42), lymphocytes (40), myelocytes (43), and platelets (44–47). Also presented in this figure are circatrigintan acrophases for certain reproductive and other hormones (22–25). Follicle-stimulating hormone (FSH) and luteinizing hormone (LH) reach greatest concentration before ovulation; as expected, both

17-OH progesterone and progesterone (22) peak later, after ovulation. The concentration as well as activity of renin (23–25) also peak following ovulation; the acrophases of these rhythms correspond quite closely in time to that for progesterone. Figure 2 also shows circatrigintan rhythms for serum vitamin A (48, 49) as well as dehydroascorbic and ascorbic acid concentrations (14)*.

Figure 3 maps circatrigintan rhythms of the cervix and endometrium (14–21, 50, 51). For the most part these variables reflect biochemical activities of the endometrium and seemingly are due to the influence on enzyme systems of reproductory hormones. As is readily apparent, most of the acrophases occur during the later half of the cycle. A few variables such as β-glucuronidase, triphosphopyridine nucleotide as (TPNH), (17), and ascorbic acid (14) reach their highest concentrations, as indicated by the circatrigintan acrophase, during the first half of the menstrual cycle.

Figure 4 reports circatrigintan acrophases for cardiopulmonary, gastrointestinal, cutaneous, as well as for several miscellaneous physiologic and biochemical variables. With regard to cardiovascular functions, highest pulse rate (9) and greatest capillary permeability to both fluid and protein (10) occur after ovulation at around the same time during the menstrual cycle. For the respiratory functions of alveolar P_{CO_2} (12) and vital capacity (13), circatrigintan acrophases occur prior to ovulation. The gastrointestinal variables of HCl concentration and parietal secretion (52) increase to greatest values just before ovulation. Also mapped in Figure 4 are circatrigintan acrophases for palmar (digital) sweat gland activity (53) and cutaneous reactivity to histamine (34). Other circatrigintan rhythmic functions documented include oral temperature and basal metabolic rate (54), as well as mammary volume (55). In addition, circatrigintan rhythms were found for the excretion of certain metabolic entities: nonprotein nitrogen (56), uric acid (56), pregnanediol (57), aldosterone (57, 58), melanocyte stimulating hormone (59), and serotonin (60, 61) among others (26, 27, 62).

DISCUSSION

The purpose of this review is to objectively analyze data, originally published in tables or graphs, by rhythmometric techniques of least

---

* Since this paper was presented the study summarized in Table 1 has been completed. Interalia statistically significant circatrigintan (menstrual) rhythms were detected in urinary excretion of 17-hydroxycorticosteroids (φ = day 19), pregnandiol (φ = day 20) and total estrogens (φ = day 19).

squares spectral and cosinor analyses (31, 32). For the most part this review encompasses those reports appearing in the literature during the last 10 years. For this span of time a great many reports are available pertaining to menstrual variation for a number of different biochemical and physiologic functions. Herein time series data from such reports have been reanalyzed so as to enable a so-called mapping, by circatrigintan acrophase, of changes in biologic processes secondary to both quantitative and qualitative changes in hormones during the menstrual cycle.

Figures 1–4 document circatrigintan rhythmicity for a great number of functions. While substantiation of rhythmicity by objective statistical techniques serves to validate previous suspected temporal variations analyzed by other methods, mapping of functions listed in these figures serves additionally to provide a pictorial overview in a more exact manner of the temporal organization of biochemical and physiologic activities subservient or secondary to reproductive functions.

For example, premenstrual tension is often characterized by irritability, weight increase, painful breasts, and gastrointestinal upset as well as general water retention. The circatrigintan maps indicate that some of these symptoms correspond temporally to an increased psychological susceptibility to irritability and depressive mood. The premenstrual syndrome also corresponds to the temporal occurrence of increased mammary volume as well as body weight gain apparently due to water retention. These latter changes seemingly reflect increased renin concentration and activity, elevated levels of aldosterone (as determined by urinary excretion rates) and progesterone, and heightened capillary permeability to both fluid and protein. Similarly, gastrointestinal upset apparently is due, at least in part, to change in HCl and parietal secretions.

Inspection of Figure 3 shows certain circatrigintan rhythmic changes in the biochemical and enzymatic complement of the cervix and endometrium apparently reflect changes due to reproductive hormones. It has long been recognized that estrogen serves to induce enzyme and coenzyme systems subservient to the metabolic processes related to proliferation of the endometrium; on the other hand progesterone serves to promote secretion of the endometrium. The endometrium metabolizes carbohydrates, lipids, and proteins not only for the nutrition of an expected implanting fertilized ovum, but also for inducing the increased rate of energy production needed for dynamic growth of endometrial tissue thereafter.

An increase in the level of energy production can take place by three metabolic pathways: the Embden-Meyerhof pathway (anaerobic metabolism), the Krebs citric acid cycle, and the hexose monophosphate cycle. As indicated in Figure 3, anaerobic glycolytic processes are greatest, as measured by concentration of lactic dehydrogenase, just after ovulation. Similarly, biochemical processes of the Krebs citric acid cycle show the greatest activity, as measured by peak concentrations of succinic and malic dehydrogenase, at the same time of the menstrual cycle.

Other enzymes, including β-glucuronidase, alkaline and acid phosphatase as well as TPNH and (DPNH), are subservient to basic metabolic processes of the endometrium. β-Glucuronidase is present in highest concentrations during the proliferative stage of the menstrual cycle. It has been suggested that this enzyme functions in cellular proliferation, conjugation of steroids, and hydrolysis of conjugated glucuronides (28). It is therefore implied that estrogenic effects upon the endometrium are associated with an increased β-glucuronidase concentration which in turn influences endometrial growth, estrogen detoxification, and carbohydrate storage (28).

Alkaline phosphatase is present in increased amounts during endometrial growth and in lesser amounts during endometrial differentiation. It is believed that alkaline phosphatase facilitates hydrolysis of phosphoric esters which participate in the intermediary metabolism of nucleoproteins, neutral fats, and glycogen (29). On the other hand, acid phosphatase is predominantly active during the progestational phase of the menstrual cycle. It has been suggested that this enzyme plays a significant role in contributing to a suitable hydrogen ion concentration favorable for survival of sperm and ovum (30). Rhythmic changes in the concentrations of DPNH and TPNH indicate that the former reaches its peak concentration during the secretory phase of the menstrual cycle; the latter attains peak concentration during the proliferative phase. Primarily, TPNH and DPNH play a role in various dehydrogenase reactions and serve as substrates for other enzyme systems.

As indicated in the introduction, experimental designs used to study menstrual rhythms in biochemical and physiologic function differ from one study to another. Table 1 serves to summarize, in a general manner, the experimental designs most commonly utilized to investigate such rhythms. Based upon the studies encountered in developing this review, it was found, as indicated in the table, that there are roughly seven (A–G) classifications of experimental designs. In study type A, different subjects were investigated on different days of the menstrual cycle. In most cases, only a single determination was obtained from each subject. The data from such studies were analyzed macroscopi-

Table 1 *Summary of Circatrigintan (Menstrual) Rhythm Study Designs*

Type of Study	No. Cycles Studied	No. Days Studied per Subject	$\bar{X}$ No. Determinations per Day	Subjects Studied Throughout		Reported Results
				Same	Different	
A	<1	1	1	+	+	Macroscopic as daily $\bar{X}$'s
B	<1	2	1	+ or	+	Before/after ovulation
C	<1	2	1	+ or	+	Before/after menstruation
D	<1	3	1	+ or	+	Macroscopic as daily $\bar{X}$'s
E	≥1	~all	1	+		Macroscopic as daily $\bar{X}$'s
F	≥1	>3	3	+		Macroscopic as daily $\bar{X}$'s
G	≥1	≥5	5 to 6	+		Microscopic as C_0, C, ϕ
		(= intervals)	(= intervals)			

cally as daily means for an assumed menstrual cycle of 28 days duration.*

In study types B and C the same or different subjects were examined on only a couple of days, usually before or after ovulation or menstruation. In such studies it was not uncommon for subjects to be studied during different stages of the menstrual cycle; nonetheless, the results were still reported for comparison of function before or after ovulation or menstruation.

In study type D the same or different subjects were studied on at least 3 days of the menstrual cycle. Usually only one determination was carried out per subject per day in such studies, and usually only one menstrual cycle was followed. The data were usually reported as daily means in tabular or graphic form. Study type E is characterized by the investigation of the same subjects at about daily intervals for at least one menstrual cycle. The results of these studies were summarized as group daily means for an idealized 28-day menstrual cycle duration.

Study type F is defined by the investigation of the same subjects at least three times daily on at least three or more days of at least one menstrual cycle. Such studies, characterized by more than two determinations per day from each subject, are not as common as the aforementioned ones. Like a majority of the previously discussed designs, results of these studies were also tabulated as daily means for an assumed 28-day menstrual cycle.

Finally, study type G is characterized by a study of the same subjects at 4-hour intervals or less during transverse 24-hour profiles scheduled every 7 (or less) days during at least one menstrual cycle. This preferred experimental design is suitable for a study of the majority of menstrual rhythms in biochemical and physiologic variables secondary to hormone changes and allows an analysis of the data by objective, statistical techniques such as least squares spectral and cosinor analyses (31, 32). These techniques enable the description of circatrigintan rhythmic phenomena in terms of a time series mean (M), amplitude (A), and acrophase (ϕ). Moreover, with the use of scheduled transverse profiles of 24-hours duration the investigator may gain additional insight into temporal structure by analyzing each of the 24-hour profiles for menstrual modulations of circadian M, A, and ϕ values. Thus, such an experimental design allows the investigator to analyze data not only as a single time series for circatrigintan

* The term macroscopic refers to those statistical techniques which serve to describe rhythmic phenomena by central tendency parameters. The term microscopic refers to analysis of time series in terms of rhythm characteristics: $M, A, \phi,$ and τ.

Table 2 *Circatrigintan Rhythms Studied by Successive 24-Hour Transverse Profiles*[a]

Physiologic Variable (units)	Rhythm Detection, p Value	Rhythm Adjusted Level M ± 1 SE	Amplitude A	Acrophase φ (days)[b]	Acrophase φ (degrees, 12° = 1 day)[b]
			(× 95% Confidence Interval)		
Skin reaction to histamine (erythema, cm²)	<.01	26.2 ± 3.0	3.8 (2.0 to 5.7)	29 (25.3 to 2.8)	−348 (−303 to −34)
Rectal temperature (°C)	<.05	37.13 ± 0.03	0.31 (0.01 to 0.60)	24.5 (18.0 to 1.0)	−294 (−216 to −12)
Mood[c] (self-rating)	>.05	0.62 ± 0.07	—	5.8	−69
Physical vigor[c] (self-rating)	>.05	0.57 ± 0.08	—	10.0	−120
Urine volume (ml)	<.05	126 ± 9.0	28.3 (2.1 to 54.5)	21.1 (16.3 to 25.8)	−253 (−196 to −310)

[a] Subjects: five healthy women; 20 to 49 years of age; length of menstrual cycle = 28.2 ± 0.67 (days ± 1 SE). No drug (particularly no oral contraceptive). February–March 1972. Physiologic variables measured every 4 hours, during 24 hours, on day 1 (when menstruation started) and again on days 8, 15, 22, and 29. Circadian M for each subject/day/variable were computed giving rise to time series for this transverse study on circatrigintan rhythms.

[b] φ reference: 00⁰⁰ (midnight) beginning of menstruation.

[c] Values for A and confidence are for φ not given when rhythms not detected.

rhythmicity but, in addition, it allows for separate analysis of data collected during each of the 24-hour profiles for determination of possible circatrigintan modulation of the circadian component through inspection of the circadian *M*, *A*, and ϕ values for different days of the menstrual cycle. For the sake of increased accuracy and completeness month of year must also be considered since circannual changes in menstrual rhythms have been reported (1, 2, 31, 34). Table 2 presents findings from a study using study type G as outlined in Table 1. Data were obtained from five healthy normally menstruating women at 4-hour intervals during successive 24-hour study profiles at 7-day intervals. Circadian rhythm adjusted time series means (*M*) for each subject, day, and variable were determined by least squares spectral analysis (31), forming a time series for this study of circatrigintan rhythms.

The data indicate circatrigintan rhythms in the cutaneous reaction to histamine, rectal temperature, and urine volume. No rhythms for self-rating of mood and physical vigor were detected for this sample. This is apparently the first reporting of a menstrual rhythm for urine volume. Findings for histamine sensitivity of skin will be reported in greater detail later in these proceedings. (33). For this rhythm, evidence of menstrual modulation in the circadian amplitude and acrophase values were obtained by utilization of transverse 24-hour profiles and rhythmometric methods of data analysis. With the present availability of objective statistical techniques to characterize circatrigintan (menstrual) rhythms it is hoped that future investigations will utilize a rhythmometric experimental design such as that suggested here (study type G).

SUMMARY

Data on biochemical and physiologic variables reported over the last decade were objectively analyzed for circatrigintan rhythmicity by the objective techniques of least squares spectral and cosinor analyses. The acrophases for rhythms validated by such techniques were graphically presented so as to form a circatrigintan (menstrual) mapping of temporal integration of biochemical and physiologic function.

A review of the type of study designs used to investigate menstrual rhythms indicates deficiencies. When appropriate it is most desirable that future studies of biochemical and physiologic circatrigintan rhythms be carried out using transverse profiles of 24-hour duration during which individual determinations be performed at least at 4-hour intervals. It is suggested that the spans between successive transverse profiles, depending upon the variable and the nature of the study, be no more than 7 days. Furthermore, it is hoped that analyses of data generated by such experimental designs will be carried out by objective techniques such as least squares spectral and cosinor techniques in order to describe rhythms in terms of rhythmometric variables—M, A, and ϕ.

REFERENCES

1. A. Reinberg and J. Ghata, *Biological Rhythms*, Walker, New York, 1964, 138 pp.

2. F. Halberg, *Ann. Rev. Physiol.*, **31**, (1969), 675.

3. E. Bünning, *The Physiological Clock*, Springer, Berlin, 1964, 145 pp.

4. G. Hildebrandt and W. Witzenrath, *Int. Z. Angew. Physiol.*, **27**, (1966), 266.

5. K. G. Doring, *Fifth Symposium of German Gesellschaft for Endocrinology*, Springer, Berlin, 1957, pp. 42–44.

6. E. V. Glanville and A. R. Kaplan, *Nature*, **205**, (1965), 930.

7. C. DeLee, Department of Development Neurology, Berchern-Antwerp, Belgium, Unpublished.

8. K. G. Doring and E. Schaefers, *Arch. Gynaekol.*, **179**, (1951), 585.

9. K. G. Doring and E. Feustel, *Klin. Woehschr.*, **31**, (1953), 1000.

10. E. M. Jones, R. H. Fox, P. W. Verow, and A. W. Asscher, *J. Obstet. Gynaecol. Brit. Commonwealth*, **73**, (1966), 666.

11. A. M. McCausland, F. Holmes, and A. D. Trotter, Jr., *Amer. J. Obstet. Gynecol.*, **86**, (1963), 640.

12. R. L. Goodland and W. T. Pommerenke, *Fertility Sterility*, **3**, (1952), 394.

13. G. K. Doring and G. Weber, *Arch. Gynaekol.*, **179**, (1951), 442.

14. J. A. Kofoed, N. Blumenkrantz, A. B. Houssay, and E. Y. Yamauchi, *Amer. J. Obstet. Gynecol.*, **91**, (1965), 95.

15. R. J. Stein and V. M. Stuermer, *Amer. J. Obstet. Gynecol.*, **61**, (1951), 414.

16. R. J. Jarrett and H. J. Graver, *Brit. Med. J.*, **2**, (1968), 528.

17. J. G. Boutselis, J. C. DeNeef, J. C. Ullery, and O. T. George, *Obstet. Gynecol.*, **21**, (1963), 424.

18. K. S. Moghissi and O. W. Neuhaus, *Amer. J. Obstet. Gynecol.*, **96**, (1966), 91.

19. J. Hochman and J. G. Schenker, *J. Clin. Endocr.*, **29**, (1969), 1120.

20. H. Schmidt, P. Berle, and K. D. Voigt, *Acta Endocr.*, **61**, (1969), 729.

21. E. C. Hughes, R. D. Jacobs, A. Rubulis, and R. M. Husney, *Amer. J. Obstet Gynecol.*, **85**, (1963), 594.

22. G. E. Abraham, W. D. Odell, R. S. Swerdloff, and K. Hopper, *J. Clin. Endocr.*, **34**, (1972), 312.

23. S. L. Skinner, E. R. Lumbers, and E. M. Symonds, *J. Clin. Sci.*, **36**, (1969), 67.

24. J. J. Brown, D. L. Davies, A. F. Lever, and J. I. S. Robertson, *Brit. Med. J.*, **2**, (1964), 1114.

25. G. Cession, *Bull. Soc. Roy. Belg. Gynecol. Obstet.,* **37,** (1967), 287.

26. M. F. Robinson and P. E. Watson, *Brit. J. Nutr.,* **19,** (1965), 225.

27. L. J. Golub, H. Menduke, S. S. Conly, *Amer. J. Obstet. Gynecol.,* **91,** (1965), 89.

28. L. D. Odell and W. H. Fishman, *Amer. J. Obstet. Gynecol.,* **59,** (1950), 290.

29. D. G. McKay, A. T. Hertig, W. A. Bardaivil, and J. T. Velardo, *Obstet. Gynecol.,* **8,** (1956), 22.

30. B. Goldberg and H. W. Jones, Jr., *Obstet. Gynecol.,* **4,** (1954), 426.

31. F. Halberg, M. Engeli, C. Hamburger, and D. Hillman, *Acta Endocr. Suppl.,* **103,** (1965), 1.

32. F. Halberg, Y. L. Tong, and E. A. Johnson, in *The Cellular Aspects of Bio-rhythms* (H. Von Mayersback, Ed.), Springer, Berlin, 1967, p. 20.

33. M. H. Smolensky, A. Reinberg, R. E. Lee, and J. P. McGovern, in *Biorhythms and Human Reproduction* (M. Ferin, F., Halberg, R. M. Richart, and R. L. Vande Wiele, Wiley-Interscience, New York, 1973, Chap. 18.

34. A. Reinberg, F. Halberg, J. Ghata, and M. Siffre, *C. R. Acad. Sci. (Paris)* **262** (1966), 782.

35. P. Proccaci, G. Buzzelli, I. Passeri, R. Sassi, M. R. Volgelin, and M. Zoppi, *Res. Clin. Stud. Headache.,* **3,** (1972), 260.

36. K. Reindl, C. Falliers, F. Halberg, H. Chai, D. Hillman, and W. Nelson, *Rass. Neur. Veg.,* **23,** (1969), 5.

37. F. Halberg, E. A. Johnson, W. Nelson, W. Runge, and R. Sothern, *Physiol. Teacher,* **1,** 4 (1972), 1.

38. T. Okazaki, *Acta Dermatol. (Kyoto),* **57,** (1962), 131.

39. G. K. Doring, *Pfugers Arch.,* **252,** (1950), 292.

40. E., Vartiainen and H. Zilliacus, *Ann. Cher. Gynaecol., Fenn.,* **57,** (1968), 287.

41. C. L. Pathak and B. S. Kahali, *J. Clin. Endocr.,* **17,** (1957), 862.

42. S. I. Dokumov, M. S. Stancheva, and S. A. Spasov, *Pol. Endocr.,* **22,** (1971), 96.

43. G. K. Doring and E. Feustel, *Arch. Gynaekol.,* **184,** (1954), 552.

44. F. J. Pohle, *Amer. J. Med. Sci.,* **197,** (1939), 40.

45. H. Pepper and S. Lindsay, *Obstet. Gynecol.,* **14,** (1959), 657.

46. J. A. McBride and C. A. Snodgrass, *J. Obstet. Gynaecol. Brit. Commonwealth,* **75,** (1968), 357.

47. J. M. Yeung, *Henry Ford Hosp. Med. J.,* **18,** (1970), 193.

48. P. A. Laurence and A. E. Sobel, *J. Clin. Endocr.,* **13,** (1953), 1192.

49. H. D. Here, Z. *Geburtshilfe Gynaekol.,* **160,** (1962), 240.

50. K. O. Hagenfeldt, L. O. Plantin, and E. Diczfalusy, *Acta Endocr.,* **65,** (1970), 541.

51. W. Carlborg, W. McCormick, and C. Gemzell, *Acta Endocr.,* **59,** (1968), 636.

52. I. MacDonald, *Gastroenterology,* **30,** (1956), 602.

53. P. C. B. MacKinnon, *J. Obstet. Gynaecol. Brit. Commonwealth,* **61,** (1954), 390.

54. B. B. Rubenstein, *Endocrinology,* **22,** (1938), 41.

55. G. K. Doring, *Arch. Gynaekol.,* **184,** (1953), 51.

56. S. E. Erikson and R. Okey, *J. Biol. Chem.,* **91,** (1931), 715.

57. M. J. Gray, K. S. Strausfeld, M. Watanage, E. A. H. Sims, and S. Solomon, *J. Clin. Endocr.,* **28,** (1968), 1269.

58. M. Reich, *Aust. Ann. Med.,* **11,** (1962), 41.
59. B. Dahlberg, *Ann. N. Y. Acad. Sci.,* **100,** (1963), 631.
60. J. Pokorny and W. Schmidt, *Z. Gynaekol.,* **84,** (1962), 169.
61. A. R. Fuchs, F. Fuchs, and S. G. Johnsen, *Int. J. Fertil.,* **9,** (1969), 139.
62. W. Z. Polishuk, H. Zuckerman, and Y. Diamant, *Fertility Sterility,* **19,** (1968), 901.

CHAPTER 16

Some Physiological and Biochemical Measurements Over the Menstrual Cycle

SHANNA H. FREEDMAN, SAVITRI RAMCHARAN, and ELIZABETH HOAG

Oral Contraceptive Drug Study, Kaiser-Permanente Medical Center, Walnut Creek, California

ALAN GOLDFIEN

Department of Medicine and Obstetrics and Gynecology, and the Cardiovascular Research Institute, University of California Medical Center, San Francisco, California

The data described in this report were originally collected in the course of an ongoing prospective study which was designed to assess non-contraceptive effects of oral contraceptive drugs (OCD) in approximately 17,000 women 18 to 54 years of age. A large number of physiological and biochemical variables were measured in oral contraceptive users and nonusers employing an automated multitest laboratory set up at the Kaiser-Permanente Medical Center in Walnut Creek, California. Although the study was not designed to collect data having to do with biorhythms, a number of physiological and biochemical measurements were found to be significantly related to time of day, day of year, and time since last menses. In fact, it was necessary to adjust for these "nuisance variables" in the analysis of OCD user-nonuser differences. It was also possible to use the data to study biological changes in relation to time since last menses, even though this had not been the primary aim of the project.

The design provided by this study was similar to type A as described by Reinberg and Smolensky (1), in which different subjects are investigated on different days of the menstrual cycle, only a single determination is obtained from each subject, and the results are reported as daily means.

Data were available not only for women with normal menstrual cycles, but also for those whose cycles were induced by oral contraceptive drugs. It was therefore possible to compare the findings for endogenous cycles with those for OCD cycles.

MATERIALS AND METHODS

Population and Sample

The women comprising the cohort for this study were members of the Kaiser Foundation Health Plan living in the Health Plan Walnut Creek area, a suburban community near San Francisco. They were predominantly white and middle class, and had been admitted into the Contraceptive Drug Study as a result of physician referral or self-referral for periodic health examinations during the period December 1968 through February 1972. For purposes of the present report, a sample of subjects was selected from this larger group according to the following criteria:

1. The only subjects included were those responding positively to the question on the medical history questionnaire: "Now and in the past year have your menstrual periods usually been regular (between 24 and 35 days)?"
2. Subjects over 50 years of age were excluded.
3. Subjects on estrogen therapy for noncontraceptive purposes were excluded.
4. Subjects should be neither pregnant nor less than 3 months postpartum.

The number of women satisfying these criteria was 8786. Past users and never users of OCD showed no differences for the variables to be presented here, and were combined to form a single nonuser group. This group accounted for just over one-half of the subjects and had a mean age of 35.4 years. The remaining patients had a mean age of 30.2 years. Ninety percent of them were users of combination type OCD (estrogen and progestin combined and ingested daily for 21 days), while 10%, or a total of 371, were users of the sequential type of OCD. Sequential users receive estrogen alone for the first 14 to 16 days of medication, and then estrogen combined with progestin for the next 5 or 6 days.

Many variables showed slight but statistically significant differences between OCD users and nonusers. We selected body weight, systolic and diastolic blood pressure, pulse rate, serum glucose, white blood cell count, red blood cell count, hemoglobin, and hematocrit for presentation.

Laboratory Methods

Weights were measured in pounds on a standard clinical scale, following a 4 to 9 hour fast, with the subject wearing only a hospital gown. Repeat measurements on random subjects were obtained twice daily and reproducibility was excellent. Monthly means ranged from 133.5 lb in June to 138.3 lb in November.

Blood pressure and pulse rate were recorded automatically on a Godart apparatus after the subjects had been in a supine position for about 5 minutes. They were taken on the left arm on a table designed to maintain a similar bend in the elbow for each subject.

Serum glucose was measured 1 hour after the ingestion of 75 grams of glucose sometime between 9:00 A.M. and 1:30 P.M. and after 4 or more hours of fasting for most of the subjects. Glucose determination was by means of the o-toluidine method.

White and red blood cell counts, hemoglobin, and hematocrit were measured electronically by means of a Coulter blood cell counter. Blood smears and differential cell counts were not made.

Data Analysis

In order to eliminate possible seasonal variation in measurements as well as fluctuations due to minor changes in technique (such as instrumentation, chemical reagents, or personnel), all data were adjusted so that the monthly mean value for each dependent variable remained constant over the 3 years for which data were available. The data were collected all within a 6-hour period—8:00 A.M. to 2:00 P.M.—and adjustment was not made for time of day.

The data were first examined by constructing plots of the mean values along with standard errors for each day of the time period representing days 4 through 30 since last menses, separately for each of 3 years. No adjustment was made for age. Peaks and troughs were found to occur consistently on the same cycle day in many instances. However, the results presented in this report are based on a three-point moving average of the combined 3-year mean.

The general hypothesis underlying the approach to the data might be stated as follows: for an individual,

$$Y = Y_0 + f(\theta) + X + Z + \epsilon + \delta,$$

where Y is the dependent variable of interest, Y_0 is the individual's mean level, θ is a measure of time (in days or radians), and X, Z are nuisance variables, for example, age, time of day, or month of year. Delta and epsilon are error terms, one representing the error in estimating θ and the other the error in measuring Y together with all "residual errors," namely, that which cannot be explained. These are usually assumed to be independent, jointly normal, and distributed about zero.

While polynomials and cosines were fitted by multiple regression analysis, they were not very informative. Even those terms which were of statistical significance accounted for only a small portion of the total variation because of the large between-subject variation which could not be removed. Therefore it was found more useful to compare daily means by Tukey's method of multiple comparisons (2). The maximum and minimum means were tested for significance and, in those cases which were significantly different, other contrasts were tested.

RESULTS

Weight

Figure 1a shows a three-point moving average of weight in pounds according to number of days since onset of last menses for OCD nonusers (normally menstruating women) and for women with cycles induced by combination type OCD. Nonusers showed an increase of about 2.5 lb in the first part of the cycles, followed by a drop of 3.5 lb in the second phase, with a gain of about 4 lb in the premenstruum.* This pattern appeared in each of the 3 years for which data were available (not shown). As depicted in Figure 1a, combination OCD users showed a slight but continuous rise over the cycle from a low of 132 lb to a high of 135 lb.† The pattern for sequential users, as shown in Figure 1b, was characterized by an increase of 3 lb in the early part of the cycle followed by a decrease of about 4 lb in the second phase, with a subsequent premenstrual rise.

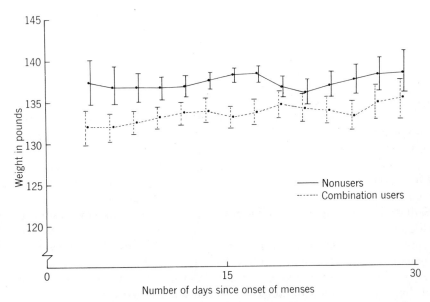

*Figure 1*a. *Mean weight (with 95% confidence intervals) of combination users and non users of oral contraceptive drugs by number of days since onset of menses.*

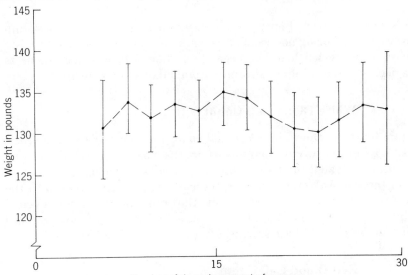

*Figure 1*b. *Mean weight (with 95% confidence intervals) of sequential users of oral contraceptive drugs by number of days since onset of menses.*

---

* Each point on the curve is an average of at least 636 subjects with a standard error of 1.1 lb at most.

† Standard error of the means for this group was at most 0.9 lb. The difference between the minimum and maximum values was significant (studentized range).

Blood Pressure and Pulse Rate

The systolic blood pressure for nonusers, as shown in Figure 2*a*, rose by about 1.5 torr at the time of midcycle, and declined steadily thereafter. The curve for the OCD users suggests a continuing rise into the premenstruum. That for the sequential users, as shown in Figure 2*b*, rose in the first part of the cycle and declined in the second phase.

Among OCD nonusers there was a gradual decline in the diastolic pressure during the latter part of the cycle* (see Figure 3*a*). A very slight decrease for combination users could be observed over the cycle. Sequential users showed a decrease in the second half of the cycle (Figure 3*b*).

As shown in Figure 4*a*, a slight rise in pulse rate can be seen over the menstrual cycle both in combination users and nonusers. Sequential users (Figure 4*b*) had a decline from 72.3 beats/minute in the first part of their cycle to a low of 70.1 beats/minute in the second part.

Serum Glucose

As is apparent from Figure 5*a*, there was a steady increase in serum glucose levels among nonusers and combination users until the last 5 to 8 days of the cycle when they began to decline. Among sequential users the decline occurred more abruptly around day 20 (Figure 5*b*).*

White Blood Cell Count

Figure 6*a* shows the white blood cell count for nonusers rising over the first half of the cycle from a low in the postmenstrual phase and remaining elevated throughout the second phase.* Among combination users the rise was more rapid in the early part of the cycle, with the level being sustained throughout the remainder of the cycle.* Sequential users showed a fall during the first phase of their cycle and a rise in the second phase starting after day 18 (Figure 6*b*).

Red Blood Cell Count

In Figure 7*a* a steady increase in the red blood cell count over the cycle for nonusers is apparent, starting from a low in the postmenstrual phase.* The combination users showed a steady decline in red blood cell count throughout the cycle, with a rise in the premenstrual phase. The curve for sequential users showed an abrupt decline around day 18 followed by a rapid rise (Figure 7*b*).

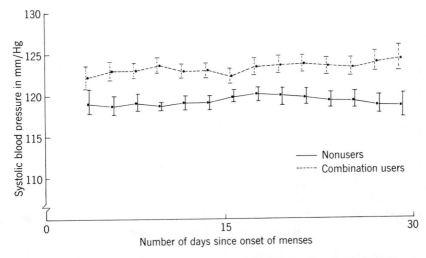

Figure 2a. *Mean systolic blood pressure (with 95% confidence intervals) of combination users and nonusers of oral contraceptive drugs by number of days since onset of menses.*

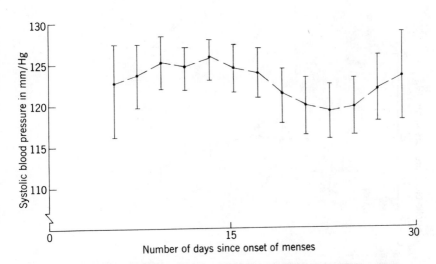

Figure 2b. *Mean systolic blood pressure (with 95% confidence intervals) of sequential users of oral contraceptive drugs by number of days since onset of menses.*

---

* The difference between the maximum and minimum values for this group was significant.

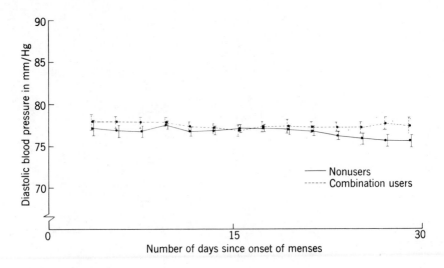

Figure 3a. Mean diastolic blood pressure (with 95% confidence intervals) of combination users and nonusers of oral contraceptive drugs by number of days since onset of menses.

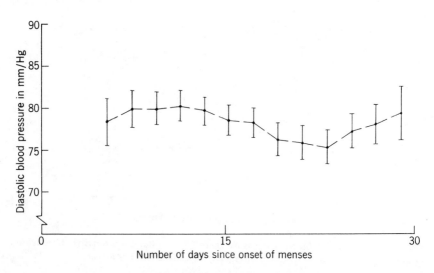

Figure 3b. Mean diastolic blood pressure (with 95% confidence intervals) of sequential users of oral contraceptive drugs by number of days since onset of menses.

266

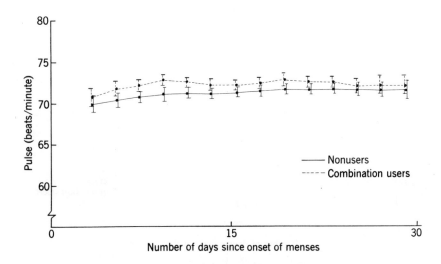

Figure 4a. **Mean pulse rate (with 95% confidence intervals) of combination users and nonusers of oral contraceptive drugs by number of days since onset of menses.**

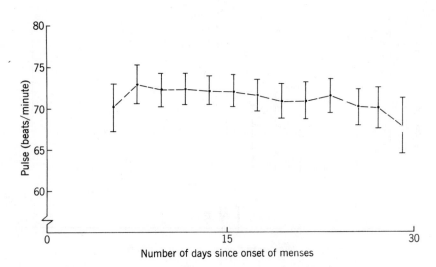

Figure 4b. **Mean pulse rate (with 95% confidence intervals) of sequential users of oral contraceptive drugs by number of days since onset of menses.**

267

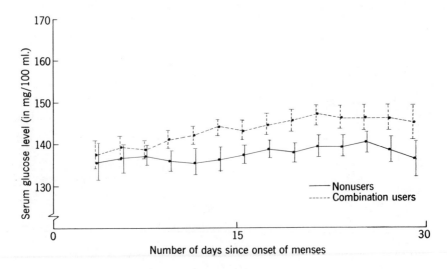

Figure 5a. *Mean serum glucose level (with 95% confidence intervals) of combination users and nonusers of oral contraceptive drugs by number of days since onset of menses.*

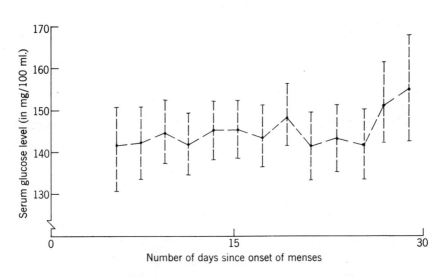

Figure 5b. *Mean serum glucose level (with 95% confidence intervals) of sequential users of oral contraceptive drugs by number of days since onset of menses.*

268

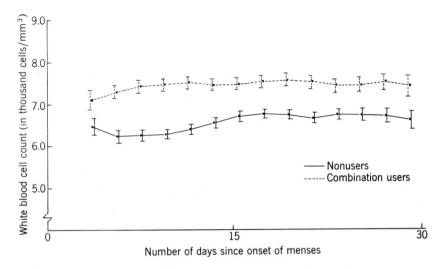

Figure 6a. *Mean white blood cell count (with 95% confidence interval) of combination users and nonusers of oral contraceptive drugs by number of days since onset of menses.*

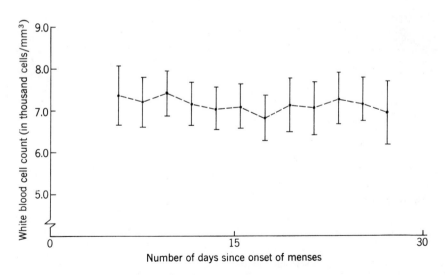

Figure 6b. *Mean white blood cell count (with 95% confidence intervals) of sequential users of oral contraceptive drugs by number of days since onset of menses.*

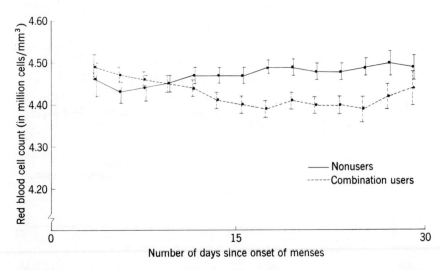

Figure 7a. *Mean red blood cell count (with 95% confidence intervals) of combination users and nonusers of oral contraceptive drugs by number of days since onset of menses.*

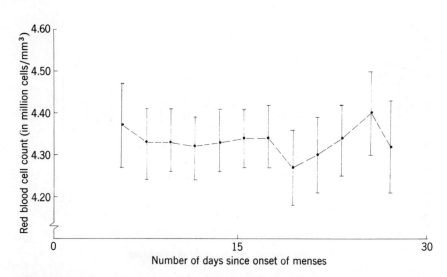

Figure 7b. *Mean red blood cell count (with 95% confidence intervals) of sequential users of oral contraceptive drugs by number of days since onset of menses.*

270

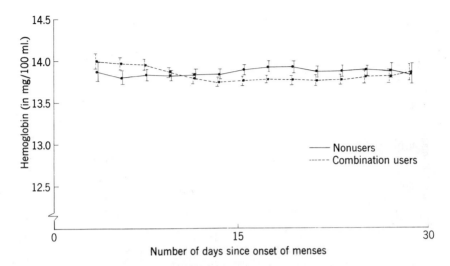

Figure 8a.   *Mean hemoglobin (with 95% confidence intervals) of combination users and nonusers of oral contraceptive drugs by number of days since onset of menses.*

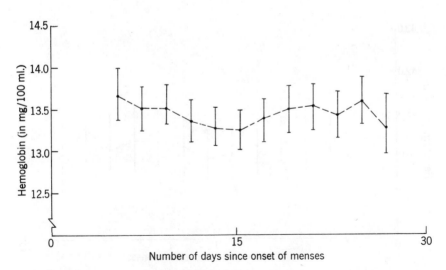

Figure 8b.   *Mean hemoglobin (with 95% confidence intervals) of sequential users of oral contraceptive drugs by number of days since onset of menses.*

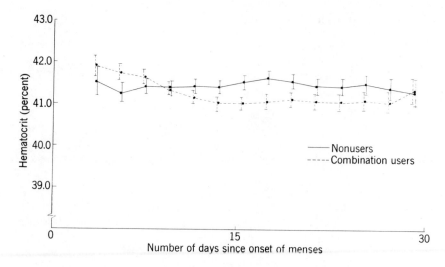

Figure 9a. *Mean hematocrit (with 95% confidence intervals) of combination users and nonusers of oral contraceptive drugs by number of days since onset of menses.*

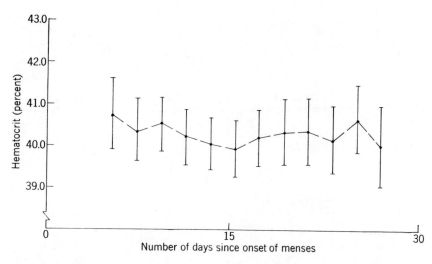

Figure 9b. *Mean hematocrit (with 95% confidence intervals) of sequential users of oral contraceptive drugs by number of days since onset of menses.*

Hemoglobin and Hematocrit

From Figures 8*a* and 9*a* it is apparent that nonusers† showed an increase in both of these measures over the cycle while combination users* showed a decline throughout, followed by a premenstrual rise. The curves for sequential users declined during the first half of the cycle and increased in the second half (Figures 8*b* and 9*b*).

DISCUSSION

The data that were available for this report consisted of cross-sectional daily means based on single observations on large numbers of women. The major drawback in the use of observations obtained in this manner is that one must use them to construct a time curve of events presumably representing a cohort of experience which should in fact be based on observations obtained through time on the same subjects. But cohort studies are difficult to carry out on more than a few women, and they suffer from the handicap that if by chance (or bad luck) there is heterogeneity within the small sample, averaging the results might obscure the time trend. On the other hand, if we make the assumption that there is a *common time trend* in the population, we can demonstrate the average pattern by sampling a large number of individuals at different points along the time axis, even though uncommon individual deviations from this pattern are thus obscured. When measurements are determined only once for each subject, an advantage is that the observations are independent. However, they do not provide information about the within-subject variation, and the spread of the measurements is much larger than if repeated observations had been made on the same woman over time.

There were some regrettable inadequacies in the information available concerning menstrual and OCD cycles:

1. Because cervical smears were taken at the examination, women were asked not to schedule their appointments during menstruation. Therefore, there were few observations made on days 1 to 4.

2. For each normally menstruating woman, the actual length of the cycle during which the observation was made was not known. Since the

---

* The difference between the maximum and minimum values for this group was significant.

† The difference between the maximum and minimum values for hemoglobin was significant.

length of the postovulatory period is reported to be more consistent between women than is the length of the preovulatory period, averaging measurements from different women would tend to mask changes that might be present at the time of ovulation or in the second phase of the cycle.

3. Whether a woman was examined in an ovulatory or an anovulatory cycle could not be establilshed. However, note that the data of Vollman suggested that the frequency of anovulatory cycles in normal women in the age group 18 to 49 years is small (3). Anovulatory cycles, if present among the nonusers, might tend to obscure menstrual changes related to progesterone secretion.

Differences were found between OCD nonusers and users, unrelated to cycle time, with respect to most of the measurements shown. Except for weight, these were associated with the ingestion of OCD. The mean weight for OCD users fell consistently below that for nonusers and this difference could be accounted for by the lower mean age of users and fewer obese subjects among them. Other variables would also be influenced by age differences. For example, in the case of diastolic blood pressure, the actual difference between OCD nonusers and users was larger than shown in the graph, since blood pressure rises with age and OCD users were, on the average, younger than nonusers.

Information was available for dose, drug brand, and duration of use of OCD, but was not used in this report since no marked variations in OCD effects were related to these factors.

Variability in the preovulatory phase of the cycle as well as the use of a three-point moving average might be expected to eliminate brief oscillations or blur sharp changes in the cycle, for example, around the time of ovulation. Since the number of observations available for sequential users in the late cycle was small, the mean values were less reliable.

The cyclic changes seen in the case of nonusers for weight and systolic and diastolic pressures showed patterns somewhat similar to those of the sequential users. Since the time of changes observed in the cycles for sequential users was closely related to the time of ingestion of estrogen, or estrogen plus progestin, one might postulate that the changes occurring among nonusers could be associated with endogenous estrogen and progesterone. The patterns shown by combination users were characterized by changes in a direction generally similar to the early cycle or "estrogen effect" demonstrated by the other two groups, but the change persisted throughout the cycle.

The patterns shown for pulse rate, serum glucose, and white blood cell count among nonusers were, in general, similar to those among combination users. The curves for sequential users were quite different from these. These results did not suggest simple hormone effects.

In studying the effects of OCD on the red blood cell count, hemoglobin, and hematocrit, it was found necessary to consider not only the effects of age and parity, but menstrual blood loss as well. Not surprisingly, red blood cell count, hemoglobin, and hematocrit were inversely correlated with menstrual blood loss. Long-term OCD users had only about one-half the menstrual blood loss of menstruating nonusers and the effect of this variable was found to be significant. No adjustment was made for menstrual blood loss in the data presented in this report. Therefore, the difference between OCD nonusers and users for these hematological values was greater than shown on the graphs.

Both combination and sequential users showed a decrease in these three measures followed by a premenstrual rise; on the other hand, nonusers showed a rise on all three. Again, no simple cyclic hormonal effect was suggested by these results.

More detailed analyses of some of these variables are in progress or planned in order to develop hypotheses relating to the effects of endogenous and exogenous gonadal hormones.

ACKNOWLEDGMENT

This project was supported by Contract No. NIH-NICHD-73-2710 with the National Institutes of Health, U.S. Public Health Service, Department of Health, Education, and Welfare, Bethesda, Maryland.

REFERENCES

1. A. Reinberg and M. Smolensky, in *Biorhythms and Human Reproduction* (M. Ferin, F. Halberg, R. M. Richart, and R. L. Vande Wiele, Eds.), Wiley-Interscience, New York, 1973, Chap. 15.

2. D. F. Morrison, *Multivariate Statistical Methods*, McGraw-Hill, New York, 1967, p. 31.

3. R. F. Vollman, in *Biorhythms and Human Reproduction* (M. Ferin, F. Halberg, R. M. Richart, and R. L. Vande Wiele, Eds.), Wiley-Interscience, New York, 1973, Chap. 11.

CHAPTER 17

Sensory Changes During the Menstrual Cycle

ROBERT I. HENKIN

Chief, Section on Neuroendocrinology, National Heart and Lung Institute, National Institutes of Health, Bethesda, Maryland

It has been commonly observed that women experience changes in perception during the menstrual cycle. However, attempts of several investigators to measure the magnitude or the timing of these changes within the cycle have not been consistently successful. Beiguelman studied taste sensitivity to phenylthiourea (PTC) during the menstrual cycle and was unable to demonstrate any change (1) whereas Glanville and Kaplan reported that changes in the taste of PTC did occur during menses (2, 3). Olfactory sensitivity to the smell of the musk-like compound exaltalide was reported to be most acute in the midcycle period by LeMagnen (4) who considered the increased acuity specific for this substance related to its supposed sexual proclivity. Köster reported that changes in olfactory acuity occurred at various times during the menstrual cycle related primarily to the length of the cycle itself (5, 6). Because these changes were demonstrable using *m*-xylol as the test stimulus, he felt these changes were based on differences in olfactory acuity rather than to any sexual relationship of the stimulus. Changes in auditory acuity during the menstrual cycle were sought by Parris and were not found (7). However, several investigators reported that changes in auditory acuity did occur during the menstrual cycle and that they not uncommonly were initiated during the middle of the cycle (8, 9). Other investigators reported that pathological changes in auditory acuity; that is, deafness, occurred during the menstrual cycle in some patients (10, 11). This phenomenon, which occurred during the luteal phase, was called premenstrual deafness and usually remitted in intensity after the onset of menses.

These studies suggested that changes in sensory acuity for several modalities occurred during menses but the nature of these changes and their relationship to the hormonal changes which occurred during the menstrual cycle were not clearly identified. The purpose of this study was to define the sensory changes which occurred during the menstrual cycle and to attempt to relate these changes to the hormonal changes that occur during this period.

MATERIALS AND METHODS

The subjects of this study were five normal female volunteers aged 19 to 23. All subjects had a normal menarche between ages 11 and 13 with regular menstrual periods ranging from 25 to 36 days. All patients remained on an air-conditioned metabolic ward and ate a normal diet during the entire study. Body weight, determined with metabolic scales daily on arising, was used to provide a gross estimate of change in the volume of body fluid. Basal body temperature was estimated daily by oral placement of a thermometer calibrated in $0.1°F$ (Ovulindex) for at least 5 minutes prior to arising in the morning.

Detection and recognition thresholds for the taste of sodium chloride (NaCl) and for the smell of pyridine in water were determined by a forced-choice, three stimulus drop or sniff technique, respectively, the methods previously described in detail (12, 13). Concentrations of NaCl used included 60, 30, 12, 6, 3, 0.8, and 0.5 mM and were expressed, for convenience, in bottle units (BU) with 60 mM called 1 BU; 30 mM, 2; and 0.5 mM, 7. Concentrations of pyridine ranged from 10^{-2}-10^{-10} M/L. Detection of auditory signals was carried out with the subject seated alone in an arm chair in an Industrial Acoustic Corporation 1204 double walled sound chamber. Acoustic signals from 125 to 8000 Hz were produced by a Beltone Clinical Audiometer and were presented to the left and right ear of each subject separately through Knight KN 848 circumaural earphones in a manner previously described (14, 15). Tactile perception on the adductor portion of the palmar surface of the left hand was measured by the application of a series of nylon filaments in a standardized manner (16) and by the measurement of two-point discrimination over the same area by the application of a two-point esthesiometer (16). Measurement of plasma luteotrophic hormone (LH) was carried out by radioimmunoassay by a method previously described (17).

All sensory measurements were made in the afternoon by one of three investigators, none of whom were aware of the menstrual status of the subject. All subjects refrained from eating or smoking for at least 1 hour prior to testing. In one subject sensory measurements were made daily over a period of 6 weeks while in four additional subjects measurements were made every 2 to 3 days. Measurements of basal body temperature was carried out daily in all subjects. In the one subject in whom LH values are reported, plasma was collected daily in the morning under fasting conditions.

RESULTS

Changes in taste, smell, hearing, light touch, and two-point discrimination and in basal body temperature and plasma LH concentration for one representative patient studied each day for a period of 6 weeks are shown in Figure 1. Increases in taste, smell, and auditory detection acuity and in light touch and two-point discrimination occurred in the follicular phase of the menstrual cycle prior to the midcycle increase in basal body temperature or in plasma LH concentration. This increased acuity was in contrast to the relative decrease in detection acuity in each of these sensory modalities which occurred during the luteal phase of the cycle. Little if any consistent change in taste or olfactory recognition acuity could be noted upon comparison of the levels observed during the follicular or luteal phases of the cycle.

The changes in acuity that occurred during the follicular and luteal phases of the cycle were within normal limits for all sensory modalities except olfaction. For NaCl, taste detection and recognition varied within the normal range of 60 to 0.5 mM, for hearing within the normal range of 0 to -15 dB (re ASO Standard for a tone of 125 Hz), and similarly for light touch and two-point discrimination (16). However, at times during the follicular phase, detection sensitivity for the smell of pyridine increased to 10^{-10} M/l, a level above the upper limit of the normal detection range which extends from 10^{-5} to 10^{-9} M/l. The noi. .al range of recognition acuity for pyridine extends from 10^{-2} to 10^{-} M/l.

Changes in taste, smell, and auditory detection acuity and in basal

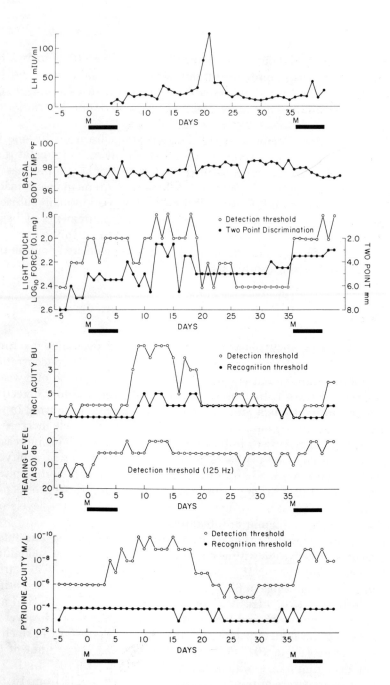

Figure 1 Daily changes in sensory function, basal body temperature and plasma LH concentration in one woman studied throughout 1 complete menstrual cycle. The

280

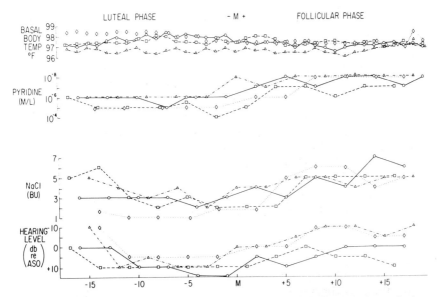

Figure 2   Changes in sensory detection acuity and in basal body temperature in four normal women studied during one menstrual cycle. Basal body temperature was measured daily whereas taste, smell, and auditory detection acuity were measured every 2 or 3 days. This figure is organized vertically in a manner similar to that of Fig. 1. On the abscissa M indicates the onset of menses in each woman, + indicates days following menses, − indicates days preceding menses. The results are similar to those observed in Fig. 1 in that increased detection of taste, smell, and auditory stimuli occur during the follicular phase of the menstrual cycle whereas a relative decrease in detection acuity occurs during the luteal phase.

abscissa of each panel is plotted in days. The onset of menses is indicated by M, the length of the menses by a solid black line. Detection and recognition thresholds for the smell of the vapor of pyridine in water (in M/l) is shown on the lowest panel. The normal range of detection is from 10^{-5} to 10^{-9} M/l, for recognition from 10^{-2} to 10^{-4} M/l. Detection acuity for a 125-Hz tone is shown on the next panel; normal levels range from 0 to -15 dB (re ASO Standard). Detection and recognition thresholds for the taste of NaCl are shown on the next panel. Thresholds are expressed in bottle units (BU), 7 referring to concentrations of 60 mM, 6 = 30 mM, 5 = 12 mM, 4 = 6 mM, 3 = 3 mM, 2 = 0.8 mM, and 1 = 0.5 mM. The normal range for detection extends from 1 to 7 BU; for recognition from 7 to 4 BU. Light touch for nylon filaments expressed as $\log_{10}$ force (0.1 mg) applied to the adductor surface of the left palm is shown in the next panel. Two point discrimination also tested on the adductor surface of the left palm is expressed in mm. Normal light touch thresholds vary from 1.8 to 2.4 $\log_{10}$ force units whereas the range of two-point discrimination extends from 3 to 7 mm. Basal body temperature in °F is plotted in the next panel and plasma LH concentration in mIU/ml is plotted on the uppermost panel. Note the detection acuity for each sensory modality tested was increased prior to the LH peak during the follicular phase of the menstrual cycle whereas a relative decrease in acuity occurred during the luteal phase following the LH peak.

281

body temperature are shown for four additional subjects studied every 2 to 3 days throughout one menstrual cycle in Figure 2. For convenience of presentation menses was designated M with the − direction designating the luteal phase, the + direction designating the follicular phase. Similar to the changes observed in Figure 1, taste, smell, and auditory detection acuity were increased during the follicular phase of the cycle with respect to the relatively decreased detection acuity observed during the luteal phase of the cycle for these same sensory modalities. Although not shown in this figure for the sake of clarity, little if any change in taste or olfactory recognition acuity occurred during either the follicular or luteal phase of the cycle. Changes in detection acuity were, in general, within the normal range of sensory acuity, as noted for the data presented in Figure 1.

DISCUSSION

These studies demonstrate that sensory detection for taste, smell, hearing, touch, and two-point discrimination is relatively more acute during the follicular phase of the menstrual cycle than during the luteal phase. These changes were observed whether the menstrual cycle was "short," that is, less than 28 days in length, or "long," greater than 28 days in length. These changes occurred for all sensory modalities measured but were more apparent for olfactory acuity. As previously noted, these changes in detection acuity were relative; this is, although changes observed were more apparent during the follicular than the luteal phase, most changes were generally within the normal limits of each sensory modality. In this sense, in order for these changes to be apparent, relatively frequent testing of sensory acuity had to be carried out in order for the trends within the normal ranges to become apparent. It is of interest that those studies in which no change in sensory acuity within the menstrual cycle was reported were those in which the least frequent testing of sensory acuity was undertaken (1, 7).

The relative increase in sensory detection acuity during the follicular phase of the menstrual cycle may be related to the effects of estrogen or progesterone on either receptor, nerve, or central nervous system activity. On the one hand there may be positive effects of estrogen producing increases in detection acuity during the follicular phase although these changes have not been isolated in any systematic manner. The increase in detection acuity to the upper limits of normal for taste, hearing, and light touch and to levels above the upper limits of normal for smell during this phase tends to support this concept. However, on the other hand, the increase of progesterone which occurs during the luteal phase coupled with the decrease in estrogen concentration may be related to the decreased detection acuity observed. Although specific sensory data are not readily available depression, accidents, admission to hospital, and suicide are more frequent during the luteal phase than during the follicular phase of the cycle (18). Progesterone has been shown to alter Na and K metabolism during the luteal phase of the cycle, and the direction of these changes have been associated with adverse effects on the metabolic balance of the fluid in the inner ear. These changes have been suggested to relate to the decreased auditory acuity seen in Meniere's syndrome. In addition, in one of the patients in whom premenstrual deafness was reported, administration of progesterone during the follicular phase, during a period of increased auditory acuity, was associated with a precipitous drop in auditory acuity to levels not different from those just prior to menses (11).

It has also been recently demonstrated that administration of progesterone to rats was associated with decreased plasma zinc concentration and that this relationship also was apparent during the estrus cycle (19). Since loss of zinc has been associated with decreased taste and smell acuity (20, 21), this relationship could provide another possible link between progesterone secretion and decreased sensory acuity.

These observations suggest that behavior in several systems may be worsened during the luteal phase of the cycle in relationship to the hormonal changes that occur during this phase. Thus the changes in sensory acuity that occur during the menstrual cycle may be a sum of the actions of both estrogen and progesterone although other factors not yet recognized may also play significant roles.

It is tempting to speculate on the meaning of the changes in sensory acuity that occur during the menstrual cycle. As noted earlier, the changes in hormonal concentration observed during the two phases of the cycle may exert specific effects on receptor, nerve, or central nervous system function which could relate to the sensory changes observed.

However, the changes that are most apparent relate to increased acuity and appear to occur prior to ovulation. Whether or not this increased acuity has any biological meaning related to increasing fertility is not known. Although there is disagreement about the level of sexual receptivity or desire prior to ovulation (22–25), there is a large body of observations suggesting that increased sexual activity occurs during the follicular phase, prior to ovulation when estrogen levels are elevated (22–24). Teleologically this activity would increase the odds of ultimate fertility. If this were the case then any increase in sensory acuity on the part of the female might be helpful particularly in relationship to mutual mate finding behavior. This concept, although difficult to recognize in man with his primary dependence upon visual acuity, is readily apparent in lower animals who must find their mates in forests, jungles, or seas where the similarity of the environment makes visual acuity less valuable than increased olfactory acuity. Certainly the presence of pheromones in insects appears consistent with this hypothesis with respect to females attracting males of the species, but mate finding does have certain mutual attraction processes. Thus, the increased sensory acuity observed during the follicular phase of the cycle in man may relate in some ultimate context to fertility. As such, these changes may provide useful information related to one aspect of ovulation.

REFERENCES

1. B. Beiguelman, *Acta Gen. Med. Gem.,* **13,** (1964), 197.
2. E. Glanville and A. R. Kaplan, *Nature,* **205,** (1965), 930.
3. E. Glanville and A. R. Kaplan, *Amer. J. Obstet. Gyncol.,* **92,** (1965), 189.
4. J. LeMagnen, *C. R. Acad. Sci.,* **226,** (1948), 694.
5. E. P. Köster, *Int. Rhin.,* **3,** (1965), 57.
6. E. P. Köster, *Olfactory Sensitivity and Ovulatory Cycle Duration,* Psychological Laboratory Report #6702, University of Utrecht, The Netherlands, 1967.
7. M. Parris, "Variations in Audiometric Thresholds Accompanying the Menstrual Cycle," Master's Thesis, Adelphi University, June 1964.
8. L. Gabrielli and D. Tarsitari, ORL, Ital. **32,** (1963), 503.
9. G. I. Matei and A. Ciovirnache, *Int. Audiol.,* **4,** (1965), 43.
10. C. W. Lloyd, "Problems Associated with the Menstrual Cycle", in *Human Reproduction and Sexual Behavior* (C. W. Lloyd, Ed.), Lea and Febiger, New York, 1964.
11. M. H. Miller and W. J. Gould, *J. Aud. Res.,* **7,** (1967), 373.
12. R. I. Henkin, P. J. Schechter, R. C. Hoye, and C. F. T. Mattern, *J. Amer. Med. Assoc.,* **217,** (1971), 434.
13. J. Marshall and R. I. Henkin, *Ann. Int. Med.,* **75,** (1971), 207.
14. R. I. Henkin, R. E. McGlone, R. Daly, and F. C. Bartter, *J. Clin. Invest.,* **46,** (1967), 429.
15. R. I. Henkin and R. L. Daly, *J. Clin. Invest.,* **47,** (1968), 1259.
16. R. I. Henkin and V. Banks, "Tactile Perception on the Tongue, Palate and the Hand of Normal Man," in *Symposium on Oral Sensation and Perception* (J. F. Bosma, Ed.), Thomas, Springfield, Ill., 1967.

17. G. T. Ross, W. D. Odell, and P. L. Rayford, *Science,* **155,** (1967), 1679.
18. P. C. B. Mackinnon and I. L. Mackinnon, *Brit. Med. J.,* **1,** (1956), 555.
19. N. Sato and R. I. Henkin, *Amer. J. Physiol.,* (1973), In Press.
20. R. I. Henkin, D. Bronzert, and H. R. Keiser, *J. Clin. Invest.,* **51,** (1972), 44a.
21. R. I. Henkin, P. J. Schechter, M. S. Raff, D. A. Bronzert, and W. T. Friedewald, "Zinc and Taste Acuity: A Clinical Study Including a Laser Microprobe Analysis of the Gustatory Receptor Area, in *Clinical Applications of Zinc Metabolism* (W. Pories and W. Strain, Eds.), Thomas, Springfield, Ill., 1973, In Press.
22. T. Benedek, *Studies in Psychosomatic Medicine: Psychosexual Functions, Woman,* Ronald, New York, 1952.
23. J. R. Udry and N. M. Morris, *Nature,* **220,** (1968), 593.
24. N. M. Morris and J. R. Udry, *Obstet. Gynecol.,* **35,** (1970), 199.
25. A. C. Kinsey, W. B. Pomeroy, et al., *Sexual Behavior in The Human Female,* Saunders, Philadelphia, 1953.

CHAPTER 18

Secondary Rhythms Related to Hormonal Changes in the Menstrual Cycle: Special Reference to Allergology

MICHAEL H. SMOLENSKY

School of Public Health, University of Texas, Houston, Texas

ALAIN REINBERG

Equipe de Recherche de Chronobiologie Humaine (CNRS No. 105) and Fondation A. de Rothschild, Laboratoire de Physiologie, Paris, France

RUFUS E. LEE

McGovern Allergy Clinic, Houston, Texas

JOHN P. MCGOVERN

University of Texas School of Biomedical Sciences, Baylor College of Medicine, McGovern Allergy Clinic, Houston, Texas

Clinic and laboratory observations on mammals suggest that ovarian secretions, among others, influence cyclic changes in sensitivity to various agents such as antigens (1, 2) and ultraviolet radiations (3–5). In animals, one of us (J.P.M.) previously reported estrus-related variations in the response of the uterine horns of either mice or guinea pigs to such chemical agents as serotonin, acetylcholine, and crystalline albumin antigen (6, 7). Using the Schultz-Dale apparatus, it was found that animals in estrus responded from five to eight times greater than controls not in estrus.

In adult women a number of physiologic as well as pathologic rhythmic changes with a period of about 30 days (circatrigintan rhythms) are thought to be secondary to hormonal changes occurring during the menstrual cycle (8, 9).* With respect to the latter, herpetic lesions are so common during menstruation that a clinical form of this skin disease is called *herpes menstrualis* (13–16). It has been suggested previously that allergy underlies this type of herpes. In addition, acne has been shown to be more or less accentuated during menstruation and unusual sensitivity to ultraviolet radiation on the first and second days of menstruation has also been reported (3–5). In regard to allergic phenomena, urticaria, angioneurotic edema, and various forms of circumscribed erythemas have been reported as occurring solely during menstruation or pregnancy.

Published reports dealing specifically with the topic of menstrual rhythms in cutaneous reactivity to antigens are few (1, 2, 17). Details of these studies are summarized in Table 1. In one study by Ozkaragoz (1) it was reported that, in 70 (23%) out of 264 skin tests, the cutaneous response to skin testing, as measured by wheal formation (to standardized intradermal injections of selected antigens), was clinically significantly higher during menstruation as compared to other times of the menstrual cycle. In a study by Hansen-Pruss (2), reactivity to various antigens was tested once daily, four and 10 days prior to and after, as well as on the first and last days, of catamenia. Using this test design it is found (Table 1) that the greatest cutaneous sensitivity to various antigens, as measured by wheal formation, occurs either 10 days before (corresponding to day 18 of the menstrual cycle when day 1 = first day of menses) and on the last day of catamenia (day 3) in a group of women previously known to be allergic, as well as in another group not previously evaluated for hypersensitivity.

In an associated study (18), Hansen-Pruss obtained blood samples once daily, following the schedule of his earlier work, from four women highly sensitive to ragweed. Passive transfer tests using prepared serums from the women were given to men (irrespective of clock hour) for bioassay of circulating allergins. As indicated in Table 1, the

---

* It is of great interest to report that for adult males, circatrigitan rhythms can be detected and quantified; urinary 17-ketosteroid excretion (10), beard growth (11), and pain threshold (12) rhythms among others are known.

Table 1   Review of Studies on Menstrual Cycle and Allergy

Biologic Variable	No. and Type of Females	No. of Cycles (No. of study days) [Tests/day]	Macroscopic Findings	First Author
Skin tests[c]	36 "Allergic"	<1 (2–3) [1]	Menses: reactivity in 23% of tests	Ozkaragoz (1)
	6 to 10 "Allergic"	1–2 (6) [1]	Peaks: Days 3 and 18	Hansen-Pruss (2)
	2 to 14 "Nonallergic"	1 (6) [1]	Peaks: Days 3 and 18	Hansen-Pruss (2)
Circulating antigens[b]	4 "Allergic"	1 (6) [1]	Peak: Day 3	Hansen-Pruss (18)

[c] Endpoint of response: Wheal.
[b] Bioassay: Passive transfer tests in males using prepared sera from females sensitive to ragweed.

289

greatest titer of antigens occurred 3 days after the onset of menstruation. The time of greatest circulating antigen titer, by an earlier study, corresponded to the timing of greatest skin reactivity to ragweed as well as to other selected antigens.

In these previous studies (1, 2, 17) on humans designed to document rhythms in skin reactivity of women during the menstrual cycle, cutaneous sensitivity was evaluated only once a day at specified but arbitrarily chosen times through the cycle. Apparently little attempt was made to standardize either subjects or times of testing for clock hour in such experiments. Even if they were considered, no prior information was available indicating whether skin reactivity to antigens was circadian rhythmic, nor were specialized electronic computer programs for rhythmometry avaiable to objectively determine and quantify circadian or circatrigintan periodicities. Work by us (19–21) indicates considerable circadian variation in the cutaneous sensitivity to histamine, house dust, and other antigens. In adult subjects standardized for periodicity study, the greatest response is expected about 20 hours after the time of midsleep. The cutaneous reactivity at the acrophase (peak of function used to approximate the rhythm) is on the average two to three, and in some subjects as much as seven, times as great as that observed 12 hours earlier at the time of the daily trough.

METHODOLOGY

Study of circatrigintan rhythms in skin reactivity was investigated at weekly intervals by means of transverse profiles of 24-hours duration. The test agent, histamine, one of the potential mediators of the allergic response, was given to two groups of eight regularly menstruating, healthy, adult women. One sample consisted of women receiving no medication (Group 1) and a second consisted of women receiving oral contraceptive medication (Group 2).

The study was carried out jointly, using similar methodology, at the physiology laboratories and Service d'Allergie Generale, Fondation A. de Rothschild in Paris, France, during the span from February 6 to April 14, 1972, and at the McGovern Allergy Clinic in Houston, Texas, during the span from April 6 to July 7, 1972.

SUBJECTS

In tests carried out in Houston, subjects consisted of nurses or paramedical personnel employed by a local clinic. This sample consisted of eight regularly menstruating, healthy, adult women receiving birth control pills and three others receiving no such medication (Table 2). All subjects were between 19 and 40 years of age and, except for CFF, all had normal menstrual histories. All the subjects were nulliparous except for M.T., M.O., and A.H. The average menstrual cycle length for those women receiving oral contraceptives was 28.0 days. The average menstrual cycle length for the subsample not taking such medication was 34.3 days.

Subjects comprising the Parisian sample were also medical or paramedical personnel. It was made up of five regularly menstruating, healthy, adult women aged from 20 to 50 years (Table 2). All had normal menstrual histories and were not receiving oral contraceptives. Except for A.G. and M.R., all were nulliparous. The average menstrual cycle duration for this sample was 28.3 days.

For at least 14 days prior to study the subjects were standardized to a regular societal routine of daily work and nightly rest. Subjects were asked to keep a diary of times of awakening and retiring. During the span of study, none of the subjects significantly altered their usual living routine. Throughout the study the mean times for retiring to and awakening from nightly rest were 2300 and 0730 for the Parisian sample and 2300 and 0615 for the Houston sample.

Table 3 outlines the protocol for this experiment. As may be seen, body temperature, mood, and activity rating were self-assessed at 4-hour intervals on menstrual days 1, 8, 15, 22, and 29. Such assessments were performed on intervening days at least daily upon awakening each morning. Oral (Houston) or rectal (Paris) temperature was determined to the nearest $0.01°F$ using basal thermometers. Mood and activity determinations were made by the placement of a mark on a speedometer-type scale. In addition to the above studied variables, urines were collected and frozen for later analysis for degradation products of estrogen and progesterone. Circatrigintan rhythms in cutaneous sensitivity to histamine were studied by testing at 4-hour intervals (commencing at 0700 of each test day) on menstrual days 1, 8, 15, 22, and 29 for at least one menstrual cycle. All skin tests were conducted on anatomically comparable areas of the flexor surfaces of the two forearms. Successive tests were carried out on similar sites of alternate forearms on each day of study.

In studies carried out in Paris a volume of 0.05 ml (1:10,000 mg/ml) histamine bichlorhydrate and in Houston 0.10 ml of histamine phosphate (1:100,000 mg/ml) was injected intradermally by means of precision syringes. It was both possible and desirable to have each subject self-administer test solutions as well as to measure the end points of response.

Table 2   Subject Composition of Groups 1 and 2

Group	Females	Age[a]	Cycle Duration[b]	Oral Contraceptives
1	M.C.	19	28	—
	S.T.	27	29	—
	A.H.	40	37[c]	—
	R.A.[d]	20	26	—
	A.N.[d]	39	28	—
	M.R.[d]	50	28[c]	—
	L.T.[d]	27	30	—
	A.G.[d]	45	29[c]	—
	$\bar{X} \pm 1$ SE	33.4 ± 4.1	29.4 ± 1.2	—
2	C.F.	21	28	Norinyl
	C.II.	23	28	Norinyl
	T.W.	23	28	Norinyl
	M.B.	20	28	Ortho Novum
	L.D.	22	28	Ortho Novum
	J.B.	26	28	Ortho Novum
	M.T.	29	28[c]	Ortho Novum
	M.O.	23	28[c]	Ovulin
	$\bar{X} \pm 1$ SE	23.4 ± 1.0	28.0 ± 0.0	—

[a] In years as of last birthday.
[b] In days.
[c] All subjects nulliparous except for M.R., A.G., A.H., M.T., and M.O.
[d] Studied at Foundation A. de Rothschild, Laboratoire de Physiologie, Paris, France.

End points of response for all tests were the area of erythema and wheal measured in a precise manner. Fifteen to 20 minutes after the standardized injection of each test material, the limits of the erythema and wheal were outlined by ball-point pen. Clear adhesive tape was placed over the tracings and transferred to paper for exact measure by use of either a KE compensating Polar Planimeter (Houston) or by a method previously used in earlier studies carried out in Paris—the correlation of area to weight of constant-weight paper.

Data collected over time were objectively quantified (14, 22) in terms of a circadian and/or circatrigintan time-series mean, denoted by $C_0{}^*$, amplitude C (equal to one-half of the within-period rhythmic

---

* The C_0 formerly termed "the rhythm adjusted level" is now referred to as mesor and abbreviated "M." Similarly, the amplitude, C, is now abbreviated as "A."

Table 3 *Protocol for Circatrigintan Study of Cutaneous Sensitivity to Histamine*

Study Day	1ᵃ	2	3	4	5	6	7	8	9	10	11	...	21	22	23	24	25	26	27	28	29
Variable																					
Skin tests	√	–	–	–	–	–	–	√	–	–	–	...	–	√	–	–	–	–	–	–	√
Temperature	√	+	+	+	+	+	+	√	+	+	+	...	+	√	+	+	+	+	+	+	√
Mood and activity rating	√	+	+	+	+	+	+	√	+	+	+	...	+	√	+	+	+	+	+	+	√
Urines	√	–	×	×	–	–	–	√	–	×	×	...	–	√	–	–	×	×	–	–	√

√ = Designated days of self-assessments and urine collections made at 4-hour intervals, commencing at 07⁰⁰ throughout an entire 24-hour span.

+ = Once-daily self-assessments of mood, activity, and body temperature make upon awakening.

× = Days of 24-hour urines.

– = No test for variable on specified day.

ᵃ Coincides to day 1 (first day of catamenia) of menstrual cycle.

variation), and acrophase ϕ (the approximated crest obtained by curve fitting) by least squares spectral and cosinor analyses, using the same methodology as described in the introductory paper of this session.

Collected data were transferred to punch cards for a subsequent two-step analysis. In the first step, a series of cosine curves, linear in frequency and each differing from one another in period by a preselected interval, was successively fit to each time series to form a spectral window. For analysis of circadian rhythms, limits of the spectral window were between 20.0 and 28.0 hours; for circatrigintan rhythms, the spectral window extended from 24 to 36 days. In the second step, C_0, C, and ϕ values for the 24.0-hour cosine fits, in the case of circadian rhythms, or for the 28.0 and 30.0-day cosine fits, for study of circatrigintan rhythms for samples of women receiving and not receiving oral contraceptive medication, respectively, were utilized for separate cosinor analysis.

For cosinor analyses, sample average C_0, C, and ϕ values along with dispersion indices were generated. The null hypothesis tested is that the C and ϕ values, determined by least squares fit of cosine curves of a given period, are random in occurrence. Statistical rejection of the null hypothesis is evidence for rhythmicity. In this paper two types of cosinor analyses are utilized. The first type is conventional and uses both the C and ϕ values determined by least squares spectral techniques. The second type utilizes the same ϕ values as in the first type; however, the amplitudes, C, are unweighted and arbitrarily set equal to 1. This so-called C unweighted cosinor allows direct evaluation of acrophase values unweighted by the amplitude.

This report deals primarily with the findings of menstrual rhythms in skin reactivity to histamine as measured by area of erythema in healthy adult, regularly menstruating women, one-half of whom had been receiving oral contraceptive medication for at least 3 months prior to study. Results of urine analyses as well as other variables investigated are to be presented in future publications.

RESULTS

Results of studies on females *not* receiving oral contraceptives are given in Figure 1. This figure graphically summarizes circatrigintan changes in the cutaneous sensitivity to histamine in five women studied in Paris as well as three women studied in Houston. Given in the figure are mean changes in the circadian C_0 (time-series average) for days 1, 8, 15, 22, and 29, relative to the overall 30-day C_0. In the graph the circatrigintan C_0 is set equivalent to 100%.

CIRCATRIGINTAN RHYTHM OF CUTANEOUS ERYTHEMA TO
HISTAMINE IN HEALTHY, REGULARLY MENSTRUATING FEMALES*

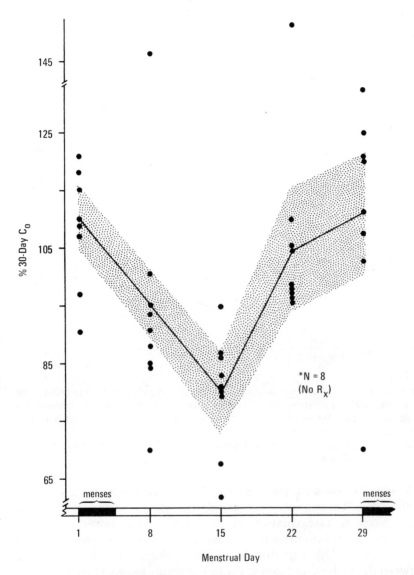

Figure 1 Shown is the individual (●) and mean (●—●) percent change in the circadian C_0 (rhythm adjusted time-series average) relative to the 30-day C_0 for the indicated days of the menstrual cycle for women not using oral contraceptives. The shaded area delineates +1 SE. The highest cutaneous sensitivity to histamine occurs on the first day of each menstrual cycle; the lowest sensitivity occurs around midcycle. Peak to trough difference equals 30%.

295

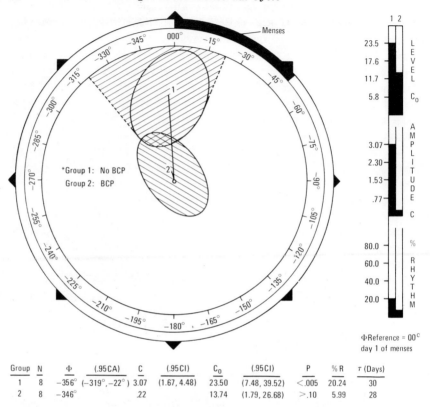

Group	N	Φ	(.95CA)	C	(.95CI)	C_0	(.95CI)	P	%R	τ (Days)
1	8	−356°	(−319°,−22°)	3.07	(1.67, 4.48)	23.50	(7.48, 39.52)	<.005	20.24	30
2	8	−346°		.22		13.74	(1.79, 26.68)	>.10	5.99	28

Figure 2   Results of cosinor analyses substantiate circatrigintan rhythmicity only for Group 1. The highest sensitivity to histamine occurs around day 1 (menses). Circatrigintan C_0 (rhythm adjusted time-series mean) and C (amplitude) for Group 2 are about ½ and ¼ compared to those for Group 1. The percent rhythm index reflects goodness of fit by a   least squares approximation of data for individual time-series comprising the sample.

As may be observed, the highest cutaneous reactivity occurs on days 1 and 29 . These days correspond, for the subjects, to the commencement of menstruation in each cycle. The lowest cutaneous reactivity to histamine occurs around the middle of the menstrual cycle—day 15. The magnitude of rhythmic change (absolute difference between the highest and lowest values) amounts to nearly 30%.

Figure 2 summarizes, by means of a polar plot, results of cosinor analyses for data obtained from the group of 8 women receiving oral contraceptives (Group 2) as well as the sample receiving no oral contraceptive medication (Group 1). On this plot the average sample acrophase is indicated by means of a directed arrow, the length of which is a measure of the rhythm's amplitude. When a rhythm is statistically validated by cosinor, at a given level of statistical significance, the error ellipse (construction of which arises directly from projections of the confidence limits for the acrophase and amplitude) will fail to overlap the center pole of the polar plot. Conversely, failure to detect a statistically significant rhythm at a preselected level of statistical significance is indicated by overlap of the error ellipse over the center pole of the polar plot.

As indicated in Figure 2, a statistically significant circatrigintan rhythm of cutaneous erythema to histamine is validated by cosinor for the sample of women receiving *no* oral contraceptive medication. As indicated both to the right and at the bottom of the polar plot, the values of the circatrigintan C_0 and C are found to be 23.5 and 3.07 cm^2, respectively. The circatrigintan acrophase occurs at $-356°$. When midnight of the first day of menstruation serves as the acrophase (ϕ) reference, its occurrence corresponds in time, by the fit of a 30-day cosine curve (1 day = 12°), to day 1 of catamenia. The 95% confidence arc of the acrophase extends from $-319°$ (3.4 days) before to $-22°$ (1.8 days) after the onset of menstruation.

As shown in Figure 2, the error ellipse for Group 2, the sample of women receiving oral contraceptives, overlaps the center pole of the polar plot. This indicates failure to substantiate a statistically significant ($p > .05$) circatrigintan rhythm for cutaneous erythema to histamine in such subjects.

Despite an inability to statistically validate circatrigintan rhythmicity for Group 2, it is of interest to consider the computed circatrigintan C_0, C, and ϕ values for comparison with those values found for the rhythm detected for females of Group 1. Although the computed sample average acrophase for Group 2 differs but little from that for the sample of women not receiving oral contraceptives, the circatrigintan C_0 and C vary considerably from those for Group 1. As graphically depicted to the right as well as indicated at the bottom of the polar plot, the circatrigintan C_0 for the sample of women taking oral contraceptives is only about one-half that determined for the sample of women not receiving oral contraceptive medication. Similarly, the circatrigintan amplitude for sample 2 approaches zero and is approximately 1/14 of that determined by cosinor analysis for Group 1.

Table 4 summarizes the characteristics of the circadian rhythm in cutaneous sensitivity to histamine for each test day of the menstrual cycle for the sample of eight healthy, regularly menstruating women not receiving oral contraceptive medication. Inspection of this table

Table 4 Circatriginitan Modulation of the Circadian Rhythm of Cutaneous Erythema to Histamine[a]

Menstrual Day	Circadian Level[b], $C_0 \pm$ SD	Amplitude[b]	Circadian Acrophase[c] (95% Confidence Interval)	Circadian Acrophase[c] C-unweighted
1	25.9 ± 3.6	4.8 (0.31 to 9.4)	0100 (1936 to 0420)[*d]	0016 (2156 to 0200)[**d]
8	22.7 ± 3.9	3.5	2320	2320 (2112 to 0248)[*d]
15	19.4 ± 4.0	1.4	1828	2036
22	21.7 ± 6.5	3.1	2236	2132
29	26.7 ± 8.8	6.1 (1.8 to 10.4)	0020 (1956 to 0404)[**d]	0024 (2132 to 0216)[**d]
Overall	23.2 ± 5.4	3.4 (1.6 to 5.3)	2344 (2144 to 0114)[***d]	2320 (2316 to 0044)[**d]

[a] Cosinor analyses of each day's data for eight healthy, regularly menstruating females. Five to six determinations/♀ per 24-hour span on menstrual days 1, 8, 15, 22 and 29 (day 1 = first day of menses). No oral contraceptives.
[b] Erythema, cm².
[c] Hours and minutes; acrophase reference = local midnight, 00^{00}.
[d] Rhythm detection: * = $p < .05$; ** = $p < .01$; all others $p > .05$.

reveals several interesting trends. As previously indicated in Figure 1, the circadian level (C_0) varies according to a circatrigintan rhythm with highest sensitivity on days 1 and 29 and lowest sensitivity around day 15.

Besides a change in the circadian C_0 throughout the menstrual cycle, both the circadian amplitude and acrophase values vary systematically during the menstrual cycle. For example, the highest values for amplitude are found at the beginning of each menstrual cycle; that is, on days 1 and 29. The lowest rhythmic within-day change (amplitude) occurs on day 15. Total alteration in the circadian amplitude through the menstrual cycle is about fourfold. When the acrophase is referenced to local midnight, for subjects strictly standardized by clock hour to a regular routine of nocturnal rest and diurnal activity, the timing of the circadian crest in cutaneous reactivity is latest on days 1 and 29 of the menstrual cycle. Earliest timing of the circadian acrophase occurs at midcycle (day 15). Parallel findings occur whether one considers the C-weighted or unweighted ($C = 1$) circadian acrophases as determined by cosinor. The magnitude of change in the circadian acrophase during the menstrual cycle is at least 4.5 hours as determined for the C-weighted cosinor. In summary, circatrigintan modulation of the circadian rhythm in cutaneous reactivity to histamine in the manner described above means that compared to day 15 the 24-hour mean cutaneous sensitivity to histamine is greatest (as revealed by the circadian C_0), while its stability (as reflected by the circadian C) is lowest around day 1. In addition, the clock hour of greatest sensitivity occurs later at the beginning of the menstrual cycle (day 1 of menses) than on any other day of the cycle.

DISCUSSION

Results of this study indicate several interesting aspects of circatrigintan temporal organization for samples of oral contraceptive users and nonusers. First, findings reveal basic differences between women taking and women not taking oral contraceptive medication. Such differences are exemplified by the failure to document, by objective statistical techniques, a circatrigintan rhythm in the cutaneous sensitivity to histamine for women receiving oral contraceptive medication. In comparison, for women taking no oral contraceptive medication, a circatrigintan rhythmicity in histamine reactivity is easily documented. Moreover, the overall time-series mean (circatrigintan C_0) for oral contraceptive users is only one-half that of nonusers.

It was pointed out earlier that this study was carried out at two locations. Since histamine dilution procedures varied between laboratories and countries, and since concentrations and volumes injected differed also, it was desirable to investigate the possibility that differences in the circatrigintan C_0 values between users and nonusers of oral contraceptives might reflect variations in histamine concentration used for testing. Results of this investigation indicate the histamine preparation used in Paris produces one and one-half times the area of erythema compared to that used in Houston. Since all eight subjects using birth control preparations were tested in Houston whereas only three of the eight nonusers were so tested, it is possible that differences in the 30-day C_0 values between the groups of subjects studied may reflect, at least in part, differences in the histamine concentrations used for testing at the separate locations. However, a comparison of the average circatrigintan C_0 values for the three nonusers and eight users of oral contraceptives studied in Houston reveals, for the latter group, a suppression in histamine reactivity (as measured by the area of cutaneous erythema) of 36%. These findings suggest that oral contraceptives may have clinical significance in the management of menstrual allergy by reduction in histamine sensitivity; however, further research is required along these lines. Furthermore, in investigations thus far carried out, only nonsequential oral birth control preparations have been studied. The effect of sequential oral contraceptives on biochemical and physiologic menstrual rhythms and in particular on histamine susceptibility and menstrual allergy will require further study also.

Detection of effects by oral contraceptive medication upon rhythm characteristics are most interesting in themselves; however, they gain additional prominence if they are compared to results obtained by Procacci (12) in studies of healthy women for pain threshold to radiant heat. As shown in Table 5, circatrigintan rhythms are validated for groups of women not receiving oral contraceptive medication, for postmenopausal women studied before or during oral contraceptive therapy, and for pregnant women. Similarly, a circatrigintan rhythm in pain threshold is detected for men. The intriguing facet of these particular results lies in the fact that for a group of seven women studied before and during oral contraceptive treatment, statistically significant circatrigintan rhythmicity is documented only for the study span in which no oral contraceptives were administered. Study of the same women during a span of oral contraceptive treatment reveals loss of the circatrigintan rhythm in pain threshold to radiant heat.

Besides basic differences in the rhythm characteristics between Groups 1 and 2, results of our studies indicate circatrigintan modulation of the circadian acrophase and amplitude values for women not using oral contraceptive medication. At the beginning of the menstrual cycle, that is, day 1 of menses, the timing of the acrophase occurs later along the 24-hour time scale as compared to midcycle (day

Table 5   Circatrigintan Rhythm in Pain Threshold (12)

Subjects		Rhythm Detection, P	Level[a] C_0	Circadian		
No.	Sex			Amplitude[a] (95% Confidence Interval)	Acrophase[b]	τ^b
8	♀[c]	<.025	1032	68.5 ± 48.2	$8.4 \genfrac{}{}{0pt}{}{+4.2}{-5.7}$	30
7	♀ d^1	<.025	977	124.4 ± 96.5	$15.5 \genfrac{}{}{0pt}{}{+2.8}{-7.3}$	25
	d^2	>.05	990	—	—	—
6	♀ e^1	<.05	854	51.3 ± 50.4	$2.6 \genfrac{}{}{0pt}{}{+4.7}{-9.1}$	29
	e^2	<.01	847	56.0 ± 34.0	2.7 ± 1.1	29
7	♀ [f]	<.05	928	39.0 ± 34.0	$12.7 \genfrac{}{}{0pt}{}{+7.7}{-4.1}$	30.5
4	♀ [g]	<.025	628	38.8 ± 25.7	$23.2 \genfrac{}{}{0pt}{}{+2.6}{-11.8}$	30
♂	(4 groups)	<.05	875	62.4 ± 60.6	2.8 ± 5.5	25

[a] In mcal/cm^2.

[b] In days.

[c] Ages 15 to 20 years. Spontaneous menstruation; no Rx. (Study span = T = 40 days.)

[d] Ages 20 to 39 years. Study before = $d1$ (T = 30 days) and during = $d2$ (T = 70 to 85 days). Oral contraceptives, Lyndiol, Organon: 5 mg − Lynestrenol + Mestranol).

[e] Ages 60 to 84 years. Postmenopausal. $e1$ = before and $e2$ = during oral contraceptive Rx. T = 85 days.

[f] Ages 55 to 68 years. Postmenopausal for at least 7 years (T = 100 days).

[g] Ages 23 to 28 years. During pregnancy (T = 45 to 90 days).

Table 6   Circatrigintan Modulation of the Circadian Rhythm of Pain Threshold to Radiant Heat[a]

Menstrual Day	Rhythm Detection P	Circadian	
		Amplitude[b]	Acrophase[c]
		(95% Confidence Interval)	
Day 7	< .001	58.7 (29.9 to 87.5)	0734 (0504 to 1011)
Day 15	< .001	49.3 (15.5 to 83.1)	0946 (0635 to 1304)
Day 23	< .003	63.7 (21.2 to 106.2)	0925 (0543 to 1236)

[a] Twenty females, ages 17 to 36; no oral contraceptives.
[b] Units mcal/cm².
[c] Hours and minutes; acrophase reference = local midnight, 00^{00}. Data from Procacci (12).

15). For the circadian amplitude, the greatest values are found for day 1 and the lowest ones are found for day 15. Although the significance of such changes in the circadian amplitude and acrophase during the menstrual cycle awaits documentation with larger samples; nonetheless they seemingly reflect underlying meachanisms of physiologic and temporal integration of the menstrual cycle. In particular, whether systemic changes in the circadian acrophase values throughout the menstrual cycle simply represent a shift in phase or whether they represent a slight alteration in the circadian period (τ) awaits further study. If the latter, changes in amplitude during the cycle may reflect such an alteration in τ.

While these findings are interesting in themselves, results obtained by Procacci (12) on circadian rhythms of pain threshold to radiant heat for different days of the menstrual cycle indicate a like phenomenon. As summarized in Table 6, the lowest circadian amplitude for pain threshold occurs around day 15 of the menstrual cycle; the highest circadian amplitude is detected on day 23. This trend in the circadian amplitude suggests that the greatest within-day change is to be expected around day 1 of the menstrual cycle. Also noteworthy is the fact that the circadian acrophase changes systematically throughout the menstrual cycle. Although there is considerable overlap of the confidence arc for the acrophases, the trend suggests that the circadian acrophase occurs later for the premenstrual phase as compared to other times of the menstrual cycle.

Earlier, Haus and Halberg (23) reported circannual modulation of the circadian rhythm of adrenocortical function in inbred C-mice. Modulation of the circadian rhythm in serum corticosterone concentration is detected for all rhythm parameters—C_0, C, and ϕ. Results of the studies given here on histamine, along with those from Procacci, further document the modulation of higher frequency components of (circadian) biologic time structure by lower frequency components (circatrigintan).

Previously, Hansen-Pruss (2), as well as Ozkaragoz (1), reported variations in the cutaneous reactivity to antigens apparently due to changes secondary to hormonal patterns associated with the menstrual cycle. In these studies, tests were carried out only once daily independent of clock hour on selected days during the menstrual cycle. Results obtained by Hansen-Pruss (2) reveal a bimodal curve with peak sensitivity occurring around midcycle as well as on day 1 of catamenia. These results differ somewhat from those reported herein. Specifically, no frequency, besides the circadian and the 30-day periodicity, was detected for cutaneous erythema to histamine. (It should be noted, however, that more frequent sampling during the menstrual cycle may reveal additional periodicities.)

It is noteworthy to point out that the experimental design used herein differed significantly from that employed by Hansen-Pruss (2) in his studies. In such investigations carried out earlier, determinations were made only once daily, either at the same or at different clock hours on different days of the menstrual cycle. However, since cutaneous reactivity to histamine is circadian rhythmic, it is not unlikely that testing on different menstrual days at the same clock hour in subjects not adhering to a regular living routine or, for that matter, at the same or different clock hour for subjects standardized to a regular routine of activity and rest, may fail to adequately reflect a representative temporal pattern for this rhythm.

In an attempt to analyze circatrigintan (menstrual) rhythms more thoroughly, we have used transverse samples of 24-hour duration with determinations at equal intervals on each test day (days 1, 8, 15, 22, and 29). This latter design enables cyclic changes during the menstrual cycle to be described through use of the rhythmometric variables of the circadian C_0, C, and ϕ rather than by conventional, central tendency measures such as the mean, x. The experimental design used here suggests that the secondary peak occurring around midcycle, as reported by Hansen-Pruss (2), probably is an artifact due to once a day testing. Apparently this artifact results from changes in the circadian acrophase during the menstrual cycle; ϕ occurs earlier around mid as compared to late cycle. Nonetheless, credit is to be given to Hansen-Pruss who, as early as 1942, pioneered study on circatrigintan rhythms in the cutaneous sensitivity to antigens.

It is most interesting that the circatrigintan rhythm in cutaneous sensitivity to histamine, with the highest reactivity occurring around day 1 of menses, seems to be temporally coordinated to various allergic phenomena. Specifically, results of one study (18) suggest a circatrigintan rhythm in the titer of circulating antigens in women not taking birth control medication. The timing of the highest titer of circulating antigens corresponds to the span of menstruation. The timing of the highest circulating antigen titer corresponds quite closely in time to the highest cutaneous sensitivity to histamine in women not receiving oral contraceptive medication.

In separate studies dealing with menstrual allergy, the occurrence of asthma has been found to vary in incidence during the menstrual cycle. For example, Wulfsohn (24), in a study of 27 adult asthmatic Bantu women, found that approximately 75% of the subjects became worse either premenstrually or menstrually. Later studies by Aoyama (25) and Chen (26) on groups of asthmatic women confirmed these findings. One of us (A.R., 27, 28) previously documented circadian rhythms in the occurrence of asthma. In these studies it was noted that the occurrence of asthma is primarily a nocturnal event which corresponds in phase, for the circadian rhythm in adrenocortical function, to the span of reduced circulating corticosteroids. Thus, it is not unusual to expect changes in bronchial asthma, perhaps secondary to hormonal change, during 24 hours; similarly, it seems reasonable to expect changes in respiratory function secondary to alterations of reproductive as well as other hormonal patterns during the menstrual cycle.

Detection of circatrigintan rhythms in the occurrence of asthma as well as in other allergic symptoms, along with the results of this study, brings to mind earlier studies dealing with the role of progesterone as a mediator of premenstrual tension—headache, irritability, anxiety, fatigue, tender breasts, and gastrointestinal upset. As noted here, the timing of the highest cutaneous sensitivity to histamine, increased incidence of asthma, and the greatest level of circulating antigens, as well as other reported allergic conditions such as premenstrual urticara, eczema, angioneurotic edema, allergic rhinitis, and coryza (29–34), all correspond to the span just prior to or during menstruation. Since the aforementioned events correspond to the late luteal phase of the menstrual cycle—the phase associated with progesterone secretion—this also suggests that increased sensitivity of women to allergy during the later portion of the menstrual cycle may, be related to increased production of this reproductive hormone.

Allergy to progesterone and its metabolites has been suggested as a probable cause of certain premenstrual symptoms comparable to premenstrual tension. Experimentally, the premenstrual syndrome has been provoked by giving progesterone to some women and relieving it in others by desensitization to progesterone (35, 36). The results of such study suggest a possible role of progesterone and other hormones in the circatrigintan rhythm of cutaneous reactivity to histamine.*

SUMMARY

Results of this study indicate circatrigintan rhythms only for women not receiving oral contraceptive medication. In women receiving such medication, no circatrigintan rhythmicity is substantiated. Moreover, the overall sensitivity to histamine for women receiving oral contraceptives is reduced as compared to those not receiving such medication.

In women not given oral contraceptives, systematic changes in the circadian C_0, C, and ϕ during the menstrual cycle suggest possible mechanisms associated with the temporal organization of reproductive function.

The experimental design used here, which should serve as a model for future studies of menstrual rhythms, enables the detection of changes in circadian C_0, C, and ϕ values during the menstrual cycle and provides a more comprehensive evaluation of menstrual rhythms as compared to studies utilizing fixed hour, single-daily determination experimental designs. At a minimum, it is suggested that transverse (24-hour) study spans be carried out at least at weekly intervals during one (or preferably more) menstrual cycle or more on the same subjects when investigating circatrigintan time structure.

ACKNOWLEDGMENT

This study was made possible by a gift from the Upjohn Pharmaceutical Company to the Texas Allergy Research Foundation.

REFERENCES

1. K. Ozkaragoz and F. Cakin, *J. Asthma Res.*, **7**, (1970), 171.
2. O. C. Hansen-Pruss and R. Raymond, *J. Clin. Endocr.*, **2**, (1942), 161.
3. H. Guthmann and W. Nagel, *Strahlentherapie*, **48**, (1933), 267.

* Since submission of this manuscript, menstrual rhythms in the females not receiving oral contraceptives have been documented for total estrogen ($\phi = 20$), and 17-hydroxycorticosteroid ($\phi = 19$) excretion. Lowest cutaneous sensitivity to histamine (around day 15) corresponds temporally to elevated circulating concentrations of these hormones.

4. F. Ellinger, *Arch. Gynaekol.*, **156**, (1934), 471.

5. G. Garrasi, *Bull. Soc. Med. Chir. Mod.*, **35**, (1935), 219.

6. J. P. McGovern, K. Ozkaragoz, A. E. Hensel, Jr., and K. L. Burdon, *Ann. Allergy*, **18**, (1960), 1342.

7. K. L. Burdon, K. Ozkaragoz, H. S. Kaufman, and J. P. McGovern, *Ann. Allergy*, **18**, (1960), 972.

8. A. Reinberg and M. H. Smolensky, Biorhythms and Human Reproduction, (Proceedings of a Conference sponsored by the International Institute for the Study of Human Reproduction), 15, In Press.

9. A. Southam and F. P. Gonzaga, *Amer. J. Obstet. Gyncol.*, **91**, (1965), 142.

10. F. Halberg, M. Engeli, C. Hamburger, and D. Hillman, *Acta Endocr. Suppl.*, **103**, (1965), 1.

11. R. B. Sothern, *Proceedings of the Interational Society Study on Biological Rhythms, 1971*, Igaku Shoin Ltd, Tokyo, In Press.

12. P. Procacci, G. Buzzelli, I. Passeri, R. Sassi, M. R. Voegelin, and M. Zoppi. Studies on the cutaneous pricking pain threshold in man. Circadian and Circatrigintan changes *Res. Clin. Stud. Headache*, 3 (1972), 260–276, Karger, Basel.

13. H. Geber and E. Rafka, *Dermatol. Wochschr.*, **90**, (1930), 737.

14. H. Bergh, in Novak, E., *Menstruation and Its Disorders*, Vol. 2 (E. Novak, Ed.), Appleton, New York, 1931, p. 308.

15. W. Amann, *Dermatol. Wochschr.*, **152**, (1966), 702.

16. W. Amann, *Dermatol. Wochschr.*, **151**, (1965), 1352.

17. K. Wisniewski-Roszkowska, *Polski Tyged. Lekar.*, **16**, (1961), 1362.

18. O. C. Hansen-Pruss and R. Raymond, *J. Clin. Endocr.*, **3**, (1943), 81.

19. A. Reinberg, *Perspectives Biol. Med.*, **11**, (1967), 111.

20. A. Reinberg, Z. Zagula-Mally, and J. Ghata, *Ann. Endocr.*, **25**, (1964), 670.

21. R. E. Lee, M. H. Smolensky, and J. P. McGovern, Manuscript in Preparation.

22. F. Halberg, Y. L. Tong and E. A. Johnson, in *The Cellular Aspects of Biorhythms*, (H. Von Mayersback, Ed.), Springer, Berlin, 1967, p. 20.

23. E. Haus and F. Halberg, *Environ. Res.*, **3**, (1970), 81.

24. N. L. Wulfsohn and W. M. Politzer, *S. African Med. J.*, **38**, (1964), 173.

25. Y. Aoyama, *Japan. J. Allergy*, **14**, (1965), 583.

26. C. Y. Chen, *J. Formosa Med. Assoc.*, **61**, (1962), 766.

27. A. Reinberg, J. Ghata, and E. Sidi, *J. Allergy*, **34**, (1963), 323.

28. A. Reinberg, J. Ghata, and E. Sidi, *Ann Endocr.*, **24**, (1963), 452.

29. B. Zondek and Y. M. Bromberg, *J. Allergy*, **16**, (1945), 1.

30. G. P. Heckel, *Surg. Gynecol. Obstet.*, **92**, (1951), 191.

31. G. P. Heckel, *Amer. J. Obstet. Gynecol.*, **66**, (1953), 1297.

32. G. P. Heckel and H. W. Scherp, *J. Clin. Endocr.*, **15**, (1955), 877.

33. G. P. Heckel, M. F. Leaky, and B. Pleger, *Amer. J. Obstet. Gynecol.*, **75**, (1958), 112.

34. G. P. Heckel, M. F. Leaky, A. L. Rose, and W. H. Strain, *Fertility Sterility*, **10**, (1959), 596.

35. J. Gillman, *J. Clin. Endocr.*, **2**, (1942), 157.

36. W. C. Rogers, *West. J. Surg.*, **70**, (1962), 100.

6 Secondary Rhythms Related to the Menstrual Cycle: Behavioral Changes

JOHN F. O'CONNOR, *Moderator*

CHAPTER 19

Behavioral Rhythms Related to the Menstrual Cycle

JOHN F. O'CONNOR, EDWARD M. SHELLEY, and
LENORE O. STERN

International Institute for the Study of Human Reproduction and the Department of Psychiatry, College of Physicians and Surgeons, Columbia University, New York, New York

For centuries, in various cultures, man has been aware of disturbances associated with the menstrual cycle. In the Greek and Roman ages such distress was ascribed to a "wandering uterus" or cycles of the moon rather than to menstruation. During the Middle Ages, psychological disorders in the female were attributed to the supernatural or, at times, to witchcraft. Writings on psychological disturbances and symptoms in females during this time of history are very similar to the literature in recent years that describes premenstrual tension (1). Many myths, superstitions, and fears have surrounded the menstrual cycle to the extent that it frequently became incorporated into religious rituals—primarily in a negative way. Women in some primitive societies were isolated from the rest of the tribe during the time of active bleeding.

In this presentation we will be dealing chiefly with psychological menstrual rhythms and sexual drive (libido) during the course of the month in the human female. There are many events that influence human activity, some known, probably most unknown. We have limited ourselves here to some of the psychological, environmental, metabolic, and hormonal factors related to the menstrual cycle. There is considerable disagreement on the part of investigators concerning the psychological and physiological phenomena that occur during the course of the cycle.

HORMONAL FACTORS

Hormonal factors have been increasingly recognized as potentially contributing to psychological distress in women. An investigation by Janiger (2) of menstrual cycle patterns in various ethnic and social groups revealed a definite parallel of physical symptoms in the different cultures. This would suggest that hormonal fluctuation may be more of a determinant in producing premenstrual and menstrual reactions than societal and familial patterns or taboos. The females he studied included literate Indians, Japanese, Nigerians, and Americans.

A description of the principal hormonal changes during the normal menstrual cycle is in order to serve as a background for the discussion of the psychological changes. Functionally the menstrual cycle is to be divided into a preovulatory and a postovulatory phase. The postovulatory phase (during which progesterone is secreted) is quite constant in length, lasting in most instances between 12 and 14 days. The proliferative phase, on the other hand, is much more variable in length and in short cycles lasts less than 14, while in long cycles it lasts more than 14 days. During the initial part of the follicular phase there is very little change in the hormonal secretions. During the later part of the follicular phase there is a rise in estrogen which reaches a peak immediately prior to luteinizing hormone (LH) release. In fact, it has recently been shown that estrogens act as the trigger for the LH surge which initiates ovulation. During the preovulatory phase, progesterone remains at very low levels, but androgens rise in parallel to the changes in estrogens. Immediately after LH release there is a rapid decrease in the levels of estrogens. However, following ovulation, there is a secondary increase in estrogens and androgens accompanied by an important increased secretion of progesterone. Concentration of the gonadal hormones reach a peak approximately 6 to 7 days after ovulation, after which there is a general decline to the baseline at the time of the onset of menses. The main androgen secreted by the ovary is androstenedione, although small amounts of testosterone are also secreted. The level of circulating testosterone in the normal female is approximately 14% to 20% of the level found in the male (3). Further details on the hormonal fluctuations during the ovarian cycle can be found in Chapter 12 by I. Dyrenfurth et al.

EXOGENOUS FACTORS INFLUENCING THE MENSTRUAL CYCLE

The reasons for the cyclic recurrence of premenstrual or menstrual symptoms, particularly the more severe ones, are basically not understood. Psychological interpretations of the activation of castration fears, confirmation of femininity, and conflicts regarding the desire to become pregnant do not provide a complete answer. Ephron (4) postulated that menstrual tension is related to emotional conditioning, life goals, and moods of the patient and that the potential to react may be related to current burdens and stresses. It is known that many external events can precipitate amenorrhea and anovulation, as well as alterations in the length of the cycle itself. Exogenous stress may well aggravate the severity of symptoms just as relief of such stress can ameliorate their intensity. Frequently women entering their first affair or at the point of marriage miss a menstrual period. Is this the wish for a child or is it a psychic disturbance effecting a hormonal imbalance? Separation and death are known to cause a temporary disturbance of menstruation (5). Similarly, cases of severe premenstrual tension and dysmenorrhea on occasion clear spontaneously.

There are many external and internal elements of a different nature which play a role in the cycle. The claim has been made that, in humans, the presence or deprivation of light influences the length and type of cycle (6). However, work with appropriate controls has yet to be reported in this area. Individuals living in red-orange rooms will tend to have a ten-point higher systolic blood pressure than those who live in blue or green rooms (7). One wonders what effect color would have on the menstrual period. We do know that in certain species green-blue light results in a limited gonadal development while, on the other hand, red-orange wavelengths will produce an increase in development (8). Changes in senses of smell, sight, hearing, and autonomic activity are generally agreed to be most marked during the premenstrual and early menstrual phase. Living in the tropics in contrast to more moderate climates is claimed to affect the regularity of the cycles. If the temperature is higher, the females will tend to have more irregular cycles (9). There appears to be a variation in birth rate which is related to the time of the year. It is known that in the Northern Hemisphere the number of children born in March is statistically greater than those born in other months of the year (8). This would bring us to the questionable conclusion that sexual activity and/or fertility increases in the midsummer period. Various other explanations have been suggested, such as summer festivals, extended daylight, and less dress in the summer exposing more of the male and female anatomy. However, Udry (10), studying seasonal coital activity, did not find any evidence in his group that would indicate intensified sexual drive during the summer months.

MOOD-BEHAVIORAL RHYTHMS

It is estimated that from 25% to 100% of women experience some form of premenstrual or menstrual emotional distrubance, depending upon the definition used (11). Eichner (12) makes the discerning point that the few women who do not admit to premenstrual tension are basically unaware of it, but one only has to talk to their husbands or co-workers to confirm its existence.

Southam and Gonzaga (13), in an extensive review of the literature, state that there is no definitive answer as to the etiology of the premenstrual syndrome. Physiological alterations such as hypoglycemia, sodium and water retention, and excess antidiuretic hormones have been considered causative, but none of these conditions are consistently manifest. Psychogenic factors are also hypothesized as being etiologic, partly due to a lack of consistent organic findings and partly because there is a high incidence of psychopathology and neurosis in women who suffer from the premenstrual syndrome. It seems likely that etiology is a combination of both physiologic and psychologic factors. The premenstrual syndrome does not occur in anovulatory females (14).

Chadwick (1), in an excellent monograph published in 1930, listed many of the symptoms experienced during the cycle but it remained for Frank (15) to introduce the term "premenstrual tension" to designate the cyclical irritability, anxiety, depression, and edema experienced by some women during the premenstrual period.

In later studies, Dalton (16–19) observed that these rhythmic physical and mood-behavioral changes were not restricted to the 4 days prior to menses, but were also present during the 4 days following onset of menstruation and, to a lesser degree, at time of ovulation (midcycle). She used the phrase "premenstrual syndrome" to describe the wide variety of physical and emotional symptoms and established the following criteria for its definition (16):

1. Existing in four consecutive menstrual cycles.
2. Incapacitating enough to warrant medication or the services of a physician.
3. Present only at specific periods during the cycle—4 days premenstrually, first 4 days of menses, or 4 days midcycle—as corroborated by calendar date.

Kramp (20) points out that many analyses have shown that a "troublesome" premenstrual syndrome is most common in the age group of 30 to 40 years, with the incidence increasing from about age 25 and peaking around 35 to 40.

The premenstrual syndrome should be differentiated from "idiopathic" or "spasmodic" dysmenorrhea (pain and cramps accompanying menstruation) (16). In dysmenorrhea the psychological concomitants are not present except for the possibility of anxiety. The pain starts on the first day of menstruation and is usually felt in the suprapubic region or the back. It tends to occur with the same intensity and duration each month. In contrast to this, the premenstrual syndrome can fluctuate from cycle to cycle and manifest itself in different symptoms (e.g., headache, depression, nausea, irritability) over a period of years. The symptom clusters are also likely to vary from individual to individual. It is speculated that females with dysmenorrhea tend to have an excess of progesterone in relation to estrogen levels (21), whereas the premenstrual syndrome has been associated with rapid progesterone withdrawal.

The findings that we present here do not primarily pertain to dysmenorrhea, but are concerned with the mood and behavioral patterns observed throughout the entire menstrual cycle. The behavioral fluctuations can be defined as a mode of activity; the emotional changes refer to feelings and mood that may or may not be translated into a behavioral unit.

The premenstrual syndrome is of considerable psychological significance and there are numerous reports in the literature discussing mood-behavior variations and the degree of psychopathology occurring throughout the menstrual cycle. Deutsch (22) and others (23, 24) feel that the perception of menstrual flow intensifies a female's pre-existing unconscious and conscious conflicts about pregnancy, sexuality, child bearing, fears of mutilation and death, uncleanliness, competitiveness, aggression, and masturbation. In the presence of a weak ego structure, neurotic and psychotic reactions can either develop or intensify during one or another phase of the cycle. Dalton (16) has listed a large number of disorders (e.g., depression, irritability, lethargy, alcoholic excess, sleep disturbances, epilepsy, gastrointestinal symptoms, migraine headaches, schizophrenic reactions, admissions to surgical and medical wards, crime rates, manic reactions) which occur cyclically during the premenstrual and menstrual phases. In a study of menstruation and acute psychiatric illness (17), she found that the highest incidence of hospital admissions occurred during the first 4 days of the menstrual cycle (25%) and the second highest occurred during the 4 premenstrual days (17%). (The normal probability for any event occurring during any 4-day period in the cycle should be about 14%.) A third, but lower peak, was observed during the ovulatory period (days 13 to 16). Dalton also noted that the influence of menstruation on psychiatric hospital admissions was greatest in females under 25 years of age (37%) (16). More work is clearly necessary to validate the statistical significance of such observations.

Jacobs and Charles (25) corroborate Dalton's findings, reporting that of 200 women seeking psychiatric help, the times of maximal contact occurred during the menstrual period, premenstrual phase, and midcycle. Mandell (26) and Wetzel et al. (27) noted an increased incidence of calls to Suicide Prevention Centers at these times.

Coppen and Kessel (28), in a study of menstruation and personality in an English population, found significant psychological symptoms, including irritability, depression, and tension states, occurring menstrually or premenstrually in over 25% of females. These symptoms correlated with the existence of other neurotic traits. Patients with psychoneuroses also tended to experience frequent menstrual irregularity. In a subsequent report, Coppen (29) noted that schizophrenics also were likely to have relapses during menses. However, their patients with affective disorders (manic-depressive, depressive states) did not evidence a similar pattern. In these females, depression was not aggravated during the premenstrual or menstrual period and, interestingly enough, two-thirds of such patients reported that their psychological symptoms were improved or unchanged at the time of menses.

Ivey and Bardwick (30), using a verbal anxiety scale, reported on patterns of affective fluctuation in the menstrual cycle by obtaining evaluations at ovulation and premenstrually. They observed increased levels of anxiety premenstrually as indicated by fears of death, mutilation, and separation.

Moos et al. (31) performed a longitudinal study of 15 women in order to determine the consistency of symptoms (both psychological and physical) reported during the course of several consecutive menstrual cycles. He examined the length and course of the cycle and related it to changes in symptoms and mood. Using his "Menstrual Distress Questionnaire" (32), it was found that the women who consistently experienced premenstrual tension had markedly high scores in anxiety, aggression, and depression. These women also rated high for the same states in other phases of the cycle. He noted that physical symptoms and the degree of sexual arousal tended to be more consistent from one cycle to another than were mood states; that is, depression and irritability. Furthermore, his sample evidenced a greater variability of psychological symptoms from cycle to cycle during the premenstrual and menstrual phases than during the intermenstrum. Moos (33) has also suggested the possiblity that specific symptom clusters exist (e.g., pain, concentration, behavioral changes, autonomic reactions, water retention, negative affect, and arousal). His hypothesis implies that the type of premenstrual syndrome may remain relatively constant in a given individual, but that it may become exaggerated at times of increasing psychological and/or physiological stress.

Paulson (34), in a study of 255 women which focused on developmental factors, found that 58% of what he referred to as the "high premenstrual tension" group had mothers who suffered painful menstrual and premenstrual dysfunction while only 27% of the "low premenstrual tension" group had symptomatic mothers. He felt that this was basically a result of maternal attitudes being transmitted to the daughter. Lamb (35) found a significant increase in intrafamilial conflict, a history of traumatic sexual experiences, and rejection of psychosexual roles in a group of women who complained of premenstrual tension.

Studies such as these, while contributory, are questionably valid, since they are subjective, basically retrospective, and rely too heavily on the daughter's memory of and relationship to the mother.

For ease of understanding, we have summarized and interpreted some of the data available in the literature on behavioral and physical phenomena observed during the menstrual cycle (16, 18, 19, 25, 27, 36).

Figure 1 represents changes that go beyond normal behavior. They include psychiatric hospitalizations, attempts at suicide or calls to Suicide Prevention Centers, arrests and convictions, and disturbances of a disorderly nature in women's prisons. As one can see, the maximum incidence of disturbances (45%) occur during eight consecutive days, four premenstrual and four menstrual; another peak occurs at the time of ovulation. As we have previously noted, the chance expectancy of

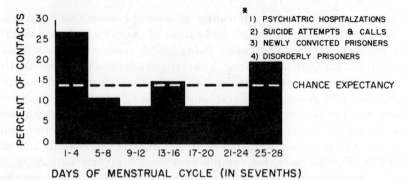

Figure 1 Menstrual cycle phase and combined incidence of psychopathologic behavior as reported in the literature (16, 18, 19, 25, 27, 36) (812 females).

any event occurring during any 4-day block of time in a 28-day cycle would be 14%. Nineteen percent of the disturbances occur premenstrually and 26% occur during the first 4 days of menstruation. The dips during days 5 to 12 and days 17 to 14 are also of interest, for it appears that in terms of probability, the midcycle is the only time where there is a "normal phase."

Figure 2 is another compilation of the literature, representing 527 acute hospital admissions for physical illness, sickness in industry, and accidents. The findings are essentially the same as the first chart, with a higher incidence during menstruation and premenstrually, and slightly lower at midcycle.

As can be seen by Figure 1, many of the studies in the literature concern patient populations who evidenced extreme psychopathology.

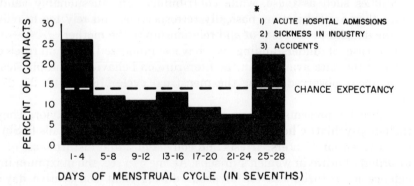

Figure 2 Menstrual cycle phase and combined incidence of morbidity in a normal population as reported in the literature (16, 18, 19, 25, 27, 36) (527 females).

At the International Institute for the Study of Human Reproduction we conducted a pilot study on three medically healthy females in their twenties who did not manifest any overtly abnormal behavioral patterns. These females undertook daily self-administered psychological testing (37) throughout the course of their menstrual cycle, and also self-administered a daily activity questionnaire which included a check list of physical symptoms and sexual activity. Basal body temperatures were recorded daily in order to establish ovulation.

In terms of the normality of the menstrual cycle, two females were ovulatory while the third had an anovulatory cycle. The two ovulatory females tended to run a parallel mood-behavior course at definite times of the month. Deviations from their individual norms occurred chiefly during three phases: premenstrual, menstrual, and postovulatory. Minor peaking was noted around the ninth and 20th days of the cycle. The smoothest part of the curve coincided with the preovulatory period; the postovulatory phase manifested more "static" (Figure 3).

The anovulatory female did not evidence any relationship between mood-behavior changes and period phase. Her cycle from start to finish consisted of continual sharp deviations (both positive and negative) from her norm. There was no indication of any predictable behavioral response in this subject (Figure 3); a not unexpected finding, since the premenstrual syndrome purportedly does not occur in anovulatory women (14). However, it seems apparent that our normal ovulatory females followed the same pattern of emotional digression that has been reported in the literature, though to a lesser degree than in women with obvious psychopathology.

In a separate study of 65 females who had sexual disorders that were psychogenic in origin (38, 39), we were unable to correlate the sexual disturbances with menstrual cycle pathology. According to psychiatric hypotheses and formulations, one could assume the existence of some relationship between severe sexual dysfunction and the premenstrual

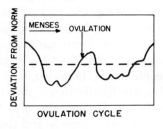 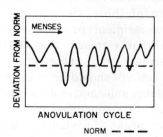

Figure 3 Representational graphs of emotional fluctuation during the menstrual cycle in two ovulatory and one anovulatory females.

syndrome. Yet this patient group did not manifest any greater incidence of menstrual symptoms, premenstrual symptoms, or dysmenorrhea than one would expect to find in an average cross-section of the female population.

LIBIDINAL FLUCTUATIONS

Whereas data on the time element of behavior response during the cycle are fairly consistent, data concerning the periodicity of libido show marked diversity. Different investigators, using different samples, report peaks of sexual arousal and activity at virtually every time during the menstrual cycle.

According to Benedek and Rubenstein (23, 24), the peak of sexual responsivity in the female is believed to occur when estrogen levels are elevated. During this phase they feel that women are more sexually interested than at other times in the cycle, as manifested by overt behavior behavior, dream content, and fantasies (although they also report evidence of increased sexual interest premenstrually). This phase culminates with ovulation and with what, in classical psychoanalytic theory, has been considered the highest level of psychosexual integration. Following ovulation the female is viewed as entering a less active, more relaxed period, with a decrease in sexual tension. Her concerns are reported as less externally oriented, more inner-directed toward her own body, and associated with the development of a passive-receptive attitude toward males. Efforts to substantiate these views have centered around the fact that progesterone is a general anesthetic in certain laboratory animals and a sedative in the human female (40). However, the dosage given to obtain these results far exceeds the normal levels in the body.

The studies of Benedek and Rubenstein (23, 24) were undertaken years ago when knowledge of hormonal factors was far less precise. The division of the cycle into the "active-estrogenic phase" and the "passive-recipient progestational phase" is no longer valid as it is known that estrogen also peaks during the progestational phase. Furthermore, while these investigators imply that the time of greatest sexual activity should be in the estrogenic midcycle period, they fail to present any substantive data concerning the distribution of intercourse.

Stopes (41) concurred with Benedek and Rubenstein, also feeling that two peaks of sexual arousal exist; one at midcycle and the other just prior to menstruation. She too, however, neglected to supply any evidence in terms of sexual behavior.

Udry (42) does provide factual data in a study of premenopausal married nonwhite working women. He found that the highest rate of intercourse (42%) occurred at about the same time as ovulation, with a sharp drop in the luteal phase and another rise (35%) in sexual activity premenstrually.

Corner (43) states that the sexual drive of the female is at its maximum just following the onset of menses. Hart (44), studying 117 British married women, reported that 22% felt the greatest sexual drive just before menstruation, 22% just after, 8% just before and after, and only 6% at midcycle. Thirty-four percent of the patients noticed no change in libido. In the same study group, 47% considered the "safe period" to be associated with menstruation, 43% had no knowledge of it, and 10% placed it at midcycle. Hart concluded that such knowledge had no influence on sexual drive.

Davis (45) and Terman (46) agree with Hart that the days just prior to and following menstruation are the times of maximal sexual arousal. Shader (47), reporting on 76 female graduate students, noted a libidinal surge premenstrually in 21%. This group also had high scores in anxiety. Conversely, the women who experienced relatively little anxiety reported no libidinal changes. McCance (48), in a study oriented toward eliciting feelings and behavior in a group of college-educated females, reports a high peak of sexual activity at the eighth day and a general increase from days 11 to 17.

Knowing that in lower mammals increased physical exercise is equated to sexual performance, Billings (49), by the use of a pedometer, measured the day to day activity of women throughout their cycles. There was a burst of activity at menstruation and preceding ovulation. Morris and Udry (50) found three such peaks—one at each end of the cycle and one coincident with ovulation. By implication the increased physical activity was equated to heightened sexual desires—a questionable premise which does, however, fit in with the work of some other investigators (48, 51).

Kinsey (52) points out that the highest level of sexual desire and arousal coincides with the premenstrual and early menstrual phase, when both estrogen and progesterone levels are markedly reduced. He found that roughly 80% of his female population preferred sexual activity at these times and also noted that orgasm occurred more frequently and more precipitously during this period than at other intervals of the month.

Masters and Johnson (51, 53) confirmed Kinsey's (52) findings on the basis of physiological parameters. They were able to demonstrate the presence of a more copious transudate and a greater degree of vasocongestion during the premenstrual-early menstrual phases and observed that the increased vascularity can intensify sexual awareness and responsivity. They also introduced a new time period of heightened sexual interest, stating that 173 of 331 women they studied (52%) favored intercourse on the last two menstrual days.

The same variability and inconsistency of findings related to periodicity of libido is apparent in data concerning the effects of oral contraception on sexual activity and desire. Bakke (54) noted a heightened sexuality in females using estrogen-progestin preparations. Others (55–57) report diminished sexual drive and decreasing levels of orgasm, and still other investigators (58–60) contend there is little or no effect on libido. Udry (61) found that the average frequency of intercourse was not affected by the use of the pill, but did observe a marked change in *periodicity* of libido among women who took oral contraceptives that inhibit ovulation, namely, the disappearance of the characteristic luteal depression of sexual activity. He attributed this to the suppression of ovulation and the resultant erasure of the luteal progestational peak. Grant and Mears (62) found that women who took strongly progestogenic oral contraceptives were more likely to experience a decrease in libido than women who were using chiefly estrogenic preparations.

Kupperman (63) also believes that progesterone significantly decreases libido in the female. Sherfey (64), on the other hand, feels that progesterone as well as estrogen can stimulate libido. However, in castrated females who are deprived of both progesterone and estrogen, there is no decrease in either sexual drive or fantasy life (51, 52, 65–68). This would lead to support of the hypothesis that androgens, rather than estrogens, particularly those that are adrenal in origin (69), are primarily responsible for determining fluctuating levels of female sexualtity (66–68, 70). Sizable amounts of exogenous androgens have frequently been observed to increase the intensity of sexual drive (71, 72) whereas there is a marked reduction of libido in androgen-deprived women (68, 73). While sexual arousal and activity are seldom diminished as a result of ovariectomy, when both ovaries *and* adrenal glands are removed, libido is greatly decreased (66–68, 74).

CONCLUSIONS

The studies previously cited indicate that the female, be she neurotic, psychotic, or a so-called "normal," undergoes rhythmic mood-hormonal-metabolic changes in the course of any given menstrual cycle. The currently existing physiologic and psychologic explanations offered cannot account completely for the findings presented. While there is increasing evidence that behavioral states and sexual response can be correlated with levels of certain hormones, particularly when exogenously administered, assertions that women are more emotional, more aggressive, or more sexually aroused at certain times during the cycle than at others need re-evaluation using a more precise methodology and research design.

This review paper is divided into two sections: (1) behavioral rhythms during the cycle that are not sexual in nature and (2) libidinal fluctuations.

Only a small portion of the findings discussed can be categorized as hard fact; these are the studies pertaining to periodicity of psychopathological behavior and incidence of morbidity as dated and verified by hospital, court, and prison records (Figures 1 and 2). Data concerning affective states, while necessarily subjective, are also fairly substantive in that adequate sized populations were studied and comprehensive psychological testing was administered. Almost all investigators agree that mood-behavior variations occur predominantly in the premenstrual and menstrual phases.

Little is known of the relationship between menstrual cycle distress and specific personality types or neuroses. Generally, the subjects are labeled as "schizophrenic," "neurotic," or "normal," but one should also take into account that basically compulsive women will react in a very different manner than women with hysterical personality structures. To the best of our knowledge, no investigation has been reported that correlates menstrual cycle symptomatology with various personality types.

Many studies concerned with behavioral patterns during the cycle also fail to consider other variables. Some workers studied only symptomatic populations. Others, who avoided selective samples, structured their studies in such a way that the subjects soom became aware of the investigator's basic interest in the cycle. For example, questionnaires that are too directly oriented toward physical and psychological symptomatology can consciously or unconsciously lead the subject to overreact to minor indispositions or mood shifts and even increase or decrease the level of sexual activity.

The material relating to libidinal fluctuations during the menstrual cycle is inconclusive and inconsistent. While there appears to be a link between sexual drive and hormonal levels, the mechanisms of interaction remain unclarified. Some reports indicate that the time of greatest female sexual interest occurs around ovulation when estrogen

322 *Behavioral Rhythms Related to the Menstrual Cycle*

levels are elevated, with interest decreasing during the luteal phase. Other investigators believe that most women do not experience significant libidinal fluctuations during the cycle, and still others contend that a maximal period of desire exists just prior to and during menstruation, or just following menstruation. While most investigators note a heightened sexual interest premenstrually (a time coincident with sharp reductions in both estrogen and progesterone secretions), some report it as the major peak of the cycle and others report it as a relatively minor manifestation.

In addition to hormonal factors, there are many external and psychological variables that can influence libidinal drive and activity in the female which have not been taken into account and which may, in part, explain the divergency of findings in the literature. Increased sexual desire just prior to menstruation may be related to loss of fear of pregnancy. Allowance should be made for the emotional and sexual behavior of the male partner. One should also obtain a life history of the subject and be aware of ethnic, religious, and socioeconomic factors, since occupation, life style, and available leisure time (e.g. holidays, weekends) can also have a profound effect on expressions of sexual activity.

To summarize, rhythmic mood-hormonal-metabolic changes occur during the human female menstrual cycle. How fluctuating endocrine levels and mood-behavior states interact has to be further defined. Improved methodology and research design will allow the investigator to correlate more precisely the many variables that exist. Such studies would surely contribute to a more complete understanding of the biological bases of human behavior.

REFERENCES

1. M. Chadwick, *The Psychological Effects of Menstruation*, Nervous and Mental Disease Publishing Co., New York, 1932.
2. F. Janiger, *Body Time* (G. Luce, Ed.), Pantheon, New York, 1971.
3. G. August, M. T. Tkachuk, and M. Grumbach, *J. Clin. Endocr.*, **29**, (1969), 891.
4. H. S. Ephron, *Psychosomatic Obstetrics, Gynecology and Endocrinology* (W. S. Kroger, Ed.), Thomas, Springfield, Ill., 1962.
5. M. Heiman, *Obstet. Gynecol.*, **7**, (1956), 3.
6. E. M. Dewan, *Amer. J. Obstet. Gynecol.*, **99**, (1967), 1018.
7. A. Schoen, Herman Miller, Inc., Personal Communication.
8. G. G. Luce, *Body Time*, Pantheon, New York, 1971.
9. N. Datta, *J. Fam. Welfare*, **4**, (1959), 38.
10. J. R. Udry and N. Morris, *Demography*, **4**, (1967), 673.

11. L. Rees, in *Psychoendocrinology* (M. Reiss, Ed.), Grune and Stratton, London, 1958.

12. E. Eichner, in *Psychosomatic Obstetrics, Gynecology and Endocrinology* (W. S. Kroger, Ed.), Thomas, Springfield, Ill., 1962.

13. A. L. Southam and F. Gonzaga, *Amer. J. Obstet. Gynecol.*, **91**, (1965), 142.

14. G. F. Melody, *Obstet. Gynecol.*, **17**, (1961), 439.

15. R. T. Frank, *Arch. Neurol. Psychiat.*, **26**, (1931), 1053.

16. K. Dalton, *The Premenstrual Syndrome*, William Heinsman Medical Books, London, 1964.

17. K. Dalton, *Brit. Med. J.*, **1**, (1959), 148.

18. K. Dalton, *Proc. Roy. Soc. Med.*, **57**, (1964), 18.

19. K. Dalton, *Brit. Med. J.*, **2**, (1960), 1425.

20. J. L. Kramp, *Acta Psych. Scand. Suppl.*, **203**, (1968), 261.

21. R. Green and K. Dalton, *Brit. Med. J.*, **1**, (1953), 1007.

22. H. Deutsch, *The Psychology of Women*, Grune and Stratton, New York, 1944.

23. T. Benedek and B. B. Rubenstein, *Psychosomat. Med.*, **1**, (1939), 245, 401.

24. T. Benedek, *Studies in Psychomomatic Medicine: Psychosexual Functions in Women*, Ronald, New York, 1952.

25. T. Jacobs and E. Charles, *Amer. J. Psychiat.*, **126**, (1970), 148.

26. A. Mandell and M. Mandell, *J. Amer. Med. Assoc.*, **200**, (1967), 132.

27. R. Wetzel, T. Reich, and J. McClure, *Brit. J. Psychiat.*, **119**, (1971), 523.

28. A. Coppen and N. Kessel, *Brit. J. Psychiat.*, **109**, (1963), 711.

29. A. Coppen, *Brit. J. Psychiat.*, **111**, (1965), 155.

30. M. E. Ivey and J. M. Bardwick, *Psychosomat. Med.*, **30**, (1968), 336.

31. R. Moos, B. Kopell, F. Melges, et al., *J. Psychosomat. Res.*, **13**, (1969), 37.

32. R. Moos, *Psychosomat. Med.*, **30**, (1968), 853.

33. R. Moos, *Amer. J. Obstet. Gynecol.*, **103**, (1969), 390.

34. M. J. Paulson, *Amer. J. Obstet. Gynecol.*, **81**, (1961), 733.

35. W. Lamb et al., *Amer. J. Psychiat.*, **109**, (1953), 840.

36. D. Janowsky, R. Gorney, et al., *Amer. J. Obstet. Gynecol.*, **103**, (1969), 189.

37. H. G. Gough and A. B. Heilbrun, *The Adjective Check List Manual*, Consulting Psychologists Press, Palo Alto, Calif., 1965.

38. J. F. O'Connor and L. O. Stern, *N. Y. State J. Med.*, **72**, (1972), 1838.

39. J. F. O'Connor and L. O. Stern, *N. Y. State J. Med.*, **72**, (1972), 1927.

40. D. A. Hamburg, *ARNMD Endocr. Cent. Nerv. Syst.*, **43**, (1966), 251.

41. M. C. Stopes, *Married Love*, 23rd ed., Putnam, London, 1937.

42. J. R. Udry and N. Morris, *Nature*, **220**, (1968), 593.

43. G. W. Corner, *Brit. Med. J.*, **2**, (1952), 403.

44. R. D. Hart, *Brit. Med. J.*, **1**, (1960), 1023.

45. K. B. Davis, *Factors in the Sex Life of 2200 Women*, Harper and Row, New York, 1929.

46. L. M. Terman, *Psychological Factors in Marital Happiness*, McGraw-Hill, New York, 1938.

47. R. Shader, A. DiMascio, and J. Harmatz, *Psychosomatics, Suppl.*, **9**, (1968), 197.

48. R. McCance, M. Luff, and E. Widdowson, *J. Hyg.*, **37**, (1937), 571.

49. E. G. Billings, *Bull. Johns Hopkins Hosp.*, **54**, (1954), 40.

50. N. Morris and J. R. Udry, *Obstet Gynecol.*, **35**, (1970), 199.

51. W. H. Masters and V. E. Johnson, *Human Sexual Response*, Little, Brown, Boston, 1966.

52. A. C. Kinsey, W. B. Pomeroy, et al., *Sexual Behavior in the Human Female*, Saunders, Philadelphia, 1953.

53. W. H. Masters and V. E. Johnson, *Human Sexual Inadequacy*, Little, Brown, Boston, 1970.

54. J. L. Bakke, *Pacific Med. Surg.*, **73**,(1965), 220.

55. A. Nilsson and L. Jacobson, *Acta Obstet. Gynecol. Scand.*, **46**, (1967), 537.

56. F. J. Kane, *J. Obstet. Gynecol.*, **102**, (1968), 1053.

57. D. Grounds, B. Davies, and R. Mowbray, *Brit. J. Psychiat.*, **116**, (1970), 169.

58. V. J. Salmon and S. H. Geist, *J. Clin. Endocr.*, **3**, (1943), 235.

59. A. L. Sopchak and A. M. Sutherland, *Cancer*, **13**, (1960), 528.

60. F. J. Kane, M. A. Lipton, and J. A. Ewing, *Arch. Gen. Psychiat.*, **20**, (1969), 202.

61. J. R. Udry and N. Morris, *Nature*, **227**, (1970), 502.

62. E. Grant and E. Mears, *Lancet*, **2**, (1967), 945.

63. H. S. Kupperman, in *The Encyclopedia of Sexual Behavior*, (S. Ellis and A. Arbanel, Eds.), Hawthorn, New York, 1961.

64. M. J. Sherfey, *The Nature and Evolution of Female Sexuality*, Random House, New York, 1972.

65. J. Bremer, *Asexualization. A Follow-up Study of 244 Cases*, Macmillan, New York, 1959.

66. S. E. Waxenberg, M. G. Drellich, and A. M. Sutherland, *J. Clin. Endocr.*, **19**, (1959), 193.

67. M. Schon and A. M. Sutherland, *J. Clin. Endocr.*, **20**, (1960), 833.

68. J. Money, in *Sex and Internal Secretions*, Vol. 2, 3rd ed. (W. C. Young, Ed.), Williams and Wilkens, Baltimore, 1961.

69. D. T. Baird, R. Horton, C. Longcope, and J. F. Taite, *Perspectives Biol. Med.*, **11**, (1968), 384.

70. G. L. Foss, *Lancet*, **260**, (1951), 667.

71. R. B. Greenblatt, *J. Amer. Med. Assoc.*, **121**, (1943), 17.

72. W. H. Perloff, *Psychosomat. Med.*, **11**, (1949), 133.

73. H. M. Bardwick, *Psychology of Women: A Study of Biodynamic Conflicts*, Harper and Row, New York, 1971.

74. S. E. Waxenberg, J. A. Finkbeiner, M. G. Drellich, and A. M. Sutherland. *Psychosom. Med.*, **22**, (1960), 435.

CHAPTER 20

Rhythmic Variations in Reaction Time, Heart Rate, and Blood Pressure at Different Durations of the Menstrual Cycle

P. ENGEL and G. HILDEBRANDT

Institut für Arbeitsphysiologie und Rehabilitationsforschung der Universität Marburg/Lahn, West Germany

It has been known for a long time that the hormonal modifications in the menstrual cycle comply with comprehensive changes in other functional systems. Lately, more interest has been taken in the subject since it has been found out that these variations also lead to changes of responsiveness, tolerance, and performance, as already shown for the circadian rhythm.

It was found, for example, that the thermic sensibility, measured at the time of acral rewarming after definite cooling, is essentially higher in the postmenstrual phase of the cycle than in the premenstrual (1), while Griefhahn (2) found the arterial pulse pressure decrease in the finger vessels following an acoustical stimulation to be smaller in the corpus-luteum phase. New investigations by Bosse and Ladebeck (3) showed that sensibility of the skin to histamine, measured by the area

325

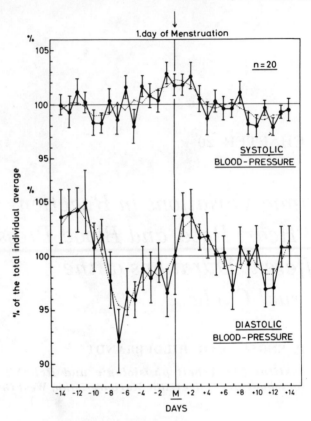

Figure 1   Mean course of systolic and diastolic blood pressure (as measured each morning immediately after waking) in the menstrual cycle. The individual cycles are synchronized to the first day of menstruation (M). The dotted curve is the result of single smoothing. Brackets indicate range of standard errors (12).

of the histamine quaddle, also undergoes a rhythmic variation with amplitudes ranging up to 100%.

There are numerous results from the last decades concerning cyclic variations of psychic and physical performance (4–11 and others) in which coordination and reaction performances, in particular, as well as different parameters of the circulatory system have been controlled.

Since these factors are of great practical importance, efforts have been made to reach a better differentiation of time in such investigations. The results of this differentiation were as follows.

In contrast to the basal body temperature, most of the functions exhibit neither a simple difference in the levels between pre- and postmenstrual phase nor a common sinus-like curve in menstrual cycle. Rather, more frequent multipeaked courses are found as can be seen in Figure 1, an example from the mean course of the morning blood pressure at first wakening. The data are synchronized to the first day of menstruation. The individual amplitudes of these variations are usually essentially higher than the average amplitude of all subjects by synchronization on the first day of menstruation. This indicates that there are interindividual differences in the phase position of the cyclic variations.

PHASE POSITION OF CYCLIC VARIATIONS

Our systematic investigations have now shown that the phase position of cyclic variations is codetermined by different factors,

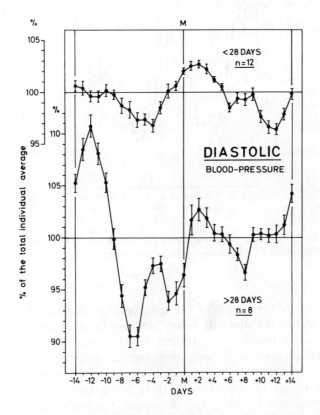

Figure 2   Mean course of diastolic blood pressure during the menstrual cycle in two groups separated after the individual cycle duration. Brackets indicate range of the standard errors (12).

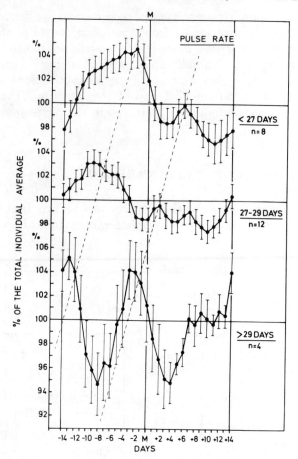

Figure 3   *Mean course of pulse rate during the menstrual cycle. Three groups with different cycle duration. Brackets indicate range of standard errors (10).*

particularly by the length of the menstrual cycle. Figure 2 shows the mean course of the diastolic blood pressure in two groups, divided according to the length of cycle and synchronized to the first day of menstruation, measured every morning immediately after waking in lying position. Apart from the amplitude of blood pressure variation in the short cycles of less than 28 days being essentially smaller, the maximum occurs in this group at around the beginning of menstruation, whereas the main maximum of the group with the longer cycle duration appears about the time of ovulation. In this group the individual amplitude of variations in diastolic pressure was significantly positive-correlated with the length of the menstrual cycle. However, there was a negative relationship between the cycle duration and the average diastolic level.

Figure 3 shows a similar behavior in the variation of pulse rate in the menstrual cycle from a further group of subjects after division into three groups according to cycle duration. Here it can clearly be seen that the main maximum of pulse rate which, in the group of shorter cycles, occurs at the end of the luteal phase, advances continuously with increasing cycle duration. After a length of 27 to 29 days this maximum is already situated before the middle of the luteal phase while, in still longer cycles, it will progress to the range of term of ovulation. Together with this, a second maximum of similar height again appears before menstruation, already marked with smaller amplitudes in the other groups.

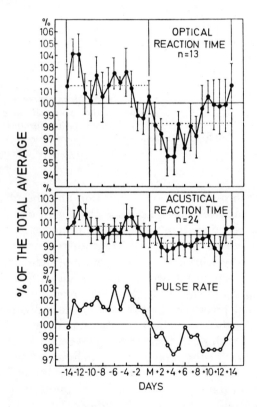

Figure 4 Mean course of the reaction time to a light signal (upper curve) and to an accoustical signal (middle curve) during the menstrual cycle. Dashed horizontal lines indicate mean premenstrual and postmenstrual niveaus. Brackets indicate standard errors. Bottom curve indicates mean course of pulse rate of the accoustically tested group (10).

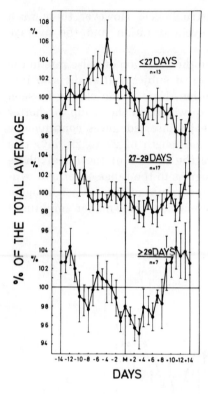

Figure 5   Mean course of the reaction time during the menstrual cycle in three groups with different cycle duration. Brackets indicate range of standard errors (10).

This relationship between cycle duration and the phase position of the menstrual variations became more obvious in our daily investigations of simple reaction time to an optical or acoustical signal, experiments carried out jointly with Witzenrath. Contrary to the earlier results obtained by Döring (5) and others, we found in the mean of a group of 37 subjects a statistically significant difference in the level between both parts of the cycle, the longer reaction times occurring in the premenstrual phase (Figure 4). Again, the individual amplitudes, calculated as the difference between the maximum and the minimum after a single smoothing of the curves, were found to be larger than in the average curves of the total group, synchronized to the first day of menstruation. Division after the cycle duration showed a considerable increase of the amplitude and a decrease in standard errors in all three groups. The systematic time shift in the position of the maximum and the minimum in the different groups can also be seen in Figure 5. In the group with shorter cycles the maximum reaction time is situated in the second part of the luteal phase, whereas with longer cycles it is already present at the end of the follicular phase. In this group the minimum is situated at the beginning of the follicular phase; in the shorter cycles, however, it is not before the end of the follicular phase. In the group with normal cycle duration the maximum and the minimum occupy an adequate middle position.

In Figure 6 the relationship between the individual length of the menstrual cycle and the temporal position of the maximum and the minimum of reaction time is presented in more detail, the means of the individual cycle duration being shown in time classes, each comprising 4 days. It is clear that with increasing cycle duration, both the maximum and the minimum advance within the menstrual cycle and vice versa. In particular the shift of the minima is more strictly correlated with the cycle duration. From the slope of the two parallel regression lines it can be concluded that the relationship between the individual cycle duration and the phase position is not only an automatic correlation, since the rate of the phase shift for a change of 1 day of cycle duration does not amount to only 1 day again but rather to three to four times as much.

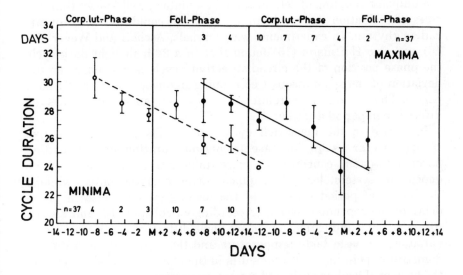

Figure 6   Correlation between the individual duration of menstrual cycle and the phase position of the minimum and maximum of the reaction time in the menstrual cycle of 37 females. Brackets indicate standard errors (10).

CONCLUSIONS

On the one hand, our results, which show a great range of variation in the phase position of the cyclic changes in reaction time—as well as similar results for the different vegetative functions—seem to be of practical importance insofar as they explain the conflicting findings in the literature concerning the optimal phase of efficiency and performance during the menstrual cycle. They should therefore provide a reason for a new revision of earlier findings related to the point of individual cycle duration.

On the other hand, within the framework of this book, the theoretical questions arising from the relationship between phase position and cycle duration are of interest. The fact that an increase in menstrual cycle duration leads to a much greater phase shift of secondary oscillations cannot account for an automatic effect, but it points to the fact that the menstrual cycle is comprised of at least two different unipolar coupled oscillating systems with relatively fixed frequencies of its own. While the leading system determines the effective period duration of the menstrual cycle, deviations from the period length of the dependent system must cause a phase angle difference of which both the direction and the magnitude will be determined by the spontaneous frequency difference of the two oscillators. Such a behavior is comparable with the unipolar coupling of self-sustained oscillators, well known from the rules of oscillation theory. A similar behavior could be shown for circadian rhythm. In experiments with animals, Aschoff and Wever (13, 14) and also Hoffmann (15) found that in a 24-hour light-dark cycle the phase position of the circadian activity cycle is dependant on the deviation of the spontaneous free running frequency from the 24-hour period. The phase shift therefore also amounts to about three times the difference in period duration.

In regard to the menstrual cycle, our results correspond to this theory insofar as an increase in period duration of the leading hormonal system, contrary to a more stable intrinsic frequency of the dependent system, leads to a phase advance of the latter, while a shortening of period duration of the leading system causes a phase delay in the dependent system. Accordingly Schlösser (16) and Döring (6) found some indications for phase angle differences in the menstrual variations of basal body temperature and the trophic state of the endometrium, pointing to further shifts in the phase relationship between the hormonal leading cycle and secondary vegetative functions.

Since the direction of the phase shift of reaction time corresponds to that of pulse rate and blood pressure variations, one can suppose that the cyclic variations of the reaction time are more strongly connected to the secondary menstrual rhythms in the vegetative system.

REFERENCES

1. H. zur Horst-Meyer and G. Heidelmann, *Schweiz. Med. Wochschr.,* **83,** (1953), 450.
2. B. Griefahn, "Extraaurale Schallwirkungen und vegetativer Tonus bei der Frau," Med. Inaug.-Diss., Göttingen, 1971.
3. K. Bosse and H. E. Ladebeck, *Z. Haut-Geschlechtskrankh.,* **47,** (1972), 365.
4. H. Brehm, *Arch. Gynaekol.,* **184,** (1953), 103.
5. G. K. Döring, *Dtsch. Med. Wochschr.,* **79,** (1954), 885.
6. G. K. Döring, *Arch. Gynaekol.,* **191,** (1958), 146.
7. L. Prokop, *Erfolg im Sport. Theorie und Praxis der Leistungssteige Steigerung,* Marathon Ed., Furlinger, Vienna, Munich, 1959.
8. E. J. Klaus and H. Noack, *Frau und Sport,* Thieme, Stuttgart, 1961.
9. G. Hildebrandt, in *Handbuch der Bäder- und Klimaheilkunde* (W. Amelung and A. Evers, Eds.), Schattauer, Stuttgart, 1962, p. 730.
10. G. Hildebrandt and A. Witzenrath, *Int. Z. Angew. Physiol.,* **27,** (1969), 266.
11. I. Bausenwein, *Muench. Med. Wochschr.,* **114,** (1972), 1325.
12. P. Engel, *Med. Welt,* **21,** (1970), 496.
13. J. Aschoff and R. Wever, *Z. vergleich. Physiol.,* **46,** (1962), 115.
14. J. Aschoff and R. Wever, *Comp. Biochem. Physiol.,* **18,** (1966), 397.
15. K. Hoffmann, *Z. Naturforsch.,* **18b,** (1963), 154.
16. W. Schlösser, *Geburtsh. Frauenheilk.,* **15,** (1955), 917.

CHAPTER 21

Sleep-Cycle Alterations During Pregnancy, Postpartum, and the Menstrual Cycle

OLGA PETRE-QUADENS and **CLAIRE DE LEE**

Department of Developmental Neurology, Born-Bunge Research Foundation, Antwerp, Belgium

Previous studies have suggested that alterations of the hormonal status could be responsible for changes in paradoxical sleep (PS) and in the other sleep stages in various conditions (1).

Endocrine changes occur spontaneously during pregnancy, postpartum, and the menstrual cycle. Combined neurophysiological and endocrinological research has already provided much information about the structures and pathways of the central nervous system involved in these changes. Hartmann (2) observed that PS percentage increased in the second half of the menstrual cycle, suggesting a relationship between PS and the secretion of hormones. However, this observation was not confirmed. According to Henderson (3), only the latency between sleep onset and the first PS epoch was decreased. But Oswald (4) reported that a decrease in PS latency always correlated an increase in PS percentage. This observation indirectly supported Hartmann's data and we thought that the hypothesis of an endocrine influence upon sleep should be tested again.

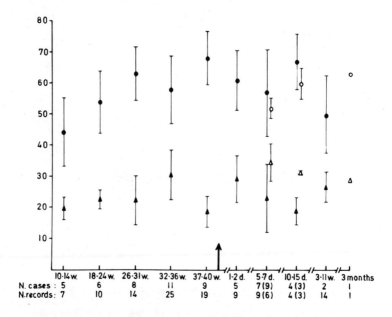

Figure 1   Percentages of PS and spindle sleep during pregnancy and postpartum. In the abscissa the age of pregnancy in weeks is given to the left to the arrow; and to the right to the arrow are the number of days, weeks, or months after delivery; and the number of cases and number of tracings are given below the arrow. The ordinate is the mean percentages and standard deviations of spindle sleep (circles) and PS (triangles). Black symbols indicate pregnancy and postpartum, and white symbols refer to the data in breastfeeding women (1).

SLEEP ALTERATIONS DURING PREGNANCY

The electro-encephalographic (EEG), -oculographic (EOG) and -myographic (EMG) patterns which characterize the sleep of the normal human adult were thoroughly modified during pregnancy.

Stage 4 was broken up and interrupted by epochs of low-voltage EEG. It did not correspond any more to the criteria of differentiation of stage 4 and resembled the alternatively slow and fast EEG of quiet sleep of the premature and the full-term neonate. The eye-movements of PS and the spindles which are characteristic of stages 2 and 3-sleep occurred simultaneously and rendered the individualization of these sleep-stages arbitrary. Therefore, we have analyzed separately the various sleep parameters.

In the EEG we have taken into account the spindles only. Their amounts increased considerably from the 20th week of pregnancy ($p <$ .05, T-test), and they were overcharged with spikes which were sometimes similar to epileptic discharges. The spindles decreased slightly between 32 and 36 weeks. They increased again between 36 and 40 weeks ($p < .05$) (Figure 1).

The EMG, which is normally abolished during PS only, varied considerably during sleep in pregnancy. It was frequently abolished out of PS (Figure 2).

The EOG showed the most striking changes. Isolated eye movements occurred during all the sleep stages, except in stage 4. They appeared as of the first weeks of pregnancy (5) and were most frequently vertical. Isolated eye movements occurred irregularly, and their relative amounts remained unchanged after the 20th week.

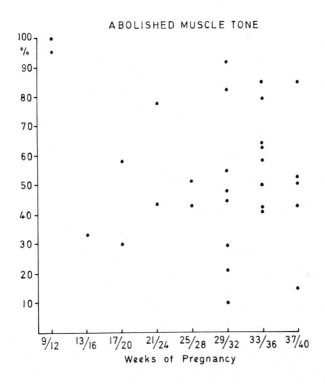

Figure 2 Muscle tone. The ordinate is the abolished muscle tone (in percentages) during stages 1, 2, 3, and 4 of sleep. The abscissa is age of preganancy (e.g., 9/12 = 9 to 12 weeks (5).

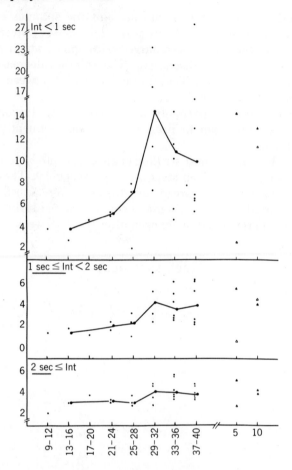

Figure 3   Evolution of the three types of eye movement intervals (I) during pregnancy and postpartum. The ordinate is the relative amount of I for 40 seconds of PS. The abscissa is the age of pregnancy (in weeks) to the left to the "break" and the age of postpartum (in days) to the right to the "break." The white triangles refer to nonbreastfeeding women; the black triangles to breastfeeding women. The dots refer to the individual values. The solid lines represent the median values.

The bursts of rapid eye movements (REM) of the pregnant women differed from those of nonpregnant subjects by an increase in eye movement density. Owing to the apparantly contradictory data of the literature, the time intervals separating consecutive eye movements have been measured. Because PS percentage is the ratio of REM sleep to total sleeping time, this method had the advantage of avoiding the confusion due to the variations in total sleeping time.

There was increased oculomotor activity related especially to time intervals shorter than 1 second. Figure 3 shows the evolution during pregnancy of the densities of the three types of eye movement intervals. The intervals (I) < 1 second gradually increased from 17 to 20 weeks of pregnancy and reached their highest values at 29 to 32 weeks. They slightly decreased from 32 weeks on, but their level remained higher than during the first months. There is still a highly significant difference indeed for the I < 1 second (p < .001, Mann-Whitney U test) between three 33-week pregnant subjects and two nonpregnant women of the same age taken as controls (Figure 4).

The increase in the amount of I < 1 second in the second half of pregnancy is significant ($2p$ = .02, Wilcoxon test) in spite of the large interindividual variations. The eye movement density increased simultaneously with the amount of spindles. The initial increase and subsequent decrease in the intensity of production of eye movements in the course of pregnancy represent a very slow rhythm covering a period of several weeks or months. It could be detected by the I < 1 second only. The 1 second $\leq I$ < 2 second and the $I \geq$ 2 second were not re-

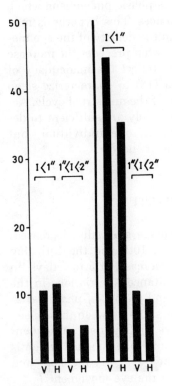

Figure 4   *Eye movement density of PS. The ordinate is the eye movement intervals (I) per minute of PS in women aged 20 to 24 years. The abscissa gives I < 1 second and 1 second ≤ I < 2 second. The left side shows nonpregnant women and the right side shows 33-week pregnant women. V = vertical eye movements. H = horizontal eye movements (1).*

lated to whatever could be responsible for this sequentially ascending and descending curve.

On the one hand, Zander (6) and Short (7) have shown that the blood progesterone level increases until 5 weeks before delivery. On the other hand, progesterone diminishes the threshold of PS in the rabbit (8). Application of progesterone directly to the forebrain of chronically implanted cats induces sleep with early appearance of PS (9). There is an increase in sleep spindles during pregnancy. This observation has been confirmed by Karacan (10). It might be the first effect of an increased progesterone level. Heuser (9) has demonstrated that injected progesterone induces sleep-spindles in cats. There are conflicting reports concerning the increase in PS percentage. On the one hand, Roffwarg (11) found a decrease in PS percentage at 7 and 9 months of pregnancy. On the other hand, Billiard (12) found a slight increase.

It may be that PS and eye movement density are two different entities. PS, defined by the simultaneous occurrence of 3 parameters— abolition of the EMG, activation of the EEG, and occurrence of eye movements—may reflect a process of physiological adaptation. The variability observed in PS percentage in different individuals may reflect their ability to integrate the variations of the phasic phenomena which seem to be more sensitive to various influences. This hypothesis may be supported by the fact that the increase in the density of the eye movement intervals shorter than 1 second somewhat precedes the increase of PS. The contradictions in the literature reflect the inadequacy of using the conventional sleep-stages criteria (13) to characterize sleep organization as a function of pregnancy and of the menstrual cycle. Accurate measurements of one sleep parameter only are sufficient to detect time-dependent variations occurring in one individual, but sequential recordings for long periods of time are needed.

SLEEP ALTERATIONS DURING POSTPARTUM

In nonbreastfeeding women, the PS percentage gradually decreased. This decrease was more abrupt between the 10th and the 15th day after delivery, and was also preceded by a decrease, at 5 to 7 days, in the density of the $I < 1$ second. The simultaneous variations in the amount of the $I < 1$ second and of stage 2 and the out-of-phase variations of PS and of stage 2 are similar to the shifts between these sleep variables observed during pregnancy (Figure 1). The new subsequent increase in PS may be considered as a rebound phenomenon which is always consecutive to PS deprivation (4). The oscillatory increases and decreases in the amounts of spindles and in the eye movement $I < 1$

second may be linked to the same mechanism. At 10 to 15 days the amount of spindles increased again ($p < .025$, T-test). The isolated eye movements progressively disappeared.

In breastfeeding women, the PS percentage remained higher. Figure 3 shows that the density in the $I < 1$ second in one breastfeeding woman was similar to the corresponding values at 6 months pregnancy and higher than in the nonbreastfeeding woman. The variations in the amount of spindles and of isolated eye movements were similar to the ones seen during pregnancy.

The increase of PS after delivery might be due to prolactine. It is doubtful, however, that these various hormones are directly acting upon the pathways responsible for PS. It is most probable that the physiological feedback mechanisms responsible for internal inhibition and sleep use the effect of several hormones on the hypothalamic structures as a link in their pathways (14–17). The changes in the amount of eye movements and of sleep spindles may possibly reflect the effect of hormones upon the central nervous system, but metabolites other than hormones may also be involved.

EVIDENCES OF NEONATAL SLEEP CHARACTERISTICS

The median values of the eye movement densities of PS corresponding to the $I < 1$ second in babies of 33 to 36 weeks, 37 to 39 weeks, and 40

Table 1 Eye Movement Intervals <1 Second during Sleep in Neonates and Pregnant Women[a]

Term (weeks)	Neonates	Pregnant Women
33 to 36	11, 80 (7)	10, 87 (6)
37 to 39	7, 07 (12)	6, 64 (3)
40	13, 17 (11)	15, 75 (3)

[a] Median values of the relative proportions of the $I < 1$ second separating the consecutive eye movements during a complete sleep cycle, that is, one interfeeding period in neonates, total night sleep in pregnant women. Values in parentheses indicate number of cases. No statistically significant differences (Mann-Whitney U and Wilcoxon tests) are found between both groups of subjects at a given term. However, the differences between the terms are significant.

weeks gestational age and in pregnant women of similar terms are indicated in Table 1. No statistically significant differences were found in these values between the pregnant women and the neonates at corresponding terms (Mann-Whitney U and Wilcoxon tests). The changes in the amount of $I < 1$ second observed in neonates of different terms are statistically significant between the 33 to 36 week and the 37 to 39 week age group, and between the 37 to 39 week and the 40 week group (.01 and .02 level of significance, respectively, Wilcoxon test). The variations were parallel in both neonates and pregnant women. However, several other similarities were detected by simple observation. From the first weeks of *pregnancy*, eye movements occurred with spindles in the EEG; an abolished EMG could be correlated with any EEG or EOG pattern; stage 4-EEG was interrupted by periods of flat tracings and resembled the "tracé alternant" (18); there was an increase in eye movement density and, simultaneously, the latency between sleep onset and the first PS epoch was decreased; a slow alternation within the high and low density eye movement epochs was detected during the night; and the sleep patterns continued to be very variable during the first weeks after delivery.

In the *premature and full-term neonate*, stage *a* or "undifferentiated sleep" (19) has been characterized by the occurrence of isolated eye movements and a number of body jerks (20); an abolished EMG has been correlated with any electroencephalographic or electrooculographic pattern (21); and quiet sleep in the neonate has been characterized by an alternatively slow and fast EEG tracing (22). The delays between the epochs with high-density REM's were twice as long in the 33 to 36 and the 37 to 39 week gestational age groups than in the babies of the 40 to 41 week gestational age; simultaneously, the latency of the first high-density eye movement period diminished with increasing gestational age (23).

The similarities in the sleep patterns between the pregnant women and the neonate might suggest that factors common to mother and foetus are responsible for them. A first hypothesis is that the alterations of the sleep patterns in the pregnant women are related to the transfer of foetal metabolites into the blood of the mother. However, it is more tempting to consider that the factor—or factors—would originate in the placenta and would be released from there simultaneously into the mother and the foetus.

CORRELATIONS BETWEEN THE SLEEP STAGES OF THE MOTHER AND THE FOETUS MOVEMENTS*

To what extent, however, are the sleep patterns in the mother and the newborn time-locked? Biological events in the pregnant women, even if their alterations are correlated with metabolic changes, remain related to the photoperiodic and synchronization effects of day and night. The simultaneity of the common sleep characteristics between mother and foetus was to be tested.

As it is not possible to collect the brain waves of the foetus through the mother's abdomen, the foetus movements have been correlated with the corresponding sleep stages of the mother.

Method

From the fourth month of pregnancy the movements of the foetus are perceived by the mother. From the seventh month these movements are frequent, sometimes very intense, and often perceptible on the mother's abdomen. We have used the external tocograph, the sensitivity of which was increased, in order to record the foetus movements. This tocograph has the advantage of being flat. It is maintained by a periumbilical rubber belt. Its axis is mobile and obturates a photoelectric cell when pressure is exerted. The changes in electric potential are proportional to the displacement of the axis. We have recorded the variations in electrical potential, first amplified in the polygraph, by means of the channels of the EEG. This allowed to obtain all the parameters perfectly synchronized on a single record.

This pressure transmitter, first mechanical, then electrical, is not sensitive to small sliding movements as an electrode would be. It records, however, the contractions of the rectus abdominis muscles of the mother when the patient is moving. The mother and foetus movements recorded by this hypersensitive tocograph may have the same speed, and since the sleep of the pregnant woman is often disturbed, the problem was to distinguish between the two types of movements. Therefore, some authors, such as Sterman (24), have fixed several restraint gauges on the mother's abdomen. This technique presents the major inconvenience to utilize several channels for recording only the foetus movements.

We have succeeded in differentiating these two kinds of contractions by recording on one lead the mother's electrocardiogram and by ampli-

* This part of the study was carried out with the assistance of Dr. J. C. Hardy.

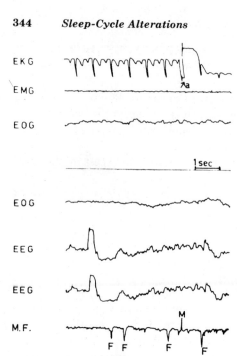

Figure 5 Recording fragment of the polygram. F = 4 foetus movements. M = 1 mother movement occurring at the precise time the artifact appears on the EKG lead (a). EKG = electrocardiogram. EMG = electromyogram. EOG = electrooculogram. EEG = electroencephalogram (39).

fying the artifacts. Movements, even of small amplitude, were communicated to the electrode fixed on the skin and entailed artifacts (fluctuation of the baseline, etc.). By using the D_2 derivation (right arm, left leg) of the mother's electrocardiogram, all the mother movements that were able to influence the tocograph also produced artifacts. Those artifacts were increased by placing electrodes at the root of the limbs and by slightly modifying their fixation system. This technique allowed us to dissociate more than 90% of the mother's movements from those of the foetus (Figure 5).

The other artifacts recorded by the tocograph were easily interpretable. They might be small uterine contractions, respiratory movements, or aortic pulsations transmitted by the uterus to the abdominal wall.

For the numbering of the foetus movements, we have established a series of criteria based on the height (minimum 0.4 cm) and on the minimum interval between two consecutive movements (1/10 second) whether the recording of the movement is mono- or biphasic. The standardization of the arbitrary criteria seems valid because repeated controls gave identical results.

The numbering of the foetus movements was done for all consecutive 20 second periods for four consecutive nights in a 33-week pregnant woman. Previous nights were supposed to habituate the subject to the experimental conditions.

Density and Periodicity of the Foetus-Movements

During four consecutive nights, 26.245 foetus movements have been numbered. Their proportion was highest—though not significantly—immediately before and after PS, namely, when a number of isolated eye movements occurred in the sleep of the mother, but when the bursts of REM's were not yet fully developed. The proportion of foetus movements was similar during PS and awakening. It was lowest during stages 1, 2, and 3—sleep without eye movements (Figure 6). The differences in the mean frequencies of foetus movements per minute were not significant whether the eye movements of the mother were isolated or grouped in bursts that indicated a continuum between them. The

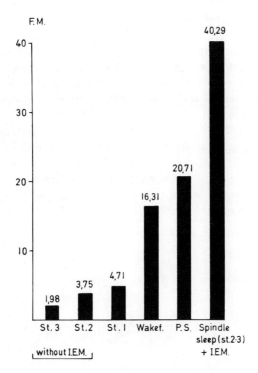

Figure 6   Frequency of the foetus movements during sleep. The abscissa is the sleep stage of the mother. The ordinate is the number of foetus movements per minute. IEM = isolated eye movements (39).

differences were significant ($p < .05$, T-test) between PS (or awakening) and sleep without eye movements. The frequency of the foetus movements increased slightly when the EEG tracing desynchronized, but it was only when the eye movements appeared that the foetus movements significantly increased.

In neonates, it is in stage a or "undifferentiated sleep" that the eye movements are isolated and the body jerks are more frequent. We are, however, unable to conclude that this 33-week old foetus is in stage a when its mother shows isolated eye movements. Indeed, a 33-week premature baby is not only asleep, but can also be awakened. We do not know what the equivalent of awakening is "in utero." It has been demonstrated that there is, in the young premature, a predominance of stage a (1). Stage a, where isolated eye movements and body jerks occur, is probably reflecting more archaic motor discharges than PS with the bursts of eye movements and the small twitches of the limbs. We may assume that the global motor discharges are predominant in the premature, because the incomplete maturation of his central nervous system does not allow him to inhibit them. Inhibition starts to appear with *bursts of REM's* and is correlated with their intensity. It has been shown that monosynaptic reflexes are depressed during PS and even more when bursts of eye movements occur (25). However, an inverse relationship between sleep and some reflexes has been demonstrated. The orbicularis oculi reflex is inhibited in slow sleep whereas it is disinhibited in PS (26).

The hypothetic factor, common to mother and foetus and responsible for the superposed activity cycles of the foetus and PS cycles of the mother, might partly account for the relative decrease of the foetus movements with bursts of REM's. In this case the decrease would be a phenomenon similar to the tonic EMG suppression of PS. It can, however, not be excluded that this decrease is also partly dependent

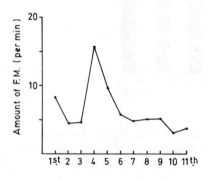

Figure 7 Frequency of the foetus movements during PS. The ordinate is the frequency of the foetus movements. The abscissa is the consecutive PS periods of the mother.

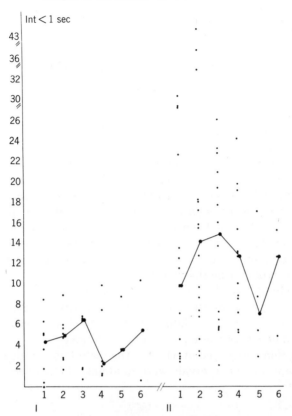

Figure 8   Density of the I < *1 second during the consecutive PS periods in pregnancy. The ordinate is the relative amounts of* I < *1 second/40 seconds of PS. The abscissa is the consecutive PS periods of total night-sleep recorded in (I) the first half and (II) the second half of pregnancy. The small dots represent individual values. The large dots represent the median values for each PS period.*

upon the intrinsic capacity of the foetal nervous system to inhibit the motor discharges. Studies on premature and newborn babies (23) provide evidence that this capacity increases with maturation and is dependent upon the integrity of the central nervous system. Sterman (27) showed that the frequency of the foetus movements is higher immediately *before* and *after* PS. He provided additional evidence that the link between mother and foetus is based upon the REM cycle and is separated from the sleep-waking cycle. Superposed upon the basic rest-activity cycle, the density of the foetus movements evidenced a slower periodicity during the night (Figure 7), similar to the periodicity of the eye movements in pregnant mothers (Figure 8). They both indicate a higher activity level during the first PS stages and a decrease toward the end of the night. Precise informations will become available only when both parameters are analyzed in a diachronic and sequential way.

THE MENSTRUAL CYCLE

We hypothesized that progesterone could be responsible for the sleep alterations observed during pregnancy. If this idea is correct, similar changes would be detected during the menstrual cycle. Data in the literature were again conflicting (3). From all the physiological parameters which have been considered in a diachronic way, we mentioned earlier that only one has been shown to be significant and periodical, namely, the bursts of eye movements of PS. The same has proven to be true during the menstrual cycle.

The density in the $I < 1$ second increased significantly during the hyperthermic phase of the ovulatory cycles (Figure 9). No changes were detected in the anovulatory cycles. The $I < 1$ second were also increased in one case where progesterone was given.

The question was raised of whether the high- and low-density periods of the eye movement intervals were randomly distributed or in a tem-

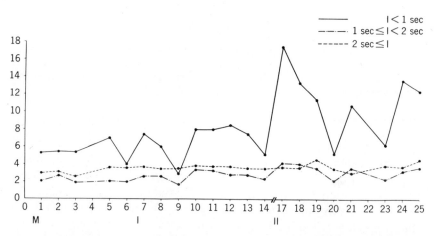

Figure 9 Evolution of the three types of eye movement intervals (I) during the menstrual cycle. The abscissa is the night of the menstrual cycle with M = menses; I = first half of the menstrual cycle; and II = second half of the menstrual cycle. The ordinate is the relative amounts of the I's for the 40 seconds of PS.

Table 2   Eye Movement Density (I < 1 second) of PS in Ovulatory Cycles as Compared to the Effects of Injected Progesterone

Subject A[a]	$M = 5.44$	$I = 7.06$	$II = 10.17$
Subject B[b]	$C = 8.59$	Prog. $= 9.36$	

[a] Median values of the $I < 1$ second of all the PS epochs of six nights during the menses (M), of 14 nights during the first (I) and of 12 nights during the second (II) half of five consecutive ovulatory cycles. The differences between I and II are significant ($2p = .05$, Wilcoxon).

[b] Median values of the $I < 1$ second of all the PS epochs of two control nights (C) and of six nights with progesterone treatment (Prog.). The control nights were insufficient to allow a statistical analysis of the data between C and Prog.

poral relationship. In other words: Is there a possibility of determining a periodicity or a rhythm in the oculomotor activity of PS during the menstrual cycle?

Statistical analysis showed that the $I < 1$ second were the only discriminative criterion. Their values were similar in all women during the first part of the menstrual cycle. This was also observed for the other sleep parameters, for the 1 second $\leq I < 2$ second and the $I \geq 2$ second. In the second part of the menstrual cycle, the latter did not change whether there had been ovulation or not. But if we considered only the $I < 1$ second, it appeared that their density was significantly higher in the ovulatory cycles than in the anovulatory cycles ($2p = .05$, Wilcoxon).

Figure 9 clearly shows a peak in the $I < 1$ second at day 17. In addition, a periodicity of 4 to 6 days may be detected by simple inspection. This figure is, however, a model constituted of the data collected during 6 consecutive months at the rate of one or two nights a week from one woman. The pregnandiol level was tested in the urines each month and the median values were 0.43 mg/24 hours at day 7 and 1.72 mg/24 hours at day 23, indicating that ovulation occurred. No significant differences were found in the $I < 1$ second between the second half of the ovulatory cycles in one subject and the nights where progesterone was given in another subject (Table 2).

The spindles also increased in the second half of the menstrual cycle and when progesterone was given.

We know that spontaneous rhythms may be artificially modified to some extent. Sawyer and his colleagues have shown that if ovulation and PS of the rat had a spontaneous periodicity, both could be triggered and their occurrence in time could be changed by modifying the light conditions. This means that the eyes are an important element in the response, and not the pineal gland as would be expected (28). A similar phenomenon was observed in humans. Dewan (29) reported that by keeping the lights on during the night at given times of the menstrual cycle, the spontaneous rhythms were changed, regularized, and the ovulation time could be controlled. The limited flexibility of this response also indicated the inherent periodicity of the organism. We have seen that only the $I < 1$ second were characteristically discriminative and periodical. Discriminative because their density was statistically higher in the cases with progesterone, whether spontaneous or injected. Periodical because their density was higher in the second half of the ovulatory cycles, their acrophase shortly following an increase in temperature (30). This infradian periodicity was superposed on a circadian periodicity since this type of eye movement burst occurred during sleep. It also overrode an ultradian periodicity. PS is a periodically recurrent phenomenon, indeed. The amount of intervals shorter than 1 second was higher toward the end of the night in the second part of the menstrual cycle (Figure 9). Weitzman (31) showed a parallel increase in the plasma-17 hydroxycortico-steroids in the PS epochs of the second half of the night. The experiments of Faure (32) and other investigators make it difficult to admit that PS is not in some way related to the hormonal status. According to Eleftheriou, the increase in oculomotor activity during PS in the human could be related to the spontaneous motor output in the ovariectomized animal after injection of sexual hormones (33). But we do not see why nor how the bursts of eye movements are selectively changed. It is known that they are generated at the level of the pons and that their patterning in bursts is dependent on the median and descending vestibular nuclei (34) and on the colliculo-mesencephalic area (35). In addition, the importance of the orbito-frontal lobes (36) for the maintenance of the bursts of eye movements is known. It has been shown that progesterone is localized in the mesencephalic structures and that the frontal cortex is involved in the neocortical mechanisms regulating hypothalamic functions (37). It can, as in the limbic system, accumulate ovarian steroids (33). Several brain structures may be involved in the functional integration of hormones. For instance, the amygdala modulate the hormonal, behavioral, and emotional processes by their regulatory function on the hypothalamus (38).

Whatever their mechanisms, changes do occur in sleep under the influence of hormones. Variations at any metabolic level represent an alteration of the "internal milieu." Further study of sleep patterns may provide valuable informations concerning the interactions between the subject and his internal or external environment. As such, they are related to wakefulness and may shed some light on the disorders of mood and attention occurring during the menstrual cycle, during pregnancy, and in postpartum psychosis.

SUMMARY

The conventional sleep-staging criteria are inadequate to characterize alterations of the phasic events as a function of the physiological changes occurring during pregnancy and the menstrual cycle, so a nonsequential analysis of the eye movements of PS was performed by measuring the time intervals separating them. The amount of spindles increased as a function of time during pregnancy, and stochastic analysis of the eye movements showed that the time intervals shorter than 1 second increased significantly in the second half of pregnancy, both events reaching a maximum at 32 weeks. It has also been shown that the activity cycles of the foetus and the eye movement cycles of the pregnant woman are superposed. Other similarities between the sleep patterns of the pregnant women and neonates are documented.

Changes similar to the ones detected in pregnancy were seen during lactation. During the menstrual cycle the eye movement density increased in the second half of the ovulatory cycles. No changes were detected in the anovulatory cycles. Since no changes are detected in the eye movement intervals longer than 2 seconds, it appears that the ultradian and infradian periodicities are detected by brain signal frequencies (spindles and eye movements) of more than 1 Hz only.

REFERENCES

1. O. Petre-Quadens, Contribution à l'étude de la phase dite paradoxale du sommeil, Acta Medica Belgica, Brussels, 1969, p. 134.

2. E. Hartmann, *J. Nervous Mental Dis.*, **143**, (1966), 406.

3. A. Henderson, G. Nemes, N. Gordon, and L. Roos, *Psychophysiology*, **2, 7** (1971), 337.

4. I. Oswald, R. J. Ashcroft, D. Berger, J. I. Eccleston, and V. R. Thacore, *Brit. J. Psychol.*, **112**, (1966), 391.

5. M. Branchey and O. Petre-Quadens, *Acta Neurol. Belg.*, **68**, (1968), 453.

6. J. Zander, *Klin. Wochschr.*, **33**, (1955), 697.

7. R. Short and B. Eton, *J. Endocr.*, **18**, (1959), 418.

8. M. Kawakami and C. H. Sawyer, *Exp. Neurol.,* **9,** (1964), 470.

9. G. Heuser, G. M. Ling, and M. Kluver, *Electroenceph. Clin. Neurophysiol.,* **22,** (1967), 122.

10. I. Karacan, W. Heine, H. W. Agnew, R. L. Williams, W. B. Webb, and Y. Y. Ross, *Amer. J. Obstet. Gynecol.,* **101,** (1968), 579.

11. H. P. Roffwarg, B. Frankel, and H. Pessah, *Psychophysiology,* **5,** (1968), 227.

12. M. Billiard, "Influence des hormones ovariennes et des gonadotrophines hypophysaires sur le sommeil chez la femme," Thèse à l'Univ. de Montpellier, 1972.

13. A. Rechtschaffen and A. Kales, *A Manual of Standardized Terminology, Techniques and Scoring System for Sleep Stages of Human Subjects,* Public Health Service, Washington, D.C., 1968.

14. E. M. Bogdanove, *Endocrinolgoy,* **73,** (1963), 696.

15. J. Faure, *World Neurol.,* **2,** (1961), 879.

16. J. Faure, in *Proceedings of the Second International Congress of Endocrinology 1964,* Excerpta Medica Foundation Series 83, Part I, Amsterdam, 1965, p. 60.

17. J. Faure, J. D. Vincent and Cl. Beusch, *Rev. Neurol.,* **115,** (1966), 443.

18. O. Petre-Quadens, A. M. de Barsy and Z. Sfaello, *J. Neurol. Sci.,* **4,** (1967), 600.

19. A. H. Parmelee and W. H. Wenner, *Develop. Med. Child Neurol.,* **9,** (1967), 70.

20. O. Petre-Quadens, *J. Neurol. Sci.,* **3,** (1966), 151.

21. O. Petre-Quadens, *J. Neurol. Sci.,* **4,** (1967), 153.

22. C. Dreyfus-Brisac and N. Monod, *Electroenceph. Clin. Neurophysiol. Suppl.,* **6,** (1956), 425.

23. O. Petre-Quadens, C. De Lee, and M. Remy, *Brain Res.,* **26,** (1971), 49.

24. M. B. Sterman, *Exp. Neurol. Suppl.,* **4,** (1967), 98.

25. F. Baldissera, G. Broggi, and M. Mancia, *Arch. Ital. Biol.,* **104,** (1966), 112.

26. J. Kimura and O. Harada, *Electroenceph. Clin. Neurophysiol.,* **33,** (1972), 349.

27. M. B. Sterman, in *Sleep and the Maturing Nervous System* (C. D. Clemente, D. P. Purpura, and F. E. Mayer, Eds.), Academic Press, New York, 1972, p. 175.

28. J. H. Johnson, N. T. Adler, and C. H. Sawyer, *Exp. Neurol.,* **27,** (1970), 162.

29. E. M. Dewan, M. F. Menkin, and J. Rock, In Press.

30. C. De Lee, *Acta Neurol. Belg.,* In Press.

31. E. D. Weitzman, H, Schaumburg, and W. Fishbein, *J. Clin. Endocr.,* **26,** (1966), 121.

32. J. Faure, J. D. Vincent, and C. Bensch, *Rev. Neurol.,* **115,** (1966), 443.

33. B. E. Eleftheriou and R. L. Hancock, *Brain Res.,* **28,** (1971), 311.

34. O. Pompeiano and A. R. Morrisson, *Arch. Ital. Biol.,* **103,** (1965), 569.

35. J. Mouret, M. Jeannerod, and M. Jouvet, *J. Physiol.,* **55,** (1963), 305.

36. J. Schlag and M. Schlag-Rey, *Brain Res.,* **22,** (1970), 1.

37. W. J. M. Nauta, *J. Neurophysiol.,* **9,** (1946), 285.

38. J. P. Gautray, Eléments de neuro-endocrinologie clinique en gynécologie et obstétrique, Masson, Paris, 1962, p. 151.

39. O. Petre-Quadens and J. C. Hardy, *Bull. Soc. R. Belge Gynécol. Obstèt.,* **38, 3,** (1968), 223.

CHAPTER 22

Rhythms in Plasma MAO Activity, EEG, and Behavior During the Menstrual Cycle

EDWARD L. KLAIBER, DONALD M. BROVERMAN,
WILLIAM VOGEL, and YUTAKA KOBAYASHI

*Worcester Foundation for Experimental Biology,
Shrewsbury, Massachusetts and Worcester State Hospital,
Worcester, Massachusetts*

The well-known cyclic rhythms of ovarian steroid hormone production during the menstrual cycle will be related in this paper to rhythms of plasma monoamine oxidase (MAO) activity, electroencephalogram (EEG) responses to photic stimulation, and certain behaviors (i.e., perceptual functioning and mood level). MAO is an enzyme believed to regulate catecholamine levels in the brain and thereby influence central nervous system adrenergic functioning (1). MAO inactivates adrenergic neurotransmitters such as norepinephrine and dopamine. Therefore, when MAO activity is elevated, adrenergic function should be depressed, and when MAO activity is reduced, adrenergic function should increase.

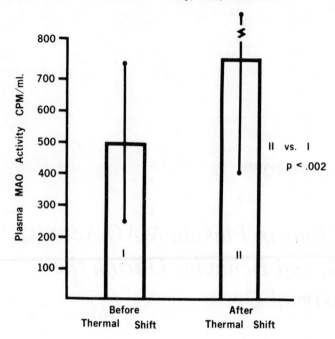

Figure 1   Means of plasma MAO activity in 19 women before and after thermal shift in basal body temperature.

OVARIAN STEROID HORMONES AND MAO ACTIVITY

Cyclic rhythms in hypothalamic MAO activity during the estrous cycle of rats have been thought to be related to adrenergic neural processes affecting the cyclic release of pituitary gonadotrophins (2, 3). Estrogens, the steroid hormones secreted by the ovary, in turn, have been shown to influence MAO activity. MAO activity in the posterior hypothalamus of the rat became significantly elevated after ovariectomy (2). The administration of estradiol benzoate returned the MAO activity toward that found in normal rats (2).

Plasma MAO activity during the menstrual cycles of normal women has also been observed to vary during different phases of the menstrual cycle (4). Plasma MAO activity was assayed using the method of Ot-suka and Kobayashi (5). Figure 1 illustrates the average plasma MAO activity of 19 regularly menstruating women before and after the time of thermal shift in basal body temperature. The average plasma MAO activity prior to the thermal shift, at a time when blood estrogen levels are known to be rising (6, 7), is significantly lower than the plasma

MAO activity occurring after the thermal shift when blood progesterone levels are known to be reaching a peak (8). These data, then, indicate a cyclic variation in plasma MAO activity coinciding with known changes in ovarian hormones.

The exogenous administration of estrogens and a progestin to amenorrheic women provides more direct evidence of the influence of ovarian hormones on plasma MAO activity (4). Plasma MAO activity was measured in 15 amenorrheic women between the ages of 18 and 36. The amenorrhea occurred without evidence of organic disease and was of 6 months to 2 years duration. Figure 2 illustrates the relationship between plasma MAO activity in these amenorrheic women and the regularly menstruating women. The mean plasma MAO activity of the amenorrheic women is significantly higher than observed during either phase of the menstrual cycle in normal women. The elevated plasma MAO activity in the amenorrheic women may be related to low blood and urinary levels of estrogen which have been reported in amenorrhea (9).

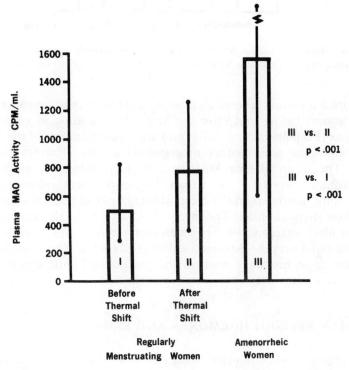

Figure 2   Means of plasma MAO activity in 15 amenorrheic women compared to 19 regularly menstruating women.

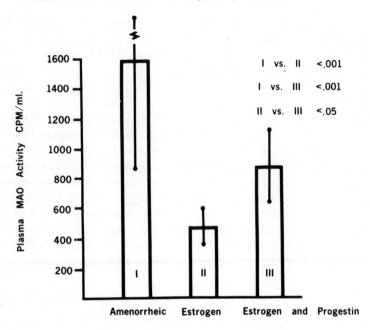

Figure 3 Means of plasma MAO activity in 7 amenorrheic women treated with conjugated estrogens (Premarin, 5 mg) alone and with a progestin (Provera, 10 ng).

Figure 3 illustrates the mean plasma MAO activity of seven amenorrheic women before and after the oral administration of conjugated estrogens (Premarin), 5 mg daily; and after a combination of the conjugated estrogens plus medroxyprogesterone acetate (Provera), 10 mg daily. The mean plasma MAO activity after estrogen treatment is significantly lower ($p < .001$) than the pretreatment values and not significantly greater than the mean MAO activity of normal women before their thermal shifts. The mean MAO activity of the amenorrheic women after estrogen plus progesterone therapy is significantly ($p < .005$) elevated over the estrogen treatment phase but is not significantly different from normal women in the postthermal shift phase of the menstrual cycle.

OVARIAN STEROID HORMONES AND EEG

The above studies clearly indicate that the ovarian hormones influence plasma MAO activity. It is possible that the ovarian hormones, in this manner, also significantly affect adrenergic neural processes in the brain. Since MAO activity is relatively low in the preovulatory phase of the cycle, heightened brain adrenergic activity would be expected at this time. When MAO activity becomes higher in the postovulatory phase of the cycle, decreased adrenergic activity would be expected. A study (10) designed to test this possibility by utilizing EEG indices is presented below.

EEG "driving" responses to photic stimulation refers to the tendency of the EEG rhythm to assume the same or a harmonic of the frequency of a bright flashing light placed before the closed eyes of the subject. EEG driving responses are known to be inhibited by drugs which act as adrenergic stimulants, for example, norepinephrine (11) and amphetamine (12); and enhanced by drugs which depress adrenergic function, for example, chlorpromazine (13). Thus the EEG driving response to photic stimulation seems to be an index of central

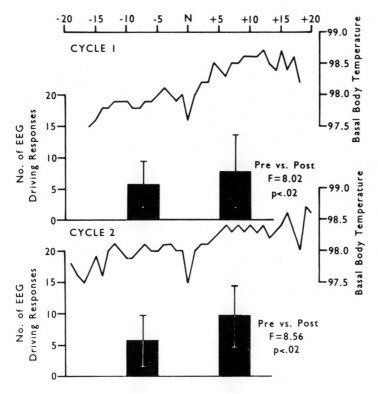

Figure 4   EEG driving responses pre- and postovulation in 14 regularly menstruating women.

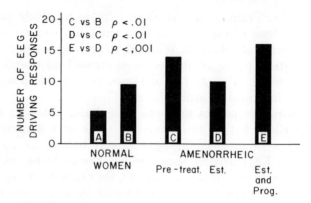

Figure 5   EEG driving responses in six amenorrheic women before and after treatment with estrogens and estrogens and a progestin.

adrenergic state. Since central nervous system adrenergic activity is expected to be greater prior to the thermal shift than afterward, fewer EEG driving responses should occur before the thermal shift than afterward.

Fourteen regularly menstruating women had their EEG responses to photic stimulation measured before and after the thermal basal body temperature shift on two successive cycles. Figure 4 illustrates the mean level of EEG driving responses at each phase of the two cycles. Significantly fewer EEG "driving" responses were observed in the prethermal shift phase of each menstrual cycle than in the corresponding postthermal shift phase. Thus these data conform to the hypothesis that low MAO activity in the prethermal shift phase of the cycle is associated with enhanced central adrenergic functioning, with the reverse true in the postthermal shift phase.

EEG driving responses were also measured in six amenorrheic women before and after the oral administration of conjugated estrogens and estrogens plus medroxyprogesterone acetate. Figure 5 illustrates that the mean level of EEG driving response of the amenorrheic women before hormone treatment is significantly ($p < .001$) elevated over that of normal women at both pre- and postthermal nadir phases of the menstrual cycle. Figure 5 also illustrates that estrogen treatment resulted in a significant ($p < .01$) suppression of EEG driving responses; while estrogen plus medroxyprogesterone acetate resulted in a return to high levels of EEG driving that were significantly ($p < .001$) greater than that observed with estrogen alone. Thus these data support the hypothesis that the ovarian hormones significantly influence central adrenergic neural processes, presumably through their influence upon brain MAO activity.

Our data suggest that central nervous system adrenergic activity is greatest during the preovulatory phase and least during the postovulatory phase of the menstrual cycle. This conclusion appears to be opposite to that of Wineman (14) who reported that peripheral autonomic indices of sympathetic arousal are highest during the postovulatory (premenstrual) phase of the cycle. The disparity between Wineman's and our conclusions may be due to the fact that we are concerned with adrenergic functioning in the brain, whereas Wineman has focused on indices known to be influenced by the adrenal medullary secretion of catecholamines. Yamori et al. (15) have recently reported data suggesting that central and peripheral adrenergic states may be negatively related to each other, which may account for the disparity between Wineman and ourselves. We are looking at central adrenergic processes while Wineman has measured peripheral indices. Elmadjian (16) has reported that urinary catecholamine levels are highest in the premenstrual phase of the cycle, a time when we believe central nervous system catecholamine levels to be reduced.

PERCEPTUAL CHANGES DURING THE MENSTRUAL CYCLE

Cyclic changes in central adrenergic functioning might be expected to manifest themselves in various behavioral phenomena. However, most attempts to measure actual changes in performances as a function of the menstrual cycle have yielded negative results (e.g., Refs. 17 and 18). A reason that performances of routine familiar tasks tend not to vary with the menstrual cycle may be that women develop internal norms or expectancies about their behavior which result in extra motivational effort at times when they feel low. If this were the case, a task which yielded no feedback to the subject and thereby prevented the formation of norms might show cyclicity over the menstrual cycle. To test this possibility, a perceptual task that yields no feedback information, the Rod and Frame Test of Perception of Verticality (19) was employed. This task is administered in a completely darkened room. The only visible stimuli are a luminescent square frame and a luminescent rod centered within the frame. The frame can be tilted 30° to either the right or left. The rod may also be tilted 30° to the right or left. Finally, the subject is seated in a chair that can also be tilted 30° left or right. The task of the subject is to tell the experimenter how to adjust the luminescent rod so that it appears to be vertical to the subject. Tilting either the frame, the body, or rod is known to influence the perception of verticality in the direction of the tilt (19, 20). Since adrenergic stimulants are known to enhance sensory responses to stimuli (21, 22), one might expect a stronger stimulus pull, that is, greater influence of frame, body, and rod tilts, in the prethermal shift phase of the menstrual cycle when we believe central nervous system adrenergic functioning to be increased. In the postthermal shift phase, when we believe adrenergic functioning is decreased, less stimulus pull should be expected.

Our initial observations were daily Rod and Frame tests on three normally menstruating women throughout one menstrual cycle. The study measured the perception of verticality under all combinations of frame, body, and rod tilts, and the data were expressed on an "activation-inhibition" dimension, that is, "activation" refers to perceptions

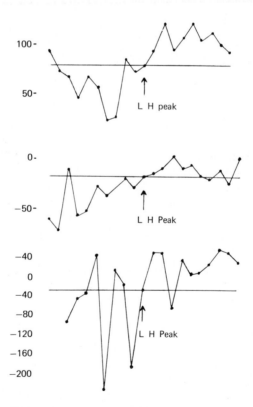

Figure 6 Daily Rod and Frame activation-inhibition scores of three regularly menstruating women throughout one menstrual cycle.

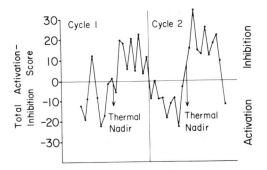

Figure 7   Daily Rod and Frame activation-inhibition scores of a regularly menstruating women during two consecutive menstrual cycles.

of verticality tilted in the direction of body, frame, or rod tilts; "inhibition" refers to verticality perceptions tilted away from frame, body, or rod tilts. Inhibition, then, represents a counterresponse to the stimuli. Figure 6 illustrates the Rod and Frame activation-inhibition scores of each of the three subjects. In each case the scores shift toward greater inhibition following the blood midcycle luteinizing hormone (LH) peak associated with ovulation. However, these changes could also be due to an adaptation to the test conditions. Hence, three more women were tested, 5 days a week, for two complete cycles. Each showed cyclicity, that is, greater Rod and Frame activation in the first half of each cycle (prethermal nadir) than in the second half. Figure 7 illustrates the Rod and Frame scores for 2 cycles of one of these women.

MAO INHIBITION AND THE ROD AND FRAME TEST

Thus the perception of verticality on the Rod and Frame Test does cycle with the menstrual cycle. Unresolved, however, is whether the perception of verticality on the Rod and Frame Test varies as a function of central adrenergic state. To test this hypothesis a manipulative study was carried out using an MAO inhibitor to increase central nervous system adrenergic activity. Six normal young adult males were tested on the Rod and Frame daily for 4 weeks. Males were selected to avoid the cyclicity we had observed in females. All subjects received placebo pills throughout the study except during the second week when threee of the subjects received an MAO inhibitor drug, isocarboxazid (Marplan), 10 mg/day. Plasma MAO activity in the three drug-treated subjects showed a marked decline during the week of treatment with the

Table 1   Average Weekly Changes in Rod and Frame Perceptions of Verticality[a]
as a Function of MAO Inhibitor[b] *or Placebo Pills*

Weeks	1	2	3	4	
MAO inhibitor Ss	0	−39.0°	−43.4°	−32.8°	$p < .001$
Placebo Ss	0	+2.1°	+13.3°	+16.9°	$p < .001$

[a] Negative scores = shift from week 1 toward greater activation. Positive scores = shift from week 1 toward greater inhibition.
[b] Isocarboxazid 10 mg daily during the second week.

MAO inhibitor drug and during the subsequent week. Plasma MAO activity levels were unchanged in the placebo-treated males.

Table 1 contains the average weekly Rod and Frame performance scores expressed in degrees departure from first baseline week. Negative values refer to a shift toward greater activation; positive values refer to a shift toward greater inhibition.

The treated subjects showed a marked shift toward activation during and immediately after drug treatment with recovery becoming apparent in the fourth week. Trend analysis indicated a statistically significant ($p < .001$) curvilinear effect in these data.

The placebo subjects show a consistent statistically significant ($p < .001$) linear trend toward increased inhibition, possibly reflecting reduced responsiveness to the various stimuli with repeated administrations.

In this study, then, low levels of plasma MAO activity induced by an MAO inhibitor drug were associated with increased activation as measured by the Rod and Frame Test. It would be assumed that the low MAO activity in the blood reflects low brain MAO activity and therefore increased brain norepinephrine levels and increased central nervous system adrenergic activity. The naturally occurring rhythms in Rod and Frame perceptions of verticality observed during the menstrual cycle, therefore, may reflect similar rhythms in brain adrenergic states as suggested by the previously described rhythms in plasma MAO activity and EEG driving.

MAO ACTIVITY IN DEPRESSED PATIENTS

Impaired central adrenergic functioning has been postulated to be the basis of mental depression, that is, Schildkraut (23) has hypothesized that mental depression is due to a catecholamine insufficiency in the brain. We would now like to present data indicating that plasma MAO activity across the menstrual cycle are associated with changes in severity of mental depression. Our data also indicate that the elevated plasma MAO activity of the depressed women can be returned toward normal by the administration of estrogen with a corresponding alleviation of depressive symptoms.

Figure 8 illustrates the relationship between pre- and postovulation plasma MAO activity levels in regularly menstruating nondepressed women versus plasma MAO activity levels of regularly menstruating, depressed women before and after oral conjugated estrogen administration. The plasma MAO activity levels of the depressed women before treatment are significantly ($p < .002$) greater than the levels found in normal women. Estrogen therapy significantly ($p < .002$) lowers the plasma MAO activity of the depressed women. These data are from an outpatient study (24) in which double blind procedures were not emp-

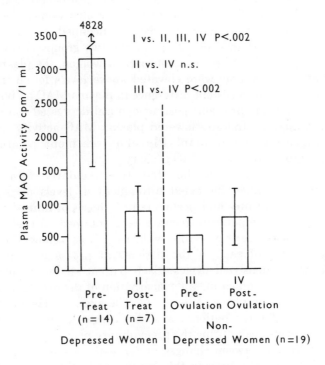

Figure 8   Comparison of plasma MAO activity in depressed women, before and after estrogen treatment (Premarin, 5 mg), versus non depressed women in pre- and postovulating phases of the menstrual cycle.

loyed. Hence, that study did not provide reliable information on the effects of estrogen administration on the symptoms of depression.

EFFECTS OF ESTROGEN ON DEPRESSIVE SYMPTOMS

A current inpatient study, however, is assessing the effect of oral conjugated estrogen administration on depressive symptoms via double blind procedures. This study also is designed to assess changes in the level of depression across the menstrual cycle in unmedicated patients. The level of depression in this study is measured weekly using the Hamilton Rating Scale (25). A level of 25 indicates severe depression. Plasma MAO activity is assessed three times weekly. Preliminary data from this study are reported below.

Table 2 presents the average Hamilton ratings of patients before and 3 months after the initiation of estrogen or placebo treatment. This group consisted of 10 premenopausal and four postmenopausal women. The mean Hamilton Rating of the estrogen-treated women declined 12 points, whereas the placebo-treated patients declined only three points. The difference in change in ratings of the two groups is statistically significant ($p < .05$). As with the outpatient women, the plasma MAO activity of these patients were elevated above levels found in nondepressed women. No consistent difference in plasma MAO activity could be seen between the pre- and postmenopausal-depressed women. Oral conjugated estrogen therapy lowered plasma MAO activity in all but one estrogen-treated woman. Placebo treatment produced no systematic effect on plasma MAO activity.

Analyses of data from the first five regularly menstruating premenopausal women indicated differences in levels of depressive symptomatology in pre- and postovulatory phases of unmedicated cycles. Basal body temperature records were obtained on these five patients to indicate time of probable ovulation. Table 3 shows the average Hamilton Rating of each of the patients at pre- and postovulation phases of their menstrual cycles. Each patient shows an increase in Hamilton Ratings, indicating an exacerbation of depressive symptoms, at the postovulatory phase of the cycle. The average plasma MAO activity of each of these patients was also higher in the postovulatory phase of the cycle than during the preovulatory phase.

The correlation between changes in depressive symptomatology and plasma MAO activity, both in this study and in the estrogen therapy study, suggest that depressive symptoms cycle with the menstrual cycle as a function of plasma MAO activity and presumably central adre-

Table 2 Average Hamilton Depression Ratings
Before and After Estrogen (Premarin, 5 to 15 mg)
or Placebo Treatment[a]

	Estrogen-Treated Patients		
	A Before Treatment	B After Treatment	C Before − After
	23.3	15.2	8.1
	36.0	18.8	17.2
	30.9	19.8	11.1
	35.2	40.0	−4.8
	28.0	17.0	11.0
	36.0	10.8	25.2
	32.0	15.8	16.2
Means	31.6	19.6	12.0
	Placebo-Treated Patients		
	D	E	F
	25.4	17.0	8.4
	28.2	23.5	4.7
	26.4	32.5	−6.1
	30.0	26.0	4.0
	22.5	19.8	2.7
	42.0	45.0	−3.0
	33.5	23.0	10.2
Means	29.7	26.7	3.0

[a] A versus B, $p < .02$. D versus E, not significant.
C versus F, $p < .05$.

nergic states. In turn, both MAO activity and central adrenergic states appear to be sensitive to gonadal steroid hormones. The well-known cyclicity of mood during the menstrual cycle in women (26) and the prevalence of premenstrual suicides (27) and premenstrual accidents (28) may similarly be a function of abnormalities in central adrenergic states accentuated at this time of the menstrual cycle. The apparent ef-

Table 3   Average Hamilton Ratings at Pre- and
Postovulatory Phases of the Menstrual Cycle in
Five Depressed Women[a]

A Pre- ovulation	B Post- ovulation	C Pre minus post
32	34	−2
28	36	−8
22	26	−4
24	34	−10
20	25	−5
Means 25.2	31.0	−5.8

[a] A versus B, $p < .03$.

fectiveness of conjugated estrogen in our depressed patients, while based on preliminary data, would suggest that these impairments in adrenergic function may be amenable to hormonal therapy.

The data presented in this paper suggest that menstrual cycle rhythms of EEG functioning, perceptual functioning, depressive symptoms, mood, and accident and suicide rates may reflect an underlying cyclicity of central nervous system adrenergic functioning which, in turn, may stem from the cyclic impact of ovarian steroid hormones on MAO activity in the brain.

ACKNOWLEDGMENT

These studies were supported, in part, by U.S.P.H.S. Grants MH-18757 and MH-11291.

REFERENCES

1. I. J. Kopin, *Pharmacol. Rev.*, **16**, (1964), 179.
2. T. Kobayashi, Y. Kobayashi, and J. Kato, in *Steroid Dynamics* (G. Pincus, T. Nakao, and J. Tait, Eds.), Academic Press, New York, 1966, pp. 305.
3. F. A. Kamberi and Y. Kobayashi, *J. Neurochem.*, **17**, (1970), 261.
4. E. L. Klaiber, Y. Kobayashi, D. M. Broverman, and F. Hall, *J. Clin. Endocr.*, **33**, (1971), 630.

5.   S. Otsuka and Y. Kobayashi, *Biochem. Pharmacol.,* **13,** (1964), 995.

6.   G. E. Abraham and E. L. Klaiber, Amer. J. Obstet. Gynecol., **108,** (1970), 528.

7.   C. S. Corker, F. Naftolin, and D. Exley, *Nature,* **222,** (1969), 1063.

8.   C. M. Cargille, G. T. Ross, and T. Yoshimi, *J. Clin. Endocr.,* **29,** (1969), 12.

9.   R. L. Vande Wiele, in *Recent Progress in Hormone Research* (E. B. Astwood, Ed.), Academic Press, New York, 1970, pp. 63–103.

10.   W. Vogel, D. M. Broverman, and E. L. Klaiber, *Science,* **172,** (1971), 388.

11.   R. Floru, A. Costin, V. Nestianu, and M. Sterescu-Volanschi, *Electroenceph. Clin. Neurophysiol.,* **14,** (1962), 566.

12.   T. Shetty, *Science,* **174,** (1971), 1356.

13.   R. S. Jorgensen and M. H. Wulff, *Electroenceph. Clin. Neurophysiol.,* **10,** (1958), 325.

14.   E. W. Wineman, *Psychophysiology,* **8,** (1971), 1.

15.   Y. Yamori, W. Lovenberg, and A. Sjoerdsma, *Science,* **170,** (1970), 544.

16.   F. Elmadjian, Fifth Pan-American Congress of Endocrinology, Lima, Peru, 1961.

17.   L. S. Hollingworth, *Functional Periodicity,* Teachers College, Columbia Univ., New York, 1914.

18.   G. H. Seward, *Psychol. Bull.,* **41,** (1944), 90.

19.   H. A. Witkin, R. B. Dyk, H. F. Faterson, D. R. Goodenough, and S. A. Karp, *Psychological Differentiation,* Wiley, New York, 1962.

20.   S. Wapner and H. Werner, *Clark Univ. Monographs in Psychology and Related Disciplines,* **2,** (1957), 1.

21.   J. E. Lebensohn and R. R. Sullivan, *U.S. Naval Med. Bull.,* **43,** (1944), 90.

22.   W. R. Thurlow, *Amer. Psycho.,* **1,** (1946), 255.

23.   J. J. Schildkraut, *Amer. J. Psychiat.,* **122,** (1965), 509.

24.   E. L. Klaiber, D. M. Broverman, W. Vogel, Y. Kobayashi, and D. Moriarty, *Amer. J. Psychiat.,* **128,** (1972), 1492.

25.   M. Hamilton, *J. Neurol. Neurosurg. Psychiat.,* **23,** (1960), 56.

26.   T. F. Benedek and B. Rubenstein, *The Sexual Cycle in Women,* National Research Council, Washington, D.C., 1942.

27.   A. Mandell and M. Mandell, *J. Amer. Med. Assoc.,* **200,** (1967), 792.

28.   K. Dalton, *The Menstrual Cycle,* Warner, New York, 1972.

7 Primary and Secondary Rhythms of the Hypothalamo-Pituitary-Adrenal Axis

ELLIOT D. WEITZMAN, *Moderator*

CHAPTER 23

Temporal Organization of the 24-Hour Pattern of the Hypothalamic-Pituitary Axis

ELLIOT D. WEITZMAN and LEON HELLMAN

Departments of Neurology and Oncology, Montefiore Hospital and Medical Center and the Albert Einstein College of Medicine, Bronx, New York

Previously well-"established" concepts have been challenged recently because of experimental studies that have emphasized the temporal organization of hormonal secretion and/or release. These studies have arisen because of the recent development and expansion of two major fields of inquiry. One, the study of biological rhythms, and the second, the multidisciplinary approach to sleep and sleep related phenomenon (1). The emphasis on frequent plasma sampling in man and other primates, associated with highly sophisticated microchemical, partially automated methods of analysis, have allowed large numbers of serially ordered hormonal assays to be correlated with a 24-hour pattern of activity. These studies have been extended to the definition of the normal pattern as well as to certain pathological (disease) entities and hold considerable promise of providing new insight into pathogenesis and treatment of a wide spectrum of neurological, endocrine and psychiatric

371

disorders. Before reviewing the recent evidence supporting the above statement, we will briefly review the historic development of our understanding of the temporal organization of neuroendocrine events.

HISTORICAL REVIEW

The early recognition of the presence of an endocrine circulating hormone system primarily emphasized a search for the detection of the hormones, the delineation of chemical specificity, the site of effect, and the endocrine glandular source. Little attention was paid to the issues of temporal relationship. The "dogma" of "constancy of the internal environment," so ably stated by Claude Bernard and Walter B. Cannon, led most investigators to search for the mechanisms of secretion and control with the explicit (and frequently implicit) assumption of their being a "feedback" mechanism which would lead to rather narrow limit constancy of hormonal concentration in the blood. A series of studies, however, carried out during the 1940s and early 1950s, raised questions about the principle of a highly controlled negative feedback "closed loop" system by the demonstration of concentration differences of certain hormones at different times of the day, month, or season (2). In 1943 it was reported by Pincus that there was a daily (24 hour) pattern of 17-OHCS in the urine of healthy adult subjects with significantly higher amounts present in the morning and lower amounts in the evening (3). This data was followed by a similar cycle demonstrated to be present in plasma 17-OHCS (4).

At about this time it was also clear that certain psychological and physiological stimuli could powerfully activate the pituitary and its controlled endocrine glands, clearly acting through the CNS with the hypothalamus and its portal-hypophyseal systems the "final common pathway" of the neuroendocrine system.

The final acceptance by the scientific community of the presence of a hormone produced in the hypothalamus, actually within neurons, which was then transported to the posterior pituitary and released there into the systemic circulation, clearly emphasized the importance and the possible (potential) role played by the CNS in the control of other pituitary hormonal systems (5). Extensive studies demonstrated the important role of brain regions in the control of pituitary hormonal release and that circulating hormones could affect multiple CNS regions. The underlying assumption generally persisted, however, that the fundamental pattern of control was a "homeostatic" mechanism, with a superimposed stimulus response system available to the organism under appropriate "stress" stimuli, or other intermittent and external sensory stimuli control (6). A number of studies, however, pointed out that "spontaneous" changes in hormone secretion and blood concentration were taking place—both with 24 hour periodicity as well as a shorter time course change. These spontaneous events and other nonpredictable temporal aspects of the hormonal pattern led to partial modification of the concept of "homeostasis" by the addition of concepts of "open" as opposed to "closed loop" system, and "variable set-point" theories (7).

During the past 8 years our research group has been engaged in the study of the temporal organization of certain hypothalamic-pituitary hormone systems as a function of the 24-hour sleep-waking cyclic pattern. The results of these studies have challenged the concept that a "steady-state" or "basal level" of most of the pituitary hormones is present during any extended time period of the 24-hour cycle and that any theory which proposes a closed feedback loop control mechanism and/or a variable regulator ("set-point") mechanism must account for the absence of a "steady-state" and the presence of episodic secretory events. In addition, recent findings in man for ACTH, gonadotropins, growth hormone, and prolactin have clearly demonstrated important temporal relationships of the hormonal secretory events to the sleep-wake cycles; in certain cases with specific correlation with sleep as a whole, as well as with specific sleep stages.

PATTERN OF ACTH—CORTISOL SECRETION

The early study of Pincus in 1943 (3) and the subsequent important work by Halberg (8) as well as others clearly demonstrated that the urine and plasma concentrations of cortisol was cyclic with a periodicity of approximately 24 hours (i.e., circadian); that the amplitude of this cycle was large (5 μg/100 ml plasma or less in the late evening, and 10–15 μg/100 ml in the early morning hours); that the major sharp rise of the 24-hours occurred during the latter third of the nocturnal sleep; and that this circadian cycle could be inverted by reversing the sleep-waking cycle in man. This inversion did not occur rapidly but took several days, actually 1 to 2 weeks.

By 1964, at the time of our first study, it was clear that the temporal pattern of normal sleep in all mammals was composed of recurrent short-term cyclic events, and that these measured physiologic changes

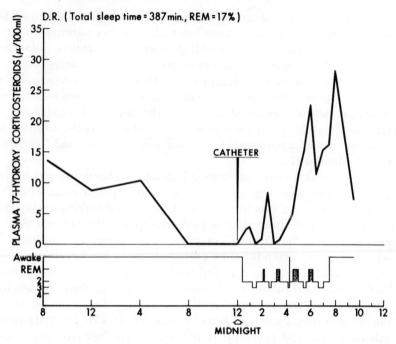

Figure 1   Graphic record of plasma concentrations of 17-hydroxycorticosteroids in a normal subject during a 26-hour period. The sleep period is graphed with the sleep stage pattern. Plasma samples were obtained every 30 minutes during the sleep period.

had associative correlative CNS active processes. In man, the latter portion of his night's sleep had been shown to consist primarily of two alternating CNS stages, rapid eye movement (REM) and Stage 2 (spindle-K-complex) sleep. Indeed, during the last 3 hours the highly active (dream associated) REM state approaches 40 to 50% of the total time (9).

The conjunction of these two major and independent lines of research led us to study the temporal pattern of hormone secretion with the development of a method utilizing a small indwelling catheter in an arm vein to obtain plasma samples at frequent intervals with a minimum of disturbance to the sleeping subject (10). The method developed allows us to define the temporal sequence of events, not only for cortisol during sleep, but is utilized by our research group for all the hormones studied, for 24 and 48 hour periods as well. The results of our early study (10) with the sampling of plasma every 30 minutes demonstrated that the rise in plasma 17-OHCS which occurred during the latter half of the night's sleep period was characterized by a series of episodic peak elevations and not by a smooth gradual rise in hormonal concentration (Figure 1). We speculated at the time that the intermittent or episodic release of cortisol might be related to recurrent changes in hypothalamic neuronal activity and that these temporal patterns might be related to the REM–non-REM sleep cycle. A subsequent study in which ^{14}C-labeled cortisol was administered intravenously just prior to the initiation of the nocturnal secretory episodes demonstrated that each episode of cortisol increment was accompanied by a proportionate decrease in cortisol specific activity, clearly demonstrating that newly formed (unlabeled) cortisol was secreted in the episodic bursts during the early morning sleep period and that between episodes, minimal or no cortisol was secreted (11). A specific one-to-one relationship between these nocturnal episodes of cortisol secretion and the REM–non-REM short-term sleep cycle, however, was not supported by other studies. When subjects were totally deprived of sleep for 1 or 2 nights, the nocturnal episodic secretion of cortisol was not prevented (12). When the sleep-waking cycle was acutely inverted, a significant delay in the reestablishment of the circadian cortisol pattern occurred and a dissociation between the sleep stage patterns and plasma cortisol concentration was present (13).

The finding of episodic secretion during the latter half of the night's sleep was extended to studies of the 24-hour pattern in normal adults who had a normal sleep-waking cycle with 7 to 8 hours of regular nocturnal sleep (14). It was found that episodic secretion was present throughout the entire 24-hour period and that the "circadian" or 24 hour pattern results from the temporal pattern (clustering) of episodic secretion (Figure 2). Utilizing a 20-minute sampling technique, an average of 9 secretory episodes occurred in a 24-hour period. Approximately 25% of the 24-hours (350 minutes) was spent in active secretion, a total of approximately 16 mg of cortisol was secreted, and a mean half-life of 66 minutes was measured. Four unequal temporal phases of episodic secretion for the 24-hour sleep-wake cycle were defined. Phase I was a 6-hour period of "minimal secretory activity" (2 minutes of secretory activity/hour, with 0.28 mg of cortisol secreted per phase), this phase beginning 4 hours before and ending 3 hours after "lights out" at night. Phase II was a 3-hour period called the "preliminary nocturnal secretory episode" (16 minutes of secretory activity/hour, with 1.7 mg secreted); the phase occurring between the third to fifth hours of sleep. Phase III was a 4-hour period, called the

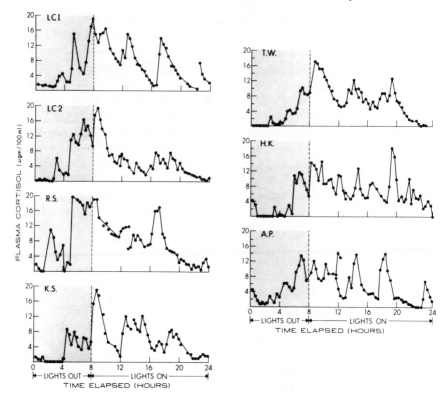

Figure 2   Plasma cortisol values of normal subjects for a 24-hour period of study. Samples obtained every 20 minutes. The period of time of "lights out" is the sleep period available at the subject's usual nocturnal sleep time.

"main secretory" phase, occurring from the sixth to the eighth hour of sleep and first hour after awakening (31 minutes of secretory activity/hour, 6.3 mg of cortisol secreted); and phase IV, the 11 hours of "intermittent waking secretory activity," occurring from 2 hours after lights on in the morning to 5 hours before sleep onset in the evening (15 minutes of secretory activity/hour, 7.2 mg secreted).

It is important to emphasize that a "basal level" or "steady-state" of cortisol concentration was not found for any extended time period of the 24-hour day. Only when the cortisol concentration falls to near zero, generally during the 4 hours in proximity to sleep onset, is there any prolonged period of constancy. This can hardly be called a "basal level," however, since essentially no cortisol is being secreted by the adrenal gland. In addition, there is considerable variability in lag time between secretory episodes and the plasma concentration at which the episodes are initiated. These findings have led us to suggest that the temporal sequence of episode initiation appears to be under CNS control as a "programmed" sequence of events, and that with a stable, repetitive daily life pattern, the association of the ACTH-adrenal secretory events with the sleep-waking cycle is part of a general program of biological rhythms. The response of the ACTH-cortisol system to acute "stress" stimuli (pain, fear, etc.) during the waking state cannot be considered to be operative in our studies as an explanation of the episodic pattern, although other possible environmental factors such as telephone calls, conversations, reading, visitors, meal time, and changes in emotional state were not rigorously defined. Their contribution, if any, to the timing of the waking secretory episodes should be evaluated by future studies.

Further studies have been carried out in our laboratories utilizing a 5-minute sampling interval in an attempt to more precisely define the temporal pattern of cortisol secretion. These studies have demonstrated that additional episodes of secretion may be missed by utilizing a 20-minute sampling frequency (Figure 3). The conclusion that we had arrived at utilizing the 20-minute sampling frequency was that the adrenal cortex was secreting cortisol with a relatively constant rate of secretion during the episodes. This was conformed with the 5-minute sampling technique, although the estimated rate of secretion was significantly different (10, 15). Since many more short episodes were noted utilizing 5 minute sampling, a rate of 105 μg/minute of cortisol was obtained, whereas 45 μg/minute was obtained utilizing 20 minute sampling. In addition, significant variations in the biologic half-life were found with 20 minute sampling. The 5-minute sampling technique demonstrated that the longer half-life estimates are the result of small episodes occurring but missed in the period between 20 minute samples. Concomitant studies of ACTH and cortisol plasma measurements have demonstrated that an appropriate temporal correlation is present for the major secretory episodes (15, 16). In addition, data from our laboratory indicates that several ACTH episodes of very short duration, if occurring in close temporal juxtaposition, may result in only one cortisol secretion episode with a smoothly rising slope.

We have carried out several studies investigating the effect of altering the sleep-waking cycle on the 24-hour pattern of cortisol secretion. In one study a group of normal subjects were subjected to a 2-week, 180° inverted sleep-wake cycle following 1 week of baseline measurements, and in a second study, normal subjects were subjected

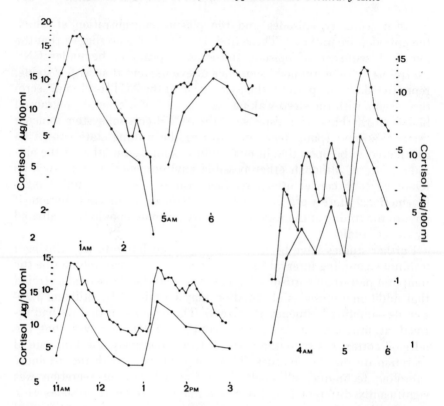

Figure 3   Graph of plasma cortisol values (log plot) with simultaneous sampling at 20 minute and 5 minute intervals compared during selected portions of the 24-hour period in normal subjects.

to a 3-week, 180° sleep-wake cycle inversion, preceded by a 3-week baseline and followed by a 3-week reinversion recovery period (13, 17). Both studies demonstrated a delay of 1 to 3 weeks before evidence of inversion of the circadian cortisol rhythm could be seen (Figure 4). Following sleep-waking reversal, the characteristic pattern of a series of peak elevations of plasma cortisol, occurring during the later part of the sleep period, was disrupted. Some subjects demonstrated a major decrease in steroid values on several reversed day sleep periods, whereas others demonstrated peak elevations very early in the daytime sleep period. A significant delay in the reestablishment of the circadian 17-OHCS occurred, and a dissociation between sleep stage patterns and plasma cortisol levels was present after sleep reversal. Although considerable variability in the 24-hour rhythm of urinary 17-OHCS took place during the 3-week inverted portion of the experiment, a rapid reinversion (return to baseline pattern) occurred within the first week.

As part of our studies of the effect of sleep-wake cycle shift on sleep and neuroendocrine patterns, we carried out a study of the effect of a prolonged 3 hour sleep-wake cycle on a group of seven normal subjects. We studied sleep stages, body temperature, and 24 hour plasma cortisol patterns (18). The subjects underwent the following 3 1/2 week (24 day) scheduled sleep-wake cycle changes while living in a hospital research unit (Clinical Research Center). Following 1 week of baseline nocturnal sleep polygraphic recording (11 P.M. to 7 A.M.), the subjects assumed a schedule for 10 days of 2 hours waking followed by 1 hour sleep (3 hour "day") throughout each 24 hour period. This was then followed by one recovery week of nocturnal sleep (11 P.M. to 7 A.M.). On days 5—6 and 13—14 an indwelling intravenous catheter was used to obtain sequential plasma samples every 20 minutes for a 24-hour period. Analysis of the pattern of cortisol secretory activity for both a baseline 24 hour period and the eighth 24-hour period of the 3-hour sleep-wake cycle revealed no significant differences in average 24-hour cortisol output, number of secretory episodes, and total secretory time. However, in the 3-hour experimental condition the secretory episodes appeared to be entrained to the 3-hour sleep-wake cycle. In this 3-hour cycle cortisol output and secretory time were maximal for the first hour after awakening, was less for the second hour of awakening, and was minimal for the third hour, that is, the hour of the next available sleep period (Figure 5). Despite the establishment of this

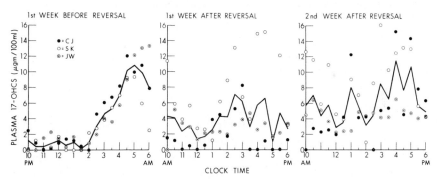

Figure 4   Graphic display of mean values of the plasma 17-OHCS for three subjects during week of nocturnal sleep (10 P.M. to 6 A.M.), and first and second week after an 180° sleep waking inversion (10 A.M. to 6 P.M.).

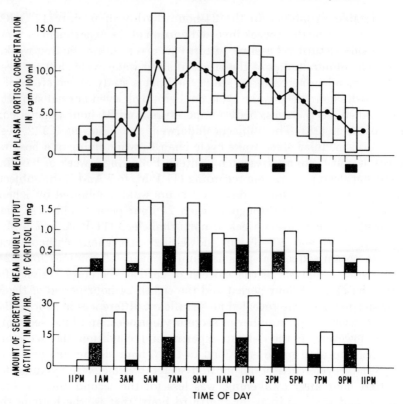

Figure 5 Top: *The mean and standard deviation of plasma cortisol concentration for a 24-hour period. The hourly mean value includes all plasma values obtained for each hour of seven studies during the eighth to ninth 24-hour period of a 3-hour sleep-wake cycle. Center: Mean hourly output of cortisol for the seven studies based on the amount secreted in the episodes or parts of episodes falling within each of the 24-hours. Bottom: Mean minutes/hour of secretory activity determined from the number and duration of episodes or parts of episodes falling within each of the 24 hours. The 8 hours of "lights out" are indicated by the shaded bars.*

3-hour pattern, a clear circadian rhythm was also present with maximal secretory activity occurring between 4 A.M. and 4 P.M. coinciding with that of the baseline rhythm and with the time of maximal sleep. However, the usual maximal secretory phase seen in baseline, between 4 A.M. and 8 A.M., was not prominent on the experimental 24-hour curves. These findings demonstrate the resistance present when altering both the 24-hour (circadian) as well as the short-term episodic secretory pattern when sleep-wake cycles are manipulated.

The 24-hour organization of the episodic secretion of ACTH-cortisol is correlated with circadian events, including light-dark changes as well as the sleep-waking cycle. In order to evaluate the role of visual perception of the 24-hour light-dark cycle in the control of these neuroendocrine temporal events, we recently studied a group of seven blind subjects, all with total absence of light perception (19). Five were totally blind from birth, one had both eyes removed during the first year of life, and one lost all vision progressively between ages 9 to 11. Utilizing our indwelling catheter technique we found that cortisol was secreted episodically in all blind subjects in patterns similar to those we previously reported in normal subjects. Five subjects had a definite circadian pattern of the cortisol episodic secretory events with a lower concentration in the evening and higher one during the early morning sleep period and subsequent waking period (Figure 6). Two of the subjects had an atypical cortisol pattern and did not clearly demonstrate a 24-hour rhythm. There was no relation between the 24-hour pattern, however, and whether the subjects had congenital or adventitious blindness. We also studied the pattern of secretion of growth hormone in the blind subjects, and these results will be described in the section on Growth Hormone.

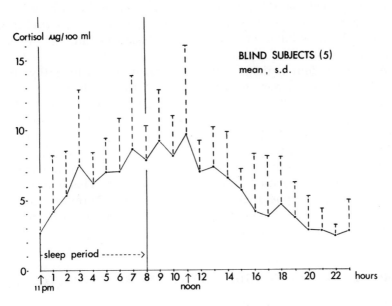

Figure 6   The mean and standard deviation of plasma cortisol values for five blind subjects for each hour of a 24-hour period.

Several clinical problems have been studied utilizing the 24-hour frequent sampling technique. Two patients with Cushing's syndrome have been studied (20, 21). One patient with bilateral adrenal hyperplasia and no evidence of a pituitary or adrenal tumor had a 24-hour pattern of cortisol secretion which was a marked exaggeration of the normal episodic pattern. The range of values was between 15 and 46 μg/100 ml. Since this pattern included major episodes of secretion with large amounts of cortisol produced in each episode, it suggests that neuronal pathways control the secretion of cortisol in Cushing's syndrome in a manner comparable to the normal subject through the mediation of the episodic secretion of ACTH and CRF. A second patient had Cushing's syndrome due to a proven adrenal cortical adenoma. Her plasma concentrations did not demonstrate major secretory episodes, but rather fluctuated in a narrow range of values between 8 and 14 μg/100 ml. Although cortisol was secreted in a series of distinct secretory episodes, the mean increase in concentration was small, suggesting that the adrenal neoplasm was relatively autonomous of hypothalamic-pituitary control. This patient had an absent circadian curve. Nine months after removal of the adrenal adenoma, a repeat study revealed that the characteristic normal 24-hour episodic pattern was now present. The findings in these two patients, as well as in a third published case (22), suggest that definition of the 24-hour pattern of plasma cortisol with frequent sampling may provide a useful means of differentiating bilateral adrenal hyperplasia from adrenal tumors in patients with Cushing's syndrome.

A study of four patients with severe spontaneous hyperthyroidism revealed that there was an increase in both the number of cortisol secretory episodes and the amount of time spent in adrenal secretory activity for the 24-hour day, and associated with a much shortened biological half-life there was an increase in production of cortisol (23). In addition, the characteristic decrease of concentration to very low values found in normal individuals at the time of sleep onset (midnight) was preserved in these patients with hyperthyroidism. This indicates that in spite of the peripheral enzymatic effects of excess thyroid hormone, as well as the direct and indirect CNS response to such metabolic changes, the circadian pattern persisted.

GROWTH HORMONE (GH) RELEASE IN RELATION TO SLEEP

During the past 5 years, it has been repeatedly confirmed that GH is secreted during the first 2 hours of sleep under normal sleep-waking conditions. At the time of our first cortisol sleep study (10), GH was measured in our nocturnal plasma samples by Dr. S. Glick to determine whether the hypothalamic-pituitary system might also release hormone during the latter part of the sleep period. We found that GH was not released at the time when cortisol was being secreted, but rather during the first hour after sleep onset (24). At that time we erroneously interpreted this sleep onset rise in GH as being related to the postprandial GH elevation, occurring 3 to 5 hours following the evening meal. Following this study, a series of investigations confirmed our findings but demonstrated that the GH release was clearly a sleep onset event (25-28). If sleep onset was shifted a few hours or prevented until 12 hours later, the release of GH was closely related to sleep onset, and no release of GH took place if the subject did not sleep at night.

Comparison of sleep stage patterns with the time of GH release suggested a relation to the electroencephalographic pattern associated with stage 3-4 sleep. In a recent study in our laboratory, plasma was sampled at 4-minute intervals during and for about 90 minutes after the transition period from waking to sleep (29). The polygraphic sleep stage scoring was carried out for both 10 and 30 second epochs during this time, and correlation was made regarding sleep stage pattern and the rise in GH concentration. The results demonstrated that the transitional events between waking and definitive sleep (stage 1 and brief

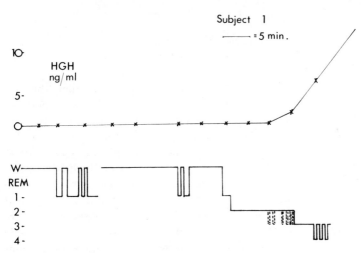

Figure 7 Graphic display of plasma GH and sleep stage at time of sleep onset at night based on sleep stage scoring using 30-second epochs. The broken line histogram indicates the first occurrence of stage 3 when the record was scored in 10-second epochs.

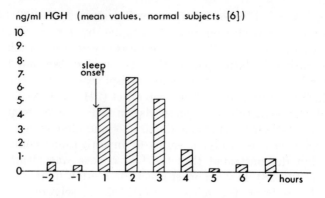

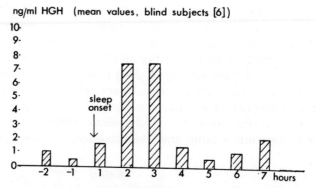

Figure 8   Comparison of mean concentration and time after sleep onset of plasma GH values for normal subjects and subjects totally blind.

periods of stage 2 sleep) were not sufficient to elicit this release. Subsequent to this transitional period, release of GH coincided with the development of definitive sleep, characterized by slow synchronous electrocortical activity (Figure 7). It should be emphasized that the presence of "slow wave" (stage 3–4) sleep is not always associated with GH release since it has been shown that especially during the latter portion of the night, stage 3–4 has been recorded without accompanying GH release (28). However, the consistent correlative temporal pattern found in the above study, as well as other reports, suggest that an underlying mechanism precipitates both the onset of "slow wave" sleep and release of GH. A number of other stimuli have been shown to produce GH release including arginine infusion, hypoglycemia, venipuncture, nonspecific stress, and L-dopa (30). In a recent study in our laboratory, it was found that the i.v. injection of 5-hydroxytryptophan induced both GH release as well as behavioral and EEG slow wave sleep in monkeys (31). The sleep onset related release of GH is not associated with changes in plasma glucose concentration, nor can it be prevented by i.v. infusion of glucose at that time (32). This later observation is different from the waking condition, since the i.v. injection of glucose will suppress the GH elevation to certain stimuli, (33) suggesting that the sleep related release utilizes a different CNS pathway.

We also studied the plasma GH concentration patterns during sleep in our studies of totally blind subjects (19). Six of the seven blind subjects had a sleep related release of GH; five had peak concentrations at sleep onset greater than 7 ng/ml, and the sixth had a small rise to a peak of 1.6 ng/ml (Figure 8). The two subjects with low or absent sleep related release of GH had the most abnormal sleep pattern. These results differ from a recent report by Krieger and Glick who reported the absence of GH release during sleep in a group of blind subjects (34). Their sleep patterns were apparently quite abnormal and may explain in part our different results. The importance of relating the GH response to a polygraphically defined sleep stage pattern is strongly emphasized.

PROLACTIN RELEASE IN MAN DURING SLEEP

During the past several years it has been demonstrated that human prolactin is distinct from growth hormone and circulates in the plasma (35). The development of an accurate radioimmunoassay technique for its measurement made it possible to study the 24-hour release patterns of prolactin in normal young men and women (36).

In a collaborative study of our laboratory group with Dr. Andrew Frantz of the Columbia College of Physicians and Surgeons, we measured prolactin and GH in plasma at 20-minute intervals for a 24-hour period in six normal adults whose sleep-waking cycles were monitored polygraphically (37). In all subjects a clear increment in the plasma prolactin concentration began 60 to 90 minutes after sleep onset (Figure 9). This initial nocturnal peak was followed by a series of larger secretory episodes resulting in progressively higher plasma concentrations during the remaining hours of sleep with peak concentrations occurring toward the end of the sleep period at approximately 5 to 7 A.M. A rapid fall in concentration occurred during the hour following awakening, with low values reached between 10 A.M. and noon. The pattern of release during the waking and sleeping

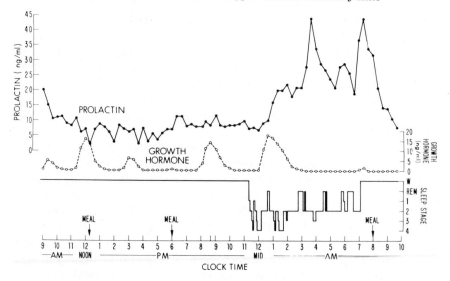

Figure 9   The 24-hour pattern of prolactin, growth hormone, and sleep stages in a normal 23-year old male.

times was clearly episodic but never fell to undetectable values. We were unable to detect readily a clear relation between the prolactin peaks during sleep and a specific sleep stage. However, one subject had a delay in sleep onset of about 2 hours, had a similar delay in the initial nocturnal sleep elevation, and had a fall at the end of sleep. Although the initial rise in prolactin after sleep onset usually began at the same time as the GH rise, the prolactin peak occurred on the average 40 minutes after the GH peak. In addition, with the exception of this sleep onset secretion episode, no consistent relation between the GH and prolactin episodes could be identified for the remaining 24-hour period. Indeed the episodic pattern of prolactin release cannot be directly correlated in temporal sequence with the other hormonal systems. Like GH and ACTH-cortisol, prolactin will be released in response to a variety of stimuli. These include stress, hypoglycemia, suckling in postpartum women, strenuous exercise, and a number of psychotropic drugs. An important additional recent finding is the elevation in concentration of plasma prolactin in patients with pituitary-hypothalamic tumors (38). Since the hypothalamic mechanism of control of prolactin has been shown to be primarily via a prolactin "inhibitory" (PIF) rather than a "releasing" factor (39, 40), the elevation of prolactin during sleep raises the possibility of a mechanism that might "inhibit" the PIF at these times. Further studies of shifts of the sleep-wake cycle and light-dark patterns are clearly necessary.

THE 24-HOUR PATTERN OF LUTEINIZING HORMONE (LH) SECRETION

The hypothalamic control of the cyclic pattern of LH secretion in adult female mammals in association with the menstrual cycle is critical for perpetuation of the species. It has been demonstrated that the timing of gonadotrophic hormone release during the ovulatory cycle in certain birds and rodents is closely linked to a 24-hour periodicity. For example, in the chicken, ovulation inducing hormone is secreted just prior to the end of the dark period, the time the chicken sleeps inducing the first ovulation in a clutch (41). Fraps has suggested that a hypothalamic threshold cycle may exist in which the lowest point may be during the dark (sleep) portion of the 24-hour day, and therefore be a "critical period" for release of gonadotrophic hormone. In the female rat it has been demonstrated that the release of LH occurs on the day of proestrus during the period of relative inactivity (lights on, sleep period) at about 2 P.M. if the rat is maintained on an approximately 14-hour light–10-hour dark schedule (time of lights on, 5 A.M.) (42). Previous studies attempting to define a possible circadian or ultradian periodicity of LH secretion in man have reported discrepant findings (43–50). Several years ago our research group undertook a study of the temporal pattern of LH secretion in man, utilizing the technique of frequent plasma sampling over a 24-hour period in association with the polygraphic definition of sleep states during the nocturnal sleep period. We have studied a group of normal young adult men, pubertal boys and girls, prepubertal children, and normal women during different segments of the menstrual cycle.

In the group of normal men (51) we found that all subjects showed discrete LH secretory episodes which were characterized by rapid rises and slower declines in concentration. The initiation and cessation of these secretory episodes occurred within narrow, well-defined LH concentration ranges. Utilizing the 20-minute sampling technique, approximately 12 secretory episodes occurred for a 24-hour period with one-third present during the sleep period. In this group of adult men we were unable to demonstrate a 24-hour rhythm nor was there evidence of a clustering of episodes as a function of the time of day. The interval between episodes was quite irregular and no clear relationship

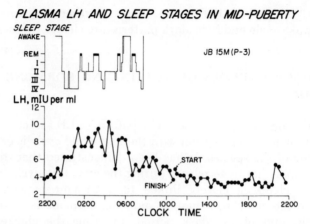

Figure 10   Plasma LH concentrations with 20-minute sampling for a 24-hour period in a 15-year old pubertal boy. The sleep stage patterns are defined during the normal time of nocturnal sleep.

could be recognized between initiation of an episode and a specific sleep stage. Rubin et al. have recently reported that LH values during sleep in men were 14% greater during REM periods when compared to other sleep stages combined (52).

A different 24-hour pattern was found in pubertal boys and girsl (53) (Figure 10). In both early and late puberty a major increment in LH concentration occurred during the sleep period. The episodes of secretion were clearly augmented during the entire sleep period. In the late pubertal group higher amplitude daytime waking LH secretory episodes began to occur, resulting in a mean LH concentration approaching the adult range. However, the mean sleep LH values were still higher than waking values in the late pubertal group. These unequivocal differences between sleep and waking portions of the 24 hours showed no sex difference. In one pubertal boy (age 15) it was demonstrated that delaying sleep onset to 5:30 A.M. resulted in a similar delay in the augmentation of LH secretion, strongly suggesting that this release is related directly to sleep such as is found with the sleep-related GH release pattern. A general relation was present between the number of LH secretory episodes during sleep and the number of sleep cycles. The secretory LH episode interval with a periodicity between 70 and 90 minutes closely matches the 80 to 90 minute non-REM–REM sleep cycle. In addition, the temporal relation to sleep stage suggested in this study that the LH secretion was initiated during non-REM sleep, with termination of the episodes in close proximity or actually during REM sleep. No such periodicity could be recognized during the waking state in these pubertal children, nor in the prepubertal and adult groups. For the prepubertal children (ages 6 to 11) no difference was found between LH concentration during the sleeping and waking states, and the values were all in a low range of concentration. These findings linking the sleep-waking pattern and the recurrent short-term sleep cycle (Non-REM–REM cycle) to gonadotrophic temporal events during puberty, suggests that certain common CNS neuronal mechanisms may be operative. The exciting developments linking the anatomy and physiology of the CNS biogenic amines in the control of both sleep and reproductive physiology emphasizes the need for cross-disciplinary programs in these areas of research interest.

Studies of the 24-hour pattern of LH secretion in adult women are complicated by the changes occurring in gonadotrophic hormones in relation to the monthly menstrual cycle. The day to day pattern of plasma LH concentration has been well documented (54, see also Chapter 12). Following menstruation, the concentration of LH in the plasma progressively rises until the midcycle ovulatory period when the "LH surge" occurs and when peak plasma concentrations two or more times greater than during other phases of the menstrual cycle are reached. During the luteal phase, LH declines several days after the LH surge and then low values are maintained through menses.

In order to study the 24-hour sleep-waking LH pattern in menstruating women, we have therefore carried out such studies during selected portions of the monthly cycle. During the early follicular phase of the menstrual cycle (first 5 days), the pattern of LH secretion was characterized by a sequence of 10 to 15 secretory episodes during the 24-hour period (55) (Figure 11). The average interval between secretory episodes was approximately 120 minutes, and no consistent difference between the inter-episode interval was found between the night and day periods. No major increment in LH occurred during sleep such as is found in pubertal children. However, there was a significant decrease in the plasma LH concentration during the first 3 hours after sleep onset in all five subjects studied (Figure 12). When the onset of the first stage 2 sleep of the night was used as a reference, a decrease of approximately 33% in mean plasma LH was present for the third hour after the onset of stage 2. There was thus a clear decrease in LH concentration in the first half with a subsequent rise in the latter half of the sleep period of the night. Comparison with an age-matched normal male group revealed that no such pattern could be recognized for the males.

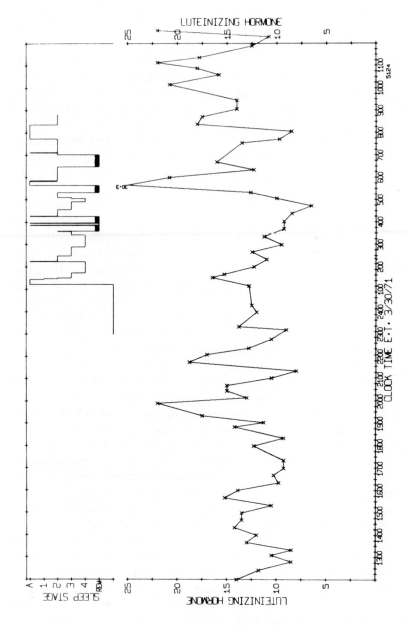

Figure 11 The 24-hour pattern of LH concentration (mIU/ml) in a normal woman (age 28) during the early follicular phase of the menstrual cycle. Sleep stage patterns are outlined for the nocturnal sleep period (12 midnight to 9 A.M.).

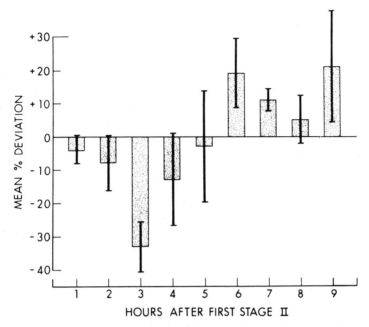

Figure 12   The mean percent deviation of LH concentration for each hour after the onset of stage 2 sleep at night for five normal women (ages 22 to 28). The "0" percent deviation value represents the mean concentration for the 24-hour period.

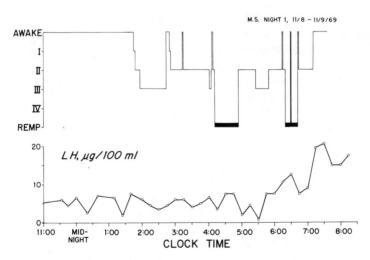

Figure 13   The nocturnal LH concentration pattern (11 P.M. to 8 A.M.) in a young women (age 33) during the late follicular phase and at the time of initiation of the presumptive preovulatory LH "surge." Sleep stage patterns are outlined above.

During the periovulatory phase of the menstrual cycle, measurement of the nocturnal (8 to 10 hours) pattern of LH secretion again revealed recurrent episodic secretory events during both the sleeping and waking periods (56). During the time of the LH surge, as defined by plasma LH concentrations greater than 40 mIU/ml, the secretory episodes were superimposed on a progressive elevation of the baseline concentration. In several subjects the presumptive onset of the LH surge could be recognized by an abrupt large increase in the plasma LH concentration. These occurred in close proximity to the end of the sleep period (Figure 13). On the succeeding night, increments in LH values to greater than 40 mIU/ml continued. These findings again emphasize the importance of short-term temporal events in the organization of hormone secretion, and the importance of frequent sampling in defining these patterns.

SUMMARY AND CONCLUSIONS

It is now clear that at least four hypothalamic-pituitary hormonal systems have temporal patterns of secretion that are closely linked to the 24-hour sleep-waking activity in man. No single principle or mechanism can explain these patterns, but rather each system has its own temporal organization and response to manipulations of the sleep-waking cycle. The ACTH-cortisol cycle is relatively resistant to change, and it can be readily dissociated from sleep in spite of a highly correlative relationship under normal stable circadian conditions. Growth hormone, on the other hand, appears to be intimately associated with a specific sleep stage in relation to the period after sleep onset, and its release can be readily shifted by shifting the time of sleep. Following the sleep onset related GH release in adults, there is usually none or only one or two small secretory episodes later in the night's sleep period. Prolactin is only beginning to be studied but there is good evidence that it is released in large quantities throughout the night in an episodic manner with initiation at the time of sleep onset and increasing in concentration throughout the sleep period with a rather abrupt termination and fall of concentration just after awakening in the morning. Preliminary evidence suggests that shift of sleep may be accompanied by a shift of the release pattern, but the evidence is too meager for firm conclusions as yet. The gonadotrophin hormone LH has a complex pattern, as expected, because of the inter-related problems of maturation (prepuberty, puberty, sexual maturity, and post maturity), sex differences, menstrual cyclicity in females, and the sleep-waking patterns. A most exciting finding is that in pubertal children there is an intimate association of LH release with sleep, as well as the suggestion of a non-REM–REM short-term cycle relationship. Evidence suggests that at certain times during the menstrual cycle in sexually mature women the release pattern of LH may also be related to sleep, with the question raised of a sleep ("critical period") associated LH episodic pattern during the "LH surge" period. All of the above hormones studied, utilizing the sequential frequent sampling technique, have been shown to have an episodic secretory pattern throughout the 24-hour sleep-wake cycle.

The importance of the biological organization of neuroendocrine systems in relation to *time* is clearly emphasized by these studies. The issues concerning the mechanisms maintaining normal temporal synchronization of hormonal release and the effect of disease upon these time-dependent processes should be pursued both from a diagnostic and pathological point of view. Indeed, the possibility that "desynchronization" of these chronobiologic events either acutely or chronically may lead to clinical disorders of functions in man should be seriously considered.

REFERENCES

1. E. D. Weitzman, "Periodicity in Sleep and Waking States," in *The Sleeping Brain* (Proceedings of the Symposia of the First International Congress of APSS) M. Chase, Ed.), B.I.S., UCLA, Los Angeles Calif., 1972, p. 193–238.

2. F. E. Yates and J. Urquhart, *Physiol. Rev., 42,* (1962), 359.

3. G. Pincus, *J. Clin. Endocr., 3,* (1943), 195.

4. F. H. Tyler, D. Migeon, A. A. Florentin, and L. T. Samuels, *J. Clin. Endocr., 14,* (1954), 774.

5. E. Scharrer and B. Scharrer, *Physiol. Rev., 25,* (1945), 171.

6. C. Fortier, *Ann. Rev. Physiol., 24,* (1962), 223.

7. E. G. Yates, The Adrenal Cortex (A. B. Eisenstein, Ed.), Little, Brown, Boston, 1967.

8. F. Halberg, *Perspectives Biol. Med., 3,* (1960), 491.

9. A. Kales, Ed., *Sleep—Physiology and Pathology,* Lippincott, Philadelphia, 1969.

10. E. D. Weitzman, H. Schaumberg, and W. Fishbein, *J. Clin. Endocr., 26,* (1966), 121.

11. L. Hellman, N. Fujinori, J. Curti, E. D. Weitzman, J. Kream, H. Roffwarg, S. Ellman, D. K. Fukushima, and T. F. Gallagher, *J. Clin. Endocr., 30,* (1970), 411.

12. F. Halberg, G. Frank, R. Harner, J. Matthew, H. Oaker, H. Granem, and J. Melby, *Experientia, 17,* (1961), 1.

13. E. D. Weitzman, D. Goldmacher, D. Kripke, P. MacGregor, J. Kream, and L. Hellman, *Trans. Amer. Neurol. Assoc., 93,* (1968), 153.

14. E. D. Weitzman, D. K. Fukushima, C. Nogeire, H. Roffwarg, T. F. Gallagher and L. Hellman, *J. Clin. Endocr.*, **33**, (1971), 14.

15. T. F. Gallagher, K. Yoshida, H. Roffwarg, E. D. Weitzman, D. Fukushima, and L. Hellman, *J. Clin. Endocr.*, **36**, (1973), 1058.

16. S. A. Berson and R. S. Yalow, *J. Clin. Invest.*, **47**, (1968), 2725.

17. E. D. Weitzman, D. G. Kripke, J. Kream, P. McGregor, and L. Hellman, *Psychophysiology*, **7**, (1970), 307.

18. E. D. Weitzman, D. Fukushima, C. Nogeire, L. Hellman, J. Sassin, M. Perlow, and T. F. Gallagher, *Psychophysiology*, **9**, (1972), 101.

19. E. D. Weitzman, M. Perlow, J. F. Sassin, D. Fukushima, B. Burack, and L. Hellman, *Trans. Amer. Neurol. Assoc.*, **97**, (1972), 197.

20. L. Hellman, E. D. Weitzman, H. Roffwarg, D. K. Fukushima, K. Yoshida, and T. F. Gallagher, *J. Clin. Endocr.*, **30**, (1970), 686.

21. M. Perlow, D. Fukushima, L. Hellman, L. Kream, J. Oppenheimer, J. Sassin, and E. D. Weitzman, In Preparation.

22. J. Tourniare, J. Orgaizyi, J. F. Riviere, and H. Rousset, *J. Clin. Endocr.*, **32**, (1971), 666.

23. T. F. Gallagher, L. Hellman, J. Finkelstein, K. Yoshida, E. D. Weitzman, H. Roffwarg, and D. Fukushima, *J. Clin. Endocr.*, **34**, (1972), 919–927.

24. S. M. Glick, *Ann. Int. Med.*, **66**, (1967), 760.

25. Y. Takahashi, D. M. Kipnis, and W. H. Daughaday, *J. Clin. Invest.*, **47**, (1968), 2079.

26. Y. Honda, K. Takahashi, S. Takahashi, K. Azumi, M. Irie, M. Sakuma, T. Tsushima, and V. Shizume, *J. Clin Endocr.*, **29**, (1969), 20.

27. J. F. Sassin, D. C. Parker, J. Mace, R. Gotlin, L. C. Johnson, and L. G. Rossman. *Science*, **165**, (1969), 513.

28. D. C. Parker, J. F. Sassin, J. W. Mace, R. W. Gotlin, and L. G. Rossman, *J. Clin. Endocr.*, **29**, (1969), 871.

29. M. A. Pawel, J. F. Sassin, and E. E. Weitzman, *Life Sci.*, **11**, (1972), 587.

30. W. H. Daughaday, in *Textbook of Endocrinology* (R. H. Williams, Ed.), W. B. Saunders Co., Philadelphia, Pa., 1968.

31. J. H. Jacoby, M. Greenstein, J. Sassin, and E. D. Weitzman, International Congress on Pharmacology, July 23–28, 1972, Abstract.

32. D. C. Parker and L. G. Rossman, *J. Clin. Endocr.*, **32**, (1971), 65.

33. S. M. Glick, J. Roth, R. S. Yalow, and S. A. Berson, *Rec. Progr. Hormone Res.*, **21**, (1965), 241.

34. D. T. Krieger and S. Glick, *J. Clin. Endocr.*, **33**, (1971), 847.

35. A. G. Frantz and D. K. Kleinberg, *Science*, **170**, (1970, 745.

36. P. Hwang, H. Guyda, and H. Friesen. *Proc. Nat. Acad. Sci. U.S.*, **68**, (1971), 1962.

37. J. F. Sassin, A. G. Frantz, E. D. Weitzman, and S. Kapen, *Science*, **177**, (1972), 1205–1207.

38. A. G. Frantz, D. L. Kleinberg, and G. L. Noel, *Rec. Progr. Hormone Res.*, **28**, (1972), 567.

39. J. Meites, in *The Neuro Endocrinology of Human Reproduction* (H. C. Mach and A. I. Sheima, Eds.), Thomas, Springfield, Ill., 1971.

40. F. G. Sulman, *Hypothalamic Control of Lactation*, Springer, New York, 1970.
41. R. M. Fraps, *Endocrinology,* **77,** (1965), 5.
42. J. W. Everett and C. H. Sawyer, *Endocrinology,* **47,** (1950), 198.
43. C. Faiman and R. J. Ryan, *Nature,* **215,** (1967), 857.
44. Franchimont, P., *Ann. Endocr.,* (Paris), **29,** (1968), 403.
45. N. T. Peterson, Jr., A. T. Midgley, Jr., and R. B. Jaffe, *J. Clin. Endocr.,* **28,** (1968), 1473.
46. H. G. Burger, J. B. Brown, K. J. Catt, B. Hudson, and J. B. Stockigt, *Excerpta Med.,* **161,** (1968), 412.
47. C. A. Strott, T. Yoshimi, and M. B. Lipsett, *J. Clin. Invest.,* **48,** (1969), 930.
48. C. Faiman and J. S. D. Winter, *J. Clin. Endocr.,* **33,** (1971), 186.
49. B. B. Saxena, H. Demura, H. M. Gandy, and R. E. Petersen, *J. Clin. Endocr.,* **28,** (1968), 519.
50. B. B. Saxena, G. Leyendecker, W. Chen, H. M. Gancy, and R. E. Peterson, *Acta Endocr.* (Copenhagen) *Suppl.,* **142,** (1969), 185.
51. R. Boyar, M. Perlow, L. Hellman, S. Kapen, and E. D. Weitzman, *J. Clin. Endocr.,* **35,** (1972), 61.
52. R. T. Rubin, A. Kales, R. Adler, T. Fagan, and W. Odell, *Science,* **175,** (1972), 196.
53. R. Boyar, J. Finkelstein, H. Roffwarg, S. Kapen, E. D. Weitzman, and L. Hellman, *New Eng. J. Med.,* **287,** (1972), 582.
54. G. T. Ross, C. M. Cargille, M. B. Lipsett, P. L. Rayford, J. R. Marshall, C. A. Strott, and D. Rodbard, *Rec. Progr. Hormone Res.,* **26,** (1970), 1.
55. S. Kapen, R. Boyar, M. Perlow, L. Hellman, and E. D. Weitzman, *J. Clin. Endocr.,* **36,** (1973), 724–729.
56. S. Kapen, R. Boyar, L. Hellman, and E. D. Weitzman, *Psychophysiology,* **7,** (1970), 337.

CHAPTER 24

A Study of Circadian Variation in Taste in Normal Man and in Patients with Adrenal Cortical Insufficiency: The Role of Adrenal Cortical Steroids

ROBERT I. HENKIN

Section on Neuroendocrinology, National Heart and Lung Institute, National Institutes of Health, Bethesda, Maryland

Circadian changes in the endogenous secretion of adrenal cortical steroids (1, 2) and of adrenocorticotropin (ACTH) (3) have been well established by many investigators. Similarly, circadian changes in phenomena related to adrenal corticosteroid or ACTH secretion have also been clearly established; for example, changes in eosinophil number or in serum copper or zinc concentration (4). In addition the relationship between adrenal cortical steroid secretion and the detection of gustatory, olfactory, or auditory stimuli has been clearly established (5-9). Whether or not a circadian variation exists for the detection of sensory signals in man has not been clearly established, and the linking of such changes to circadian changes in the endogenous secretion of adrenal cortical hormones has also not been sought.

The purpose of this study is to demonstrate (*1*) the existence of a circadian variation in the detection of the taste and smell of NaCl in

397

normal man, (2) the abolition of this circadian variation in patients in whom endogenous secretion of adrenal cortical hormone was absent (those with adrenal cortical insufficiency), and (3) the reinstitution of the circadian pattern of variation coincident with the institution of cyclic exogenous administration of carbohydrate-active steroids to patients with adrenal cortical insufficiency.

METHODS

The subjects of this study were 17 normal volunteers, nine men and eight women aged 21 to 65; two patients with anterior pituitary insufficiency, aged 21 and 36; and three patients with Addison's disease, aged 25 to 65. All patients with Addison's disease had clinical features of this disease, urinary 17-hydroxycorticosteroids that were below 2 mg/24 hours, and did not increase with 40 U ACTH given intravenously over 8 hours each day for 5 days. Both patients with anterior pituitary insufficiency had hypothyroidism, hypogonadism, and adrenal insufficiency, with urinary 17-hydroxycorticosteroids in J.M. and S.P. of 1.1 and 2.2 mg/day, respectively, rising to 27 and 10.9 mg/day, respectively, with ACTH. All patients remained in an air-conditioned metabolic ward and ate a normal diet, which was well tolerated even when they were not receiving treatment. Sodium intake was 100 to 200 meq/day. Body weight, determined with metabolic scales daily on arising, was used to provide a gross estimate of changes in the volume of body fluids. The patients and volunteers were studied under two conditions: (*1*) untreated for 4 or more days and (*2*) treated with prednisolone (Δ_1F), 20 mg per day in four divided doses given orally every 6 hours for 2 to 5 days.

Taste was evaluated by measuring detection and recognition thresholds. A detection threshold is defined as the lowest concentration of test solution that can be consistently distinguished from distilled water. A recognition threshold is defined as the lowest concentration of test solution that can be consistently recognized appropriately; for example, salty would be the appropriate response for NaCl. The solutions of test substance or glass-distilled water were administered as single drops in a forced-choice three stimulus drop technique previously described (5, 10). For this study only sodium chloride (NaCl) was used as the test substance. NaCl solutions used to test taste acuity in normal subjects were, in mM, 0.5, 0.8, 3, 6, 12, 30, and 60. For taste acuity in patients with adrenal cortical insufficiency NaCl concentrations tested were, in mM, 10^{-7}, 10^{-6}, 10^{-5}, 10^{-4}, 10^{-3}, 10^{-2}, 0.1, 0.5, 0.8,

3, 6, 12, 30, and 60. For convenience, changes are reported in bottle units (BU) with the value at 3 A.M. equal to 1 BU, this value always being the lowest concentration detected by the subjects studied. Concentrations greater than that representing 1 BU were scaled as 2, 3, 4, or higher BU, representing the higher concentrations relative to that concentration of 1 BU. Only median detection or recognition thresholds of the group of subjects under the specific conditions studied are presented. Responses were obtained independently by four observers every 3 hours over a period of 30 hours. At the time of testing, subjects had not smoked or eaten for at least 1 hour.

Olfaction was evaluated by measuring detection thresholds by a forced-choice, three stimulus sniff technique in a manner previously described (6, 11). For this study, detection of the vapor above a series of solutions of NaCl was carried out (6). Responses were obtained independently by one observer every 3 hours at the same time intervals at which taste acuity was measured. Measurement of olfactory acuity always followed measurement of taste detection and recognition acuity. NaCl solutions used to test olfactory acuity in normal subjects were, in mM, 60, 150, and 300. Concentrations used to test olfactory detection in patients with adrenal cortical insufficiency were the same used to test taste acuity except that more concentrated solutions of NaCl were also used, including 150 and 300 mM. As for taste, all changes were reported in BU for convenience, the value at 3 A.M., the lowest concentration detected, being equal to 1 BU. If subjects could not detect any difference between water and the 300 mM solution of NaCl, the value was listed as greater than the highest bottle unit tested. All values are given as median detection thresholds for the specific group of subjects studied.

RESULTS

Circadian Variation in Taste Thresholds in Normal Volunteers

The circadian pattern of change in taste detection acuity for NaCl in 17 normal volunteers in shown in Figure 1. Detection acuity was greatest (0.8 mM) at 12 and 3 A.M. followed by a relatively sudden decrease in detection acuity to the lowest acuity level at 6 A.M. (30 mM) which persisted through 9 A.M. This was followed by a gradual increase in acuity over the next 12-hour period. Little if any change in taste recognition acuity occurred over the period of time studied, the median level remaining at 30 mM over the entire study. All changes in

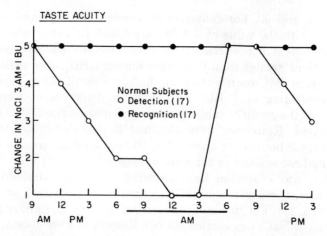

Figure 1   Circadian variation in median detection and recognition thresholds for the taste of NaCl in 17 normal volunteers. 1 BU equals 0.8 mM, 2 = 3, 3 = 6, 4 = 12, and 5 = 30. Time in hours is noted on the abscissa. The dark line on the abscissa denoted darkness.

acuity were within the normal limits of detection and recognition acuity for NaCl previously observed (detection, 0.5 to 60 mM; recognition, 6 to 60 mM).

Comparison of the circadian pattern of change in taste detection acuity for normal men and women is shown in Figure 2. Although the overall pattern for the two sexes is similar in general outline, there are

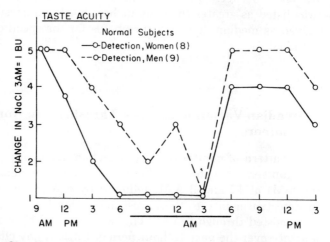

Figure 2   Circadian variation in median detection thresholds for the taste of NaCl in eight normal women and in nine normal men. 1 BU equals 0.8 mM, 2 = 3, 3 = 6, 4 = 12, and 5 = 30.

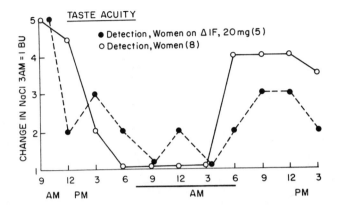

Figure 3   Circadian variation in median detection threshold for the taste of NaCl in normal women before and after administration of prednisolone (Δ,f) for 2 days. 1 BU = 0.8 mM, 2 = 3, 3 = 3, 3 = 6, 4 = 12, and 5 = 30.

specific differences. For women, detection acuity increased to maximum levels (0.8 mM) at 6 P.M. and remained at that level for 9 hours, followed by the relatively sudden fall to lower acuity (12 mM) at 6 A.M. where it remained for 6 hours, followed by the gradual increase in acuity previously noted. For men, detection acuity was at maximum level only at 3 A.M. (0.8 mM), followed by the relatively sudden decrease in acuity to 60 mM where it remained for 9 hours. This was followed by the gradual increase in detection acuity previously observed. As previously noted, changes in detection acuity for NaCl were all within the normal limits previously observed.

Comparison of the circadian pattern of change in taste detection acuity in normal women before and after administration of prednisolone is shown in Figure 3. Little if any change in circadian variation occurred following administration of prednisolone for 2 days to normal women.

Circadian Variation in Smell Thresholds in Normal Volunteers

The circadian pattern of change in smell detection acuity for NaCl in 10 normal volunteers is shown in Figure 4. Detection acuity was greatest (150 mM) between 9 P.M. and 3 A.M., fell relatively quickly to lowest levels (>300 mM) at 6 A.M., and remained at this level until 6 P.M. when there was a relatively rapid rise to the greatest detection acuity observed over the time period studied. As with taste detection

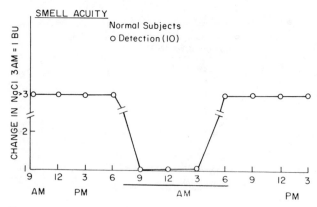

Figure 4   Circadian variation in median detection threshold for the smell of NaCl in 10 normal volunteers. 1 BU equals 150 mM, 2 = 300, and 3 = > 300.

and recognition acuity, all changes in olfactory detection were within the normal limits previously observed (150 to >300 mM).

Comparison of the circadian pattern of change in olfactory detection acuity for normal men and women is shown in Figure 5. Although the overall pattern for the two sexes is similar in general outline, there is a difference with respect to the degree of responsiveness over the time period studied. For women, detection acuity increased to maximum levels (150 mM) at 9 P.M. and remained there for the next 6 hours, followed by a relatively rapid decrease in acuity to lowest levels (>300 mM) at 6 P.M. where levels remained until 6 P.M. After this time values increased to their maximum levels. For men, the pattern of change was similar but the increase in acuity which occurred at 12 midnight

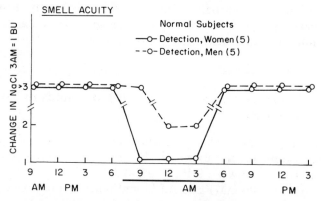

Figure 5   Circadian variation in median detection threshold for the smell of NaCl in five normal men and in five normal women. 1 BU equals 150 mM, 2 = 300, and 3 = > 300.

and persisted through 3 A.M. was only to 300 mM. This was followed by a return to the lowest level of acuity (>300 mM) at 6 P.M. where levels remained until 9 P.M.

Comparison of the circadian pattern of change in olfactory detection acuity in normal women before and after administration of $\Delta_1 F$ is shown in Figure 6. Little if any change in circadian variation occurred following administration of $\Delta_1 F$ for 2 days to normal women.

Circadian Variation in Taste Thresholds in Patients with Adrenal Cortical Insufficiency

The circadian pattern of change in taste detection acuity for NaCl in patients with adrenal cortical insufficiency treated with the carbohydrate-active steroid $\Delta_1 F$ (20 mg given in four divided doses orally every 6 hours for 2 days) is similar to that observed in normal subjects (Figure 7). Off therapy for more than 4 days detection acuity increased significantly as previously noted (5, 9). However, little, if any circadian pattern of change in NaCl detection acuity was observed during the period over which this increased detection acuity occured. These phenomena occurred both in the patients with adrenal cortical insufficiency and in those with panhypopituitarism.

Comparison of the circadian pattern of variation in detection acuity for NaCl in patients with adrenal cortical insufficiency treated with oral $\Delta_1 F$ in divided doses with that of normal subjects is also shown in Figure 7. The similarity of circadian pattern of change between the treated patients and the normal subjects can be easily observed. However, the magnitude of the changes in the treated patients was

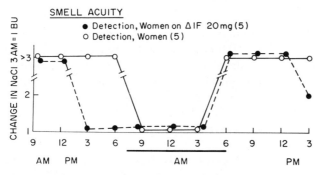

Figure 6   Circadian variation in median detection threshold for the smell of NaCl in normal women before and after administration of prednisolone ($\Delta_1 F$) for 2 days. 1 BU = 150 mM, 2 = 300, and 3 = > 300.

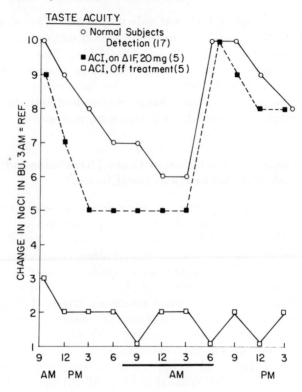

Figure 7 Circadian variation in median detection threshold for the taste of NaCl in normal volunteers and in patients with adrenal cortical insufficiency (ACI) without treatment and after treatment with the carbohydrate-active steroid prednisolone ($\Delta_1 F$). This hormone was administered in divided doses, orally, 5 mg given every 6 hours for 2 days. 1 BU = 10^{-4} mM, 2 = 10^{-3}, 3 = 10^{-2}, 4 = 10^{-1}, 5 = 0.5, 6 = 0.8, 7 = 3, 8 = 6, 9 = 12, and 10 = 30.

somewhat greater than in the normal subjects, (0.5 to 30 mM for the patients, 0.8 to 30 mM for the normal subjects). Also, the time over which the maximum level of sensitivity occurred was greater in the patients than in the normal subjects. Return to a lower level of sensitivity occurred with a relatively rapid fall of acuity at 6 A.M. in both the treated patients and the normal subjects, with a gradual rise in sensitivity to maximum levels thereafter.

Circadian Variation in Smell Thresholds in Patients with Adrenal Cortical Insufficiency

The circadian pattern of change in smell detection acuity for NaCl in patients with adrenal cortical insufficiency treated with the carbohydrate-active steroid $\Delta_1 F$ (20 mg given in four divided doses orally every 6 hours for 2 days) is similar to that observed in normal subjects (Figure 8). Off therapy for more than 4 days detection acuity increased significantly as previously noted (6, 9). However, no circadian pattern of change in NaCl detection acuity was observed during the period over which this increased detection acuity occurred. These phenomena occurred in each of the patients with adrenal cortical insufficiency and panhypopituitarism.

Comparison of the circadian pattern of variation in detection acuity for the vapor above a series of solutions of NaCl in patients with adrenal cortical insufficiency treated with carbohydrate-active steroids and in normal volunteers is also shown in Figure 8. The circadian pattern of variation in the treated patients and normal subjects is essentially the same.

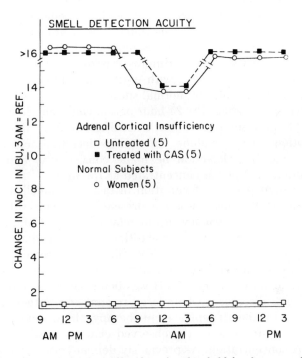

Figure 8   Circadian variation in median detection threshold for the vapor above a series of solutions of NaCl in normal volunteers and in patients with adrenal cortical insufficiency without treatment and after treatment with the carbohydrate-active steriod (CAS) $\Delta_1 F$. This hormone was given orally in divided doses, 5 mg/dose, every 6 hours for 2 days. 1 BU = 10^{-6} mM, 2 = 10^{-5}, 3 = 10^{-4}, 4 = 10^{-3}, 5 = 10^{-2}, 6 = 10^{-1}, 7 = 0.5, 8 = 0.8, 9 = 3, 10 = 6, 11 = 12, 12 = 30, 13 = 60, 14 = 150, 15 = 300, and > 16 = > 300.

DISCUSSION

The results of this study demonstrate that there is a specific circadian pattern of variation in taste and smell detection acuity for NaCl in normal men and women. This circadian pattern of variation closely follows the well-known endogenous secretion of cortisol over a 24-hour period and suggests a relationship between these phenomena. The circadian pattern of variation for women for olfactory detection acuity extends over a wider range of sensitivity than for men although the pattern for the two sexes is similar. The observed changes in sensitivity over a 24-hour period are within the normal limits of sensitivity previously reported (9). Circadian changes have been observed for detection acuity but not for recognition acuity.

While the circadian variation for the taste of NaCl may be easily understood, that for the detection of the vapor above a series of solutions of NaCl may not. Previous studies have demonstrated that the stimulus above the NaCl solutions which is most probably detected is a minute concentration of chlorine gas generated by the action of light on the NaCl solutions (6). This stimulus was detected with ease as different from water by patients with untreated adrenal cortical insufficiency but with difficulty by normal subjects. This was obvious in that most subjects throughout the 24-hour period could not detect the vapor above a 300-mM solution of NaCl as different from that above two other solutions of water alone. Only when detection acuity was most sensitive could this detection be made consistently, and then only above solutions of NaCl as concentrated as 150 mM.

The circadian pattern of variation in taste and smell detection observed in normal subjects was not present in patients with untreated adrenal cortical insufficiency, again relating the circadian pattern of variation in these two sensory modalities to the circadian variation in the endogenous secretion of cortisol. Since the circadian pattern of variation was abolished both in patients with panhypopituitarism, in whom little or no pituitary ACTH was being secreted, and in patients with Addison's disease, in whom large amount of pituitary ACTH were being secreted while off replacement therapy, there appears to be little or no relationship between the observed changes in sensory detection and ACTH concentration. As previously demonstrated, it is during the period after withdrawal of replacement therapy for 3, 4, or more days that the well-known increased detection acuity for both taste and smell detection acuity occurs (5, 6, 9). However, upon resumption of therapy with the carbohydrate-active steriod $\Delta_1 F$ a circadian pattern of variation similar to that seen in normal volunteers was observed, the detection acuity for both taste and smell returning to normal levels. It is unlikely that the cyclic nature of the replacement therapy was important for the return of the circadian variation since the pattern was present in other patients, not reported here, treated with dexamethasone, another carbohydrate-active steroid given orally once daily in doses of either 0.5 or 0.75 mg.

The circadian pattern of variation in taste and smell detection was not abolished in normal subjects after treatment with large doses ofΔ_1F for 2 days; the amount given was large enough to suppress endogenous secretion of ACTH. Once again, this suggests that ACTH secretion is not a critical factor in the initiation of these circadian changes in sensory detection acuity. Indeed, other circadian patterns of change related to cortisol secretion, for example, serum copper or zinc concentration, were not abolished following administration of exogenous carbohydrate-active steroids (4). However, it is possible that in patients who receive large amount of exogenous carbohydrate-active steroids for long periods of time or in patients with untreated Cushing's syndrome, the circadian pattern of variation in detection acuity for taste and smell may be dampened by the significant decrease in detection acuity which is produced by the presence of large concentrations of carbohydrate-active steroids for long periods of time (9).

The mechanism by which carbohydrate-active steroids influence detection acuity for both taste and smell is not clearly understood. Indeed, whether this hormone affects receptor, nerve, central nervous system function, or some combination of the three to produce the effects observed is not known. The limited information available suggests that carbohydrate-active steroid effects can be shown in receptor, nerve, and central nervous function so that localization of specific effects may be difficult. However, the changes in circadian variation due to carbohydrate-active steroid activity offer a fruitful area of investigation in which to learn more about the interaction of hormones on several aspects of neural function.

REFERENCES

1. G. Pincus, *J. Clin Endocr.*, **3**, (1943), 195.
2. F. H. Tyler, D. Migeon, A. A. Florentin, and L. T. Samuels, *J. Clin. Endocr.*, **14**, (1954), 174.
3. M. M. Martin and D. E. Hellman, *J. Clin. Endocr.*, **24**, (1964), 253.
4. M. Lifschitz and R. I. Henkin, *J. Appl. Physiol.*, **31**, (1971), 88.
5. R. I. Henkin, J. R. Gill, Jr., and F. C. Bartter, *J. Clin. Invest.*, **42**, (1963), 727.

6. R. I. Henkin and F. C. Bartter, *J. Clin. Invest.*, **45,** (1966), 1631.

7. R. I. Henkin, R. E. McGlone, R. Daly, and F. C. Bartter, *J. Clin. Invest.*, **46,** (1967), 429.

8. R. I. Henkin and R. L. Daly, *J. Clin. Invest.*, **47,** (1968), 1259.

9. R. I. Henkin, in *Perception and Its Disorders* (D. Hamburg, K. Pribram, A. Stunkard, Eds), Association for Research in Nervous and Mental Diseases (Vol. 48) Williams and Wilkins Inc., Baltimore, 1970.

10. R. I. Henkin, P. J. Schecter, R. C. Hoye, and C. F. T. Mattern, *J. Amer. Med. Assoc.*, **217,** (1971), 434.

11. J. Marshall and R. I. Henkin, *Ann. Int. Med.*, **75,** (1971), 207.

CHAPTER 25

The Effect of Hypokinesis on Plasma
ACTH and Cortisol Concentrations

CAROLYN S. LEACH and PAUL C. RAMBAUT

*Biomedical Research Division, National Aeronautics and
Space Administration, Manned Spacecraft Center,
Houston, Texas,*

J. VERNIKOS-DANELLIS and C. M. WINGET

*Biomedical Research Division, National Aeronautics and
Space Administration, Ames Research Center, Moffett
Field, California,*

B. O. CAMPBELL

*Physiology Department, Baylor College of Medicine,
Houston, Texas*

ABSTRACT

The effects of prolonged bed rest on the pituitary-adrenal system was
assessed in five healthy young males who were submitted to 56 days of
bedrest in a 14L:OD (lights-on at 9:00 A.M.) environment. Circulating
levels of cortisol and ACTH were determined in blood samples drawn

at four hourly intervals for 48-hour periods before 10, 20, 30, 42, and 54 days during and at 10 and 20 days post-bed rest. Significant fluctuations in the circulating levels of both hormones occurred, with cortisol peaking in the early morning and ACTH in the late evening. Mean daily ACTH levels increased sharply after the 30th day of bed rest and showed a threefold increase by the 54th day. Mean daily cortisol rose during the early part of bed rest and showed a significant decrease by the 54th day. These changes in the daily mean were a reflection of the change in the amplitude of their respective circadian rhythms. The possible mechanisms involved in these changes are discussed.

Adaptability to virtually every type of environmental change remains one of man's outstanding characteristics. However, each of these changes has occurred with the force of gravity remaining constant. In the presence of gravity, man has evolved the necessary anatomic structure and physiologic mechanism to permit him to live and function effectively. These mechanisms have also undergone intensive study only in a 1 *g* environment. In general, the logistics of physiological studies in space flight (0*g*) have proved most difficult to surmount. A careful examination of the nature and duration of most physiological changes can be more readily investigated using ground-based hypokinetic experiments in which bed rest is employed as an analogue of weightlessness. Bed rest has been used to simulate weightlessness and confinement in numerous investigations (1–3). The skeletal and cardiovascular reactions have been extensively investigated, but little attention has been given to the endocrine system.

We have studied five healthy men, ambulatory as well as during a 56-day period at bed rest. Other physiological data from this study have been previously reported in this conference (4).

MATERIALS AND METHODS

Six healthy males, ages 20 to 26, and weighing approximately 55 $\pm$ 1 kg were studied. These subjects followed an experimental procedure which included a 20-day ambulatory pre-bed rest control period, 56 days of absolute bed rest, and a 20-day post-bed rest recovery period. The question was raised in a previous study (5) whether some of the observed changes, particularly those around day 20, may have been due to the bleeding schedule. Therefore, one subject was not bled in this study, and only temperature, heart rate, and urine was collected from this subject. The results indicated that the changes we had previously observed under these conditions (6) were not due to excessive blood loss. This report includes data from only the five bed-rested subjects. These subjects were selected out of 60 applicants based on interviews and psychological tests. All subjects were on a 14-hour light:10-hour dark regimen (lights on at 9:00 A.M.) and fed a balanced diet of 2500 cal/day.

The subjects were bled by repetitive venous punctures every 4 hours for a 48-hour period at each of nine points throughout the study; these included two 48-hour periods before bed rest, at 10, 20, 30, 42, and 54 days after confinement to bed, and 13 and 20 days after the subjects had again become ambulatory. Fifteen milliliters of blood was removed at each bleeding time to obtain approximately 5 ml plasma and 2.5 ml serum. It was kept cold in crushed ice during collection and separation, frozen promptly, and stored frozen.

Once each 48-hour bleeding period (8:00 A.M. of second day) hemoglobin, hematocrit, RBC, and WBC's were determined in all subjects. No appreciable changes were observed throughout the study.

Analysis of the blood plasma samples included, among others, hydrocortisone determined by Murphy's competitive protein-binding technique (7) and adrenocorticotrophic hormone (ACTH) using the method of radioimmunoassay (8).

RESULTS AND DISCUSSION

The diurnal rhythms of plasma ACTH and cortisol levels have been studied in depth (9, 10). Besides documenting the existence of rhythms in these circulating hormones, studies have been conducted with reference to the effect of changes in work-rest or light-dark cycles on these rhythms (11, 12).

Our experiment studied the effect of hypokinesis (bed rest) on the rhythms of ACTH and cortisol in healthy individuals. The mean hormone concentrations for the six samples analyzed each 24 hours are shown in Figure 1. The mean 24-hour secretion of ACTH increases gradually from the beginning of bed rest but rises more sharply after 30 days. The increase measured by the 54th day of bed rest was threefold. There was a gradual return post-bed rest to control levels. In contrast the mean cortisol level rose significantly during the early part of the study, almost doubling after 20 and 42 days of bed rest, but decreased sharply thereafter at a time when ACTH was still rising.

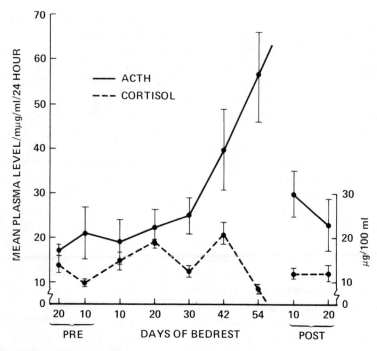

Figure 1   Mean circulating cortisol and ACTH per 48 hour sampling period during 56 days of bedrest. Vertical lines represent SE (n = 5).

These changes in the daily mean of circulating ACTH and cortisol concentrations are a reflection of the change in the amplitude of their respective circadian rhythms.

Figure 2 shows these circadian rhythms in *cortisol* for the five subjects at various intervals during the study. Bed rest had little effect on the circadian rhythmicity of this hormone. A significant fluctuation in plasma cortisol was evident with peak levels occurring around 7:30 A.M. Throughout the experiment. However, progressive bed rest initially increased and by the 54th day reduced the amplitude of the steroid rhythm.

The daily rhythm in circulating ACTH (Figure 3) was considerably less stable than the cortisol rhythm, showing several peaks and phase shifts as the bed rest progressed. These results confirm previous reports of significant diurnal fluctuations in ACTH and cortisol. The rhythmicity of both hormones is generally variable with evidence of several peaks. Numerous spikes in these hormonal secretions throughout the day have been reported recently to occur with a frequency of approximately 20 minutes (13, 14). Since in this study sampling was at four hourly intervals, some of the variability in our results may be related to this phenomenon. Individual variability was considerably greater for the ACTH than the cortisol rhythms. However, progressive bed rest reduced the amplitude of the steroid rhythm and increased the amplitude of the ACTH secretion. During bed rest and in the post-bed rest period a secondary peak around 1600–2000 hours was observed in hydrocortisone. This has been reported previously by

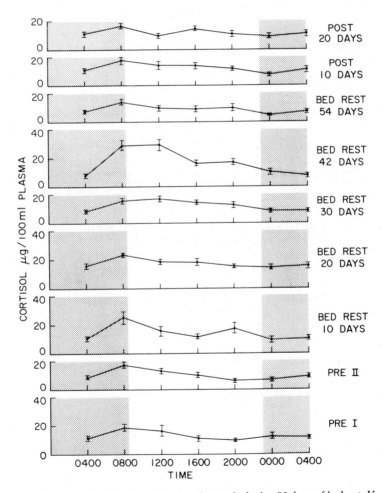

Figure 2   Plasma cortisol rhythm at various intervals during 56 days of bedrest. Vertical lines represent SE (n = 5). Stippled area represents lights off period.

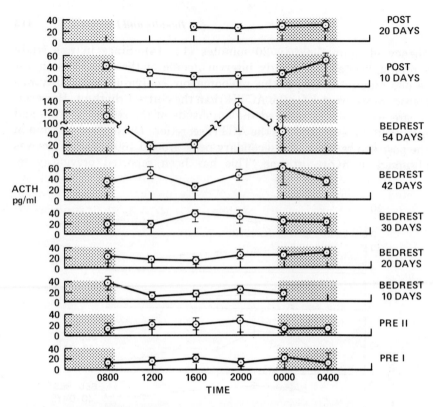

Figure 3 Plasma ACTH rhythm at various intervals during 56 days of bedrest. Vertical lines represent SE (n = 5). Stippled area represents lights off period.

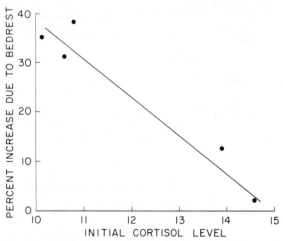

Figure 4 Correlation between mean 24-hour pre-bed rest cortisol level with the mean 24-hour level during bed rest for each of the five subjects.

Perkoff et al. in studies conducted on 51 normal subjects (15). As for cortisol, the most impressive change was also in the amplitude of the ACTH rhythm which started to increase after the 30th day and fluctuated from 10 to 120 pg/ml (a 12-fold change) over a 24-hour period by the 54th day of bed rest. This large amplitude in the rhythm decreased promptly when the subjects got out of bed. Thus an inverse relationship between the changes in the mean daily circulating levels of cortisol and ACTH throughout bed rest did exist.

The rise in daily cortisol levels early in the study may be partly attributed to the reduction in plasma volume that is known to occur within the first week of bed rest (16) and partly to the rise in ACTH observed since relatively small changes in circulating ACTH are sufficient for maximal cortisol secretion. That the well-known negative feedback mechanism regulating ACTH secretion was functional at this time is indicated by the data in Figure 4 which shows a correlation between mean 24-hour pre-bed rest cortisol levels and the mean 24-hour level during bed rest for each of the five subjects. It may be seen that those subjects with the lowest initial levels showed the greatest increase during bed rest and vice versa. The decrease in measurable free cortisol in the plasma is thought to be a reflection of an increase in the urinary excretion of the hormone. This is supported by previous studies in which an increase in 24-hour excretion of cortisol was noted in individuals bed-rested for prolonged periods of time (17).

It is possible that the greater increase in circulating ACTH that occurs with progressive bed rest may be in response to the decreasing plasma cortisol levels. It is also possible that this large increase in ACTH measure by radioimmunoassay may not reflect levels of the biologically active hormone. These possibilities are being investigated.

Information acquired from this study and others has led researchers in space-life sciences to examine more closely the projected effect of long duration space flight on man's homeostatic mechanisms. The mission planners are now aware of significant metabolic changes occurring in a space-flight crew during their exposure to the weightless environment. Considerations such as these profoundly affect the way in which mission schedules are designed.

REFERENCES

1. J. E. Deitrick, G. D. Whedon, and E. Shorr. *Amer. J. Med.,* **4,** (1948), 3.

2. P. B. Miller, R. L. Johnson, and L. E. Lamb. *Aerospace Med.,* **35,** (1964), 1194.

3. C. L. Donaldson, S. B. Hulley, J. M. Vogel, R. S. Hattner, J.H. Bayers, and D. E. McMillan. *Metab. Clin. Exp.,* **19,** (1970), 1071.

4. C. M. Winget, J. Vernikos-Danellis, C. W. DeRoshia, S. E. Cronin, C. S. Leach, and P. C. Rambaut, in *Biorhythms and Human Reproduction* (M. Ferin, F. Halberg, R. M. Richart, and R. L. Vande Weile, Eds.), Wiley-Interscience, New York, 1973, Chap. 38.

5. J. Vernikos-Danellis, C. S. Leach, C. M. Winget, P. C. Rambaut, and P. B. Mack, *Appl. J. Physiol.,* **33,** (1972), 644.

6. C. M. Winget, S. E. Cronin, P. C. Rambaut, and P. B. Mack, *J. Appl. Physiol.,* **33,** (1972), 640.

7. B. E. Murphy, *J. Clin. Endocr.,* **27,** (1967), 973.

8. B. O. Campbell, C. S. Leach, and H. S. Lipscomb. *Physiologist,* **14,** (1971), 118.

9. F. C. Bartter, C. S. Delea, and F. Halberg, *Ann. N. Y. Acad. Sci.,* **98,** (1962), 969.

10. G. W. Clayton, L. Librik, R. L. Gardner, and R. Guillemin. *J. Clin. Endocr.,* **23,** (1963), 975.

11. R. T. W. L. Conroy, A. L. Elliott, and J. N. Mills., *Brit. J. Ind. Med.,* **27,** (1970), 170.

12. F. Halberg, G. Frank, R. Harner, J. Matthews, H. Aaker, H. Gravem, and J. Melby, *Experientia,* **17,** (1961), 282.

13. D. T. Krieger, W. Allen, F. Rizzo, and H. P. Krieger, *J. Clin. Endocr.,* **32,** (1971), 266.

14. F. Ceresa, A. Angeli, G. Boccuzzi, and G. Molino, *J. Clin. Endocr.,* **29,** (1969), 1074.

15. G. T. Perkoff, K. Eik-Nes, C. A. Nugent, H. L. Fred, R. A. Nimer, L. Rush, L. T. Samuels, and F. H. Tyler, *J. Clin. Endocr.,* **19,** (1959), 432.

16. P. C. Johnson, T. B. Briscoll and W. R. Carpertier, *Aerospace Med.* **42,** (1971), 875.

17. C. S. Leach, S. B. Hulley, P. C. Rambaut, and L. F. Dietlein. Space Life Sci., In Press.

CHAPTER 26

Long-Term Changes in Corticosteroid Excretion

GEORGE C. CURTIS

*Department of Psychiatry, University of Michigan Medical
Center, Ann Arbor, Michigan*

The physiological range of adrenal cortical secretion as assessed by
various techniques has been studied extensively, and several sources of
its normal variability identified. Notable among these are circadian
variations (1), intracircadian secretory bursts (2, see also Chapter 23),
and changes due to stressful physical or psychological stimuli (3).

Measures of total daily urinary corticosteroid excretion filter out circadian and intracircadian changes and lend themselves to chronic
studies of adrenal function, although they may be affected to some
extent by factors such as the metabolism and disposal of corticoid hormones. It has been reported that urinary corticoid excretion remains
relatively constant, at least over periods of several weeks within a given
individual, even despite chronic and acute emotional stress such as
that associated with the fatal illness of an offspring; this relatively
stable level has been termed the "chronic mean" and may be related in
part to personality variables (4, 5). The "chronic mean" value is of
considerable interest as, for example, a possible correlate of prognosis
in cancer (6–8).

There are relatively few data to show just how stable or how chronic the "chronic mean" is. A single individual, reported by Fox et al. (9), collected consecutive 12-hour urine specimens almost continuously for 3 years. Over periods of several months his corticosteroid excretion levels showed marked shifts, both in mean level and in variability, related apparently to the course of his psychoanalytic treatment. Studies of this subject, if they had spanned several weeks only, could have led to the conclusion that his chronic corticosteroid excretion rate was high and labile, high and stable, or low and stable, depending upon when they were carried out.

In a study of even more heroic proportions, Hamburger collected his own urine almost continuously for 16 years for determination of total daily 17-ketosteroid excretion, a measure derived partly from testicular androgens and partly from adrenal corticosteroids. Computer analysis of these data by Halberg et al. (10) by least squares fitting of cosine functions revealed superimposed low frequency, low amplitude but statistically significant cycles with periods of about a week, about 20 days, about 30 days, and about a year.

The present report adds to the small number of studies in which adrenal function was studied in the same subject over several years.

METHODS

A healthy male subject collected his urine more or less continuously, but with a number of interruptions, from August 22, 1961, when he was 34 years old, through September 8, 1965, when he was 38. A total of 795 samples were collected, mostly of 24-hours duration, but occasionally of 48 or 72 hours. He ordinarily slept from around 2300 or 2400 hours til around 0700 and worked a 50 to 60 hour week, consisting of weekdays from about 0830 to 1730, Saturdays from about 0900 till noon, and about 1815 and 2200 on Monday and Thursday evenings. The only illnesses during this interval were head colds, which were rare and never disabling.

Urinary 17-hydroxycorticosteroids (17-OHCS) were measured by the method of Glenn and Nelson (11). Standard solutions and two quality control urine specimens were run with each batch of unknown samples. The coefficient of variation due to analytical error was 3 to 4% when replicate aliquots from the same urine sample were run at the same time, but around 7% when replicates were run on different days. Statistical analysis of quality control data always revealed that the between-runs component of measurement error was significantly greater than the within-runs component.

RESULTS AND DISCUSSION

The mean and standard deviations of all 795 samples were 6.9 $\pm$ 1.2 mg/24 hours, yielding a coefficient of variation of 17.4 and 95% confidence limits of 5.7 to 8.1 mg/24 hours. The highest and lowest single values were 20.1 and 2.2 mg/24 hours. The mean of 6.9 mg/day is in the lower midrange of values for normal males in five previously reported studied employing the same analytical method (4, 5, 12–14). The extreme values were also similar to previously reported extreme values for male subjects.

Figure 1 depicts the mean and standard deviation month-by-month throughout the study, together with the grand mean and the standard deviation around it. No progressive trend was apparent, though the monthly means tended to be higher in the middle years of the study (1963 and 1964) than in the first two and final years. This pattern conceivably could be part of a cycle with a period of several years, but such a speculation is untestable without a much longer data series. Excluding the last 3 months when sampling was very limited, monthly means ranged from a high of 9.0/24 hours in June 1964 to a low of 5.0 in October 1964. The coefficient of variation ranged from 3.4 to 40.6%. Thus, like the subject reported by Fox et al. (9), the chronic level and variability estimated from measurements over an interval of a few weeks were subject to rather marked shifts.

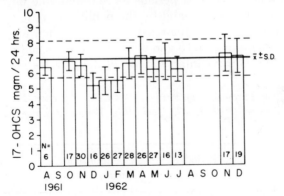

Figure 1   Month-by-month 17-hydroxycorticosteroid excretion rates by a healthy adult male over a 4-year span.

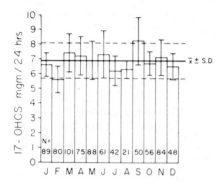

Figure 2 Monthly averages of 17-hydroxycorticosteroid excretion rates by a healthy adult male.

In order to test for seasonal changes, all data from each calendar month were averaged across years. These results are summarized in Figure 2. Though analysis of variance showed a highly significant difference between calendar months, the pattern was not clearly seasonal, suggesting, if anything, several cycles per year rather than one.

To test further for significant cyclical changes, the data were kindly analyzed by the Chronobiology Laboratory of the University of Minnesota by the least squares fitting of cosine curves as described by Halberg et al. (15). Spectra linear in frequency were done with several intervals among trial periods. Several statistically significant components or "candidate periods" in the low frequency domain of biological rhythms were thus detected. Since this method sometimes yields pseudosignificant candidate periods, nonlinear least squares analyses were also carried out (16). Out of eight candidate periods initially obtained, two, with periods of 18.0 and 45.6 days, were confirmed as mathematically valid by the nonlinear approach. Estimates of their parameters are summarized in Table 1. Their amplitudes were quite small, amounting to 3.8 and 6.8% respectively, of the grand mean or mesor value. Circavigintan (about 20 day) cycles may be charac-

Table 1   Rhythm Parameters of Urinary 17-OCHS Excretion Estimated by Nonlinear Least Squares Method. All Variability Estimates Expressed as Standard Errors

Mesor (mg/24 hours)	Period (days)	Amplitude (mgm/24 hours)	Phase (degrees)
6.87 ± 0.054	18.00 ± 0.038	0.266 ± 0.076	-60 ± 33
	45.6 ± 0.134	0.473 ± 0.077	-186 ± 18.5

teristic of the human male, having been detected in the present study and in 17-ketosteroid excretion by Halberg et al. (10). Neither of the periodicities detected in the present study correspond closely to any obvious environmental or social cycles, such as the week or the year, and if confirmed must probably be regarded as endogenous unless eventually shown to be driven by some heretofore unrecognized environmental cycle. The major significance of infradian (longer than a day) physiological cycles may be as mechanisms underlying the periodic recurrence of some physical and mental illnesses.

The finding of cycles of such low amplitude must be interpreted with caution, however, despite their statistical significance. Although the difference between trough and crest of the best fitting curve is twice the amplitude, the changes in question are not greatly different in magnitude from the analytical error of the biochemical method which, as stated above, was between 3 and 7%, with a between-runs component detectably greater than within-runs. If the between-runs component of the analytical error were rhythmic, then very low amplitude laboratory rhythms *might* become detectable by powerful statistical techniques when applied to large series of data. This contribution, however, would be difficult to reconstruct since it would depend on the period of the laboratory rhythm, the frequency of runs, and the number of unknown samples per batch.

On the other hand the amplitude of the possibly real biological cycle and its statistical significance may in fact be greater than the mathematical analyses suggest. Biological systems often oscillate, not with the fixed frequency of theoretical cosine functions but with shifting, wobbling, accelerating, or decelerating frequencies, which give poor least squares fits with curves of fixed frequency. For example, a least squares analysis with trial periods at fixed frequencies performed on almost any musical composition might lead to the conclusion that the piece was not significantly rhythmic, the reason being the frequent shifts in rhythm and tempo. The issue could be resolved, both for the musical composition and for the present data, by temporal amplitude and phase diagrams (10) designed to detect shifts and drifts of phase or frequency.

SUMMARY

Total 17-hydroxycorticosteroid excretion (17-OHCS) was determined daily, with interruptions, for approximately 4 years in a healthy male subject. The coefficient of variation of all the data was 17.4%, supporting previous reports that 17-OHCS excretion is *relatively* constant over time in individual subjects. However, if serial measurements had been made over only several weeks at different parts of the total time span, markedly differing conclusions would have been drawn regarding mean levels and variability.

Rhythmometry by electronic computer detected statistically reliable superimposed low amplitude cycles with periods of 18.0 and 45.6 days. The former was similar in period (i.e., in the circavigintan region) to cycles of urinary 17-kestosterone excretion and of androgenic functions observed by other investigators. Thus circavigintan physiological cycles may be characteristic of the human male.

ACKNOWLEDGMENTS

The author is indebted to Stephen Brumberg, R. Leon Joshlin, Sarah Perry, and G. Mark Spitalny for technical assistance, to Max L. Fogel for statistical consultation, to Donald McEvoy for biochemical consultation, and to Franz Halberg and the Chronobiology Laboratories of the University of Minnesota for least squares analyses of the data.

The work was carried out while the author was attached to the Eastern Pennsylvania Psychiatric Institute, Philadelphia, and to the Department of Psychiatry, University of Pennsylvania. Supported in part by Grants K3-MH-7723 and MH-08806 of the National Institute of Mental Health, U.S.P.H.S. and by research funds of the Commonwealth of Pennsylvania.

REFERENCES

1. G. C. Curtis, *Psychosomat. Med.*, **34**, (1972), 235.
2. E. D. Weitzman, D. Funkushima, C. Nogeire, H. Roffwarg, T. F. Gallagher, and L. Hellman, *J. Clin. Endocr.*, **33**, (1971), 14.
3. J. W. Mason, *Psychosomat. Med.*, **30**, (1968), 576.
4. C. T. Wolff, S. B. Friedman, M. A. Hofer, and J. W. Mason, *Psychosomat. Med.*, **26**, (1964), 576.
5. S. B. Friedman, J. W. Mason, and D. A. Hamburg, *Psychosomat. Med.*, **25,** 4 (1963), 364.
6. R. D. Bulbrook, F. C. Greenwood, and J. L. Hayward, *Lancet*, **1**, (1960), 1154.
7. J. Marmorston, P. J. Geller, and J. M. Weiner, *Ann. N.Y. Acad. Sci.*, **164**, Art. 2 (1969), 483.
8. F. D. Moore, S. I. Woodrow, M. A. Aliapoulios, and R. E. Wilson, *New Eng. J. Med.*, **277**, (1967), 411.

9. H. M. Fox, B. J. Murawski, G. W. Thorn, and S. J. Gray, *Arch. Intern. Med.*, **101**, (1955), 859.

10. F. Halberg, M. Engeli, C. Hamburger, and D. Hillman, *Acta Endocr.* **50**, Suppl. 5 (1964), 103.

11. E. M. Glenn and D. H. Nelson, *J. Clin. Endocr.* **13**, (1953), 911.

12. E. J. Sachar, J. W. Mason, J. R. Fishman, D. A. Hamburg, and J. H. Handlon, *Psychosomat. Med.*, **27**, (1965), 435.

13. G. C. Curtis, M. L. Fogel, D. McEvoy, and C. Zarate, *Psychosomat. Med.*, **28**, (1966), 696.

14. G. D. Curtis, M. L. Fogel, D. McEvoy, and C. Zarate, *J. Clin. Endocr.*, **28**, (1968), 711.

15. F. Halberg, Y. L. Tong, and E. A. Johnson, in *The Cellular Aspects of Biorhythms* (H. von Mayersbach, Ed.), Springer, Berlin, 1967, p. 20.

16. J. Rummel, J. K., Lee, and F. Halberg, in Biorhythms and Human Reproduction (M. Ferin, F. Halberg, R. M. Richart, and R. L. Vande Wiele, Eds.), Wiley-Interscience, New York, 1973, Chap. 5.

8 Rhythms of the Male Hypothalamo-Pituitary-Testicular Axis

A. VERMEULEN, *Moderator*

CHAPTER 27

Rhythms of the Male Hypothalamo-Pituitary-Testicular Axis

A. VERMEULEN, L. VERDONCK, and F. COMHAIRE

Department of Endocrinology and Metabolism, Medical Clinic, Akademisch Ziekenhuis, Gent, Belgium

The gonadal function in mammals is subject to a complex rhythmicity. Although it is more obvious in the female, there also exists in males a gonadal rhythmicity in which we can distinguish a developmental rhythm, a circadian rhythm, short-term variations, and eventually a seasonal cyclicity.

In this report we will discuss the rhythmicity in the pituitary-gonadal axis and in androgen secretion, mainly in terms of hormone levels in plasma. Only occasionally will data concerning urinary excretion of hormones and their metabolites be mentioned.

RHYTHMICITY IN THE FUNCTION OF THE PITUITARY GONADAL AXIS IN RELATION TO SEXUAL DEVELOPMENT

Leydig cells appear in the *fetal testes* when the embryo has a length of 29 mm and display 3β-ol steroid-dehydrogenase activity in 30 mm fetuses (1). This differentiation of fibroblasts into Leydig cells seems to occur independently of any pituitary stimulation and is, moreover, probably independent of extragonadal factors since it also occurs in the explanted testicular tissue (2). The initial hormonal secretion by the Leydig cell, responsible for the differentiation of the genital tract which occurs before the ninth or 10th week, seems to be independent of pituitary stimulation, as masculinization of the genital tract is complete even in the absence of hypothalamo-pituitary tissue. Maternal choriongonadotropins may be involved, however. Kaplan et al. (3) detected immunoreactive luteinizing hormone (LH) and follicle-stimulating hormone (FSH) in the fetal pituitaries and sera as early as the end of the first trimester. During the second half of pregnancy the fetal testes undoubtedly need pituitary stimulation for normal functioning (4) since in anencephalic fetuses testes are hypoplastic and Leydig cells reduced in number. After the sixth month of fetal life the testicular Leydig cells normally decrease in size, suggesting that both the placental and the pituitary gonadotropins become insufficient to maintain these cells; as the FSH secretion by the fetal pituitary also decreases after the 30th week of life (5), it would then appear that gonadotropin stimulation by the pituitary during fetal life goes through a maximum.

Mizuno et al. (6) observed that *at birth* plasma levels of androstenedione, but not of testosterone, are higher in male than in female babies; this difference persists for some days after birth. Rivarola et al. (7), however, did not observe this difference. None of the authors (6–9) who determined testosterone in cord blood observed any sexual difference. Recently Younglai (10) reported a statistically significant difference in "17β-hydroxyandrogen" concentration of amniotic fluid between a group of male and female babies, but sexes could not be distinguished on this basis in the majority of individual cases.

Leydig cells degenerate and testicular secretion becomes quiescent soon after birth, and it is only in the prepubertal stage that a reactivation of the pituitary-gonadal axis occurs.

However, in recent years it has become evident that even in the young child the pituitary-gonadal axis is active and that there exists an *active prepubertal feedback mechanism.* Indeed, immunoassayable LH and FSH levels have been detected in the *circulation* of sexually infantile children of any age group studied (11–13). Moreover, Laron and Zilka (14) reported a compensatory hypertrophy of the descended testis of prepubertal males with unilateral cryptorchidism, and Blizzard et al. (15) reported a significant increase in plasma LH and FSH levels 7 days after gonadectomy of a 4-month old child. Finally, Frasier and Horton (16) observed that plasma testosterone and androstenedione levels are higher in boys than in girls, suggesting an active tes-

ticular secretion during childhood, and we (17) observed lower testosterone levels in anorchid children than in P_1 prepubertal children.

Even in the prepubertal child the testes can easily be stimulated to
an androgen secretion as high as in normal adults (18); it is generally
assumed that the high sensitivity of the hypothalamus for the feedback
by sex steroids during infancy precludes the release of adult levels of
gonadotropins.

In the *prepubertal period* the sensitivity of the hypothalamus for the
feedback effect of sexual hormones decreases, and consequently the
plasma levels of gonadotropins increase progressively.

In prepubescent boys a gradual increase of *LH and FSH* levels with
progressing puberty occurs (19–22), the mean age at which FSH and
LH begin to rise being, respectively, 9.8 and 10.9 years in girls, and
11.5 and 12 years in boys. The subsequent time course of plasma FSH
and LH levels varies somewhat from one study to another. Postpubertal values for plasma FSH are approximately twice prepubertal
values in both sexes; postpubertal LH levels are three to four times
prepubertal values, the increasement being somewhat higher in girls.

As far as the *androgen secretion* in prepubertal boys is concerned,
during childhood the main androgen in plasma is *androstenedione* (23).
Its mean concentration during stage I is 62 ± 8.9 ng/100 ml and this
does not change significantly during pubertal development. Although it
has been shown that in the calf (24), the piglet (25), and the
prepubescent rat (26) this androstenedione is an important secretory
product of the testis, this has not been proved in man. As androstenedione levels do not increase upon HCG stimulation (16) but do
increase upon ACTH stimulation (27), we may accept that androstenedione has an adrenal origin in children.

Plasma testosterone levels, on the contrary, show a progressive
increase with the stage of pubertal development (22, 23, 27, 28). We observed a rise in plasma testosterone concentration from a mean value of
33.9 ng/100 ml (16.7–68 ng) in the P_1 pubertal stage (n = 35), to 90.6
ng/100 ml (50.9–157) in the P_2 group (n = 50), 181 (118–278) in the P_3
group (n = 30), 387 (287–663) in the P_4 group (n = 19) and finally
648 ng/100 ml (496–896) in the P_5 pubertal stage (n = 15); the rise
in the free testosterone fraction is even more impressive (Figure 1). In
anorchic children of similar age plasma testosterone levels are significantly lower than in the P_1 group; this shows that in the P_1 stage Leydig
cells are actively secreting androgens in accordance with the observation
that in the child testosterone levels are much lower than in the P_1 prepubertal stage (29). As androstenedione levels remain constant whereas
testosterone levels increase, the testosterone/androstenedione ratio

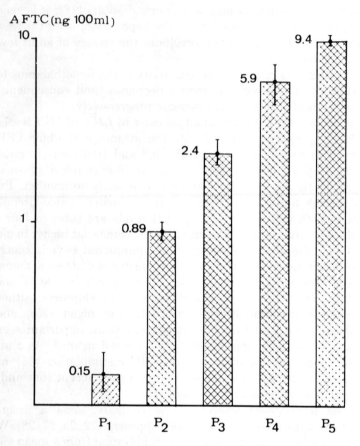

Figure 1 Apparent free testosterone levels (AFTC) as a function of pubertal development.

shows an important shift during pubertal development. *5α-Dihydrostesterone,* on the other hand, was shown to increase in parallel with plasma testosterone during sexual maturation (30) (Figure 2).

Dehydroepiandrosterone levels appear to be rather constant in prepubertal male subjects and do not vary as a function of pubertal stage (31). It is generally considered as being mainly an adrenal secretory product; Lipsett (32) calculated that only about 6% would have a testicular origin. On the other hand, Saez et al (3) observed an increase in plasma levels from 138 $\pm$ 40 ng/100 ml to 394 $\pm$ 204 ng/100 ml after HCG stimulation, suggesting a testicular origin. Kirchner et al, (33), however, did not find any increase in dehydroepiandrosterone after HCG stimulation of adult males.

Dehydroepiandrosterone sulfate levels, on the other hand, rise progressively until puberty (31) from very low levels before the age of 7. This DS probably originates mainly in the adrenals since this rise was also observed in a patient with anorchia. Nevertheless, it seems that it may also be secreted by the testes as Saez (31) observed a 50% suppression of DS levels after fluoxymesterone and a threefold increase in young children after HCG stimulation for 14 days. This stimulation did not, however, increase DS levels in older children or in adults (34).

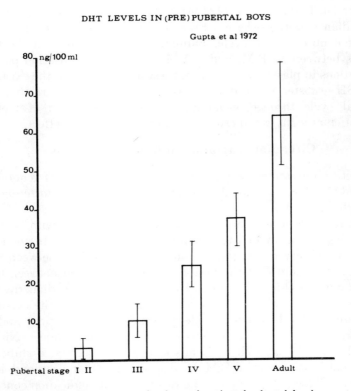

DHT LEVELS IN (PRE) PUBERTAL BOYS

Gupta et al 1972

Figure 2   Plasma dihydrotestosterone levels as a function of pubertal development (30).

THE HYPOTHALAMO-HYPOPITUITARY TESTICULAR AXIS IN ADULTS

Circadian Rhythm in Gonadotropin Levels

Several authors (35–37) observed a *nycthemeral rhythm* in *LH plasma levels*, the latter being generally lowest between 2 and 6 p.m. This is in agreement with the findings of Fraschini and Motta (cited in Ref. 38) who found a diurnal cycle in the amount of LH and FSH stored in the pituitary, with highest levels being in the afternoon. According to Burger (36), the nadir of the LH levels precedes the nadir of the testosterone levels, whereas LH levels start to increase just before an increase in testosterone levels in observed. On the other hand, Swerdloff et al. (39), Peterson et al. (40), Nillius and Wide (41), Faiman and Winter (42), Boyar et al. (43) did not observe any 24-hour rhythm in LH levels.

The existence of a *diurnal cycle in FSH levels* is even more controversial. Faiman et al. (42, 44) and Saxena et al. (45) did observe a circadian variation whereas Franchimont (5) and Peterson et al. (40) did not observe any cycle. Faiman et al. (42) observed the highest levels between 11 P.M. and 7 A.M.; as the FSH variation paralleled variations in plasma testosterone, these authors suggest the existence of a FSH-testosterone feedback mechanism, eventually related to the pineal cycle through melatonin secretion. Inhibition of melatonin secretion at night would relieve the gonadal inhibitory effect.

Circadian Variation in Androgen Levels

Despite observations to the contrary (33, 46, 47), it is generally accepted that there exists a diurnal cyclicity of *plasma testosterone levels* (26, 28, 42, 47–53).

Diurnal variations have generally been studied with a plasma sampling frequency of 4 to 6 hours. With this sampling frequency it is generally observed that testosterone levels are highest between 8 A.M. and 12 noon and that the lowest levels are observed between 6 P.M. and midnight (Figure 3). In distinction with the circadian rhythm of plasma cortisol, which is very similar in all subjects, the cyclicity of testosterone levels shows considerable individual variations and often important day to day variations exist in the same subject. Nieschlag and Ismail (52) as well as Evans et al. (54) observed a shift of the maximal levels toward 8 P.M. by disturbing regular sleep and activity periods. These authors do not, however, give any indication concerning the duration of sleep reversal before disturbances in testosterone rhythmicity were observed. The amplitudes of the variations are much smaller than for cortisol, and the mean ratio of the highest to the lowest values does not exceed 1.5, whereas for cortisol and ratio of 4 or 5 is quite usual. Similar variations in *urinary testosterone* and epitestosterone were observed by Okamoto et al. (53). In 1943 Pincus was the first to describe (55) a circadian rhythm in urinary 17-ketosteroid excretion. As urinary rhythms are discussed in Chapter 30, we will not go further into this topic.

Circadian variations in other plasma androgens have been less extensively studied.

We have no data concerning *dihydrotestosterone*. In view of its low concentration in plasma and the rather important methodological errors in its determination, it would be rather difficult to detect a rhythm of low to moderate amplitude.

As *17OH-progesterone* in the male seems to originate 90% from the testes and for 10% from the adrenals, it is not surprising that, as Strott and Lipsett showed (56), 17OH-progesterone shows an circadian rhythm with large amplitude, the 8 P.M. plasma level representing

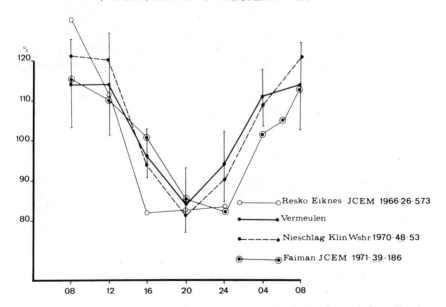

Figure 3 Circadian variations in plasma testosterone levels (% of mean). Sampling frequency: 4 hours.

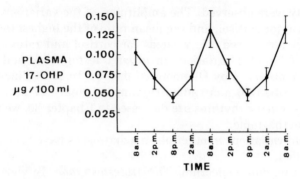

Figure 4   Circadian variations in plasma 17OH progesterone levels ± 1 SE, measured every 6 hours in nine normal men (56).

only 40% of the 8 A.M. value. The importance of this variation is perhaps rather surprising (Figure 4).

De Jong and Van der Molen (57) reported that unconjugated *dehydroepiandrosterone* levels in plasma show important fluctuations with 5 P.M. values about 50% of 8.30 A.M. values, whereas Rosenfield et al. (58) using 20-minute sampling, reported that DHEA is secreted episodically and synchronously with cortisol by normal man.

As far as *androstenedione* levels are concerned, about 1/3 of which has a testicular origin, Crafts et al. (51) reported a nycthe meral rhythm synchronous with the cortisol rhythm with an amplitude of 40 to 65%. Recently we have been determining androstenedione in normal males with a sampling interval of 30 minutes and observed a rhythmicity which paralleled cortisol rhythmicity.

Laatikainen and Vihko (59) observed a diurnal variation in plasma levels of most *sulfoconjugated androgens* (DHEA, androsterone, epiandrosterone, androst-5-ene-3β,17β-diol) with the highest values in the afternoon and the lowest between midnight and 8 A.M. McKenna and Rippon (60) as well as De Jong and Van der Molen (57) observed a similar pattern for dehydroepiandrosterone sulfate in male subjects. The difference in acrophase with respect to most other steroid hormones may be related to the long half-life of sulfoconjugated steroids. Migeon et al. (61) observed, however, a distinctly different pattern with the highest dehydroepiandrosterone sulfate levels in the morning.

Ultradian Variations in the Activity of the Pituitary Gonadal Axis

Nankin and Troen (62) were the first to report a repetitive pattern of *abrupt elevations of LH levels* throughout the day; this was confirmed by Boyar et al. (43), Santen (63), and Naftolin et al. (64). Further details can be found in Chapter 23. The ratio of the highest to the lowest LH value in each subject varies between 2 and 4.4. Boyar et al. (43) as well as Santen (63) suggest that between spikes there is a cessation of secretion. Boyar et al. (43) calculated a "low set point" for LH secretion (initiation) whereas cessation occurred within a narrow, well-defined high LH concentration range. This suggests a short feedback mechanism, although testosterone feedback cannot be excluded.

The relationship between pulsatile LH release and REM sleep is controversial. Whereas Rubin et al. (65) observed that LH is secreted in episodes during sleep, with LH levels 14% higher during REM sleep, Boyar et al. (43) did not observe more frequent LH spikes during sleep.

Figure 5   Short-term variations in plasma testosterone levels in two normal males (AV & VDW). Sampling frequency: 5 minutes.

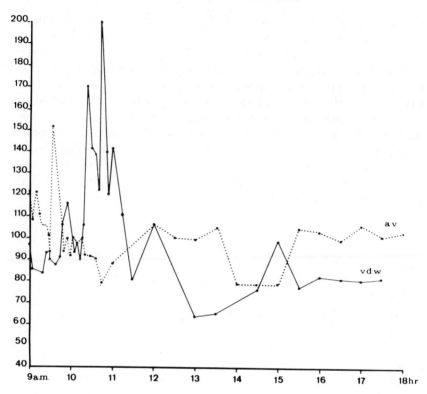

*Figure 6   Ultradian variations in plasma testosterone levels in two normal males (AV &
VDW) (percent of mean of all values).*

We found no data in the literature concerning episodical variations
in plasma testosterone levels. In two normal volunteers we determined
testosterone levels every 5 minutes for 2 hours between 9 and 11 A.M.,
and subsequently every 30 minutes for 8 hours. It appears (Figure 5)
that the plasma testosterone pattern is characterized by a few irregularly occurring peaks, with sudden rising and a slower descending
limb.

All determinations were performed in quadruplicate and only variations greater than can be accounted for by lack of assay precision (6%)
are considered. The ratio of the lowest to the highest value was 1.9 in
the first subject and 3.7 in the second (Figure 6). We attempted to determine the half-life of testosterone from the semilogarithmic plot of
the testosterone values whenever there were three points in the
descending limb. The mean half-life was 11 minutes.

Plasma LH values were determined simultaneously. Irregular oscillatory variations were observed, but a correlation with the testosterone secretory peaks was not directly evident.

Evans et al. (54) studied rapid changes in plasma testosterone levels during sleep, using a sampling interval of 5 to 20 minutes. On the basis of their results the authors suggest an association of the individual fluctuations in testosterone level with periods of REM sleep. These authors report, moreover, that administration of long-acting ACTH severally depresses REM sleep and minimizes fluctuations in plasma testosterone. We (17) observed that long-acting ACTH depresses testosterone levels in the male for at least 24 hours. This may be of interest in view of the fact that Kreuz et al. (66) observed a significant decrease in plasma testosterone levels during psychological stress.

Concerning pulsatile variations in plasma *androstenedione*, we are not aware of data in the literature. We determined plasma androstenedione in the same samples on which short-term variations in testosterone were studied. As expected, peaks were not synchronous with those of testosterone.

As far as the other androgens are concerned, Rosenfield et al. (58) observed that *dehydroepiandrosterone* was secreted episodically and synchronously with cortisol.

Long-Term Variations in the Activity of the Pituitary-Gonadal Axis

Peterson et al. (40) determined daily plasma LH levels for 3 months in four subjects and did not observe any long-term rhythmicity. However, the results were not analyzed by sophisticated statistical methods that might have uncovered rhythm that were not apparent in the crude data.

As far as long-term variations in androgen secretion are concerned, available data are relatively scarce. In a study on *urinary 17 KS excretion* by Hamburger (67) and analyzed by Hallberg et al. (68), cycles of about 1 week, about 20 days, about 1 month, and about 1 year were observed. The excretion was highest in September and lowest in May; however, the amplitude of the cycle was low. As urinary 17 KS have mainly an adrenal origin, it is evident that variations in their excretion do not necessarily reflect any variation in gonadal activity. See also Chapter 26.

Manson (69) reported cyclic variations (cycles of 4 weeks) in the frequency of neutrophil leucocytes with "androgen induced" nuclear appendages. Eik-Nes (70) in dogs observed a trend toward higher secretion of *testosterone* in early spring compared to late winter months.

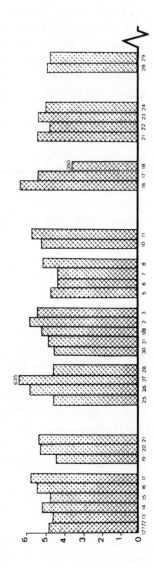

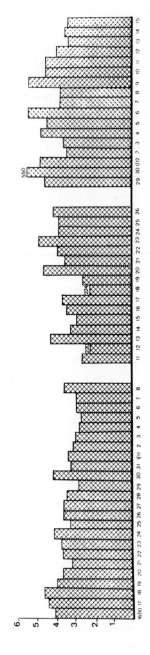

Figure 7 *Daily variations in plasma testosterone levels determined at 11 A.M. over 6 weeks in Summer and 2 months in Winter in a normal male, age 43.*

Longitudinal studies concerning testosterone excretion were first reported by Ismail and Harkness in 1966 (71 and Chapter 30). These authors made two short serial studies of 14 days and one of 26 days. They reported intervals of 6 and 12 days between peaks of excretion. Corker and Exley (72) studied testosterone excretion for 42 days in two male subjects and found a coefficient of variation of about 30% in excretion values but little evidence for a 8 to 10 day cycle as had been observed in estrone excretion (73). Burger et al. (50) determined testosterone plasma levels thrice weekly for 7 weeks. They observed wide variations but no apparent rhythmicity. Fox et al. (74) studied plasma testosterone levels in the same subject daily at 10 A.M. and correlated the values with sexual activity. Testosterone levels sampled daily varied between 145 and 565 ng/100 ml. Peak values were detected at irregular intervals which bore no apparent relationship to sexual activity. Incidently, testosterone levels were invariable higher during coitus or immediately after orgasm. Moreover, evening levels were lower than 8 A.M. levels in 40 out of 43 samples.

In a limited study we determined plasma testosterone levels in a male subject daily at 11 a.m. for 40 days in summertime (July–August) and for 60 days during winter (15 October–15 December) (fig. 7). During

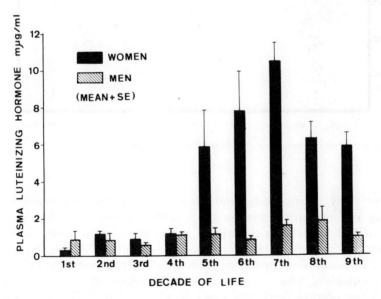

Figure 8   Variations of plasma LH levels as a function of age. Each group consists of between five and 12 normal subjects. All women in the fifth decade were postmenopausal (76).

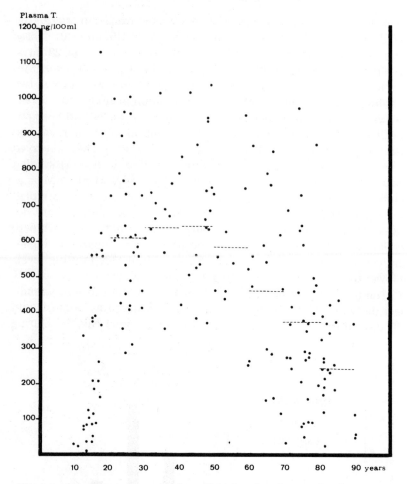

Figure 9   Plasma testosterone levels as a function of age in normal males.

summer daily plasma testosterone levels varied, with one exception (360 ng/100 ml) varied between 430 and 630 ng/100 ml, with a mean value of 550 ng. During the winter period testosterone levels varied between 220 and 550 ng/100 ml with a mean value of 363 ng/100 ml, about 25% lower than the mean summer value. It appears that at least in this individual, who was in good general condition during the study, plasma testosterone levels are lower in winter than during the summer. Daily variations were more important in winter than during the summer period, but nevertheless much smaller than the variations observed by Fox et al. (74). In the view of the circadian variation in plasma testosterone, urinary testosterone levels may be more appropriate to use in the study of long-term variations in gonadal activity. However, as many substances may influence the enzymatic hydrolysis of the testosterone glucuronide, hydrolysis should be checked for completeness, as variations in hydrolysis might be taken for variations in testicular function. Moreover, variations in testosterone metabolism may also affect the testosterone excretion.

EVOLUTION OF TESTICULAR FUNCTION WITH AGE IN ADULTS.

Although there is no abrupt fall in the testosterone secretion in old age, there is nevertheless ample clinical evidence that androgenicity decreases with aging.

Several authors (39, 75–77) observed slightly higher *LH and FSH levels* in plasma in patients over 50 years old with, however, a wide range of individual values overlapping partly the range in younger individuals, in distinction to the levels observed in postmenopausal women

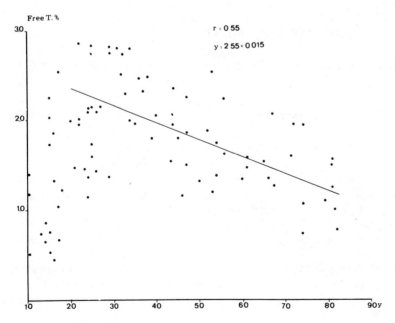

Figure 10   Free plasma testosterone fraction as a function of age.

(Figure 8). Coppage and Cooner (78), Kent and Acone (79), as well as Gandy and Peterson (29) did not observe any decrease in *plasma testosterone* levels with advancing age. Hollander and Hollander (80), Kirchner and Coffman (81), and our group (82), however, observed in a large group of subjects that notwithstanding a wide range in individual values, there is a definite decrease in testosterone levels in males over 50 years old (Figure 9). The decrease was more obvious when only the free testosterone fraction was considered (Figure 10). This is a consequence of the fact that the testosterone binding capacity of the testosterone binding globulin increases with decreaseing testosterone levels. Moreover, the HCG stimulation test revealed that the testicular capacity to secrete testosterone is decreased. This decrease is consistent with the well-known decrease in urinary testosterone excretion in old age. Moreover the metabolic clearance rate also decreases, resulting in a decreased blood production rate. This is, however, to a large extent the consequence of the decreased free testosterone fraction, as increasing this fraction to values found in younger adults increases the MCR (82).

Finally, in old age a shift in the metabolism of testosterone is also observed; less 17β-hydroxyl metabolites are excreted and 5β metabolites become relatively more important (82).

ACKNOWLEDGMENT

This work was supported in part by Grant No. 1214 of the F.W.G.O.

REFERENCES

1. L. J. Pelliniemie and M. Niemi, *Z. Zellforsch.*, **99**, (1969), 507.

2. R. Picon, *Arch. Anat. Microscop. Morphol. Exp.*, **56**, (1967), 281.

3. S. L. Kaplan, M. M. Grumbach, and T. H. Shepard, *Pediat, Res.*, **3**, (1969), 512.

4. A. Jost, in *The Pituitary Gland*, Vol. 2 (G. W. Harris and B. T. Donavan, Eds.), Butterworths, London, 1966, p. 299.

5. P. Franchimont, *Ann. Endocr.*, **29**, (1968), 403.

6. M. Mizuno, J. Lobotsky, C. W. Loyd, and T. Kobayski, *J. Clin Endocr.*, **28**, (1968), 1133.

7. M. A. Rivarola, M. G. Forest, and C. J. Migeon, *J. Clin. Endocr.*, **28**, (1968), 34.

8. H. H. Simmer, M. V. Frankland, and M. Greipel, *Steroids*, **19**, (1972), 215.

9. G. P. August, M. Tkachuk, and M. M. J. Grumbach, *J. Clin. Endocr.*, **29**, (1969), 891.

10. E. V. Younglai, *J. Endocr.*, **54**, (1972), 513.

11. R. B. Rifkind, H. E. Kulin, and G. T. Ross, *J. Clin. Invest.*, **46**, (1967), 1925.

12. R. B. Rifkind, H. E. Kulin, P. L. Rayford, C. M. Cargille, and G. T. Ross, *J. Clin. Endocr.,* **31**, 1970), 517.

13. P. A. Lee, A. R. Midgley, and R. B. Jaffe, *J. Clin. Endocr.,* **31**, (1970), 248.

14. Z. Laron and E. Zilka, *J. Clin. Endocr.,* **29**, (1969), 1409.

15. R. M. Blizzard, R. Penny, T. P. Foley, Jr., A. Baghdassarian, A. Johanson, and S. S. C. Yen, "Pituitary Gonadal Interrelationships in Relation to Puberty," in *Gonadotropins* (B. Saxena, C. C. Beling, and H. M. Gandy, Eds.), Wiley-Interscience, New York, 1972, p. 502.

16. S. O. Frasier and R. Horton, *Steroids,* **8**, (1966), 777.

17. A. Vermeulen, *Verhandelingen van de Koninklijke Vlaamse Akademie voor Geneeskunde van Belgie* **35**, (1973), 95.

18. J. M. Saez and M. Bertrand, *Steroids,* **12**, (1968), 754.

19. S. Raiti, A. Johansson, C. Light, C. J. Migeon, and R. M. Blizzard, *Metabolism,* **18**, (1969), 234.

20. A. J. Johansson, H. Guyda, C. Light, C. J. Migeon, and R. M. Blizzard, *Pediatrics,* **74**, (1969), 416.

21. I. Burr, P. Sizonenko, S. Kaplan, and T. Grumbach, *Pediat. Res.,* **4**, (1970), 25.

22. C. P. August, M. M. Grumbach, and S. L. Kaplan, *J. Clin. Endocr.,* **34**, (1972), 319.

23. D. S. Frasier, F. Gafford, and R. Horton, *J. Clin. Endocr.,* **29**, (1969), 1404.

24. H. R. Lidner, *J. Endocr.,* **23**, (1961), 139.

25. H. R. Lidner, *J. Endocr.,* **23**, (1961), 171.

26. J. A. Resko, H. H. Feder, and R. W. Goy, *J. Endocr.,* **40**, (1968), 485.

27. R. L. Rosenfield, B. J. Grossman, and M. Ozoa *J. Clin. Endocr.,* **33**, (1971), 249.

28. A. Vermeulen, T. Stóica, and L. Verdonck, *J. Clin. Endocr.,* **33**, (1971), 709.

29. H. M. Gandy and R. E. Peterson, *J. Clin. Endocr.,* **28**, (1968), 949.

30. D. Gupta, E. McCafferty, and K. Rager, *Steroids,* **19**, (1972), 411.

31. J. M. Saez and M. Bertrand, *Steroids,* **12**, (1968), 749.

32. M. B. Lipsett, in *The Human Testis* (E. Rosemberg and C. A. Paulsen, Eds.), Plenum Press, New York, 1970, p. 407.

33. M. A. Kirchner, M. B. Lipsett, and D. R. Collins, *J. Clin. Invest.,* **44**, (1965), 657.

34. T. Yamaji and H. Ibayashi, *J. Clin. Endocr.,* **29**, (1969), 273.

35. P. Franchimont, *Ann. Endocr.,* **27**, (1966), 273.

36. H. G. Burger, In *Protein and polypeptide hormones* (M. Margoulies, Ed.), Excerpta Medica Foundation, I.C.S., **161**, (1969), III–729.

37. B. B. Saxena, G. Leyendecker, W. Chen, H. Gandy and R. E. Peterson, *Proc. Karol Symp. Res. Methods Reprod.,* **142**, (1969), I.

38. L. Martini, F. Fraschini and M. Motta, *Rec. Progr. Hormone Res.,* **24**, (1968), 439.

39. R. S. Swerdloff and W. D. Odell, in *Gonadotropins* (E. Rosemberg, Ed.) Geron X, Inc., Los Altos, Calif., 1968, p. 155.

40. M. T. Peterson, A. R. Midgley, and R. B. Jaffe, *J. Clin. Endocr.,* **28**, (1968), 1473.

41. S. J. Nillius and J. Wide, *Acta Endocr.,* **65**, (1970), 583.

42. C. Faiman and J. S. D. Winter, *J. Clin. Endocr.,* **31**, (1971), 186.

43. R. Boyar, M. Perlow, L. Hellman, S. Kapen, and E. Weizmann, *J. Clin. Endocr.,* **35**, (1972), 73.

44. D. Faiman and R. J. Ryan, *Nature,* **215,** (1967), 5103.

45. B. B. Saxena, H. Demura, M. Gandy, and R. E. Peterson, *J. Clin. Endocr.,* **28,** (1968), 519.

46. B. Hudson, J. P. Coghlan, A. Dulmanis, and M. Wintour, *Excerpta Med. Found., I.C.S.,* **83,** (1964), 1127.

47. T. Kobayashi, J. Lobotsky, and C. W. Lloyd, *J. Clin. Endocr.,* **26,** (1966), 610.

48. R. Dray, A. Reinberg and J. Sebaoun, *C. R. Acad. Sci., Paris,* **261,** (1965), 573.

49. A. L. Southren, S. Tochimoto, N. C. Carmody, and K. Isurgi, *J. Clin. Endocr.,* **25,** (1965), 1441.

50. H. G. Burger, J. B. Brown, K. J. Catt, B. Hudson, and J. R. Stockigt, *Excerpta Med. I.C.S.,* **161,** (1968), 412.

51. R. Crafts, A. Llerena, A. Guevara, J. Lobotsky, and C. W. Lloyd, *Steroids,* **12,** (1968), 151.

52. E. Nieschlag and A. A. A. Ismail, *Klin. Wochenschr.,* **48,** (1970), 53.

53. M. Okamoto, C. Setaishi, K. Nakagawa, Y. Horiuchi, K. Moriya, and S. Itoh, *J. Clin. Endocr.,* **32,** (1971), 846.

54. J. I. Evans, A. M. Maclean, A. A. A. Ismail, and D. Love, *Proc. Roy. Soc. Med.,* **64,** (1971), 841.

55. G. Pincus, *J. Clin. Endocr.,* **3,** (1943), 195.

56. Ch. A. Strott, T. Yoshimi, and M. B. Lipsett, *J. Clin. Invest.,* **48,** (1969), 930.

57. F. H. De Jong and H. Van der Molen, *J. Endocr.,* **53,** (1972), 461.

58. R. S. Rosenfield, L. Hellman, H. Roffwarg, E. D. Weitzman, D. K. Fukushima, and T. F. Gallagher, *J. Clin. Endocr.,* **33,** (1971), 87.

59. T. Laatikainen and R. Vihko, *J. Clin. Endocr.,* **28,** (1968), 1356.

60. J. McKenna and A. E. Rippon, *Biochem. J.,* **95,** (1965), 107.

61. C. J. Migeon, A. R. Keller, B. Lawrence, and T. H. Shepard, *J. Clin. Endocr.,* **17,** (1957), 1051.

62. H. R. Nankin and P. Troen, *J. Clin. Endocr.,* **33,** (1971), 558.

63. R. J. Santen, *Abstracts of the 4th International Congress of Endocrinology, Washington 1972,* Excerpta Medica I.C.S. No. 256, Abstract No. 254.

64. F. Naftolin, S. S. C. Yen, and C. C. Tsai, *Nature, New Biol.,* **23,** (1972), 92.

65. R. T. Rubin, A. Kales, R. Adler, T. Fagan, and W. Odell, *Science,* **175,** (1972), 196.

66. L. E. Krenz, R. M. Rose, and J. R. Jennings, *Arch. Gen. Psychiat.,* **26,** (1972), 479.

67. C. Hamburger, *Acta Endocr.,* **17,** (1954), 116.

68. F. Hallberg, M. Engeli, C. Hamburger, and D. Hellman, *Acta Endocr. Suppl.* **103,** (Vol 50), (1965).

69. J. C. Manson, Life Sci., **4,** (1965), 329.

70. K. Eik Nes, *Rec. Progr. Hormone Res.,* **27,** (1971), 517.

71. A. A. A. Ismail and Harkness, *J. Endocr.,* **34,** (1966), XVII.

72. C. S. Corker and D. Exley, *J. Endocr.,* **40,** (1968), 255.

73. D. Exley and C. S. Corker, *J. Endocr.,* **35,** (1966), 83.

74. C. A. Fox, A. A. A. Ismail, D. M. Love, K. E. Kirkham, and J. A. Loraine, *J. Endocr.,* **52,** (1972), XLI.

75. R. J. Ryan and Ch. Faiman, in *Gonadotropins* (E. Rosemberg, Ed.) Geron X, Inc., Los Altos, 1968, p. 333.

76. D. S. Schalch, A. F. Parlow, R. C. Boon, and S. Reichlin, *J. Clin. Invest.*, **47**, (1968), 665.

77. P. Franchimont, *Secretion normale et pathologique de la somatotrophine et des gonadotrophines humaines*, Masson, Paris 1968.

78. W. S. Coppage and A. E. Cooner, *New Engl. J. Med.*, **273**, (1965), 773.

79. J. Z. Kent and A. B. Acone, in *Androgens in Normal and Pathological Conditions (A. Vermeulen and D. Exley, Eds.)*, *Excerpta Medica Foundation, I.D.S.* **101**, (1966), 31.

80. N. Hollander and V. P. Hollander, *J. Clin. Endocr.*, **18**, (1958), 966.

81. M. A. Kirchner and G. D. Coffman, *J. Clin. Endocr*, **28**, (1968), 1347.

82. A. Vermeulen, R. Rubens, and L. Verdonck, *J. Clin, Endocr.*, **34**, (1972), 730.

CHAPTER 28

Short- and Long-Term Rhythms in Testicular Function in the Bull

MACIEJ KRZANOWSKI

Institute of Zootechny, Experimental Station at Grodziec Slaski, Laboratory of Applied Biochemistry at Gumna/Cieszyn, Poland

Two different studies showing periodicity in Bull testicular function have been reported.

FLUCTUATIONS OF MALE FERTILITY

The most interesting, and from the practical point of view, the most promising of the discoveries in the field of the male sexual rhythm seems to be that of Kihlström (1) who reported that bull fertility was rhythmic. The fertility of the bulls increased with an average period of 3 weeks; that is, similar to that of the cow's sexual cycle. The cycle was more regular in younger bulls. The phasing of the rhythm of fertility and of the number of spermatozoa ejaculated differed, and this might lead to the conclusion that the male fertility cycle depends on the rhythm of qualitative changes in spermatozoa (Kihlström and Hultnäs, quoted in Ref. 2). The abundance of spermatozoa in the bull ejaculate and the widely applied artificial insemination in cows make cattle an especially advantageous species for such investigations. The great

447

number (50 to 150) of cows inseminated with spermatozoa from the same ejaculate allows one to consider female fertility as a statistical constant for a given population in a given season of the year. It permits one, therefore, to accept the differences in NR (nonreturn to heat in 30 to 60 days) percentages of groups of cows inseminated with different ejaculates as the result of differences in the quality of spermatozoa.

One aim of this study was to repeat Kihlström's work. The studies were carried out on the basis of recorded 60 days-NR percentages after inseminations with the semen of 28 bulls kept at the Animal Insemination Center at Drogomysl (Poland) in the years 1963–1970. The records of 14 individuals covered a period of 3 or more years. Time analysis was carried out on a total of 3000 items of data, each of which concerned the percentage of NR after insemination with the individual's semen collected in 1 day. The semen was not collected at regular time intervals since the material used for analysis was mostly retrospective. However, research on the time of occurrence of spermatozoa marked with isotopes in the ejaculate proved that this time was almost independent of the frequency of semen collectd (3); it seems, therefore, that this irregularity was not of major importance. Analysis of the data was performed by the cosinor method (4) in the Chronobiology Laboratories of the University of Minnesota. Eighteen of the 26 animals demonstrated a statistically significant rhythm in fertility at less than the .05 level; the average period of the cycles investigated was 18.6 days, which is in agreement with the earlier results of Kihlström. The average differences between the results in the more and less fertile half of the estimated cycles amounted to 1 to 4%. There was also reported in 16 young bulls (less than 2.5 years old at the time the observations began) a statistically significant circannual rhythm in fertility, which attained its maximum in January. Courot, Goffaux, and Ortavant (5) also found (in similar geographic and climatic conditions) the NR percentage of cows inseminated with fresh semen in the autumn–winter to be the highest. However, using frozen spring sperm or frozen autumn sperm for insemination in spring or autumn demonstrated that not males but females were responsible for this winter crest of fertility. They found that the male and female annual crests of fertility in the cattle in French Jura were opposite and the best results in fertility were obtained with cows inseminated in autumn by semen collected in spring.

FLUCTUATIONS OF DNA SPERMATOZOAL CONTENT

The second part of this study deals with the DNA rhythm in spermatozoa. The rule of a constant amount of DNA in all somatic cells and half of it in the haploid gametes within the individual and within all the individuals of the same species is, as yet, a textbook axiom. The high variability in spermatozoa DNA content was considered by some authors as the reason for reduced male fertility (6, 7). However, for several years opinions have been expressed that the DNA content may not be constant in healthy fertile males' spermatozoa (8); that it may differ in different breeds (9) or in different individuals (10–12). The aim of this investigation was to check whether the scatter of spermatozoa DNA content within different ejaculates was due to the method of analysis and whether the fluctuations seen were rhythmic.

This study is discussed in detail because of its general importance and because attempts to revise the principle of the constancy of DNA content in spermatozoa meet with attacks of methodology (13). The investigation was carried out on three groups of ejaculates:

1. *Ejaculates from different bulls* (collected and examined simultaneously every 2 to 5 days). Eight regularly fertile bulls of Black-and-White Lowland Breed in two subgroups (1a and 1b) of four were collected and examined simultaneously for 6 months. Subsequently the investigation on four of these bulls (subgroup Ic) was prolonged for another 5 months. There were 120 tests (of four, in seven times of three ejaculates). All ejaculates, from which some portions of semen were taken for these tests, were subjected to routine laboratory tests in the above mentioned A.I.C. at Drogomyśl (at least 80% of spermatozoa had a progressive movement) and succesfully used to inseminations (at least 70% NR).

2. *Successive ejaculates of an individual bull* (collected within a few minutes). They were collected from different bulls. The variation in four successive ejaculates from each was studied. There were 22 tests of four ejaculates.

3. *Different dilutions of one ejaculate* (control group, particular samples prepared in concentrations: $1 + 0$, $1 + 1$, $1 + 3$, and $1 + 7$ parts of diluent). There were 36 tests of four samples.

During one analytical run, four samples were tested simultaneously. There were four ejaculates from different bulls (Group 1), or four ejaculates from the same bull (Group 2), or four dilutions of the one ejaculate (control group). The scatter of the results within one analytic run, if due only to the errors of the method, should be similar in every one of these groups.

The total amount of DNA in a sample was tested by Schneider's method by determining the DNA quantity in the hydrolyzate on the basis of the extinction value in ultraviolet light at a wavelength of 260 nm as in the previous work (9). The concentration of spermatozoa was calculated in a Bürker's chamber. The average DNA content in a spermatozoon was obtained by dividing the total DNA content by the number of spermatozoa. Chemical analyses were repeated three times and counts of spermatozoa 20 times in both control and tested groups. DNA was determined in 60 samples in parallel on the basis of the sugar component by Dische's reaction with diphenylamine. A comparison of the results in pairs showed a good correlation ($r = +0.92$). Admixture of other cells as the source of DNA was less than 1/300; therefore, they did not significantly influence the total DNA amount as compared to the degree of its fluctuations.

The ratio of DNA to spermatozoon, the average from several dozen ejaculates collected at the same period of the year, did not differ significantly in the majority of the individuals investigated. In eight bulls tested during 6 months this average amount calculated from 38 or 39 ejaculates was (in arbitrary units): 5.54, 5.64, 5.64, 5.65, 5.67, 5.69, 5.75, and 6.03. The coefficients of variance of these average means ranged between 9.1 and 11.9%.

All these results, except the last one, do not differ significantly. Within particular ejaculates, however, the differences were high. An analysis of variances (Table 1) showed that the scatter of the results in simultaneously tested ejaculates of different bulls was rather similar to that in successive ejaculates of one individual. The scatter of the results in both these groups, however, was significantly ($p < .01$) higher than that in the control group (i.e., due to the method). The coefficient of variance of the real content of DNA (average per one spermatozoon) in consecutive ejaculates of the same individual approximated 7.3% (with allowance for variance due to analysis error estimated on the basis of variance in the control group).

The results of DNA tests of the ejaculates collected from the particular bulls over a period of 6 to 11 months were analyzed by the cosinor method (4). The analysis demonstrates that there was a statistically significant circannual rhythm in the DNA content in spermatozoa with a period of increase in autumn and a period of decrease in spring (Figure 1). The average amounts of DNA per one spermatozoon in autumn surpassed the ones in spring by 12 to 13%. In one individual a systematic increase of DNA content in spermatozoa (Table 2) was observed 2 months earlier than in the others investigated in parallel (bull with an average DNA content in 6 months of investigations equal to 6.03). This shift in DNA rhythm in one bull, tested simultaneously with individuals whose DNA did not increase, indicates that the reaction to synchronizing factors may be individualistic. It also supports the conviction that the observed circannual rhythm of DNA content in spermatozoa was not due to a "rhythm in the method"

Table 1 Variances in the Series of Analyses Carried out in Parallel and the Significance of Differences[a]

	df	S2	Group Ia	Ib	Ic	II	III
The series of ejaculates from different bulls	Group Ia 114	0.2265	*	$F = \dfrac{0.3329}{0.2265} = 1.97$ $p < .05$	$F = 1.6$ $p < .01$	$F = 1.15$ NS	$F = 2.7$ $p < .01$
	Ib	114	0.3329		*	$F = 1.2$ NS	$F = 1.27$ NS	$F = 4.0$ $p < .01$
	Ic	111[b]	0.3974			*	$F - 1.5$ $p < .05$	$F = 4.73$ $p < .01$
The series of consecutive ejaculates from one bull	II	66	0.2611				*	$F = 3.11$ $p < .01$
The series of portions from one ejaculate (control of the method)	III	108	0.0839					*

[a] The variance in all the groups is significantly higher than in the control. Also, between the subgroups of different ejaculates (Ia/Ib; Ia/Ic) and between the subgroups Ic and the group of consecutive ejaculates of one bull, the differences of variances are significant. This is because the subgroups Ib and Ic contained the bull with early increasing DNA spermatozoa content (see Table 2).

[b] The difference between the number of observations in this section and df is due to the fact that seven series, where not all of the bulls had been included, were not taken into account.

Table 2 Average Content of DNA (in Arbitrary Units per one Spermatozoon) Dependent on the Period of the Year in these bulls who were tested over 11 months. Individual Differences[a]

	March 20 to May 10, 1968	May 13 to June 24, 1968	June 26 to August 9, 1968	August 12 to September 30, 1968	October 8 to October 31, 1968	November 4 to November 26, 1968	November 29 to December 21, 1968	December 24, 1968 to January 20, 1969	January 23 to February 11, 1969
Bull No. 6, average values from 9 to 10 ejaculates	5.59	5.81	6.28	6.46	6.67	7.23	6.70	6.17	Not examined (slaughtered)
Bulls No. 2, 5, and 8, data of combined average values from 27 to 30 ejaculates	5.49	5.71	5.82	5.64	6.20	6.65	6.06	6.17	6.24
p	NS	NS	NS	$<.001$	$<.01$	$<.05$	$<.05$	NS	

[a] There is about 1 month's difference between the acrophase of Figure 1 and the crest of DNA fluctuation in this table. The figure is based on eight bulls; that is, on the data of four bulls presented in this table and also on the data of the four bulls examined only in spring and summer when the average level of DNA was lower. Their presence probably shifted the acrophase of the rhythm.

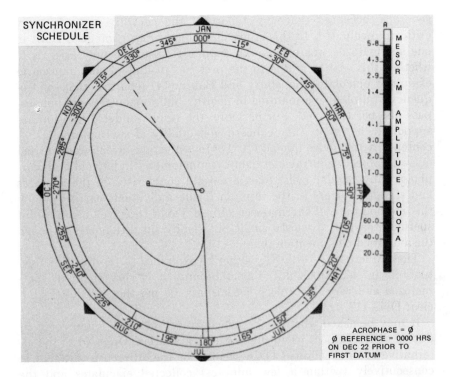

Figure 1   DNA in arbitrary units per 1 spermatozoon from 8 bulls (Ref = Dec 22). The ellipse dose not cover the pole; therefore, the statistical significance of the rhythm is established.

of analysis; that is, to any systematic error. Time analysis of the collected data also suggested the possibility of the existence of 5-, 7-, and 30-day cycles. However, when investigations carried out on ejaculated semen, containing the spermatozoa differing in age within the range of 10 days (3), any attempt to find a possible cycle of changes in spermatozoal DNA content with a shorter period (e.g., circaseptan) is very difficult. Large differences in DNA within successive ejaculates allow us to guess whether there are higher frequency rhythms (circadian and ultradian).

No correlation was found between DNA and fertility. The experiments of Courot et al. (5) quoted earlier, allow one to guess that male fertility, if tested independently of female fertility, would be negatively correlated with DNA.

The axiom concerning the constancy of DNA content in all the cells of a species, supported primarily by karyological studies, is so generally accepted that any report to the contrary must be very critically evaluated. In this study only the average content of DNA in one spermatozoa was calculated; therefore, suspicion may arise as to whether this scatter is due to the presence of diploid spermatozoa in examined ejaculates. Salisbury and Baker (14) reported that the frequency of diploid spermatozoa in healthy bulls' ejaculates did not surpass one per thousand. Furthermore, the results under discussion are supported, as far as they concern nonequality of spermatozoal DNA content, by the above-quoted (10–12) investigations which were carried out with objective ultraviolet light cytophotometry in haploid and morphologically regular cells. Also, a careful comparison of the averages and standard errors in the tests on fertile individuals carried out by this method on bull spermatozoa (15, of Table 1) as well as on rabbit spermatozoa (16) discloses small but significant differences although the authors do not mention them.

These fluctuations of DNA cannot be explained as representing fluctuations in its mitochondrial pool because it is very small in spermatozoa and very likely does not surpass .05 per thousand of the nuclear DNA (17, 18).

It may be asked whether the nonconstancy of DNA found could be due to differences in its formation or degradation. Opposed to the alternative of the decline of DNA seem to be the differences in DNA in consecutively (within a few minutes) collected ejaculates and the opinion that DNA-ase does not degrade DNA in living spermatozoa. Indeed, we have not found any decrease of the DNA in spermatozoa during 4 days of storage at $+4°C$. The synthesis of DNA in spermatozoa is known to take place at least 40 days before its ejaculation and at no latter stage of its development that in its mother spermatocyte during the zygotene phase. The fluctuations under discussion should, therefore, have their origin in the cells of the seminiferous epithelium.

It is impossible to reconcile the results presented with our current knowledge of the role and fate of DNA in cleaving cells. The scatter of its total content values is due perhaps to differences in the quantity of the hypothetical, genetically not active, "luxury" DNA (8). Without going into the biological character of the DNA found, it is my conclusion that its total content in healthy bulls spermatozoa is not equal but oscillates around some average mean value and only this value, determined from a number of ejaculates, is equal in the majority of individuals tested over the same period of the year. There also seems to be independent of these higher frequency oscillations, a circannual fluctuation in their level, which in Poland reached its crest in autumn in this 1-year study.

SUMMARY

Kihlström's observation that there is a rhythm of bulls' fertility was confirmed. Its average period was 18.6 days in this study. A circannual rhythm was also found. It may, however, be the result of fertility fluctuations of either the male or female or of both.

It was found that the mean DNA content in regularly fertile bull spermatozoa varied within succeeding ejaculates of the same male in a significantly higher degree than within the different samples taken from one ejaculate examined as a control. The coefficient of variability of the real DNA content was approximately calculated. In the succeeding ejaculates of the same male it amounted to 7.3%. However, the DNA contents averaged from a number of ejaculates collected in the same period of the year were similar in different individuals. There were also found seasonal differences—the highest values of DNA appeared in autumn, their average means exceeded those from spring by 12 to 13%. This suggests that the DNA content in spermatozoa is not stable, but oscillates around some value which also fluctuates seasonally. Only this average value from a number of ejaculates collected in the same period of time for the majority of individuals is rather constant.

REFERENCES

1. J. E. Kihlström, *Acta Physiol. Scand.*, **59**, (1963), 370.
2. J. E. Kihlström, *Experientia*, **22**, (1966), 630.
3. D. Amir and R. Ortavant, *Ann. Biol. Animale Biochim. Biophys.*, **8**, (1968), 195.
4. F. Halberg, Y. L. Tong and E. A. Johnson, in *The Cellular Aspects of Biorhythms (H. V. Mayersbach, Ed.), Springer, Berlin, 1967, p. 20.*
5. M. Courot, M. Goffaux, and R. Ortavant, *Ann. Biol. Animale Biochim. Biophys.*, **8**, (1968), 209.
6. C. Leuchtenberger, F. Schräder, F. Weir, and D. P. Gentile, *Chromosoma*, **6**, (1953), 61.
7. M. Parez, J. P. Petel, and C. Vendrely, *C. R. Acad. Sci., Paris*, **251**, (1960), 2581.
8. W. Sandritter and K. D. Grosser, *Symp. Biochem. Hung.*, **4**, (1964), 63.
9. M. Krzanowski and R. Mackowska, *Folia Biol.*, **17**, (1969), 357.
10. H. Berchtold, *Zuchthygiene*, **1**, (1966), 22.

11.   W. Leidl and R. Stolla, *J. Reprod. Fert.*, **18,** (1969), 173.

12.   R. Stolla, *Giessenere Beitr. Erbpathol. Zuchthyg.*, **43,** (1970) 43.

13.   M. F. Kramer. *Ann. Biol. Animale Biochim. Biophys.*, **8,** (1968), 217.

14.   G. W. Salisbury and F. W. Baker, *J. Reprod. Fert.*, **11,** (1966), 477.

15.   B. L. Gledhill, M. P. Gledhill, R. Rigler, Jr., and N. R. Ringertz, *Exp. Cell Res.*, **41,** (1966), 652.

16.   R. Bouters, Esnault, R. Ortavant, and G. W. Salisbury, *Nature,* **213,** (1967), 181.

17.   T. Mann, *Bull. Soc. Chim. Biol.*, **40,** (1958), 745.

18.   G. F. Bahr, and W. F. Engler, *Exp. Cell Res.*, **60,** (1970), 388.

CHAPTER 29

Oscillatory Changes in LH Secretion in Men

HOWARD R. NANKIN and PHILIP TROEN

Endocrine Service, Department of Medicine, Montefiore Hospital, and the University of Pittsburgh School of Medicine, Pittsburgh, Pennsylvania.

As part of our research in reproductive endocrinology we have been interested in characterizing circulating levels of luteinizing hormone (LH) in men. Double antibody radioimmunoassay has been utilized to assay hormone concentrations and the method has been reported (1–3). The National Pituitary Agency and the Endocrine Study Section of the National Institute for Arthritis and Metabolic Diseases kindly supplied LH and follicle-stimulating hormone (FSH) pituitary reference preparation LER-907, immunochemical grade LH LER-960, and the anti-hCG used in the background studies (see later). Multiple sampling assays characterizing the variations of LH were done with an anti-hCG obtained from the Wellcome Co. This antibody was chosen because it gave results which were quite similar to the NPA-NIH anti-hCG (3). All specimens from a given study of a patient were determined in the same assay. Reported values were the means of two or three (12-hour studies) determinations. Aliquots of a single normal serum were added to 20 consecutive tubes in the same assay. Calculating these as 10 pairs of tubes gave a mean of 76.6 ng/ml (approximating our average for

457

men) and an intra-assay variation of $\pm 10\%$ (2 SD). Interassay variation ranged from ± 12 to 32% (2 SD) (3).

BACKGROUND

Before discussing the oscillatory changes, a review of some background data which suggested that oscillations occurred and prompted us to perform multiple sampling studies is in order:

1. From 6 to 8 daily specimens were obtained from five male volunteers (24 to 33 years). LH values from these specimens revealed the following ratios when highest to lowest daily concentrations were compared for each man, respectively: 1.7, 1.8, 2.1, 2.3, and 3.1. Thus levels of LH did not appear to be tonic but rather demonstrated fluctuations.

2. Sera were obtained from 79 healthy new male hospital employees (20 to 50 years). The mean LH was 75.5 ng/ml $\pm$ 76 (2 SD), with a range of 30 to 210 ng/ml. Half of the values clustered in the lower 26% of the range, while there was a splaying of the other half of the specimens between the mean and the upper limit of the LH range. Similar non-Gaussian results were reported by others (4). This distribution would not fit with normally distributed tonic levels.

3. Reports from other laboratories demonstrated that growth hormone (5) and ACTH (6) were both released in pulsatile fashion in normal humans.

4. The half-time for LH after an intravenous bolus (7, 8) or squeezing the pituitary at hypophysectomy (9) was found to range between 21 and 69 minutes—thus suggesting a short effective life for circulating LH.

MULTIPLE SAMPLING STUDIES

To determine the frequency and regularity of the LH elevations, and to see if a diurnal rhythm was present, blood specimens were obtained via indwelling venous catheters at 15-minute intervals from the same four normally active men (23 to 26 years) between 6:00 A.M. and 6:00 P.M. on one occasion (2), and from 7:00 P.M. to 7:00 A.M. on the second occasion (3). In 42 instances, equally divided between the day and night studies and occurring, on the average, every 2.3 hours, decreasing concentrations of LH were followed by abrupt elevations of hormone which reached peak concentrations and declined. By our definition, elevations occurred if two or more constant or decreasing concentrations (troughs) were followed by two or more higher levels (Figure 1). The highest concentration of LH during an elevation, providing one or more subsequent samples had lower LH levels, was called the peak. No regular periodicity to elevations was apparent in our volunteers or in those men reported by Krieger et al. (10). However, the day and night LH patterns in D.P. and in J.S. were remarkably characteristic for each person (Figure 2). D.P. (fertile) had infrequent elevations and ran low LH levels, while J.S. (normal sperm count) had frequent elevations. The other two volunteers had gonadotropin patterns that appeared more variable (Figure 3). During the daytime studies no intervals with increased frequency of elevations or higher concentrations at peaks were apparent. However, in the overnight study, peaks were lower in early evening and reached their highest concentrations in all four men between 3:00 and 7:00 A.M. while they were asleep. Ratios of each man's highest to lowest LH concentration ranged from 2.1 to 2.9 and from 2.3 to 4.8 in the daytime and nighttime studies, respectively.

We also decided to assess a shorter multiple sampling technique which might be utilized for diagnostic clinical problems. These studies were carried out between 9:00 A.M. and 7:00 P.M. with the patients reclining on cots. Smoking was not permitted and the subjects were 1

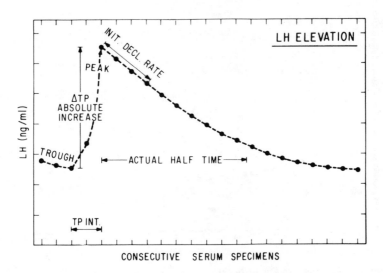

Figure 1   Schematic representation of LH elevation. Absolute LH increases were peak values minus preceding lowest trough concentrations. Initial rate of LH decline was calculated using peak and the next three specimens. TP INT. stands for lowest trough to peak time interval.

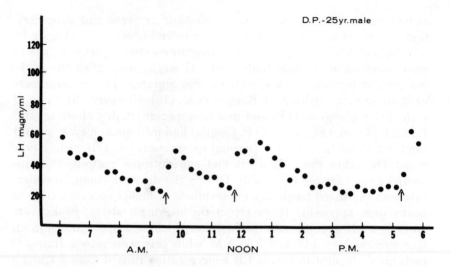

Figure 2a   *The patterns to LH oscillations at the daytime study (see also Figure 2c) and overnight study (see Figures 2b and 2d), done 5 months later, were remarkably similar in D.P. and J.S., respectively (2).*

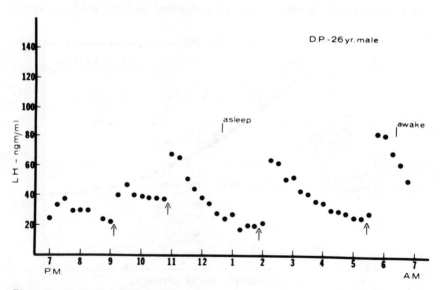

Figure 2b   *See legend to Figure 2a (3).*

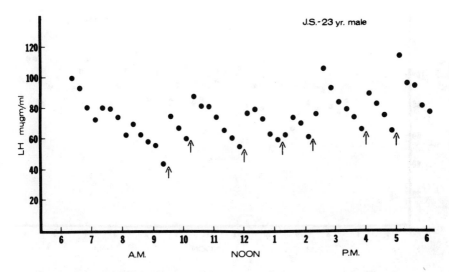

Figure 2c   See legend to Figure 2a (2).

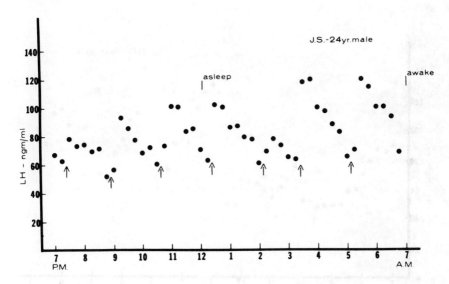

Figure 2d   See legend to Figure 2a (3).

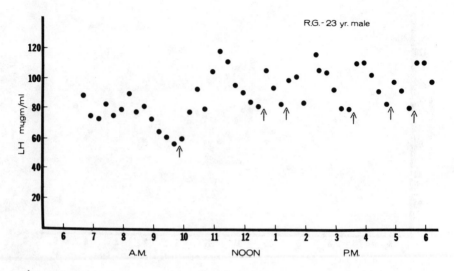

Figure 3a   *The LH fluctuations in the two 12-hour studies in daytime (see also Figure 3c) and overnight (see Figures 3b and 3d) were variable in subjects R.G. and L.K. (2).*

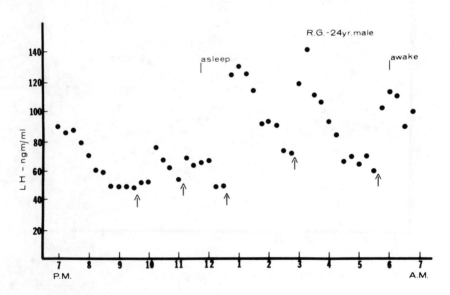

Figure 3b   *See legend to Figure 3a (3).*

462

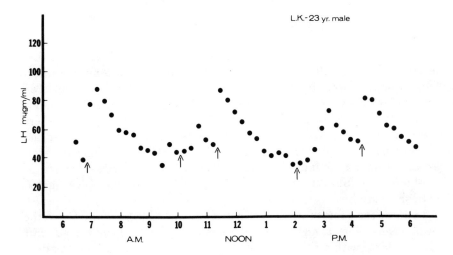

Figure 3c   *See legend to Figure* 3a *(2).*

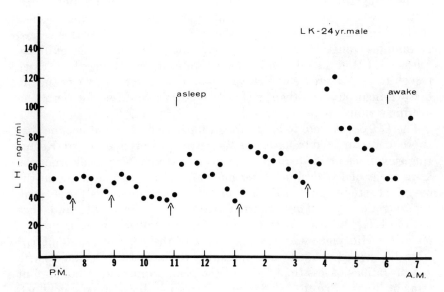

Figure 3d   *See legend to Figure* 3a *(3).*

or more hours postprandial. Initially, sampling was done for 1.5 to 4 hours. Subsequently, the short studies were done for 2.5 hours in an attempt to document at least one complete set of trough and peak LH levels in subjects. Although specimens were obtained at 5-minute intervals in one subject, 10-minute sampling was used for the others. Ten men (23 to 36 years) were studied on 15 occasions. Five men had a second study a week after their initial one. In 14 short studies there were 18 LH elevations. The first 110-minute study of one volunteer disclosed only a progressive LH decline from 46.2 to 30.4 ng/ml; however, an elevation was found in a second study. LH elevations occurred at 1.9-hour intervals during the 33.5 hours that men were investigated with short studies.

In 17 of the short study elevations it was possible to characterize the peak levels. The trough to peak increases (Figure 1) averaged 28.1 ng/ml (48%) with a range of 15.6 to 57.3 ng/ml (19 to 178%). Analysis of the overnight studies revealed that the mean trough to peak increase was 40.2 ng/ml (87%) with a range of 11.5 to 81.2 ng/ml (26 to 224%).

The trough to peak intervals (Figure 1)averaged 17.6 minutes (range 10 to 40) when specimens were obtained at 10-minute intervals in short studies. In the overnight studies the trough to peak interval averaged 30 minutes (range 15 to 60). It is interesting to speculate that pituitary release of LH occurs at relatively fixed rates in men. Since the average trough to peak increase of LH was greater (40.2 versus 28.1 ng/ml) in the overnight studies, then it should require more time for the mean overnight trough to peak interval.

The LH peaks were followed by an initially more rapid decline that subsequently tended to flatten as the rate of LH decline slowed. Recurring elevations or inadequate time for sampling after a peak limit the assessment of LH decline. After only four of the 21 elevations in the overnight studies did LH fall to levels less than half the peak concentrations (3). This occurred twice in volunteer D.P. and once each in L.K. and R.G. With semilog plots most of the tapering present on arithmetic plots was straightened and the concentrations of LH from the peaks, down to and including the next two specimens with concentrations of less than one-half the peaks, appeared to fall on four straight lines (Figure 4). Semilog regression lines were calculated (Monroe 1785 Calculator) and the four $T_{1/2}$'s were: D.P., 70 and 114 minutes; R. G., 93 minutes; and L.K., 108 minutes (mean 96).

Another approach is to determine the rate of LH fall during the interval that included the peaks and the subsequent three specimens (total intervals, 30 minutes in short or 45 minutes in overnight studies) (Figure 1). More prolonged sampling was infrequently possible in the short studies. Eleven (of 17) peaks from nine short studies of eight men fulfilled the criteria of having a documented elevation, a peak, and at least three subsequent 10-minute specimens. Since this initial decline appeared linear, all of the data from the 11 peaks and subsequent three specimens were appropriately averaged and a linear regression line was calculated (Olivetti-Underwood 101 Calculator). The initial rate of fall after the peaks was equivalent to a half-time of 78 minutes (correlation coefficient 0.995) (Figure 5). Similar estimates could be determined from 20 of the 21 overnight elevations where the initial rate of LH fall was equivalent to a mean half-time of 105 minutes. Since the postpeak decline of LH flattens, a slower rate of decrease was anticipated in the overnight studies because the four specimen interval was 45 minutes instead of 30 minutes with 10 minute sampling. The four actual $T_{1/2}$'s (mean 96 minutes) averaged 23% longer than the four corresponding initial postpeak rates of LH decline (mean 74 minutes). Of interest is the fact that usually LH declined to a certain level before another elevation occurred (Figures 2 and 3). There were instances where LH remained at this lower level for a prolonged time without further decline being evident—which suggests a basal release of LH by the pituitary. $T_{1/2}$'s reported here after spontaneous oscillations were similar to those found after LHRH (11). The initial rates of LH decline

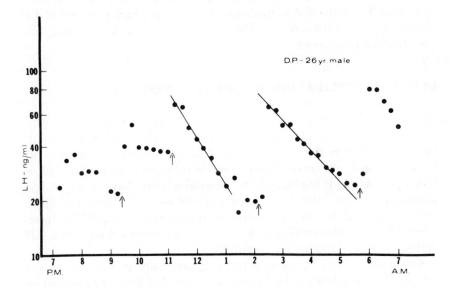

Figure 4   With semilog plots most of LH decline appeared to fall on straight lines and actual $T_{1/2}$'s were calculated (3).

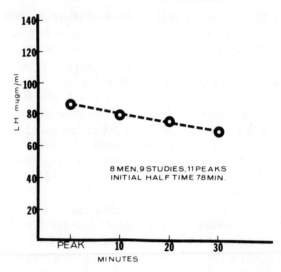

Figure 5 The initial rate of LH decline was calculated using averaged data from 11 elevations during short studies and was equivalent to a half-time of 78 minutes.

and the actual half times found after spontaneous LH oscillations were longer than the corresponding results found after an i.v. bolus (7, 8) or by squeezing the pituitary at hypophysectomy (9). More gradual endogenous LH decline is probably due to a gradual tapering of LH release by the pituitary, possibly to delayed systemic mixing, and possibly to a basal release.

CLINICAL APPLICATION OF OSCILLATIONS

Short (2.5 hours) multiple sampling sudies have become our method for appraisal of circulating LH. Considering the total range of LH values obtained in the 15 short studies of 10 volunteers as normal, we have recorded instances where men with partial testicular failure (oligospermia and low serum testosterone) may have some values which fall within, and have some LH values well above, the upper limits of normal. Studies of infertile men reporting "normal" LH levels when FSH was abnormally high (12) may have missed abnormal LH values because of inadequate sampling.

Two patients with hypogonadotropic hypogonadism (Kallmann's and Prader-Willi syndromes) had abnormally low LH levels and lacked oscillations.

In 232 boys (11 to 15 years) we found LH to rise gradually between Tanner genital stages I to V. The means, ranges, and standard deviations, particularly in stages II to V, were similar to the distributions we noted in our series of healthy men and suggested the occurrence of LH oscillations. Boyar et al. have recently reported episodic increases of LH in pubertal males (13). Oscillation patterns may prove useful in assessing and following pubertal development.

SUMMARY AND CONCLUSION

In men LH is released form the pituitary in repetitive bursts throughout the day and night. These hormonal oscillations occurred, on the average, every 1.9 to 2.3 hours but there was no constant pe-periodicity. Serum levels of LH were highest between 3:00 and 7:00 A.M. (sleeping) when compared to levels detected earlier in the evening in four normally active men. During oscillations, this gonadotropin reached peak levels within time intervals that averaged from 17.6 to 30 minutes. The data suggested that the trough to peak time interval varied directly with the absolute trough to peak increase of LH. Once the peak was reached, LH promptly started a more gradual decline at a rate equivalent to a half-time approximately four times the preceding trough to peak interval. The decline of LH appeared to taper. Gradual pituitary shut off, delayed mixing, or basal LH release may contribute to the "slow" LH decline. With 2.5-hour multiple sampling studies gonadotropin titers could be evaluated, oscillations could be characterized, and individuals could be detected with LH levels that varied from normal to high.

Mathematical models (14) of sporadic endocrine-CNS oscillations have been suggested. That the pituitary may release all of its hormones by sporadic bursts and that homeostatic mechanisms may function better with such systems are possibilities. Preliminary data suggest that although testosterone and LH fluctuate episodically, these fluctuations were not significantly correlated with each other (15). While there is no doubt that the testes and testosterone feed back on the central nervous system, the relationship between the sporadic LH elevations and testicular function has not been precisely delineated.

ACKNOWLEDGMENTS

The authors wish to express sincere appreciation to Shu-Wen Ke, Marilyn McCartney, Carol Phillips, Kathy Smith, Nancy Sweeney, and Eloise Swenson for assistance with the described studies.

Supported in part by the B. F. Anathan Fund.

REFERENCES

1. H. R. Nankin, T. Yanaihara, and P. Troen, *J. Clin. Endocr.,* **33,** (1971), 360.

2. H. R. Nankin and P. Troen, *J. Clin. Endocr.,* **33,** (1971), 558.

3. H. R. Nankin and P. Troen, *J. Clin. Endocr.,* **35,** (1972), 705.

4. R. J. Ryan, M. D. Cloutier, A. B. Hayles, J. Paris, and R. V. Randall, *Med. Clin. N. Amer.,* **54,** (1970), 1049.

5. H. Quabbe, E. Schilling, and H. Helge, *J. Clin. Endocr.,* **26,** (1966), 1173.

6. D. T. Krieger, W. Allen, F. Rizzo, and H. P. Krieger, *J. Clin. Endocr.,* **32,** (1971), 266.

7. P. O. Kohler, G. T. Ross, and W. D. Odell, *J. Clin. Invest.,* **47,** (1968), 38.

8. D. S. Schalch, A. F. Parlow, R. C. Boon, and S. J. Reichlin, *J. Clin. Invest.,* **47,** (1968), 665.

9. S. S. C. Yen, D. Llerena, B. Little, and O. H. Pearson, *J. Clin. Endocr.,* **28,** (1968), 1763.

10. D. T. Krieger and M. Fogel, IV International Congress of Endocrinology, Washington, D.C., June 1972 (Abstract 464).

11. S. S. C. Yen, R. Rebar, G. VandenBerg, F. Naftolin, Y. Ehara, S. Engblom, K. J. Ryan, and K. Benirschke, *J. Clin. Endocr.,* **34,** (1972), 1108.

12. S. W. Rosen and B. D. Weintraub, *J. Clin. Endocr.,* **32,** (1971), 410.

13. R. Boyar, M. Perlow, L. Hellman, S. Kapen, and E. Weitzman, *J. Clin. Endocr.,* **35,** (1972), 73.

14. N. Rashevsky, *Bull. Math. Biophys.,* **34,** (1972), 65.

15. F. P. Alford, H. W. G. Baker, B. Hudson M. W. Johns, J. P. Masterton, and Y. C. Patel, IV International Congress of Endocrinology, Washington, D.C., June 1972 (Abstract 465).

CHAPTER 30

Variations in Testosterone Excretion by Man

R. A. HARKNESS

Department of Paediatric Biochemistry, Royal Hospital for Sick Children, and Department of Clinical Chemistry, University of Edinburgh, Edinburgh, Scotland

The excretion of testosterone in urine by the adult human male provides a measure of testosterone secretion by the testes. About 1% of injected testosterone is excreted as testosterone glucuronide and this excretion can be used in physiological and clinical studies (1) despite some variations (2). There is general agreement on the actual amounts of testosterone excreted as determined by a number of different methods (3).

Serial urinary studies have long been used to study variations of ovarian function but serial studies of testicular function have been limited. The most marked physiological variation in testosterone excretion by normal males is the rise in mean level associated with sexual activity (4); this has now been confirmed by studies in rats (5), rabbits (6, 7), and bulls (8), as well as by further work in man (9). The importance of central control mechanisms in this response is indicated by the rise in testosterone levels before coitus in several of the above studies as well as in studies on beard growth (10). Beard growth is triggered and in part maintained by testosterone in men (11). However, it

469

should be noted that considerable beard growth continues even after orchidectomy, adrenalectomy, and hypophysectomy (12).

There is evidence for low frequency, small amplitude rhythms in male androgen secretion although much of the evidence is indirect. In 1956 it was found that the sperm count in rabbit ejaculate had a 5-day periodicity (13). Similar periodicity was found independently in the volume of daily ejaculates (14). These observations were extended to man who also showed a 5-day periodicity (15, 16). The cell cycle in the epithelium of the seminiferous tubules lasts 16 days (17) and, therefore, appears to have little effect on the changes shown in semen which have been thought to be due to alterations in androgenic stimulation. Studies in beard growth in man have shown a 33-day rhythm as well as a difference between a Wednesday minimum and Sunday maximum which was just significant (18). Similar weekly changes have been found in another study (10) and would appear to be related to sexual activity.

More direct evidence of variations in androgen secretion has been provided by studies of 17-oxosteroids and estrone excretion by four men which showed an 8 to 10 day cycle (19), although these compounds may originate from the adrenal cortex as well as from the testes (20, 21). The most extensive series of observations on 17-oxosteroid excretion by one healthy man have been analyzed by Halberg and his colleagues (22) who found rhythms with frequencies of about 1 week, 20 ± 3 days, 1 month, and 1 year.

The most direct evidence for periodicity in the endocrine function of the testes has been provided by studies of urinary testosterone. Regular peaks of urinary testosterone were noted in five subjects studied over periods of from 14 to 29 days. Four subjects showed peaks at 4 to 6 day intervals and one man at about 12 day intervals (4). Further studies of two men over 42 days by another group using a similar method showed an average interval between peaks of 9 days in one subject, but no regular pattern in the other man (23).

In an extension of the original series (4), three further subjects were found who also showed no clear pattern in their excretion (24). These tentative conclusions on possible rhythms in urinary testosterone were derived from inspection of the data in histogram form. After further accumulation of data, a more definitive mathematical analysis (25) has now been performed in collaboration with Professor Halberg and Dr. Simpson. The rhythms were fitted by the method of least squares using periods of 30 to 3 days in 1-day increments.

The principal rhythms of testosterone excretion by nine normal men are shown in Table 1. These rhythms were found in 12 series of consecutive 24-hour urine collections lasting from 10 to 45 days. The principal rhythms appear to fall into two groups, one group of 3 to 5 days and another group of 12 to 18 days. Eight less marked rhythms which appeared to be related to the main rhythm were also found. In these less marked rhythms the average period length in days varied from one-third to twice the length of the principal rhythm. It would appear advisable to concentrate on those components for which p is 0.05 or less. Six men showed such rhythms. Three subjects showed significant 5-day rhythms, one of whom in a later study showed a 17-day rhythm with a 4-day component which was not significant (p 0.11). One subject showed significant 12- and 18-day rhythms in a 45-day study; two other men showed significant 15-day rhythms. The effect of sexual activity is shown in studies C1 and 2 and E1 and 2; in subject C it did not alter the main 5-day rhythm but in subject E the average period length was reduced from 15 to 8 days. In both studies the level of significance was reduced (Table 1). The amplitude of the rhythms ranges from 3.6 to 18.8 μg/24 hours, which is larger than might

Table 1   *Rhythms in Urinary Testosterone Excretion by Nine Normal Men*

Subject	Average Period Length (days)	Amplitude (in μg/24 hours)	Standard Error	Prob-ability (p)	Level (in μg/24 hours)	Standard Error	Serial 24 hour collections, (N)
N	15	15.9 ± 2.5		<.01	51.1 ± 1.7		17
E (1)[a]	15	5.9 ± 1.4		<.01	48.0 ± 1.0		28
E (2)	8	7.7 ± 3.7		.14	47.5 ± 2.6		21
C (1)[a]	5	8.5 ± 2.1		.01	54.7 ± 1.5		13
C (2)	5	16.9 ± 6.4		.07	54.8 ± 4.6		14
C (3)	17	9.1 ± 2.8		.01	51.9 ± 2.0		33
	4			.11			
H	5	14.5 ± 5.2		.05	80.8 ± 3.7		17
J	14	15.8 ± 6.8		.10	105.6 ± 4.6		17
ED	5	11.0 ± 4.1		.05	58.3 ± 2.9		18
A	4	6.1 ± 2.5		.08	49.2 ± 1.8		28
W	3	18.8 ± 7.5		.10	50.8 ± 5.1		10
P	18	3.5 ± 1.1		.01	13.9 ± 0.8		45
	12			.01			

[a] No sexual activity.

reasonably be derived from an adrenal cortical contribution, urinary testosterone in castrate men being, about 5 μg/24 hours or less (4). The levels in micrograms per 24 hours in Table 1 are generally very close to the mean excretion determined with this and other methods (3). However, levels in subjects H and J were somewhat high, although this is sometimes found in men of their age (20 years). Subject P showed a low level which may be due to his very lean body build.

The 5-day periodicity in urinary testosterone (Table 1) agrees with studies of semen changes in man (15, 16) and the rabbit (13, 14). It has been suggested that an approximately 10-day rhythm may be due to basic sexual control mechanisms in the central nervous system (14, 19). The demonstrated 5-day rhythm may be compatible with these suggestions.

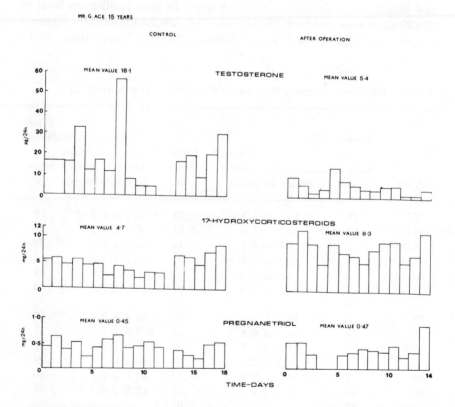

Figure 1   *The effect of amygdalotomy on urinary testosterone, 17-hydroxycortico-steroids, and pregnanetroil in a male aged 15 years.*

CLINICAL CONDITIONS ASSOCIATED WITH INCREASED VARIABILITY IN URINARY TESTOSTERONE EXCRETION BY MEN

A study of the clinicopathological correlations of variations in urinary testosterone has shown that there is an association between acne vulgaris and increased variability in urinary testosterone (26). Since only limited studies were possible for a maximum of 5 days, no patterns were detectable, and the results were expressed as a ratio of minimum to maximum. However, on this basis it was possible to predict correctly that the patient whose urinary testosterone levels are shown in Figure 1 had acne vulgaris. Increased variability of plasma testosterone levels during male puberty has also been found (27, 28).

MECHANISMS WHICH MAY BE INVOLVED IN DAY TO DAY VARIATIONS IN TESTOSTERONE EXCRETION

The approximately weekly rhythms in beard growth (10,18) and 17-oxosteroids (22) have been suggested to arise from environmental changes at the weekend. It is difficult to account for the rhythms in urinary testosterone shown in Table 1 in this way; it seems more likely that these rhythms may be related, at least in part, to basic sexual rhythms in the central nervous system.

Central Nervous System

There is little information on the mechanisms which may be important in small amplitude variations in testosterone secretion, although the central nervous system must be important, especially those structures which modify hypothalamic function, such as the amygdaloid nucleus. The effect of amygdalotomy in man on urinary testosterone, studied in collaboration with Mr. E. R. Hitchcock and Dr. L. Murray, is shown in Figures 1 and 2. The methods used were those employed by Harkness et al.(29). Studies were conducted in the same environment before operation and after discharge from the surgical wards. The reduction in urinary testosterone was also associated with a reduction in urinary oestrogens. This reduction in levels of gonadal steroids is consistent with the effects of amygdalotomy in mature male cats and rats (30) and with the suggestion that the amygdaloid nucleus facilitates hormone release in the adult (31).

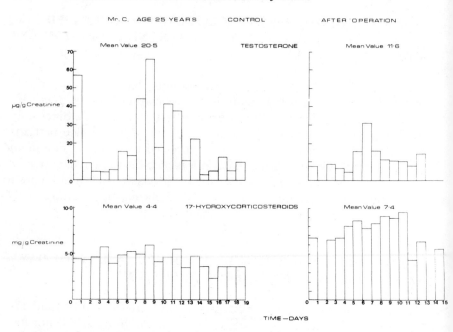

Figure 2 The effect of amygdalotomy on urinary testosterone and 17-hydroxycortico-steroids in a male aged 25 years.

Peripheral Mechanisms

Peripheral mechanisms seem to be important in the relatively long-term changes of urinary testosterone extending over periods of several days. Changes in the metabolism of steroids in liver and other tissues will be especially important (32, 33). Alterations in food intake could operate directly on hepatic steroid metabolism (34). However, even complex regulatory enzymatic interactions occur within minutes (35); longer term variations extending over several days may involve the lengthier processes of "induction" of enzymes by their substrates, especially those enzymes involved in the hydroxylation of steroids and drugs (see Ref. 33). Long-term testosterone administration increases its metabolic clearance rate by processes which occur slowly (36). Similar increases occurred after barbiturate administration, which also caused a persistent increase in 6β-hydroxycortisol excretion (37). Alterations in endogenous hormone production have now been shown to be associated with alterations in the 6β-hydroxylation of steroids. The metabolism of a fixed dose of a testosterone analogue to its 6β-hydroxy metabolite has been studied in women during the menstrual cycle in which there are large alterations in endogenous hormone production. During the luteal phase there was a marked increase in the conversion of a single oral dose of 5 mg of the anabolic steroid 17β-hydroxy-17-methyl-1,4-androstadien-3-one, methandrostenolone, to its 6β-hydroxy metabolite (Table 2) as determined by high-temperature catalytic reduction (38) after thin-layer chromatographic separation of the urinary metabolites. These metabolites are excreted almost entirely as free steroids (39), thus avoiding the effects of alterations in conjugation mechanisms. Similar but less marked increases in 6β-hydroxycortisol excretion have been noted in the luteal phase of the menstrual cycle (40) and after drug administration (41).

The above findings may account for the absence of a sex difference in the metabolic clearance rate of testosterone in some studies (42) but not in others (43, 44), despite agreement on values in normal men. The available evidence thus suggests that long-term variations in steroid metabolism, especially hydroxylation, occur in response to endogenous as well as exogenous stimuli.

Alterations in the rate of removal of testosterone could cause variations in testosterone production and urinary testosterone excretion (45). In addition, there is related evidence for mechanisms which could cause desynchronization.

Table 2   Variation in Metabolism of the Anabolic Steroid Methandrostenolone during the Menstrual Cycle

		Percentage of the 5-mg Dose Recovered from Urine as	
Age in Years	Day of Menstrual Cycle	17-Epimer of Methandrostenolone	6β-Hydroxy-Methandrostenolone
25	10	6.0	0.3
	20	0.7	1.6
38	10	1.7	0.1
	20	1.4	0.8
37	10	2.8	0.5
	25	0.8	1.1
20	10	1.2	0.4

The clearance of testosterone and related steroids from the circulation is associated with the conversion of small amount of these compounds to oestrogens which are potent hypothalamic inhibitors of gonadotrophin release (46, 47). Estrogens can be synthesised in adrenalectomized, ovariectomized subjects (48 to 50, 46) although the adrenal cortex appears essential for increased estrogen excretion to be readily detectable after gonadal stimulation (29). The importance of the liver in the control of gonadal function is also indicated by the occurrence of hypogonadism and gynecomastia in men with hepatic cirrhosis. In this condition, testosterone levels are very variable (51) and estrogen excretion is high (52) as well as 6β-hydroxycortisol output (40).

From the preceding evidence it has seemed justifiable to suggest (24) that this peripheral "biosynthesis" of potent inhibitors of central control mechanisms occurs by an "inducible" mechanism. Changes in this mechanism would be slower than the rapid response of the gonads to stimulation and could, therefore, be responsible for some interference in normal control mechanisms or desynchronization.

CONCLUSION

Significant variations in testosterone excretion by men have provided evidence for rhythms in the hypothalamo-pituitary testicular axis and functionally related systems. Further studies of such changes may provide insight into some of the disorders of the endocrine and reproductive systems.

SUMMARY

Urinary testosterone can be used as an index of testosterone secretion by the testes in the adult human male. Twelve serial studies of urinary testosterone from nine normal men lasting from 10 to 45 days have shown significant rhythms when subjected to definitive mathematical analysis. The principal rhythms appear to fall into two groups, one group with an average period length of 3 to 5 days and another group with an average period length of 12 to 18 days. Central and peripheral mechanisms involved in the control of gonadal steroid secretion have been investigated. In two subjects, amygdalotomy reduced the level of testosterone excretion. In addition, the metabolism of a testosterone analogue was found, in three women, to alter during the normal menstrual cycle.

ACKNOWLEDGMENTS

My thanks are due to F. Halberg and H. Simpson for their help in the analysis of the rhythms in the excretion of urinary testosterone. The help of E. R. Hichcock and L. Murray, as well as that of L. G. Whitby and J. A. Strong, is also gratefully acknowledged.

REFERENCES

1. A. M. Camacho and C. J. Migeon, *J. Clin Invest.*, **43,** (1964), 1083.

2 J. W. Goldzieher and M. C. Williams, *Acta Endocr.* (Copenhagen), **67,** (1971), 371.

3. A. A. A. Ismail and R. A. Harkness, *Biochem. J.*, **99,** (1966), 717.

4. A. A. A. Ismail and R. A. Harkness, *Acta Endocr.* (Copenhagen), **56,** (1967), 469.

5. Z. Herz, Y. Folman, and D. Drori, *J. Endocr.*, **44,** (1969), 127.

6. M. Saginor, and R. Horton, *Endocrinology,* **82,** (1968), 627.

7. G. C. Haltmeyer and K. B. Eik Nes, *J. Reprod. Fertil.*, **19,** (1969), 273.

8. C. B. Katongole, F. Naftolin, and R. V. Short, *J. Endocr.*, **50,** (1971), 457.

9. C. A. Fox, A. A. A. Ismail, D. N. Love, K. E. Kirkham, and J. A. Loraine, *J. Endocr.*, **52,** (1972), 51.

10. Anon., *Nature,* **226,** (1970), 869.

11. J. B. Hamilton, in *The Biology of Hair Growth* (W. Montagna and R. A. Ellis, Eds.), Academic Press, New York, 1958, p. 400.

12. J. A. Strong, J. Bruce. and C. W. A. Falconer, *Lancet,* **2,** (1959), 1055.

13. V. C. Doggett, *Amer. J. Physiol.*, **187,** (1956), 445.

14. J. E. Kihlström, *Experientia,* **22,** (1966), 630.

15. V. C. Doggett and R. K. Keillers, *Anat. Rec.*, **142,** (1962), 227.

16. S. J. Zimmerman, M. B. Maude, and M. Moldawer, *Fertility Sterility,* **16,** (1965), 342.

17. Y. Clermont, *Physiol. Rev.*, **52,** (1972), 198.

18. J. E. Kihlström, *Life Sci.*, **10,** (1971), 321.

19. D. Exley and C. S. Corker, *J. Endocr.*, **35,** (1966), 83.

20. D. T. Baird, A. Uno, and J. C. Melby, *J. Endocr.*, **45,** (1969), 135.

21. R. P. Kelch, M. R. Jenner, R. Weinstein, S. L. Kaplan, and M. M. Grumbach, *J. Clin. Invest.*, **51,** (1972), 824.

22. F. Halberg, M. Engeli, C. Hamburger, and D. Hillman, *Acta Endocr. Suppl.*, **103,** (1965), 50.

23. C. S. Corker and D. Exley, *J. Endocr.*, **40,** (1968), 255.

24. R. A. Harkness, in *Control of Gonadal Steroid Secretion*, (D. T. Baird and J. A. Strong, Eds.), Edinburgh Univ. Press, Edinburgh, 1971, p. 146.

25. F. Halberg, *Ann. Rev. Physiol.*, **31,** (1969), 675.

26. R. A. Harkness, A. A. A. Ismail, G. W. Beveridge, and E. W. Powell, in Testosterone (J. Tamm, Ed.), Thieme, Stuttgart, 1968, p. 192.

27. S. D. Frasier, F. Gafford, and R. Horton, *J. Clin. Endocr.,* **29,** (1969), 1404.

28. G. P. August, M. M. Grumbach, and S. L. Kaplan, *J. Clin. Endocr.,* **34,** (1972), 319.

29. R. A. Harkness, E. T. Bell, J. A. Loraine, A. A. A. Ismail, and W. I. Morse, *Acta Endocr.* (Copenhagen), **58,** (1968), 38.

30. A. Kling, J. Orbach, N. B. Schwartz, and J. C. Towne, *Arch. Gen. Psychiat.,* **3,** (1960), 391.

31. B. T. Donovan, in *Control of Gonadal Steroid Secretion* (D. T. Baird and J. A. Strong, Eds.), Edinburgh Univ Press, Edinburgh, 1971, p. 1.

32. H. Schriefers, *Vitamins and Hormones,* **25,** (1967), 271.

33. B. N. La Du, H. G. Mandel, and E. L. Way, Eds., *Fundamentals of Drug Metabolism and Drug Disposition,* Williams and Wilkins, Baltimore, 1971.

34. A. L. Herbst, F. E. Yates, D. W. Glenister, and J. Urquhart, *Endocrinology,* **67,** (1960), 222.

35. C. Frieden, *J. Biol. Chem.,* **245,** (1970), 5788.

36. A. L. Southren, G. G. Gordon, and S. Tochimoto, *J. Clin. Endocr.,* **28,** (1968), 1105.

37. A. L. Southren, G. G. Gordon, S. Tochimoto, E. Krikun, D. Krieger, M. Jacobson, and R. Kuntzman, *J. Clin. Endocr.,* **29,** (1969), 251.

38. P. M. Adhikary and R. A. Harkness, *Anal. Chem.,* **41,** (1969), 470.

39. P. M. Adhikary and R. A. Harkness, *Acta Endocr.* (Copenhagen), **67,** (1971), 721.

40. F. H. Katz, M. M. Lipman, A. G. Frantz, and J. W. Jailer, *J. Clin. Endocr.,* **22,** (1962), 71.

41. K. Thrasher, E. E. Werk, Y. Choi, L. J. Sholitan, W. Mayer, and C. Olinger, *Steroids,* **14,** (1969), 455.

42. R. Horton and J. F. Tait, *J. Clin. Invest.,* **45,** (1966), 301.

43. C. W. Bardin and M. B. Lipsett, *J. Clin. Invest.,* **46,** (1967), 891.

44. A. L. Southren, G. G. Gordon, S. Tochimoto, G. Pinzon, D. R. Lane, and W. Stypulkowski, *J. Clin. Endocr.,* **27,** (1967), 686.

45. D. L. Berliner and T. F. Dougherty, *Pharm. Rev.,* **13,** (1961), 329.

46. C. Longcope, T. Kato, and R. Horton, *J. Clin. Invest.,* **48,** (1969), 2191.

47. P. C. MacDonald, J. M. Grodin, and P. K. Siiteri, in *Control of Gonadal Steroid Secretion* (D. T. Baird and J. A. Strong, Eds.), Edinburgh Univ. Press, Edinburgh, 1971, p. 158.

48. C. D. West, B. L. Damast, S. D. Sarro, and O. H. Pearson, *J. Biol. Chem.,* **218,** (1956), 409.

49. E. Chang and T. L. Dao, *Biochim. Biophys. Acta,* **57,** (1962), 609.

50. D. T. Baird, R. Horton, and J. F. Tait, *Perspectives Biol. Med.,* **11,** (1968), 384.

51. W. S. Coppage and A. E. Cooner, *New Engl. J. Med.,* **273,** (1965), 902.

52. J. B. Brown, G. P. Crean, and J. Ginsburg, *Gut,* **5,** (1964), 56.

9 Secondary Rhythms Related to Testicular Function

ROBERT B. SOTHERN, *Moderator*

CHAPTER 31

Rhythms in Cognitive Functioning and EEG Indices in Males

**EDWARD L. KLAIBER, DONALD M. BROVERMAN,
WILLIAM VOGEL,**
and **EDMUND J. MACKENBERG**

*Worcester Foundation for Experimental Biology,
Shrewsbury, Massachuseets, and Worcester State Hospital, Worcester Massachusetts*

A cognitive index termed the Automatization Cognitive Style is thought to be influenced by testosterone in the human male (1). The Automatization Cognitive Style is defined as greater or lesser ability to perform simple repetitive tasks than expected from the person's general level of performance on a heterogenous battery of tasks (2). Strong Automatizers perform relatively better on simple repetitive tasks, such as speed of naming color hues, than expected from their general level of proficiency. Strong Automatizers are also proficient at developing skills on repetitive tasks, thereby making them automatic; hence the term automatization (3). Thus the Strong Automatizer tends to be a faster reader, speaker, or typist than the Weak Automatizer (4). Strong Automatizers have also been found to be less susceptible to mental fatigue than Weak Automatizers, that is, the rate of continuous performance on simple coded addition problems deteriorated more over time in Weak Automatizers than in Strong Automatizers (3).

AUTOMATIZATION ABILITY AND TESTOSTERONE

One indication that automatization ability might be dependent on variations in testosterone activity are the positive correlations of the automatization index with physical parameters thought to reflect testosterone stimulation, that is, pubic hair development and chest and bicep circumferences (5). A positive correlation also has been observed between automatization and 24-hour urinary 17-ketosteroid excretion (5). These studies (5) were carried out with 50 male college students whose average age was 21.65 years. Chest and biceps circumferences, height, and weight were measured on each subject. Assessments of pubic hair growth were made from photographs. Raters assigned each picture to one of five standards of development and intercorrelations of the three raters ranged from .88 to .93. Two 24-hour urine specimens were collected from each subject and assayed for 17-ketosteroid levels by the method of Drekter et al. (6). The pubic hair, chest and bicep circumferences, weight, and 24-hour urinary 17-ketosteroid excretions were all positively correlated with the automatization index as shown in Table 1.

The above correlational study tends to support the hypothesis that automatization is influenced by testosterone activity. Unpublished data from our laboratory on 33 young adult human males give further support to the testosterone hypothesis. Measurements of testosterone

Table 1   *Intercorrelations of the Automatization Cognitive Style with Anthropometric Measures and Urinary 17-Ketosteroid Excretions in 50 Normal Males*

	Index of Automatization Cognitive Style
Pubic hair	.410**
Chest circumference	.388**
Right biceps circumference	.356*
Weight	.286
Height	.104
Urinary 17-ketosteroids	.303*

* $p < .05$.
** $p < .01$.

Table 2 *Testosterone Metabolic Clearance and Production Rates in Weak and Strong Automatizers*

	Age	Height, inches	Weight, lb	Testo, MCR 1/24 hours	Testo, MCR/M²	Plasma Testo, µg/100 ml	Testo, PR mg/24 hours	Testo, PR/M²
Strong Automatizers ($n = 17$)	21	70.6 ±2.3	175.9 ±18.1	1876 ±383	948 ±183	0.75 ±0.20	14.0 ±3.6	7.1 ±1.9
Weak Automatizers ($n = 16$)	21	70.6 ±1.8	156.4 ±16.0	1196 ±302	642 ±156	0.75 ±0.18	9.1 ±4.3	4.9 ±2.2
Significance	NS	NS	$p < .01$	$p < .01$	$p < .01$	NS	$p < .01$	$p < .01$

metabolic clearance (MCR) and production rates (PR) utilizing the constant infusion technique as conceptualized by Tait (7) and applied by Levin et al. (8) were made on these subjects. Plasma testosterone levels were measured by the method of Riondel et al. (9). Cognitive testing established that the subjects consisted of 16 Weak and 17 Strong Automatizers. Plasma testosterone levels were not different in the two groups. However, testosterone metabolic clearance rates and production rates were significantly greater in the Strong Automatizers ($p < .01$). See Table 2.

Since the Strong Automatizers weighted significantly more than the Weak Automatizers, MCR and PR were corrected for body surface area. Comparison of the corrected MCR and PR differences between Weak and Strong Automatizers remained significant ($p < .01$). See Table 2. Again, this study supports a relationship between automatization and testosterone.

AUTOMATIZATION AND EEG

Testosterone's association with automatization may be related to the reported ability of testosterone to counteract fatigue in nerve-muscle preparations (10, 11). However, this possibility leaves unanswered the question of whether the hormone is acting upon nerve or muscle. While the functions tested in automatization, that is, rapidity of verbal response rate, usually are interpreted as representing central nervous system (CNS) functioning, it is possible that what is being tested is peripheral muscle fatiguability, for example, tongue and jaw muscles.

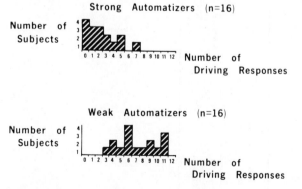

Figure 1   Frequency distributions of driving responses of Strong and Weak Automatizers.

An automatization -EEG study (12) gives support to the hypothesis that differences in automatization ability are related to differences in CNS functioning. This study was carried out with 32 normal male college students. The group was equally divided between Strong and Weak Automatizers. The mean ages of the 16 Strong and 16 Weak Automatizers were 21.2 and 21.4 years, respectively; their mean Wechsler Adult Intelligence I.Q. scores were 127.0 SD 8.0, and 131.0 SD 12.0. There were no significant age or I.Q. differences between the two groups.

The EEG measure employed was EEG "driving" in response to photic stimulation. When a subject whose eyes are closed is exposed to a bright flickering light, the EEG occipital rhythm tends to be "driven," that is, the EEG rhythms may assume the frequency of the flickering light. An EEG driving response is defined as two consecutive seconds of EEG waves at the frequency or a subharmonic of the frequency of the photic stimulation.

We observed that the EEG driving response to photic stimulation was much less in the 16 Strong Automatizers compared to the 16 Weak Automatizers. The distribution of driving responses of Strong and Weak Automatizers is shown in Figure 1. Half of the 32 subjects made four or less driving responses; the other half five or more responses. A chi-square analysis comparing Strong and Weak Automatizers showed that 13 Strong and three Weak Automatizers were below the group median for all 32 subjects; three Strong and 13 Weak Automatizers were above the group median. The chi-square of 12.50 ($p < .001$) indicates that significantly more Weak than Strong Automatizers gave greater than the median number of responses.

The mean number of driving responses of the Strong Automatizers was 2.25 SD 2.1; the mean of the Weak Automatizers was 7.25 SD 2.7. An analysis of variance indicates that the difference between the two means is statistically significant ($p < .001$).

EFFECTS OF TESTOSTERONE ON EEG AND AUTOMATIZATION

The EEG data support the hypothesis that automatization ability is associated with central nervous system differences. The study does not, however, bear directly on the relationship of testosterone to CNS functioning as reflected by the EEG. To examine this question, still another study was undertaken.

EEG driving was measured weekly over a period of months in three hypogonadal males before and after long-term androgen therapy (13). EEG driving during photic stimulation occurred at high levels of frequency in all three subjects prior to androgen treatment. The three subjects then received varying doses of testosterone cypionate in oil at 4 week intervals. All three subjects showed a definite reduction of EEG driving responses following testosterone therapy.

In two subjects a simple repetitive automatization task, that is, the speed of naming of repeated objects, was administered prior to each EEG testing session. Both subjects performed this automatization task very slowly prior to androgen treatment. The administration of testosterone clearly accelerated their performances of the automatization task, and the changes in EEG driving and in the speed of object naming were significantly ($p < .001$) correlated in each subject.

An interesting cyclicity of both EEG driving and speed of object naming was noted in relationship to time of the administration of testosterone. During the week following testosterone administration, EEG driving was markedly reduced and speed of object naming enhanced. In the subsequent 3 weeks the EEG driving tended to increase and the speed of object naming slowed. This pattern regularly repeated itself. We attributed these variations to changes in the patient's testosterone levels; the highest levels occurring just after, and the lowest levels occurring just before the hormone injections.

Figure 2 illustrates this cyclicity in relation to time of testosterone administration in one of the subjects.

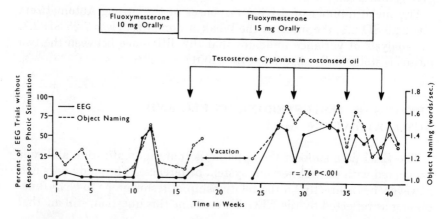

Figure 2   Percent of EEG trials without response to photic stimulation and speed of naming repeated objects as a function of testosterone treatment in a late maturing boy.

Table 3   Mean A.M. and P.M. Performances of Color Naming in Males[a]

A.M. (10:00)	52.67 seconds	
P.M. (4:00)	53.87 seconds	$p < .05$
Difference	1.20 seconds	

[a] $n = 56$; ages from 12 to 38.

DIURNAL VARIATIONS IN TESTOSTERONE, AUTOMATIZATION AND EEG

The preceding study indicates that testosterone does directly affect both automatization ability and EEG driving. Natural variations in testosterone levels, then, ought to be associated with corresponding variations in automatization ability and EEG driving with photic stimulation.

Diurnal variations in plasma testosterone concentrations have been reported (14), that is, plasma testosterone levels at 9 A.M. are approximately 20% higher than at 4 P.M. It would be expected, then, that automatization ability would be better in the A.M. compared to the P.M., and that EEG driving should occur less often in the morning than in the afternoon.

A study (15) was designed to test these hypotheses and also to test whether the subject's age and state of physical maturation are related to the magnitudes of the predicted diurnal changes. Physical maturation was indexed by pubic hair development and chest and biceps circumferences. Twenty normal male subjects were obtained for each of two age groups, 24 to 38 years and 18 to 23 years; and 16 subjects were obtained in a 12 to 16 years age group.

Half of the subjects from each age group were first tested for automatization ability and EEG driving in the morning at 10 A.M., and then again at 4 P.M. of the same day. The other half of the subjects were tested first at 4 P.M. and then again at 10 A.M. of the following day. This procedure was employed to balance the gains in performance due to practice that are known to occur on tasks measuring automatization ability. A failure to control for this known practice effect could prevent the expected A.M.-P.M. effect from being observed.

Table 3 shows the mean performance of all subjects of the speed of color naming in the A.M. and P.M. The mean A.M. performance is slightly but significantly ($p < .05$) faster than the mean P.M. performance. The statistical analysis employed a mixed Latin square,

Table 4   Mean Number of EEG Driving Responses in the A.M. and P.M.

	Number of EEG Driving Responses	
A.M. (10:00)	9.67	
P.M. (4:00)	10.89	$p < .01$
Difference	1.22	

repeated measures analysis of variance design which extracted variance due to order of task administration as well as hour of task administration. The reason statistical significance was reached with such a slight difference between means is that random variations in performances of automatization tasks tend to be minimal. Much of the variation that does occur usually can be assigned to known effects, for example, practice or age. These results, then, indicate that the predicted A.M. to P.M. shift in performances of automatization tasks does occur. The present study was not designed to demonstrate repeated A.M. to P.M. rhythms. However, Hollingworth, in 1914 (16), reported that similar A.M. to P.M. shifts in speed of color naming did occur repeatedly on successive days. Hollingworth did not associate these changes to hormonal phenomena.

In the EEG data only 48 of the subjects were usable since muscle artifacts on the records prevented scoring in eight cases. Table 4 shows the mean number of driving responses to photic stimulation in the A.M. and P.M. of these 48 subjects. There are significantly ($p < .01$) fewer EEG photic driving responses in the A.M. compared to the P.M.

The above results support the hypothesis that natural variations in testosterone are associated with A.M. to P.M. changes in EEG photic driving. Once again, the data indicate that changes in automatization

Table 5   Means of Color Naming in A.M. and P.M. as Function of Age

	A.M. (seconds)	P.M. (seconds)	P.M. − A.M. (seconds)	
Young (12 to 16 years) $n = 16$	60.68	64.00	3.32	
Middle (18 to 23 years) $n = 20$	50.20	50.00	−0.20	$p < .05$
Oldest (24 to 38 years) $n = 20$	48.75	49.65	0.90	

Table 6   *Means of Color Naming in A.M. and P.M. as Function of Androgenicity*

	A.M. (seconds)	P.M. (seconds)	P.M. − A.M. (seconds)	
High androgenicity $n = 28$	52.28	51.89	−0.39	
Low androgenicity $n = 28$	53.07	55.85	2.78	$p < .01$

ability and EEG photic driving tend to covary as a function of testosterone levels. In this case the variation in testosterone is due to a natural daily cycle.

The A.M. to P.M. shifts in automatization ability as a function of age were examined next. Table 5 indicates that the largest A.M. to P.M. shift occurs in the 12 to 16 year age group and secondarily in the 24 to 38 year age group. A slight nonsignificant shift occurs in the opposite direction (P.M. better than A.M.) in the 18 to 23 year group. It is noteworthy that the peak of testosterone production rates in males have been reported to occur in this age range (17), suggesting that a high testosterone production rate may tend to protect against or to minimize A.M. to P.M. shifts in automatization ability.

While we do not have plasma testosterone levels or production rate data on these subjects, the anthropometric maturation indices may reflect the level of testosterone stimulation. Each subject was assigned an androgenicity rating based on his combined relative standings on each of three parameters: pubic hair development, chest circumference, and biceps circumference. The subjects within each of the three age groups were then dichotomized into categories of high or low androgenicity.

Table 6 shows the means of speed of coloring naming in the A.M. and P.M. as a function of these androgenicity ratings.

The entire A.M. to P.M. shift is apparently due to the less physically mature (low androgenicity) subjects. The more physically matured, high androgenicity subjects actually show a slight, nonsignificant reversal of the expected A.M. to P.M. shift. The difference between the two groups of subjects is statistically significant ($p < .01$).

The final breakdown of the A.M. to P.M. data involves the interaction of age and androgenicity. Table 7 indicates that this interaction is highly statistically significant ($p < .001$). The source of this effect appears to be the accentuated relationship of differences in androgenicity in the youngest age group to A.M. to P.M. shifts in au-

Table 7   Means of Color Naming Speed in A.M. and P.M. as Function of Age and Androgenicity

	A.M. (seconds)	P.M. (seconds)	P.M. – A.M.	(seconds)
Youngest age group (12 to 16 years)				
Low androgenicity $n = 8$	65.50	70.62	5.12	
High androgenicity $n = 8$	55.87	57.37	1.50	
Middle age group (18 to 23 years)				
Low androgenicity $n = 10$	48.70	49.90	1.20	$p < .001$
High androgenicity $n = 10$	51.70	50.10	−1.60	
Oldest age group (24 to 38 years)				
Low androgenicity $n = 10$	47.50	50.00	2.50	
High androgenicity $n = 10$	50.00	49.30	−0.70	

tomatization performance. The greatest variation in androgenicity undoubtedly occurred within this young group.

Unfortunately, the eight subjects with unusable EEG records occurred to a large extent in one sequence of one of the three age groups which prevented analyses of the EEG data in relation to age or level of physical maturation.

EFFECTS OF TESTOSTERONE ADMINISTRATION ON DIURNAL CHANGES IN AUTOMATIZED BEHAVIORS

A.M. to P.M. variations in automatization ability, then, may be a function, to a large degree, of testosterone levels. One way to test this hypothesis would be counter the A.M. to P.M. variation of testosterone by exogenously administering the hormone in the P.M. to prevent the afternoon decline.

The next study (18) reports such an investigation. Unfortunately, at the time that this study was initiated, measurements of EEG photic driving were not available in our laboratories. Therefore, only automatization ability changes were studied.

Sixty-seven normal adult male subjects had automatization ability (serial subtraction performances) measured in the morning and again in the afternoon. Two different serial subtraction tasks were employed to minimize practice effects.

Sixty-seven subjects were processed in order to obtain 54 subjects who could be divided into 27 pairs matched for A.M. serial subtraction performances.

An intravenous saline infusion, using a constant infusion pump, had been started prior to the A.M. testing. Following the morning testing session, one subject from each pair had testosterone added to the solution being infused in amounts which supplied 0.8 mg of testosterone per hour for the next 4 hours. The other half of the pair continued to receive saline. To control for possible psychological factors, all subjects were told that they were receiving testosterone throughout the infusion.

In Table 8 the average number of serial subtraction problems solved correctly in a 3-minute time period in the A.M. and P.M. are shown.

Because of the matching, there is no difference between the two groups in the A.M. An A.M. to P.M. decline occurs in both groups, but the decline is significantly ($p < .05$) greater in the saline group. This suggests, then, that the A.M. to P.M. decline in automatization ability can be partially reversed by the administration of testosterone.

To determine more precisely the relationship of testosterone to the A.M. to P.M. shift, plasma testosterone levels were measured in 28 subjects who received testosterone and 20 saline infused subjects. The A.M. to P.M. changes in plasma testosterone are presented in Table 9. The control subjects had a 21% decline in their plasma testosterone levels, while the testosterone infused subjects had a 44% rise in their plasma testosterone.

The morning to afternoon change in plasma testosterone concentrations of the control subjects represents, presumably, the normal diurnal cyclicity of testosterone. Hence, the relationship of these data

Table 8   Mean Number of Correct Serial Subtraction Problems in A.M. and P.M. of Matched Testosterone and Saline Infused Subjects

	A.M.	P.M.	P.M. − A.M.
Testosterone subjects $n = 27$	38.44 ± 14.11	35.74 ± 14.02	-2.70
Saline subjects $n = 27$	38.29 ± 13.86	32.92 ± 12.00	-5.37
			$F = 4.24$ $p < .05$

Table 9   A.M. and P.M. Plasma Testosterone Concentrations in Testosterone Infused and Saline Infused Subjects

	Means in $\mu g/100$ ml		
	A.M.	P.M.	% Change
Testosterone infused subjects $n = 28$	$0.73 \pm .37$	$1.05 \pm .51$	$+44\%$
Saline infused subjects $n = 20$	$0.70 \pm .21$	$0.55 \pm .22$	-21%
Testosterone infusion A.M. versus P.M.	$p < .001$		
Saline infusion A.M. versus P.M.	$p < .01$		

to changes in automatization ability were examined. Those control subjects with the highest P.M. plasma testosterone concentrations were found to have the least deterioration in performance; those subjects with the lowest P.M. plasma testosterone concentrations had the greatest deterioration. This phenomenon is most clearly seen when the 20 control subjects are divided into three groups, high, middle, and low, based on their P.M. plasma testosterone concentrations. The differences between the means of the changes in performance, as shown in Table 10, are significant ($p < .02$). These observations support the hypothesis that testosterone levels are related to the magnitude of A.M. to P.M. shifts in automatization ability.

In summary, the data presented here suggest that natural variations in testosterone levels such as occur from the morning to afternoon are

Table 10   Relationship of P.M. Plasma Testosterone Concentrations to Changes in Serial Subtraction Performances in Control Subjects

Plasma Testosterone Concentration	Mean Change in Number of Correct Problems, P.M. − A.M.
Low (0.27 to 0.38 $\mu g/100$ ml) $n = 6$	-11.67
Middle (0.40 to 0.72 $\mu g/100$ ml) $n = 8$	-7.82
High (0.74 to 0.92 $\mu g/100$ ml) $n = 6$	-2.66
	$F = 8.38$
	$p < .02$

associated with variations in a specific ability, automatization; and with variations in EEG photic driving. Also, administration of testosterone acts to counter the A.M. to P.M. shifts in automatization ability and to influence EEG photic driving.

ACKNOWLEDGMENT

This work was supported in part by Public Health Service grant MH-18757 from the National Institute of Mental Health.

REFERENCES

1. D. M. Broverman, I. K. Broverman, W. Vogel, R. D. Palmer, and E. L. Klaiber, *Child Develop.* **35,** (1964), 1343.

2. D. M. Broverman, *J. Consult. Psychol.,* **28,** (1964), 487.

3. D. M. Broverman, I. K. Broverman, and E. L. Klaiber, *Percept. Motor Skills,* **23,** (1966), 419.

4. D. M. Broverman, *J. Pers.,* **28,** (1960), 167.

5. E. L. Klaiber, D. M. Broverman, and Y. Kobayashi, *Psychopharmacologia,* **11,** (1967), 320.

6. I. J. Drekter, A. Heisler, G. R. Scism, S. Stern, S. Pearson, and T. H. McGavach, *J. Clin. Endocr.,* **12,** (1952), 55.

7. J. F. Tait and S. Burstein, in *The Hormones* (G. Pincus, K. V. Thimann, and E. B. Astwood, Eds.), Academic Press, New York, 1964.

8. J. Levin, C. W. Lloyd, J. Lobotsky, and E. H. Friedrich, *Acta Endocr.* (Copenhagen), **55,** (1967), 184.

9. A. Riondel, J. F. Tait, M. Gut, S. A. S. Tait, E. Joachim, and B. Liddle, *J. Clin. Endocr.,* **23,** (1963), 599.

10. M. Gans and R. G. Hoskins, *Endocrinology,* **10,** (1926), 56.

11. E. H. Herrick and R. S. Storer, *Trans. Kansas Acad. Sci.,* **52,** (1948), 203.

12. W. Vogel, D. M. Broverman, E. L. Klaiber, and K. J. Kun, *Electroenceph. Clin. Neurophysiol.,* **27,** (1969), 186.

13. P. G. Stenn, E. L. Klaiber, W. Vogel, and D. M. Broverman, *Percept. Motor Skills,* **34,** (1972), 371.

14. J. A. Resko and K. B. Eik-Nes, *J. Clin. Endocr.,* **26,** (1966), 573.

15. E. J. Mackenberg, Unpublished Ph.D. Dissertation, Clark University, 1972.

16. H. L. Hollingworth, *Psychol. Rev.,* **21,** (1914), 473.

17. H. Persky, K. D. Smith, and G. K. Basu, *Psychosomat. Med.,* **33,** (1971), 265.

18. E. L. Klaiber, D. M. Broverman, W. Vogel, G. E. Abraham, and F. L. Cone, *J. Clin. Endocr.,* **32,** (1971), 341.

10   Synchronizers of Reproductive Function

MICHAEL MENAKER, *Moderator*

CHAPTER 32

Synchronizers of Reproductive Function in Men and Women

IRWIN H. KAISER

Department of Gynecology and Obstetrics, Albert Einstein College of Medicine, Bronx, New York

There are a large number of specific events in reproduction in man and only a few have been studied from the viewpoint of chronobiology. In the search for possible synchronizers the furthest anyone has gone are the relationships of a few phenomena to day and night and to social time. Beyond this there exists only speculation, and that at a very superficial level. The critical experimental models suitable for studies in Homo have not yet been proposed.

In this review, mainly reports of the past 5 years have been considered in detail. There do not appear to be any studies in males. In the female the studies have been directed at three matters: the timing of the menarche, the timing of ovulation and menses, and the timing of birth.

Although there is a consensus that the age of menarche has decreased as the size of adolescent women has increased in urban populations in recent years (1), it is by no means clear why this is so. An historical review, however, suggests that the age of menarche in classic Rome and Greece was not markedly different from the present rural age (2). It has been proposed, by analogy to effects of prolonged lighting in rodents, but without experimental evidence, that increased artificial lighting, particularly in the temperate zone, might produce an earlier menarche (3). It does appear to be established that blinded girls experience an earlier menarche (4, 5) and that this is proportional to the sight deficit. A negative report on this matter (6) unfortunately does not sort out proposita by type of blindness. It is based on hospitalized girls and employs as controls sighted girls who are also hospitalized. The authors do not specify the reasons for institutionalization. The deviations from ordinary behavior among blinded pubescents is so complex that it cannot be concluded that this is an effect of light deprivation alone, especially when institutionalized subjects are used.

The recent development of effective ovulation inducers and of suitable assay methods for gonadotropins has provided tools to study synchronizers of ovulation and menstruation but as yet no studies have been published. Slivka et al., observing three women for one cycle each and doing assays three times a day for plasma follicle-stimulating hormones and luteinizing hormones found no evidence of within-day cyclicity during follicular or luteal phases. They eliminated the values obtained on the day before, day of, and day after the midcycle peaks (7).

It is known that severe emotional and nutritional stress affect the menstrual cycle and that the menstrual cycle correlates with psychologic and motor phenomena. But the efforts to identify more specific environmental synchronizers have been meager. Udry and Morris have demonstrated, in selected populations, a maximum in coital activity at midcycle in the female, as well as a maximum in orgasm (8). They have also shown a corresponding maximum of activity as measured by a pedometer (9).

Clark (10) has reviewed the literature which suggests that ovulation may be induced in Homo, as in several other mammalian species, by copulation. The best evidence brought to bear derives from instances of rape, a very unsatisfactory experimental model. Prill et al. (11) observed shortening of the period of bleeding and prolongation of menstrual interval among women on holiday at relatively increased altitudes—from 300 to 1500 meters above control—as compared with those on holiday without altitude changes. But Cameron observed no clear evidence of changes in menstrual function or fertility among air hostesses serving on jet aircraft, except during the initial months of the experience (12, see also Chapter 33). Although a group of women have been studied in underground isolation, the report does not include menstrual data (13). The period of observation was brief: the variability of menses is such that prolonged observation and rigid controls are needed.

The last observation bears on the suggestion by Dewan that he could influence menstrual patterns in the direction of regularity by exposing the subjects to illumination during sleep (14) and to that of Zondek et al. that ovulation might be induced by the ringing of a bell (15). In both cases the interpretation is affected by the fact that the subjects were aware of the purpose of the experiment.

Halberg and the author reviewed the considerable literature on the timing of the onset of labor and of birth (16). There is strong evidence of a daily rhythm of onset of labor whether by initiation of contractions or spontaneous rupture of membranes. Furthermore, the duration of labor is related to its time of onset and there is therefore a rhythm of delivery as well. This rhythm may be absent from premature and abnormal births. Techniques for the induction of labor can completely obscure these phenomena. Rippmann has since reported the absence of a daily rhythm but did not correct for artifacts (17). Orban et al., who used a large population of contemporary urban deliveries, found birth maxima at 0500 to 0700 hours and at 1100 to 1200 and minima at 2000, 2200, and 0200 hours. There were unchanged from year to year and season to season (18). There is no obvious explanation of the difference of these results from those reported earlier (16). Orban does confirm the absence of a pattern for premature deliveries and still births. Rippmann also searched for an ultradian relationship of birth with phases of the moon but found none (19). MacDonald, using a much smaller population in rural South Carolina, observed more births at the time of new and full moons. Confronted with a bimodal distribution, he lumped the two high quarters and the two low quarters and asserted statistical significance (20).

COMMENT

There is, of course, a substantial body of information on synchronizers of reproduction in species other than man. A number of investigators have been unable to resist argument by analogy to events in Homo. But it is known that social phenomena seriously affect individual reproductive function—for example, the results of crowding of rats. Homo is certainly the most complex social animal.

It is well established that both violent crimes and suicide vary in incidence among women in relation to the menstrual cycle (21). There has been no reference to these or related events among their sexual or social partners, nor whether the relationships of menses to suicide and crime are heightened or diminished by social isolation.

The correlations observed between rape and pregnancy have been used to propose that copulation induces ovulation. A number of other explanations are possible and caution is mandatory if these data are to be used to suggest the operation of physiological synchronizers.

The correlations of birth, menarche, and menses with a limited spectrum of environmental clues are as yet at a superficial level. There is no useful evidence in Homo either of the way in which *intrinsic* synchronizers of reproductive functions work nor of any specific way they are linked to external influences.

REFERENCES

1. G. M. Skutsch, *Lancet,* **2,** (1970), 571.
2. D. W. Amundsen and C. J. Diers, *Human Biol.,* **41,** (1969), 125.
3. N. A. Jafarey, M. Y. Khan, and S. N. Jafarey, *Lancet,* **2,** (1970), 471.
4. K. Magee, J. Basinska, B. Quarrington, and H. C. Stance, *Life Sci.,* **9,** (1970), 7.
5. L. Zacharias and R. Wurtman, *Obstet. Gynecol.,* **33,** (1969), 603.
6. J. B. Thomas and D. J. Pizzarello, *Obstet. Gynecol.,* **30,** (1967), 507.
7. J. Slivka, C. M. Cargille and G. T. Ross, *J. Appl. Physiol.,* **30,** (1971), 905.
8. J. R. Udry and N. M. Morris, *Nature,* **220,** (1968), 593.
9. N. M. Morris and J. R. Udry, *Obstet. Gynecol.,* **35,** (1970), 199.
10. J. H. Clark and M. X. Zarrow, *Amer. J. Obstet. Gynecol.,* **109,** (1971), 1083.
11. H. J. Prill and H. Maar, *Med. Klin.,* **66,** (1971), 986.
12. R. G. Cameron, *Aerospace Med.,* **40,** (1969), 1020.
13. M. Apfelbaum, A. Reinberg, P. Nillus, and F. Halberg, *Presse Med.,* **77,** (1969), 879.
14. E. M. Dewan, *Amer. J. Obstet. Gynecol.,* **99,** (1967), 1016.
15. B. Zondek, C. J. Meyer, and J. P. Hagedoorn, *Proceedings Sixth World Congress on Fertility and Sterility,* Israel Academy of Sciences and Humanities Press, Tel Aviv, 1970, pp. 165–170.
16. F. Halberg and I. H. Kaiser, *Ann. N. Y. Acad. Sci.,* **98,** (1962), 1056.
17. E. T. Rippmann, *Gynaecologia,* **158,** (1964), 31.
18. G. Orban and E. Czeizel, *Gynaecologia,* **163,** (1967), 173.
19. E. T. Rippmann, *Amer. J. Obstet. Gynecol.,* **74,** (1957), 148.
20. R. L. McDonald, *J. Genet. Psychol.,* **108,** (1966), 81.
21. A. Mandell and M. Mandell, *J. Amer. Med. Assoc.,* **200,** (1967), 132.

CHAPTER 33

The Effects of Flying and of Time Changes on Menstrual Cycle Length and on Performance in Airline Stewardesses

F. S. PRESTON and S. C. BATEMAN

Air Corporations Joint Medical Service, Speeedbird House, Heathrow Airport, London, England

R. V. SHORT

A.R.C. Unit of Reproductive Physiology, and Biochemistry and Department of Venterinary Clinical Studies, Cambridge, England

R. T. WILKINSON

M.R.C. Applied Psychology Unit, Cambridge, England

There is now an extensive literature on human circadian rhythms (1), and in recent years people have become increasingly concerned with the way in which these rhythms may be upset by time changes (2, 3). A number of studies have been carried out on students (4–6), civilian and military pilots (7–10), and astronauts (9, 11), particularly with respect

501

to sleep deprivation, but there have been few investigations of cabin crews.

There are also few reports on the effects of flying on the menstrual cycles of stewardesses. The Russians claimed that menstrual cycle length was undisturbed, although flying increased the menstrual discharge and exacerbated dysmenorrhoea (12). In a retrospective questionnaire-type investigation undertaken by Swissair, no long-term deleterious effects of flying on any aspect of the menstrual cycle could be found (13).

It is known that a variety of environmental stresses, including travel, can either postpone or suppress ovulation in women (14). When the environmental change occurs in the preovulatory phase, ovulation is inhibited or delayed and menstruation is postponed. But when the same environmental changes occurs in the postovulatory phase, menstruation occurs at the expected time. However, the subsequent ovulation may also be postponed if the environmental change still persists. There are therefore theoretical grounds for supposing that the menstrual cycle could be prolonged by the stress of repeated time-zone changes, and two of the authors have noted that a number of British Overseas Airways Company (BOAC) stewardesses do complain of irregular cycles after they start flying on long transmeridian routes.

The purpose of the present investigation was first to determine the extent of the disturbance in menstrual cycles of such stewardesses, and second to test the effects of time-zone changes on performance in a controlled environment isolation unit.

MATERIAL AND METHODS

The Effects of Flying on the Menstrual Cycle

Log books for recording menstrual cycles were issued to 119 stewardesses on first joining BOAC. None of these girls had complained of irregular cycles, and none was on oral contraceptives. They were asked to record cycle lengths during their first year of flying on transmeridian routes; at the end of that time 29 log books were returned for analysis, with details of 225 menstrual cycles.

The Effects of Time Changes on the Menstrual Cycle and on Performance

Seven BOAC stewardesses from the preceding study and one British European Airways (BEA) stewardess volunteered for this investigation.

They were aged between 21 and 32 years, and were taken off flying duties for a complete month, with the exception of the BEA girl who was employed only on short routes within Europe. The volunteers were asked to record oral temperatures at 0700, 0855, 1100, 1355, 1600, 1755, 2100, and 2255 hours. These times were specifically arranged to avoid the main meal times. Each girl also kept a personal log of the number of hours of sleep taken each night. Recording started on April 7, 1971 and ended at least a week after the period of the isolation unit. On April 21 all the volunteers were brought to London Airport for a day's briefing and practice in performance tests. On the basis of these tests (see later) the girls were subsequently divided into two matched groups of four, and one person in each group was subsequently designated as team leader.

Following this pre-experimental period, each of the two groups of girls spent 4 days in the controlled environment isolation unit at Risley, the property of the Department of Physiology, Manchester University (3). Temperature and humidity levels were kept constant; there was no natural lighting, nor auditory or visual contact with the world outside,

Table 1 Daily Routine

Time	Duties
0700	Main lights on, body temperature before rising, addition and reaction time
0705	Breakfast
0855	Body temperature
0900–1100	Work period
1100	Body temperature
1100–1230	Free time
1230–1330	Lunch
1355	Body temperature
1400–1600	Work period
1600	Body temperature
1600–1800	Tea, free time
1755	Body temperature
1800–2000	Work period
2000–2100	Dinner
2100–2300	Free time
2255	Body temperature, addition and reaction time
2300	Retire to bed, main lights out

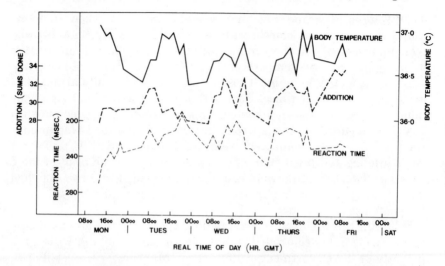

Figure 1   Mean body temperatures, Addition, and Reaction time scores for the four control (NTZ) subjects in the isolation unit.

apart from a two-way speaker for use in emergencies. The main lighting was controlled by an external time-switch, and could also be turned off at will from within. Subsidiary reading lights were available throughout the 24 hours. Radios, television, and newspapers were withheld, but a record player was provided. The unit was well provisioned with a variety of dried, tinned, and deep-frozen foods, and some alcoholic drinks were also available in the evening, after the last work session.

The first group entered the isolation unit at 1300 hours on May 3 and left 96 hours later. They were not subjected to any time alterations, and served as the control, no time zone change (NTZ), group. The daily routine is outlined in Table 1.

The time zone change (TZ) group entered the isolation unit at 1300 hours on May 10 and left 96 hours later. On entry, the time was immediately advanced 8 hours to 2100 hours, simulating a prolonged flight in an Easterly direction. Thereafter the regime was the same as for the NTZ group (see Figure 1 and Table 1) until 1300 GMT on May 12 (2100 hours in the isolation unit), when the clock was advanced a further 8 hours, bringing the time within the isolation unit to 0500 hours on May 13. The established routine was then followed for the next 2 days, and the group was released from the unit at 1300 GMT on

May 14, when their apparent time was 0500 hours on May 15. Thus they were in effect subjected to a third 8 hour time change in order to resynchronize with GMT.

Performance Tests

The workload tasks were carried out in sessions of 2 hours duration in each of which the program was: 2 minutes Addition, 2 minutes Reaction time, 20 minutes Short-term memory, 20 minutes Vigilance, 20 minutes Visual search, 2 minute Addition, and 2 minutes Reaction time. All of these continued without a break except Reaction time.The order of the three 20-minute tasks was balanced as follows:

Addition. columns of 5 two-digit numbers were totalled, the score being the number of sums completed.

Reaction time. A key was pressed as quickly as possible after the onset of a light, which occurred 10 times at quasi-random intervals.

Short-term memory. Runs of eight digits were read out from a tape-recorder, the interval between digits being 0.5 second and between runs 6 seconds. In the latter interval the girls recalled and wrote down the numbers; the score was the number of errors and omissions.

Vigilance. This was similar to proof checking. Each girl had a sheet of numbers, 50 lines of 50 digits in each. A tape-recorded voice read the numbers at a speed of 2.4/second and each girl checked her sheet for errors, of which there were from seven to nine randomly distributed on each set of sheets and different for each girl. The score was the number of errors detected.

Visual search. Each girl had a book containing 40 pages of letters of the alphabet, 50 rows of 50 letters on each page. Fifteen letters of the alphabet were randomly distributed on each page. The other 10 letters (the letter "I" being excluded) appeared only five times each, randomly among the rest. At the head of each page one of these 10 letters was indicated as the target letter. The girls had to find and encircle each of the five target letters before passing on to the next page, where a different target letter was assigned. The score was the number of pages completed.

The 2-hour working sessions were carried out each day at 0900, 1400, and 1800 hours, except for the first and last days (see Table 1). The short Addition and Reaction time tests were carried out at the start and end of each work session, and also when body temperature was taken. Thus these two brief assessments of performance were taken at roughly

2-hour points throughout the hours of wakefulness. The work-load tasks were designed so that they could be self-administered. The complete material and scoresheets were passed out of the unit for marking at the end of each workday. Except for the Reaction time tests the girls did not know their scores, although they were told that they would be given an overall indication of them at some later date.

Two or three weeks before entering the isolation unit, all the girls were given instructions and practice on the tasks and also on the methods of administering them. On the basis of their practice score in Short-term memory, Vigilance, and Visual search, the girls were matched in pairs and one member of each pair allocated either to the TZ or the NTZ group.

RESULTS

The Effect of Flying

The menstrual cycle. Table 2 shows the mean cycle lengths and their standard deviations for each of the 29 BOAC stewardesses. Taking account of the number of cycles recorded, chi-square was used to compare these standard deviations with a normal value of 3.94 at the age of 20, calculated from a large population of women (15). The irregularity of cycle length was significantly greater than normal ($p < .05$) in eight of the girls. Over the group as a whole, chi-square shows this figure to b much larger than would be expected by chance ($p < .01$). In only one case (No. 18) was there evidence of abnormally short cycles.

The Effect of Time Changes

Performance. The results of the workload tasks are shown in Table 3, averaged over the girls in each group and over (*1*) all sessions, and (*2*) the last 24 hours in the isolation unit. Over all sessions the efficiency of the TZ group was lower than that of the NTZ group; this difference was significant (using a *t*-test on the log transformed scores) in Visual search ($p < .05$) and Reaction time ($p < .02$). By the last day in the unit, although the performance of both groups improved with practice, the inferiority of the TZ group became more marked in all tasks but Visual search, so that their efficiency was now significantly lower in Visual search, Short-term memory, and Reaction time ($p < .05$ in each case).

Table 2 *Menstrual Cycle Lengths of BOAC Stewardesses*

Subject No.	No. of Cycles	Mean Length (days)	±SD
1	7	29.0	2.8
2	4	29.8	3.6
3	3	30.0	—
4	4	46.3	12.5**
5	13	27.2	2.4
6	6	28.0	3.4
7	9	27.6	1.1
8	11	26.5	2.5
9	11	29.9	2.7
10	4	32.3	5.3
11	11	30.1	6.7**
12	11	28.6	3.5
13	7	29.4	3.1
14	4	31.8	2.4
15	10	29.5	4.1
16	10	26.4	1.4
17	6	37.5	13.1**
18	8	21.1	10.1**
19	9	30.8	5.5
20	11	28.6	2.4
21	5	27.6	0.9
22	4	26.5	4.1
23	12	26.1	0.9
24	9	25.3	6.8**
25	5	29.8	4.4
26	5	43.8	13.3**
27	6	29.5	4.0
28	7	34.9	11.6**
29	13	30.1	5.8*

* $p < 0.05$
** $p < 00.1$.

Scores of Addition, Reaction time, and Body temperature were averaged for each group at each of the 2-hour points of measurement. Figures 1 and 2 show the trends of these averages in the NTZ and TZ groups, respectively, over the whole period in the isolation unit, the scale along the lower abcissa representing Greenwich mean time (GMT), and the upper abcissa in Figure 2 representing the time in the isolation unit. The usual diurnal cycle of body temperature is evident in the NTZ group and the pattern of the Addition and Reaction time scores matches this cycle, particularly after the first day when the girls had settled down to the routine. Product moment correlations of the levels of Body temperature, Addition, and Reaction time during Thursday and Friday (real time) (see Figures 1 and 2) were as follows: Body temperature versus Addition, $r = +0.69$ ($p < .01$); Body temperature versus Reaction time (where a low score indicates good performance), $r = -0.41$ ($p < .1$); and Addition versus Reaction time, $r = -0.57$ ($p < .02$). In the TZ group (Figure 2) no clear diurnal cycles were apparent in any of these measures and their patterns over time do not appear to show related trends, the corresponding correlations being respectively $r = +0.22$, -0.24, and -0.28, none of which is significant.

Sleep. The girls kept sleep logs during their isolation in the unit and for control periods of 33 days before and 11 days afterward. Average control sleeping times were 7.61 hours for the NTZ group and 7.97 hours for the TZ group. These fell to 6.54 and 4.99, respectively, in the unit. Thus the average daily loss of sleep in the unit was 1.07 hours for the NTZ group and 2.99 hours for the TZ group. The difference between the two was significant ($p < .05$).

Table 3   The Effect of Time Changes on Performance

	Over all sessions		Last day	
	NTZ	TZ	NTZ	TZ
Short-term memory (errors)	22.6	31.2	15.0	27.3
Vigilance (signals detected)	4.9	4.2	5.3	4.5
Visual search[a] (pages completed)	4.8	3.7	6.2	4.6
Reaction time[b] (milliseconds)	141	172	139	167
Addition (sums done)	31	25	32	27

[a] Difference between NTZ and TZ significant at $p < .05$ level.
[b] Difference between NTZ and TZ significant at $p < .02$ level.

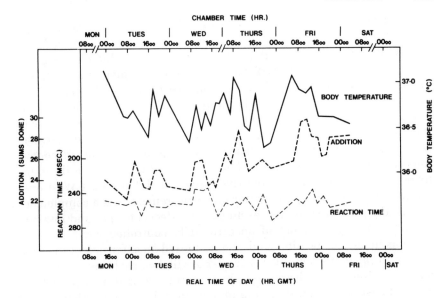

Figure 2 Mean body temperatures, Addition, and Reaction time scores for the four subjects who underwent time-zone changes (TZ). Upper abscissa indicates time in isolation unit.

DISCUSSION

The results of this study leave no doubt that a high proportion of stewardesses flying transmeridian routes (28%) have irregular menstrual cycles. This irregularity seems to be mainly due to occasional prolonged cycles, and it is tempting to conclude that they are caused by the stress of repeated time-zone changes. The life of the corpus luteum in nonpregnant women seldom exceeds 14 days (16), so extended cycles are likely to be due to a prolongation of the preovulatory phase.

Although there were only four girls in each of the groups, the time-zone change produced a significant impairment of efficiency both over the performance tasks as a whole and in particular in the ability to react quickly, memorize, and search. Overall, the results appear to offer a clear indication that some aspects of performance are impaired by transferring people across time zones.

In previous studies of the effects to time-zone transition on performance, subjects have been tested for a preflight baseline period, physically transported from one part of the world to another, and then retested for a number of postflight days in the new time zone. Typically the time change has been a single one of 8 hours followed eventually by a reversal on the return journey. Clear shifts in the phase of the 24-hour rhythm of performance have been demonstrated (8), but changes in the overall level of performance have usually been of the order of about 2 to 8% and at borderline levels of significance (4–6, 8, 17). Hauty and Adams (17) sought to control for the effect of the flight itself by comparing the results of flights along the East-West meridian with North-South flights of about the same duration, but, of course, imposing no time-zone change. The North-South flight gave the same increase in the reports of subjective fatigue as did the East-West ones but failed to show the same performance decrements. As a complement to this, the present study has excluded the effects of flight and retained the time change. The use of an artificially controlled environmental unit has made this possible and also permitted an extension of the time change to the equivalent of two West-East flights, each involving an 8-hour transition. The decrements in performance in this setting have been statistically significant and considerably larger than those reported previously for a single 8-hour transition. Across tasks they range from 17 to 38% when averaged over the whole period of transition and from 18 to 82% when measured on the last day.

The fact that we observed performance decrements at all supports the conclusion of Hauty and Adams (17) that it is the time-zone transition itself and not the mere fatigue of flying which is mainly responsible for impaired performance following flights along the East-West meridian. The size of the decrement as compared with that in previous studies may well be due to the severity of a double time change within the course of 4 days. The possibility that the present tasks were particularly sensitive is discounted by consideration of reaction time. This test was similar in both its form and in the frequency and duration of its administration to that used by Klein and his co-workers (5, 8), yet it gave decrements of about 20% in our setting as against an average of 6% in their most comparable single 8-hour transition (5).

The data from the NTZ group reveal a good relationship between Body temperature and Performance such that when temperature is highest during the 24 hours, performance is most efficient. This agrees with previous findings (18, 19). However, the present daytime correlations were higher than some of those previously noted, due perhaps to more stable measurement of body temperature and performance in the controlled environment of the unit. Time-zone changes destroyed this relationship and a more random fluctuation of body temperature was accompanied by a lower level of performance; whether these two outcomes are casually related, however, cannot be assessed.

In the chamber the daily sleep of the experimental TZ group was reduced to about 5 hours, a degree of sleep deprivation which has been shown to lower performance (20, 21). The control NTZ group achieved nearly 7 hours sleep. Thus it seems likely that the inferior performance of the group subjected to time-zone changes may have been partly due to loss of sleep as well as to the disturbance of the circadian rhythm.

In operational situations it would seem reasonable to regard disruption of the body temperature cycle as an indication that peak performance during the 24 hours is likely to be lower than normal and that sleep may well be disrupted, leading to lower overall efficiency. There are various lessons to be learned from this work in the scheduling of cabin crew in commercial airlines., so that they remain at full operational efficiency during rostered duty.

ACKNOWLEDGMENTS

We are particularly indepted to the eight volunteer stewardesses who took part in this experiment and gave up so much of their time with such enthusiasm. We would also like to thank J. N. Mills of the Department of Physiology, University of Manchester and his staff for permission to use their Isolation Unit at Risley. We are indebted to the laboratory staff at the Air Corporations Joint Medical Service Medical Centre at Speedbird House for technical assistance.

REFERENCES

1. R. T. W. L. Conroy and J. N. Mills, *Human Circadian Rhythms*, Churchill, London, (1970).
2. P. V. Siegel, S. J. Gerathewohl, and S. R. Mohler, *Science*, **164**, (1969), 1249.
3. A. L. Elliott, J. N. Mills, D. S. Minors, and J. M. Waterhouse, *J. Physiol.*, **221**, (1972), 227.
4. G. T. Hauty and T. Adams, *Aerospace Med.*, **37**, (1966), 668.
5. K. E. Klein, H. M. Wegmann, and B. I. Hunt, *Aerospace Med.*, **43**, (1972), 119.
6. S. A. Lewis, G. A. Christie, J. R. Daly, J. I. EVans, and M. Moore-Robinson, in *Aspects of Human Efficiency: the Diurnal Rhythm and Loss of Sleep* (W. P. Colquhoun, Ed.), English Univ. Press, London, 1972, p. 307.
7. A. N. Nicholson, *Aerospace Med.*, **41**, (1970), 626.
8. K. E. Klein, H. Bruner, H. Holtmann, H. Rehme, J. Stolze, W. D. Steinhoff, and H. M. Wegmann, *Aerospace Med.*, **41**, (1970), 125.

9. C. A. Berry, *Aerospace Med.*, **41**, (1970), 500.

10. F. S. Preston and S. C. Bateman, *Aerospace Med.*, **41**, (1970), 1409.

11. A. N. Nicholson, *Aerospace Med.*, **43**, (1972), 138.

12. V. F. Shmidova, *Gig. Truda Prof. Zabol.* (Moscow), **10**, (1966), 55.

13. R. G. Cameron, *Aerospace Med.*, **40**, (1969), 1020.

14. S. Matsumoto, M. Igarashi, and Y. Nagaoka, *Int. J. Fert.*, **13**, (1968), 15.

15. A. E. Treloar, R. E. Boynton, B. G. Behn, and B. W. Brown, *Int. J. Fert.*, **12**, (1967), 77.

16. R. Vande Wiele, J. Bogumil, M. Dyrenfurth, R. Ferin, R. Jewelewicz, M. Warren, T. Rizkallah, and G. Mikhail, *Rec. Progr. Hormone Res.*, **26**, (1970), 63.

17. G. T. Hauty and T. Adams, Federal Aviation Agency Report No. AM-65-30 (1965).

18. N. Kleitman and D. P. Jackson, *J. Appl. Physiol.*, **3**, (1950), 309.

19. W. P. Colguhoun, M. J. F. Blake, and R. S. Edwards, *Ergonomics*, **12**, (1969), 856.

20. R. T. Wilkinson, *Progr. Clin. Psychol.*, **8**, (1968), 28.

21. R. T. Wilkinson, *Proc. Roy. Soc. Med.*, **62**, (1969), 903.

CHAPTER 34

The Control of Brain Serotonin
Concentration by the Diet

JOHN D. FERNSTROM

*Laboratory of Neuroendocrine Regulation, Department of
Nutrition and Food Science, Massachusetts Institute of
Technology, Cambridge, Massachusetts*

INTRODUCTION

The concentrations of tryptophan and of most other amino acids in
human plasma undergo characteristic and parallel daily oscillations (1,
2). Tryptophan levels are lowest between 2 and 4 A.M., and rise 50 to
80% to attain a plateau late in the morning or early in the afternoon.
The amplitude of the rhythm exhibited by any particular amino acid
tends to vary inversely with its availability in the body: Concentrations
of relatively scarce amino acids (e.g., cysteine, methionine) rise and fall
by as much as twofold during each 24-hour period, while the more
abundant amino acids (e.g., glycine, glutamate) vary only 10 to 30%.
Rhythms in the plasma concentrations of tryptophan and other amino
acids are also observed in rats (3) and mice (4) (Figure 1); however,
peak tryptophan levels, 35 to 150% above those at the daily nadir, occur about 8 to 10 hours later than in humans, a phenomenon which

513

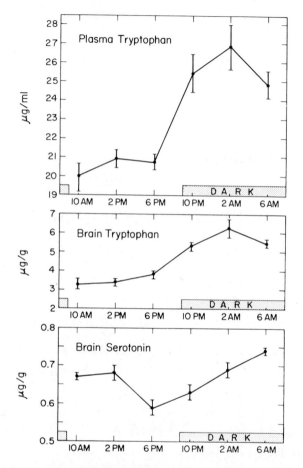

Figure 1   Daily rhythms in plasma tryptophan, brain tryptophan, and brain serotonin. Groups of 10 rats kept in darkness from 9 P.M. to 9 A.M. were killed at intervals of 4 hours. Vertical bars represent standard errors of the mean. (3)

probably reflects the rat's tendency to consume most of its food during the night.

Plasma amino acid rhythms appear to be influenced at least in part by the diet. Thus, for example, the daily concentration curves for tryptophan, tyrosine, and phenylalanine are significantly altered when human subjects shift from consuming normal foods which contain protein to a diet essentially devoid of protein (1). Moreover, the rhythm in plasma tryptophan can be abolished by placing subjects on a total fast (5).

The mere existence of plasma amino acid rhythms does not establish that such variations are of any physiologic consequence. To explore their possible significance, we have attempted to determine whether the naturally occurring daily fluctuations in the plasma concentration of a particular amino acid can actually influence its metabolic fate. The amino acid whose plasma concentration seems most likely to influence its metabolic fate is tryptophan: The quantities of free and peptide-bound tryptophan present in mammals and in most foods are the lowest of all the amino acids (3); moreover, evidence has already been obtained that daily rhythms in the ingestion of tryptophan-containing proteins (and presumably in the portal venous concentration of tryptophan) cause parallel rhythms in hepatic polysome aggregation (9), and in the activity and synthesis of at least one specific liver protein, tyrosine aminotransferase (2).

DISCUSSION

The compound synthesized from tryptophan which we chose for study was serotonin, a putative neurotransmitter in the mammalian central nervous system. This amine appeared to be a promising candidate with which to demonstrate a physiologic relationship between the availability of precursor amino acid and the rate of product formation, in that evidence already existed suggesting that the concentration of tryptophan in brain influenced serotonin synthesis. For example, Lovenberg et al. (10) showed that tryptophan hydroxylase, the enzyme that catalyzes the rate-limiting step in serotonin biosynthesis, has an unusually large K_m relative to normal brain tryptophan concentrations; thus this enzyme might be unsaturated *in vivo* and respond to increases in brain tryptophan by becoming more saturated with amino acid and synthesizing more serotonin. Consistent with this hypothesis was the observation that raising brain tryptophan levels by injecting the amino acid into rats does elicit rapid increases in brain serotonin and its principal metabolite, 5-hydroxyindole acetic acid (8). Finally, it was known that a daily rhythm occurred in the concentration of serotonin in brain (6), and it seemed conceivable that this oscillation might in part be due to the daily variations in plasma and brain tryptophan (3).

Initial experiments were designed to determine whether brain serotonin concentrations could be increased by injecting very small doses of tryptophan around the time of day (noon, 3 hours after the beginning of the light period) when plasma and brain tryptophan, and

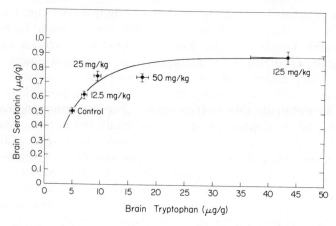

Figure 2   Dose-response curve relating brain tryptophan and brain serotonin. Groups of 10 rats received tryptophan intraperitoneally at noon, and were killed 1 hour later. All brain tryptophan and serotonin levels were significantly higher than control levels (p < .01) (10).

brain serotonin concentrations are known to be low (Figure 1). We hoped that a dose of tryptophan could be identified that (1) raised brain serotonin, but (2) was smaller than the amount of the amino acid normally consumed by the animal (rat) each day, and (3) did not elevate plasma and brain tryptophan levels beyond their normal daily peaks. All of these goals were met by injecting intraperitoneally a 12.5 mg/kg dose of the amino acid to male rats weighing 150 to 200 g. This dose, which constitutes less than 5% of the amount of tryptophan that rats would be expected to ingest each day in 10 to 20 g of standard rat chow, produced peak elevations in plasma and brain tryptophan which were well within the ranges that occurred normally in untreated animals, and caused brain serotonin levels to rise by 20 to 30% ($p < .01$) within 1 hour (7). A dose of 25 mg/kg caused further elevations in *both* brain tryptophan and serotonin, but larger doses (50, 125 mg/kg), while inducing even greater rises in brain tryptophan, did not further elevate brain serotonin (Figure 2). [Studies by others (8) have shown that further increments in tryptophan dose continue to produce increases in brain 5-hydroxyindoleacetic acid (but not serotonin), and thus presumably in serotonin synthesis. The failure of brain serotonin content to continue to rise when plasma and brain tryptophan are elevated beyond their normal dynamic ranges thus most likely reflects a limitation in the ability of serotonin-producing neurons to store the amine.]

DISCUSSION

The observed increase in brain serotonin caused by injecting very small doses of tryptophan was thought to be compatible with the hypothesis that the nocturnal rise in brain serotonin in normal rats is related to the daily rhythms in plasma and brain tryptophan (see Figure 1). However, it does not seem necessary to conclude that a substrate-induced rhythm in serotonin synthesis is the *only* factor responsible for the daily rhythm in brain serotonin content. For example, it is also possible that the serotonin rhythm reflects changes in the rate at which the monoamine is released from neurons or metabolized intraneuronally.

Now that small increases in plasma tryptophan had been shown to cause parallel changes in brain tryptophan and serotonin, it became of interest to determine whether physiological *decreases* in the plasma concentration of the indoleamino acid also lowered brain serotonin content. We attempted to lower plasma tryptophan by injecting rats with insulin. It was not known whether exogenous insulin lowered

Table 1   The Effect of Insulin on the Concentrations of Tryptophan, Other Amino Acids, and Glucose in Rat Plasma (12)[a]

Amino Acid	Concentration (μg/ml)		Percent Change
	Control	Insulin	
L-Tryptophan	11.1 ± 0.6	16.6 ± 0.9	+50*
L-Tyrosine	11.4 ± 0.8	9.4 ± 1.0	−18
L-Phenylalanine	11.8 ± 0.3	10.8 ± 0.9	−9
L-Serine	24.6 ± 1.2	17.0 ± 1.3	−31*
L-Glycine	29.6 ± 1.4	18.3 ± 2.0	−38*
L-Aline	25.3 ± 0.9	11.7 ± 0.7	−54*
L-Valine	21.5 ± 1.3	17.3 ± 1.8	−20
L-Isoleucine	13.7 ± 0.8	7.1 ± 0.4	−48*
L-Leucine	21.3 ± 1.2	16.6 ± 2.1	−22
	(mg/100 ml)		
Glucose	96.0 ± 2.5	43.0 ± 4.8	−55*

[a] Groups of five to 10 fasting 150 to 200 g rats were killed 2 hours after receiving insulin (2 units/kg, i.p.).
* $p < .01$.

plasma tryptophan in rats, probably because a simple assay for the amino acid had only recently become available. However, there was abundant evidence that insulin exerts this effect on almost all other amino acids examined, largely by increasing their uptake into striated muscle (11).

Rats were fasted overnight and then injected with a dose of insulin (2 units/kg, i.p.) known to lower blood glucose levels. To our surprise, the hormone did not lower plasma tryptophan, but rather *increased* its concentration by 30 to 50% (12). This effect was independent of the route of administration; it was associated with 55% decline in plasma glucose and with major reductions in the plasma concentrations of most other amino acids studied (Table 1), including the neutral amino acids generally believed to compete with tryptophan for uptake into the brain (13, 14). Two hours after rats received insulin, brain tryptophan levels were elevated by 36% ($p < .001$), and brain serotonin by 28% ($p < .01$) (15).

The increase in brain serotonin content observed in rats receiving insulin could have resulted not from increased availability of substrate, but from reflexes activated by the accompanying hypoglycemia. To determine whether the physiologic secretion of insulin in normoglycemic animals also increases plasma and brain tryptophan and brain serotonin concentrations, these indoles were measured in rats given access to a carbohydrate diet after overnight fasting. In a typical experiment, animals ate approximately 5 g of food the first hour, and 2 g each hour during the second and third hours (12, 15). Plasma tryptophan levels were significantly elevated one, two, and three hours after food presentation; tyrosine concentrations were depressed at all three times studied (Table 2). Brain tryptophan rose 22% during the first hour, reached a peak 65% above control values ($p < .001$) after 2 hours, and remained significantly elevated at 3 hours (Table 2). Brain serotonin concentrations rose during the first hour, became significantly elevated by the end of the second hour, and remained so after 3 hours (Table 2) (15).

On the basis of these observations, a model was proposed to explain the mechanisms by which dietary inputs affect brain serotonin. According to this model, carbohydrate consumption elicits insulin secretion, thereby raising plasma tryptophan levels (by mobilizing the amino acid from as yet undefined pools); this elevation, in turn, causes a corresponding increase in brain tryptophan, which increases the saturation of tryptophan hydroxylase, increases serotonin synthesis, and, ultimately, increases brain serotonin levels. With this model we predicted that the consumption of a diet containing both carbohydrates

Table 2 *The Effect of Carbohydrate Ingestion on Brain Serotonin Concentrations and on Plasma and Brain Tryptophan* (15)[a]

	Time after Presentation of Food (hours)			
	0	1	2	3
Plasma tryptophan (μg/ml)	10.86 ± 0.55	13.56 ± 0.81**	14.51 ± 0.70***	13.22 ± 0.65**
Brain tryptophan (μg/g brain, wet weight)	6.78 ± 0.40	8.32 ± 0.63*	11.24 ± 0.52***	9.81 ± 0.50***
Brain serotonin (μg/g brain, wet weight)	0.549 ± 0.015	0.652 ± 0.046	0.652 ± 0.012***	0.645 ± 0.017***
Plasma tyrosine (μg/ml)	13.03 ± 0.29	9.55 ± 0.34***	8.67 ± 0.26***	9.03 ± 0.21***

[a] Average animal weight was 160 g.
* $p < .05$ differs from 0-hour group.
** $p < .02$ differs from 0-hour group.
*** $p < .001$ differs from 0-hour group.

and protein would cause an even greater rise in brain serotonin. In addition to elevating plasma tryptophan via insulin secretion, the tryptophan molecules in the dietary proteins would also contribute directly to plasma tryptophan; hence plasma tryptophan concentrations would increase even more than after ingestion of a protein-free diet, and brain tryptophan and serotonin would show similar amplifications in response. When this model was tested by giving fasted rats access to diets containing either casein or a synthetic amino acid mixture similar in composition to 18% casein, it was immediately apparent that it was in need of major revision. As expected, protein consumption was followed by a major increase (about 60%, $p <$ .001) in plasma tryptophan; however, neither brain tryptophan nor brain serotonin was at all increased (16).

Other investigators, using brain slices (13) or animals treated with pharmacological doses of individual amino acids (14), had shown that groups of amino acids (e.g., neutral, acidic, basic) are transported into brain by specific carrier systems, and that within a group the member amino acids compete with each other for common transport sites. Since protein ingestion introduces variable amounts of all of the amino acids into the blood, it seemed possible that brain tryptophan failed to increase after protein ingestion because the plasma concentrations of other, competing amino acids increased even more than that of tryptophan. To test this hypothesis, we allowed groups of animals to eat either a synthetic diet containing carbohydrates plus all of the amino acids in the same proportions as present in an 18% casein diet, or this diet minus five of the amino acids thought to share a common transport system with tryptophan (i.e., tyrosine, phenylalanine, leucine, isoleucine, and valine). Both diets significantly increased plasma tryptophan levels above those found in fasted controls. However, only when the competing neutral amino acids were deleted from the diet did large increases occur in brain tryptophan, serotonin, or 5-hydroxyindoleacetic acid (Figure 3).

To rule out the possibility that the increase in brain 5-hydroxyindoles observed in rats consuming the latter diets was simply a nonspecific consequence of the omission of any group of amino acids from the diet, we repeated the above experiment omitting aspartate and glutamate instead of the five neutral amino acids. (These two amino acids comprise approximately the same percent of the total alpha-amino nitrogen in casein as the five competing amino acids. Because they are charged at physiologic pH, they are transported into the brain by a carrier system different from that transporting tryptophan (13). Hence, their absence would not be expected to alter the postprandial

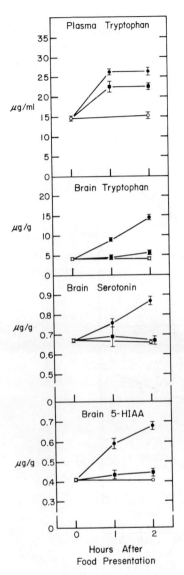

Figure 3 The effects of ingestion of various amino acid-containing diets on plasma and brain tryptophan, and brain 5-hydroxyindole levels. Groups of eight rats were killed 1 or 2 hours after diet presentation. Vertical bars represent standard errors of the mean. Fasting controls, O; complete amino acid mix diet, ■; mix diet minus tyrosine, phenylalanine, leucine, isoleucine, and valine, ●. One and 2 hour plasma tryptophan levels were significantly greater in animals consuming both diets (p < .001) than in fasting controls. All brain tryptophan, serotonin, and 5-hydroxyindoleacetic acid levels were significantly greater in rats consuming the diet lacking the five amino acids than in fasting controls (p < .001 for all but 1 hour serotonin, p < .01 (16).

competition between tryptophan and other amino acids within its transport group for uptake into the brain). At 1 and 2 hours after presentation of this diet or the complete amino acid mixture, plasma tryptophan concentrations again increased 70 to 80% above those of fasted controls ($p < .001$). However, neither diet caused increases in brain tryptophan, serotonin, or 5-hydroxyindoleacetic acid.

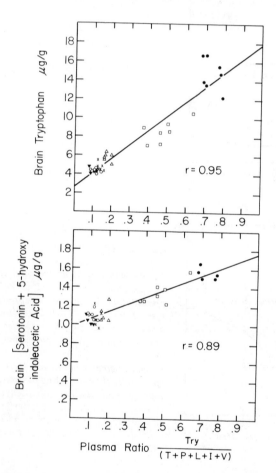

Figure 4 Top: Correlation between brain tryptophan concentration and the plasma ratio of tryptophan to the five competing amino acids in individual rats studied in the experiment described in Figure 3. r = 0.95 (p < .001 that r = 0). Bottom: Correlation between the sum of brain serotonin and 5-hydroxyindoleacetic acid, and the plasma ratio of tryptophan to the five competitor amino acids in individual rats studied in the experiment described in Figure 3. r = 0.89 (p < .001 that r = 0).
Controls: 0 hours, ○; 2 hours, ▼. Complete amino acid mix diet: 1 hour, ×; 2 hours, △.
Complete mix diet minus five competing amino acids: 1 hour, □; 2 hours, ● (16).

These results were interpreted as showing that brain tryptophan and 5-hydroxyindole levels do not simply reflect plasma tryptophan, but also depend upon the plasma concentrations of other neutral amino acids. This relationship was illustrated by a correlation analysis comparing the brain tryptophan level and the ratio of plasma tryptophan to the five competing amino acids among individual rats given various diets that contain differing amounts of each amino acid. This analysis yielded a correlation coefficient of 0.95 ($p < .001$ that $r = 0$), whereas the correlation between brain tryptophan and plasma tryptophan alone was less striking ($r = 0.66$; $p < .001$ that $r = 0$). Similarly, the correlation coefficient for brain 5-hydroxyindoles (serotonin plus 5-hydroxyindoleacetic acid) versus the plasma amino acid ratio was 0.89 ($p < .001$), whereas that of 5-hydroxyindoles versus tryptophan alone was only 0.58 ($p < .001$) (Figure 4). Thus the brain concentrations of both tryptophan and the 5-hydroxyindoles more nearly reflect the ratio of plasma tryptophan to competing amino acids than the plasma tryptophan concentration alone. The reason that brain tryptophan and serotonin appeared in our earlier formulation to depend upon plasma tryptophan alone was that all of the physiological manipulations tested at that time (i.e., tryptophan injections, insulin injections, carbohydrate consumption) raised the numerator in the plasma tryptophan:competitor ratio while either lowering the denominator or leaving it unaltered. Only when the rats consumed protein were both the numerator and the denominator elevated.

SUMMARY

The effect of food consumption on 5-hydroxyindoles in rat brain may now be modeled as in Figure 5. Since carbohydrate ingestion elicits insulin secretion, it simultaneously raises plasma tryptophan and lowers the concentrations of the competing neutral amino acids in rats (12); hence the ratio of plasma tryptophan to competing amino acids increases, leading to elevations in brain tryptophan and serotonin. In contrast, protein consumption provides the plasma with an exogenous source of all of the amino acids; however, the ratio of tryptophan to its competitor amino acids is almost always lower in dietary proteins than it is in plasma. Probably for this reason, protein ingestion increases the plasma levels of tryptophan less than it does the concentrations of competing amino acids and thereby decreases the tryptophan:competitor ratio. The insulin secretion elicited by protein consumption will, by itself, produce an opposite change in this ratio. Thus brain tryptophan and 5-hydroxyindole levels can decrease, increase, or remain un-

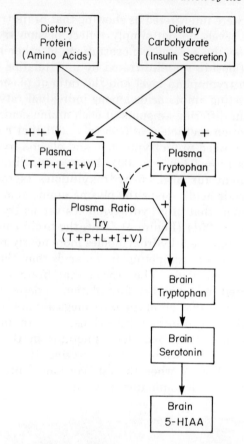

Figure 5 Proposed sequence describing diet-induced changes in brain serotonin concentration in the rat. The ratio of tryptophan to (tyrosine + phenylalaline + leucine + isoleucine + valine) in the plasma is thought to control the tryptophan level in the brain (16).

changed after eating, depending on the proportion of protein to carbohydrates in the diet and the amino acid composition of the particular proteins.

The extent to which these observations on rats also apply to humans and other mammals awaits clarification. Insulin does not seem to raise total plasma tryptophan in humans as it does in rats; however, it does depress the concentrations of the other neutral amino acids and thus elevates the ratio of tryptophan to its competitors (17). In both species the more relevant index to tryptophan availability may be plasma *free* tryptophan (i.e., tryptophan not bound to albumin) (18). Plasma-free tryptophan concentrations do fall precipitously in human subjects who consume a glucose load. When a method becomes available for estimating brain serotonin synthesis in humans, it will be interesting to determine whether such synthesis correlates best with free or total plasma tryptophan concentrations, or with their ratios to the concentrations of other neutral amino acids.

Serotonin-containing neurons have been implicated in the control of ovulation. Spontaneous ovulation in the rat can be prevented by elevating brain serotonin concentrations just prior to the "critical period" [with a monoamine oxidase inhibitor (19), or with 5-hydroxytryptophan (20)]. This effect is probably related to the inhibitory influence of serotonin on LH secretion. The intraventricular injection of serotonin decreases the concentration of LH in serum within 1 hour of monoamine administration (21). Inasmuch as serotonin concentrations in brain are rapidly influenced by the diet, it is tempting to speculate that nutritional factors (e.g., the type of food consumed or the time pattern of food ingestion) may participate in the control of ovulation.

ACKNOWLEDGMENTS

These studies were supported in part by grants from the John A. Hartford Foundation and the National Aeronautics and Space Administration.

REFERENCES

1. R. J. Wurtman, C. M. Rose, C. Chou, and F. Larin, *New Engl. J. Med.,* **279,** (1968), 171.

2. R. J. Wurtman, in *Mammalian Protein Metabolism,* Volume IV (H. N. Munro, Ed.). Academic Press, New York, 1970, p. 445.

3. R. J. Wurtman and J. D. Fernstrom, in *Perspectives in Neuropharmacology,* (S. H. Snyder, Ed.)., Oxford Univ. Press, Oxford, 1972, p. 145.

4. M. I. Rapoport, R. D. Feigen, J. Bruton, and W. R. Beisel, *Science,* **153,** (1966), 1642.

5. E. B. Marliss, T. T. Aoki, R. H. Unger, J. S. Soeldner, and G. F. Cahill, *J. Clin. Invest.,* **49,** (1970), 2256.

6. P. Albrecht, M. B. Visscher, J. J. Bittner, and F. Halberg, *Proc. Soc. Exp. Biol. Med.,* **92,** (1956), 702.

7. J. D. Fernstrom and R. J. Wurtman, *Science,* **173,** (1971), 149.

8. A. T. B. Moir and D. Eccleston, *J. Neurochem.,* **15,** (1968), 1093.

9. B. Fishman, R. J. Wurtman, and H. N. Munro, *Proc. Nat. Acad. Sci. U.S.,* **64,** (1969), 677.

10. W. Lovenberg, E. Jequier, and A. Sjoerdsma, *Advan Pharmacol.,* **6A,** (1968), 21.

11. I. G. Wool, *Fed. Proc.,* **24,** (1965), 1060.

12. J. D. Fernstrom and R. J. Wurtman, *Metabolism,* **21,** (1972), 337.

13. R. Blasberg and A. Lajtha, *Arch. Biochem. Biophys.,* **112,** (1965), 361.

14. G. Guroff and S. Udenfriend, *J. Biol. Chem.,* **237,** (1962), 803.

15. J. D. Fernstrom and R. J. Wurtman, *Science,* **174,** (1971), 1023.

16. J. D. Fernstrom and R. J. Wurtman, *Science,* **178,** (1972), 414.

17. H. N. Munro and W. S. T. Thomson, *Metab. Clin. Exp.,* **2,** (1953), 354.

18. R. H. McMenamy and J. L. Oncley, *J. Biol. Chem.,* **233,** (1958), 1436.

19. C. Kordon, F. Javoy, G. Vassent, and J. Glowinski, *Eur. J. Pharmacol.,* **4,** (1968), 169.

20. A. Psychoyos, *C. R. Acad. Sci., Paris,* **263,** (1966), 986.

21. I. A. Kamberi, R. S. Mical, and J. C. Porter, *Endocrinology,* **87,** (1970), 1.

11 Characteristics of Biorhythms

FRANZ HALBERG, *Moderator*

CHAPTER 35

Circadian and Lower Frequency Rhythms in Male Grip Strength and Body Weight

J. F. W. KÜHL, J. K. LEE, F. HALBERG, E. HAUS,
R. GÜNTHER, and E. KNAPP

*Chronobiology Laboratories, Department of Pathology,
University of Minnesota, Minneapolis, Minnesota, and Department of Medicine, University of Innsbruck, Innsbruck,
Austria*

Ten healthy male medical students, living under "usual" if somewhat standardized conditions in Austria (1–3) and self-measuring various body functions, provide a store of information of interest to students of biorhythms as well as to those involved in the study of reproduction.

This study demonstrates how a relatively few self-measurements requiring modest instrumentation (1) suffice to detect and quantify parameters of circadian rhythmic changes in psychomotor performance. Thus we shall describe the temporal relation between human performance variables and body temperature at circadian and other frequencies. Particularly germane for the student of human reproduction are the lower-than-circadian frequency changes in body temperature and certain performance data.*

The body temperature which changes in the course of the female menstrual cycle deserves scrutiny for possible changes with similar frequency in the human male. With very extensive data covering several years, a 30-day component in the male human body temperature already has been demonstrated (4). The question examined herein is whether systematically collected body temperature and/or performance data, measured with a fixed sampling scheme over a few months, allow the detection of spectral components in both the circadian region and in a region of periods longer than a day yet shorter than a month. This question is prompted by the search for a cycle in the male reproductive system (5). The extensive literature on body temperature and performance rhythms (6–8) has been critically reviewed.

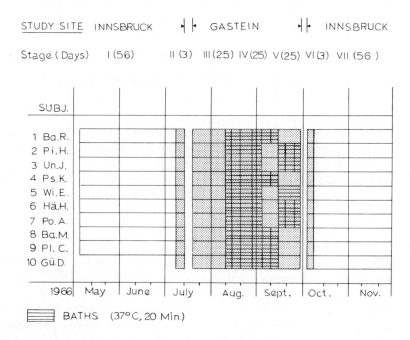

Figure 1 Study plan—site and duration. Sampling for performance functions nine times a day, as indicated in Figure 2, during stages II to VI, in dotted sections.

* Supported by the United States Public Health Service (5–K6–GM–13,981 and 1 R01–CH–14445–01), and NASA

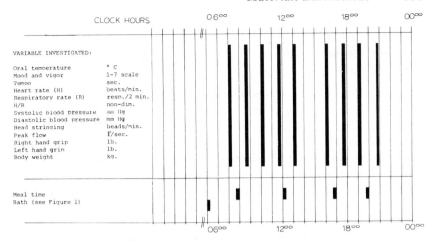

Figure 2   Partial list of variables examined and sampling schedule.

MATERIALS AND METHODS

Ten male medical students, 23 years of age on the average (range: 22 to 26 years), were each given a physical examination, including X-ray of the chest, electrocardiogram, white and red blood cell counts, differential blood smear examination, erythrocyte sedimentation rate, and urine analysis. There were no pathologic findings.

The subjects recorded their routine of retiring and awakening during a first stage of 56 days in Innsbruck, Austria, as shown in Figure 1. The study's subsequent stages—II to IV, Figure 1—involved the self-measurements indicated in Figure 2. Herein we shall focus primarily upon data collected in stages III, IV, and V, each of 25-day duration. During these stages, the subjects were invited to awaken at 07^{00} (except on days with "bath," see later) and to retire at 22^{00}; meals were served before 08^{00}, after 12^{00}, before 17^{00}, and before 20^{00}, at the times indicated in Figure 2. Figure 2 also indicates the frequency and times of sampling and the variables examined—only some of them discussed in this report.

The geographic displacement from Innsbruck to Bad Gastein, Austria, preceding Stage III, did not involve a time-zone shift. Stages IV and V but not stage III involved baths for all of the subjects on the schedule apparent from Figure 2. The morning bath at 05^{00} involved immersion for 20 minutes in water at 37°C containing on the average 10×10^{-9} Curie of radon per liter. The ratios of radon to radium A, B, and C in the water were 1:0.8:0.4:0.3. The thoron content in the water was estimated at about 10 picocurie (pc)/liter (9).

Among the numerous functions studied, the following variables will be considered in this report:

1. Oral temperature, measured with calibrated clinical thermometers kept sublingually 8 minutes for each measurement.

2. Grip strength, measured by the standing subject with a Jamar Dynamometer manufactured by the Asimov Engineering Company, Los Angeles, California. The needle follower on the dynamometer was turned to zero before each test. The subject stood up and assumed a military stance with eyes forward, shoulders back, and feet about a foot apart. With dynamometer held in the right hand first, he extended the arm downward at about a 30°-angle from the body and squeezed quickly with full effort and without changing position or approaching hand to thigh. The grip thereafter was released, and the best value indicated by the follower recorded. The procedure was then repeated with the left hand.

3. Body weight, measured on a calibrated scale with a precision of 0.1 kg.

Data analysis was carried out with methods described elsewhere (5, 10, 11) or introduced at this meeting (12, 13).

DISPLAY OF RESULTS

In mathematics the degrees on a polar coordinate are measured in counterclockwise direction. In introducing degrees measured clockwise with a negative sign for a biologic acrophase, the concepts of a delay and an advance in phase (φ) are presented by a graphical method along both polar and rectangular coordinates in Figure 3. For the given function, say, with a period $\tau = 24$ hours, the phase, φ, is referred to as the lag from the starting time or zero phase to the first crest in a single cosine function best fitting the data. The phase reference or zero phase is indicated as 0° at the top of each circle or at the origin of the rectangular coordinates.

When $\varphi = 0°$, the first crest of the function occurs at the beginning of the time series (i.e., $t_i = 0$). This zero phase is indicated as a dashed reference function in each graph. Solid lines represent functions with phase differing from zero (i.e., $\varphi \neq 0°$) and are compared with the dashed function. In the second and third rows delays with $\varphi = -45°$ and $-135°$ are shown on polar (left) and rectangular (right) coordinates, where φ indicates the distance from the start to the first crest in absolute values.

The phases with positive signs in the fourth and fifth rows are read as advances. This means that the first crest of a function has occurred before the time series is started, before $\varphi = 0°$. In most cases, in the study of biologic rhythms, we present phases, called acrophases (10,

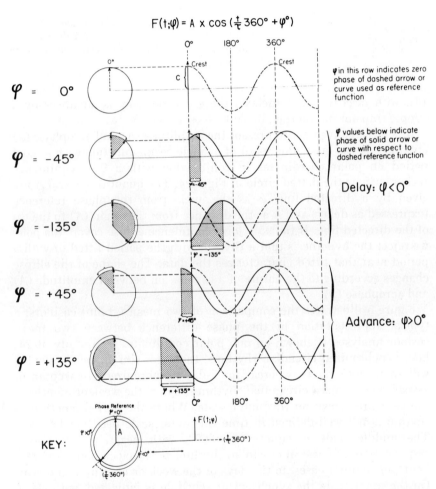

Figure 3 *Phase relation between 2 cosine functions. Timing of a rhythm shown on two equivalent coordinate systems in order to introduce the concepts of delay in negative degrees and advance in positive degrees for the continuous line representation of a rhythm. One of the 2 functions, that is, the dashed arrow or curve, serves as the reference function with $\varphi = 0°$.*

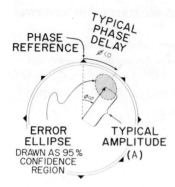

*Figure 4   Cosinor, from cosine and vector, repre-
sents a demonstration of amplitude and acrophase
by the length* (A) *and angle* (φ) *of a directed line
with a joint 95% confidence region given as an
ellipse.*

14), with negative signs (delay); these can be read as advances by a simple transformation (i.e., by $360° + \varphi$) (14).

It is often convenient to present the amplitude (A) and acrophase (φ) obtained by the least squares fit of a single cosine function with a fixed period on polar coordinates in conjunction with a 95% confidence region such as the dotted circle in Figure 4. The quantities φ and A are given by a directed line, φ as the angle from the phase reference (expressed as delay) and A as the distance from the origin (A) to the tip of the directed line. If the 95% joint confidence ellipse covers the pole, we reject the hypothesis that a rhythm with the period fitted or with a period near that fitted characterizes the data. The shape of the ellipse changes according to the nature of a correlation between amplitude (A) and acrophase (φ).

Figure 5 illustrates the comparison of two mean (group) cosinors: a typical representation of the phase difference between two mean cosinor analyses is shown on one polar coordinate. In a study of 24-hour-synchronized circadian rhythms, the outer rim of such a display will show clock hours, whereas for 7-day-synchronized circaseptan or 1-year-synchronized circannual rhythms days of the week or months of the year are shown on this outer scale. Where the period length of a rhythm is not well-defined in time this outer scale need not be used. The middle scale is equated to 360°, with negative values corresponding in the case of circadian rhythms to the clock hour ($15° \equiv 1$ hour), or in other cases, to the days of the week or months of the year. On the inner scale the synchronizer schedule is indicated, with sleep, for example, shown as a black arc.

Let us now assume that we compare two sets of time series. Here we shall compare time series obtained from the same group of subjects under two different conditions. For this set comparison the time series provided by each subject are analyzed by the least squares method with the fit of a 24-hour period, and the amplitudes and acrophases thus derived are used for a mean group cosinor. By the mean cosinor analysis (14), a mean amplitude and acrophase are obtained and presented along with the joint 95% confidence region of these

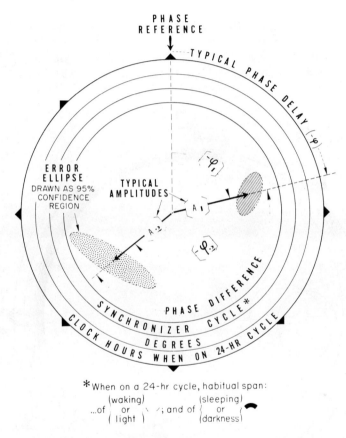

*When on a 24-hr cycle, habitual span:
...of {(waking)/(light)} or 〳 〵; and of {(sleeping)/(darkness)} or 〳

Figure 5   *Internal timing may be assessed as the phase angle difference or, briefly, phase difference between two rhythms with the same frequency. Internal timing can be expressed for a 24-hour-synchronized circadian rhythm in clock hours as well as in degrees, and it is desirable in this case to add information on the routine of waking and sleeping such as the times of turning on the lights in the bedroom on awakening, and turning them off on retiring, when such events describe a given subject's routine. Internal timing in degrees, as phase difference, can be given for any rhythm, whatever its frequency, so long as the rhythms compared have the same frequency.*

parameters. For the second set of series the same procedure is applied and again the results are graphed. Figure 5 then allows a rough comparison of the two sets in terms of mean amplitudes and acrophases. In the abstract Figure 5 a large difference in acrophases is depicted. Both acrophases, φ_1 and φ_2, are indicated as delay in negative degrees (clockwise) starting from the phase reference at the top of the figure.

RESULTS ON CIRCADIAN RHYTHMS

The cosinor summary of temperature time series obtained from the groups of subjects in stages III and IV of the study, without (set *A*) and

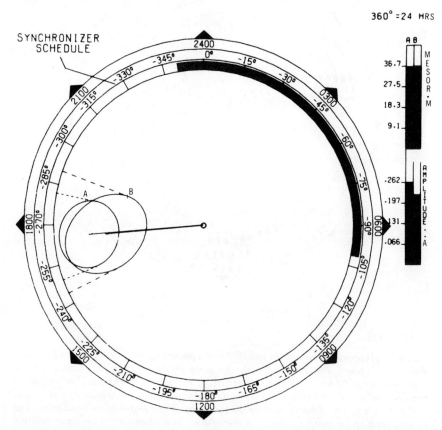

Figure 6   Remarkable similarity in acrophase of circadian rhythm in oral temperature during two 25-day spans, one without baths (A), the other with baths (B) in 10 healthy males.

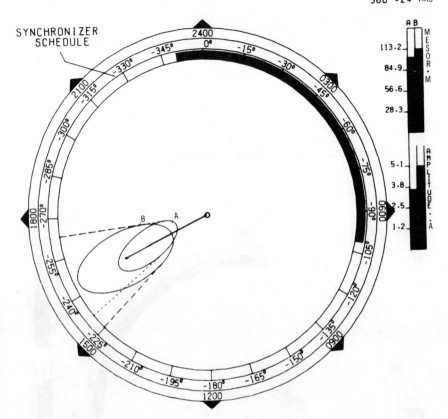

Figure 7   Remarkable agreement in timing of circadian rhythms in grip strength of right hand during spans without baths (A) and with baths (B).

with (set *B*) hydrotherapy, is shown in Figure 6. After data from each subject were analyzed by fitting a single cosine function with a 24-hour-synchronized period, the group characteristics—amplitude and acrophase—were computed by mean cosinor analysis. On the outer rim of the graph clock hours are indicated with midnight at the top (phase reference = 0 hour), and in the middle scale the degrees, computed as 360°/24 for 1 hour, are provided clockwise with negative signs corresponding to clock hours. On the inner scale the black arc shows the synchronizer schedule, in this case the average sleep span. The amplitudes and acrophases are similar for sets *A* and *B*, but set *B* has a larger error ellipse than set *A*. This means that a better estimate of mean rhythm characteristics is obtained from set *A* in terms of standard errors.

At the right of the polar graph, black bars indicate the rhythm-adjusted mean or mesor (*M*) and the amplitude (*A*) for sets *A* and *B* in precise units.

Figures 7–9 show comparable results for grip strength of the right hand and the left hand and for body weight. It can be seen in each summary that there is little difference in the acrophases obtained from the two sets of time series. For grip strength of the right hand the results show that sets *A* and *B*, with similar acrophases, have possibly different amplitudes, also indicated at the right.

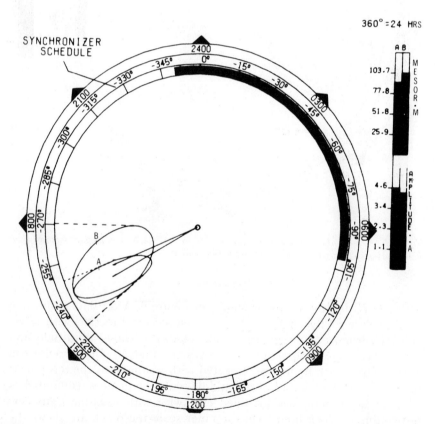

Figure 8   Extent of agreement in circadian rhythm parameters for grip strength of left hand in the absence as well as in the presence of baths.

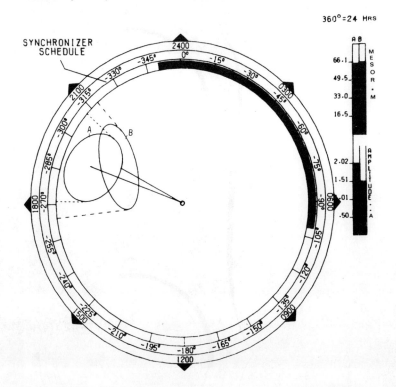

Figure 9 Circadian rhythm in body weight of ten medical students during a span without baths (A) and in the presence of baths (B).

Thus, by the mean cosinor technique, the occurrence of circadian rhythms is validated for the variables displayed in Figures 4–7 and for all other variables investigated during the three stages of study, each covering 25 days (cosinors not shown; 1–3). Actually, as Figure 8 indicates, an estimation of timing without the weighting by the amplitude (done in the cosinor) yielded comparable results for grip strength of either hand or for the sum of grip strength in both hands.

RESULTS FOR RHYTHMS WITH A FREQUENCY LOWER THAN CIRCADIAN

The spectral region from 2 to 25 days was examined by fitting, one at a time, cosine functions in increments of 0.5 days by the linear least

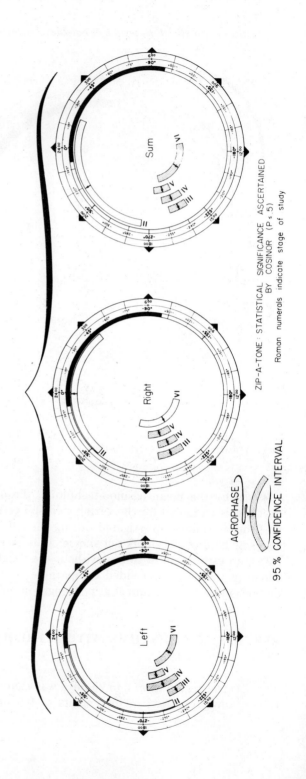

ZIP-A-TONE: STATISTICAL SIGNIFICANCE ASCERTAINED
BY COSINOR (P ≤ 5)

Roman numerals indicate stage of study

ACROPHASE

95 % CONFIDENCE INTERVAL

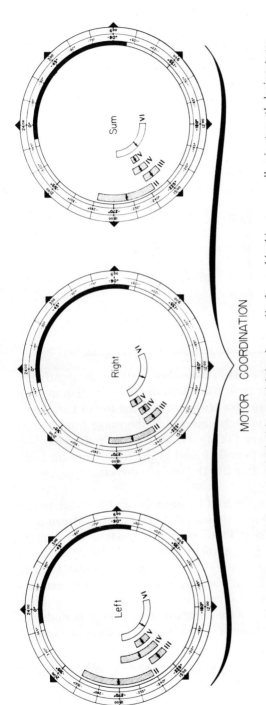

MOTOR COORDINATION

Figure 10 Acrophase estimates may be summarized without weighting for amplitudes, and in this case as well, grip strength during stages III and IV without and with baths, respectively, shows good agreement for data from the 10 healthy medical students, summarized separately by cosinor in Figures 7 and 8. Agreement of such acrophases computed from other stages is further apparent for stage V of 25 days' duration (as were stages III and IV) and for stage VI of 3 days' duration. Stage II, also of 3 days' duration, did not allow a statistically significant rhythm description for either hand or the sum of the two hands.

Comparable data in the bottom half of the figure are given for eye-hand coordination. Again, data from 10 subjects during three spans, each of 25 days' duration, agree extremely well in acrophase, whereas shorter spans yield less reliable and not invariably statistically significant results.

541

Table 1   *Periods Detected by Linear Least Squares Window for Grip Strength of Right and Left Hand*

τ_i	y_i	τ_i	y_i	τ_i	y_i
2.0	0	10.0	2	18.0	0
2.5	0	10.5	1	18.5	0
3.0	0	11.0	0	19.0	0
3.5	0	11.5	3	19.5	0
4.0	0	12.0	3	20.0	0
4.5	2	12.5	3	20.5	0
5.0	0	13.0	7	21.0	1
5.5	2	13.5	1	21.5	0
6.0	2	14.0	1	22.0	0
6.5	1	14.5	1	22.5	1
7.0	2	15.0	1	23.0	0
7.5	1	15.5	0	23.5	0
8.0	3	16.0	0	24.0	0
8.5	2	16.5	0	24.5	0
9.0	3	17.0	1	25.0	0
9.5	1	17.5	1		

squares method (5). Such an analysis described statistically significant periods with p-values smaller than .05. Moreover, distribution of such periods within the examined sample was not the same for all variables.

Thus for grip strength the periods detected in the range from 2 to 25 days clustered around 13 days (Table 1). The same impression was gained from inspection of the periods found in the time series for body weight. For oral temperature, in turn, the periods detected appeared to be randomly distributed.

To check these impressions, the null hypothesis can be postulated as the occurrence of randomly distributed periods; the alternative hypothesis is that the frequency distribution of periods in the domain from 2 to 25 days will show a maximum at 13 days.

Cox and Stuart's test for a trend (11) constitutes a nonparametric instrument for this purpose. It is a modification of the sign test and can be used to test for an upward or downward trend in a data set. Thus this test can be used to study the frequency distribution of periods detected in original physiologic time series by the linear least squares fit of a single cosine function.

For the test data, consisting of the frequency of detected periods, a scale with an increasing order of trial periods seems appropriate. We write: (y_i, τ_i), $i = 1$, $\cdots, n$, where y_i is the frequency of a given period, τ_i, whereas n is the total number of periods analyzed by the linear least squares method in order to detect a given period. In our case τ_i ranges from 2 to 25 days by increments of half days. Thus, for the given data (y_i, τ_i), it is desired to test the randomness of occurrence, along the scale, of trial periods extending from 2 to 25 days against the alternative hypothesis that the frequency of periods tends to be minimal at 2 days, increasing to a maximum at 13 days, and decreasing again toward 25 days. This assumption is made for the purpose of applying a simple, quick, nonparametric sign-test equivalent as a first, yet in itself insufficient, step toward describing a physiologic state of affairs.

In order to test these hypotheses, the given data, such as the pooled candidate periods (12) on grip strength shown in Table 1, are rearranged into pairs as shown in Table 2.

In Table 2 the first half of the results (listed in the first column) is paired with the second half of results (in the second column). Two frequency findings in a given row form a pair; that is, $(7, 0)$, $(1, 4)$, $\cdots$, $(1, 0)$. Let us denote the sequence of pairs as (a_i, b_i) $i = 1, \cdots, 46$ (from 2 to 24.5) and replace each

Table 2   Rearrangement of Table 1 Data as Pairs

τ_i (days)	y_i	τ_i	y_i	Sign
13.0	7	19.0	0	+
13.5	1	7.0	4	−
12.5	3	19.5	0	+
14.0	1	6.5	1	0
12.0	3	20.0	0	+
14.5	0	6.0	1	−
11.5	2	20.5	0	+
15.0	2	5.5	1	+
11.0	1	21.0	1	0
15.5	0	5.0	0	0
10.5	2	21.5	1	+
16.0	0	4.5	0	0
10.0	0	22.0	0	0
16.5	0	4.0	1	−
9.5	1	22.5	0	+
17.0	1	3.5	0	+
9.0	4	23.0	0	+
17.5	1	3.0	0	+
8.5	2	23.5	0	+
18.0	0	2.5	0	+
8.0	4	24.0	0	+
18.5	0	2.0	0	0
7.5	1	24.5	0	+

Table 3   Periods Detected (τ_i) in Certain Body Functions and Cox-Stuart Test of Randomness for Frequency Distribution of Rhythms with a Period Mode Near 13 Days

Variables[a]	$\tau_i{}^b$ Mean $\pm$ SE	N''	N'	p-Value
Body weight	12.30	14	12	0.0065
Grip strength				
Left + right	10.77	17	14	0.0064
Left	11.15	13	11	0.0112
Right	10.56	12	8	0.1938
Oral temperature	10.97	11	5	0.7256

[a] Number of trial periods tested, 47 in each case; see text for definitions of N'' & N'.

[b] See Figures 1 and 2 for the more representative mode.

pair (a_i, b_i) with a "+" if $a_i > b_i$, "–" if $a_i < b_i$, and "0" for ties as shown in the last column.

Define N' as the total number of "+" signs and N'' as the total number of "+" and "–" signs.

For $N'' \leq 20$, we can use a table (3, in Ref. 11) to obtain the critical region for a given significance level. For $N'' > 20$ a method for approximation also is shown (at the end of Table 3 in Ref. 11). In our case, $N' = 14$ and $N'' = 17$. Thus the table shows for $N'' = 17$ that a p-value of .0064 corresponds to values of N' equal to (or greater than) 14. Therefore, H_0 is rejected.

The alternative hypothesis may then be accepted for the case of grip strength data of both hands (Tables 1–3). However, for dynamometry of the right hand, the Cox-Stuart test for trend does not reject the null hypothesis of randomly distributed periods. At first such a result seems surprising. However, for right-handed individuals—and all subjects were right handed—the hand used for a variety of purposes throughout the day may be subject to local fatigue and other perhaps not fully-defined factors. By contrast, the hand not used in everyday activity may reflect more directly spontaneous rhythms, less complicated by factors such as use during the minutes or hours preceding the test.

It is pertinent that a difference between the two hands also was found by the use of Downs' rotational correlation coefficient (13). A statistically significant correlation was found between body temperature and eye-hand coordination of both hands at the 5% level and very near the 5% level for coordination of the right hand, but not between oral temperature and coordination of the left hand.

Table 1 also reveals that for the case of body weight the Cox-Stuart test rejects the null hypothesis with a *p*-value of .0065. By contrast, for oral temperature a corresponding *p*-value is .72, suggesting that temperature does not reflect in the subjects studied any prominent periods in the domain from 2 to 25 days.

Since the linear least squares method can sometimes show periods that represent side-lobes of other components, a more rigorous method, the nonlinear least squares technique, also was applied to these admittedly limited and noisy data. It was anticipated that if the results of a linear least squares analysis were spurious, the analysis with a nonlinear method would not reveal a maximum frequency at 13 days. It is interesting, therefore, that for grip strength of the left hand the periods actually found in the 25-day time series by nonlinear windows were at a length near 13 days, namely at 11.1, 12.2, 12.3, 12.5, 12.7, 13.2, 14.2, and 15.0 days, respectively. For body weight, nonlinear least squares windows revealed periods of 11.5, 12.3, 12.5, 12.6, 12.7, and 12.8 days, respectively, in the data of eight of the subjects, each studied for three 25-day spans (stages III and IV). For grip strength of the right hand, the nonlinear window revealed three periods of 12.7, 12.9 and 14.4 days. It is pertinent further than when four of the oral temperature series were scrutinized by the more rigorous nonlinear window, the single period validated was of 7.1-days length.

The findings on rhythms in grip strength of the left hand are the more interesting since they could represent a gauge of male sex gland activity—a point yet to be proved. Data from Harkness (15) on testosterone levels of men analyzed by our methods also reveal periods averaging about 13 days. However, until such correlations between biochemical variables and behavior are established concomitantly during the same cycle of a given individual, the possibility remains that the agreement of 13-day periods could be coincidental.

In summary, then, circadian rhythms and rhythms with a period of about 2 weeks were detected in time series of presumably healthy men. A systematic scrutiny of candidate rhythms for gauging a possible male sex gland cycle should include monitoring body weight and the grip strength of the left hand.

FUNCTIONAL DOWNS CORRELATIONS AT THE CIRCADIAN FREQUENCY

Relationships among the acrophases of rhythmic physiologic variables can now be explored by the computation of a rotational correlation introduced by Downs at this meeting (13). The test statistic, the so-called Downs coefficient, R, among circadian acrophases has been computed and discussed from a methodologic viewpoint by Downs (13). R values equal to or higher than $+.75$ or lower than $-.75$ are regarded by the original author as significant at or below the 5% level. Results for several variables of the 10 medical students who self-measured temperature and other functions nine times a day for 25 days are particularly pertinent to this discussion.

A correlation between acrophases for oral temperature and dynamometry with the right and/or left hand is not detected, the rotational correlation coefficients being $-.39$ and $-.34$, respectively. This result is reminiscent of differences between oral temperature and grip strength series of the same subjects in the spectral region of frequencies lower than 1 cycle/day. Such findings lead us to qualify the merits of the temperature rhythm's circadian acrophase as a gauge of performance in general. However, it is pertinent that the oral temperature acrophases, so often related to all kinds of performance, correlate with the sum of the coordination of both hands at the 5% level and very near the 5% level (with $R = .74$) with coordination of the right hand. The R for circadian acrophases in oral temperature and coordination of the left hand is .70.

As to other variables for which a circadian rhythm has been earlier demonstrated, a positive correlation is found ($R = .82$) between acrophases for mental state and coordination of the left hand tested by the number of beads strung in one minute. No correlation is detected between acrophases for mental state and the coordination of the right hand (invariably tested before the left hand) ($R = .58$). Acrophases of coordination of the right hand correlate with those for coordination of the left hand ($R = .81$) and also with those for the sum of the coordination values for the two hands ($R = .94$). Mental state acrophases correlate positively with those for vigor, the R being .88. $R = -.80$ for a correlation of acrophases for tempo (2-minute estimation by counting from 1 to 120) with those for diastolic blood pressure; a confirming correlation of $-.80$ is found for acrophases of tempo and a second diastolic blood pressure measurement series. Without claiming that such correlations in themselves provide a physiologic basis for a better understanding of performance, it should be emphasized that the tools are now available to explore physiologic mechanisms that may underlie any relation between changes in body temperature and certain variations in performance occurring with the same frequency and with some defined time relation.

Table 4   *Concomitant Description of a Circadian and an Infradian Rhythm in Grip Strength (lb) of a Healthy Man (EH) by Simultaneous Linear Fit of Two Cosine Curves[a]*

Hand Grip	P	Rhythm (%)	Mesor, M (SE of M)	Period, τ	Amplitude A (SE of A)	Acrophase, ϕ (95% Confidence Arc)
Right	< .005			16 days	2.26 (0.7)	$-84°$ (-50, -117)
		33	130.2 (0.5)			
	< .001			24.2 hours	8.00 (0.7)	$-214°$ (-204, -223)
Left	< .001			13 days	3.65 (0.7)	$-126°$ (-105, -147)
		27	117.9 (0.5)			
	< .001			24.2 hours	6.56 (0.7)	$-227°$ (-125, -239)

[a] Initiated from results of a chronobiologic window with single cosine fitting; followed by tests of the model consisting of the sum of two cosine curves revealing that the model is statistically significant below the 5% level; continued by iteration for optimizing the fit of the sum of two cosines with computation of the probability (P) for rejecting a zero amplitude in each cosine component when this hypothesis of "no rhythm" is true.

On the one hand it is quite clear from the results of this study that a broad parallelism between body temperature and performance more generally, notably as the latter varies around the 24-hour scale, will have to be qualified for the kinds of performance and for the frequency of any rhythms in the variables investigated (16, 17).

SPOTCHECK ON INFRADIAN GRIP STRENGTH RHYTHM

Limited data, self-measured by one of us (E.H.) for only 26 days provide another example of a 13-day period for grip strength of the left hand but not for right hand grip strength or oral temperature. The 13-day period found by the linear least squares fit of separate single cosine functions was validated by the simultaneous fit of a two-component cosine function (Table 4).

REFERENCES

1. R. Günther, F. Halberg, and E. Knapp, *Kurverlaufs und Kurerfolgsbeurteilung, Symposium II* (W. Teichmann, Ed.), Sanitas, Bad Wörishofen, 1968, pp. 106–111.

2. F. Günther, E. Knapp, and F. Halberg, *Z. Ang. Bäder Klimaheilkunde,* **16** (1969), 123.

3. R. Günther, F. Halberg, and E. Haus, *Proceedings of the International Society for the Study of Biological Rhythms, Little Rock, Arkansas, November* 8–10, 1971.

4. F. Halberg, *Physiological Problems in Space Travel* (J. D. Hardy, Ed.), Thomas, Springfield, Ill., 1964, pp. 298–322.

5. F. Halberg, M. Engeli, C. Hamburger, and D. Hillman, *Acta Endocrinol. Suppl.* **103** (1965), 54.

6. J. N. Mills, *Transmission Processes Between Clock and Time Manifestations*, In Press.

7. W. P. Colquhonn, *Aspects of Human Efficiency: Diurnal Rhythm and Loss of Sleep*, English Univ. Press, 1972, pp. 344.

8. N. Kleitman, *Sleep and Wakefulness*, Univ. Chicago Press, Chicago, 1965, 552 pp.

9. Analysis by Forschungsinstitut Gastein, Badgastein, Austria, 1966.

10. F. Halberg, E. A. Johnson, W. Nelson, W. Runge, and R. Sothern, *Physiol. Teacher,* **1** (1972), 1.

11. W. J. Conover, *Practical Nonparametric Statistics*, Wiley, New York, 1971.

12. J. Rummel, J.-K. Lee, and F. Halberg, in *Biorhythms and Human Reproduction* (M. Ferin, F. Halberg, R. M. Richart, and R. L. Vande Wiele, Eds.), Wiley-Interscience, New York, 1973, Chap. 5.

13. Downs, T., in *Biorhythms and Human Reproduction* (M. Ferin, F. Halberg, R. M. Richart, and R. L. Vande Wiele, Eds.), Wiley-Interscience, New York, 1973, Chap. 7.

14. F. Halberg, Y. L. Tong, and E. A. Johnson, *The Cellular Aspects of Biorhythms* (H. von Mayersbach, Ed.), Springer, Berlin, 1967, pp. 20–48.

15. A. Harkness, in *Biorhythms and Human Reproduction* (M. Ferin, F. Halberg, R. M. Richart, and R. L. Vande Wiele, Eds.), Wiley-Interscience, New York, 1973, Chap. 30.

16. A. Fort, J. A. Gabby, R. Jackeet, M. C. Jones, S. M. Jones, and J. N. Mills, *J. Physiol.* **219,** (1972), 17.

17. A. Fort, M. T. Harrison, and J. N. Mills, *J. Physiol.* **231,** (1973), 114.

CHAPTER 36

Menstrual Changes of the Circadian Temperature Rhythm in Women

HUGH W. SIMPSON and ERNA E. HALBERG,

Chronobiology Laboratories, University of Minnesota, Minneapolis, Minnesota

INTRODUCTION

Several analogies with physics come to mind in the case of a woman's temperature series, wherein both the circadian and menstrual components can be of a similar order of amplitude. One might expect as a minimum a superposition phenomenon. Moreover, the mechanisms underlying two or more biological rhythms with different frequencies in the same system might interact. Frequency, phase, amplitude or mesor modulation may result.

This communication is a beginning investigation into the possibilities of endocrine interactions revealed by modulation or superposition (1). It is a study of data collected for other research projects and in this respect is not ideal; nevertheless the results provide a rather more dynamic picture of the time structure in human temperatures than has previously been published.

MATERIALS AND METHODS

Five hundred and fifty body temperature measurements were made over a 7-month span at various times of day by a woman, 39 years of age. Each reading was obtained after voiding urine directly onto the bulb of an Ovulindex mercury thermometer designed for "safe period" family planning practice. This method has an advantage over oral methods in that it is quick—the time taken to void—and difficulties arising from recent meals or hot drinks are minimized.

The data were originally collected as part of a plan to study rhythms during an attempt to ski to the North Pole and later while living in a static base camp in the Northwest Territories. The subject had previously participated in several polar expeditions which did not overtly affect her menstrual cycle. Moreover, inspection of the raw data plot reveals that measurements of the first two cycles obtained while living in Minneapolis, Minnesota, and then in the Northwest Territories of Canada are grossly similar to those from cycles during the Arctic trek.

The data were transferred to I.B.M. punch cards and subjected to linear least squares spectral analysis (2). The trial periods of the cosine model fitted extended from 936 to 18 hours, with 12-hour decrements between successive trials in the range of longer periods and 0.1-hour decrements in that of shorter periods. In this way, parameters of both the circadian and the menstrual rhythms were obtained.

RESULTS

The results of these analyses, and a chronogram of the raw data with the "best-fitting" sinusoid superimposed, are seen in Figure 1. The menstrual temperature rhythm stands out more clearly than the circadian one. It is interesting that the mesors are similar to the ones obtained by the same subject in the temperate zone. An unexpected finding from gross inspection is the relatively sinusoidal form of the menstrual rhythm rather than the step function representation seen in some standard texts.

In order to be able to resolve any mesor, amplitude, or phase modulation, the eight menstrual cycles seen in the chronograms of Figure 1 were "folded," that is, superimposed upon each other, in Figure 2. The period estimate derived from the least squares spectral analysis, 26 days, was used to form a single idealized cycle. More specifically, Figure 2 was obtained by pooling for each consecutive 3-day

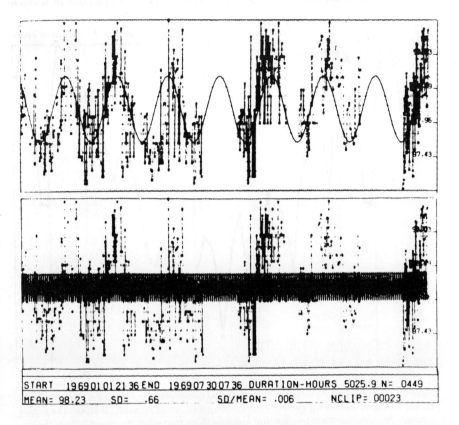

START	19 69 01 01 21 36 END	19 69 07 30 07 36	DURATION-HOURS 5025.9 N= 0449
MEAN= 98.23	SD= .66	SD/MEAN= .006	NCLIP= 00023

MENSTRUAL RHYTHM   CIRCADIAN RHYTHM

τ = 636·h = 26·5·d τ=24·h

P (SE/C) =	.01 (	.07)		.01 (	.21)	
MESOR(.95CI) =	98.25 (	98.22	98.28)	98.19 (	98.13	98.26)
AMPLITUDE(.95CI) =	.50 (	.46	.53)	.21 (	.12	.30)
ACROPHASE(.95CA) =	-332	-323	-341)	-260(	-236	-284)

Figure 1 Top: Least squares spectral analysis. Computer microfilm plots of raw data overprinted with "best-fitting" cosine waves: on the top the menstrual component = 636 hours = 26.5 days; in the middle the circadian component = 24 hours = 1 day (components found by analyses in spectral range from 936 to 18 hours).

Bottom: Summary of parameters for 2 "best-fitting" components in spectral range examined. Note that data points read N = 449 whereas the text indicates N = 550 (also Figure 2). The difference is due to the fact that the plot program is only able to accept up to N = 450; hence some data points had to be randomly removed.

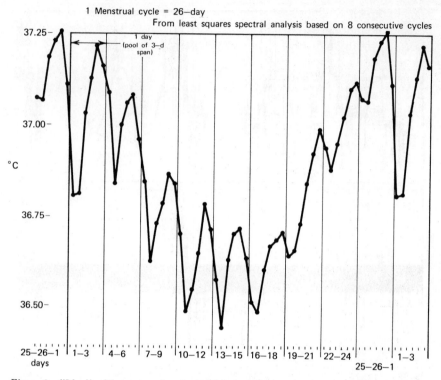

Figure 2   "Idealized" menstrual cycle (26 days) constructed from original data shown in Figure 1. Eight "menstrual cycles" are superimposed, using for such "folding" the period estimate from the least squares spectral analysis. Note that when the mesor is high, the circadian amplitude tends to be high. Also note that during the proliferative stage the peaks on the chronogram are invariably at 20^{00} whereas in the secretory stage they average out at 00^{00}.

span of each 26-day period observations made during similar 4-hour spans (00^{00} to 04^{00}; 04^{00} to 08^{00} and so on). Thereby it was hoped to remove much random interference that was not consistent during consecutive menstrual cycles.

DISCUSSION

The idealized "menstrual" temperature cycle shown in Figure 2 represents data from only one woman. Much further work remains to be done before deciding whether any one of its features is reasonably representative of an average "normal" cycle.

In Figure 2, high resolution of the circadian and menstrual components is afforded by the large number of observations and the pooling method. Thus we gain a more dynamic perspective of several—circadian and menstrual—aspects of a temperature-time structure than is conveyed by the approaches in standard texts (3–5). The slow drop in temperature mesor after menstruation is unexplained; at this time blood progesterone concentration is usually rather low (3) but it was not determined in this case. Blood progesterone is usually held responsible as the thermogenic agent in the menstrual cycle.

Interesting differences in the timing of the circadian rhythm seem to occur in different stages of the menstrual cycle. During the presumed "proliferative" stage, the circadian rise of temperature extends from 08^{00} to 20^{00}, whereas in the "secretory" stage the rise begins at 04^{00} or 08^{00} and extends to 00^{00}. If one wanted to predict—in this case from two temperature readings 4 hours apart—the current stage of the cycle, then readings taken at 20^{00} and 00^{00} would provide the answer. On the average, a fall between these times would indicate a "proliferative" stage and a rise "secretory," presumptive evidence that ovulation had occured. Alternatively a prediction might be by observing the relative temperatures at 04^{00} and 08^{00} on the same day; again, on the average, a fall of say 0.2 to 0.25°C would indicate the "secretory" phase whereas a lesser fall or a rise would suggest the "proliferative" phase.

A consistent movement of the daily peaks from 20^{00} (first half of chronogram) to 00^{00} (second half) suggests a 4-hour phase modulation. To investigate this possibility further, a 24-hour cosine wave was fitted by least squares to the idealized data, in a chronobiologic serial section (6) with 2-day intervals and 12-hour increments. While only a slight phase change was found, this same computer-prepared chronobiologic serial section revealed a mesor modulation. In this case the circadian mesor was high near menstruation and low near midcycle.

Much more extensive analyses of the concomitantly evaluated circadian and menstrual components of human temperature variation are indeed warranted in the light of the result here presented. It should be realized that variance spectra have earlier been used (7) in a noisy situation to resolve menstrual and circadian components. By comparison to such variance spectra, the linear least squares approach (Figure 1) provides an improved resolution. Such analyses are further greatly enhanced by the combination of linear and nonlinear least squares fitting methods presented at this meeting (8).

Another series of oral temperatures collected by one of us (E.H.) serves to introduce the combined linear-nonlinear chronobiologic window presented in Table 1. These oral temperatures are first analyzed by the linear least squares method to obtain candidate parameter values that serve as input for another series of computations yielding the final estimates. The values in Figure 1 are such candidate

Table 1 Circadian and Menstrual Components Detected Concomitantly by the Nonlinear Least Square Method in Oral Temperature of a Healthy Woman (E.H.), 48 Years of Age (December 22, 1968, to March 21, 1969)[a]

Least Squares Analysis[b]	Mesor (M) (°F)	Period, $\tau_i \pm$ SE hours (days)	F° Amplitudes (A_i) ($\pm$SE)	Acrophases, ϕ_i; 95% Confidence Arc (degrees)
Linear	98.1_7	$\tau_1 = 24$ $\tau_2 = 576(24)$	$A_1 = 0.22$ $A_2 = 0.39$	$\phi_1 = -241; -219, -263$ $\phi_2 = -58; -46, -70$
Nonlinear	98.1_4	$\tau_1 = 23.9_4 \pm 0.028$ $\tau_2 = 570.1_4\ (23.7_6) \pm 9.8_8\ (\pm 0.4_1)$	$A_1 = 0.22 \pm 0.046$ $A_2 = 0.38 \pm 0.039$	$\phi_1 = -277; -235, -319$ $\phi_2 = -66; -41, -91$

[a] Lower decimals of dubious significance.

[b] Set of candidate values from linear least squares method serves as input for obtaining final estimates from nonlinear squares method.

parameter values. The results in the bottom half of Table 1 indicate that the nonlinear approach actually validates the results obtained with respect to both period and amplitude by the linear analysis (top half). Particularly relevant to data collection complementing methods of analysis is the development of a compact thermograph by K. Lange in the Chronobiology Laboratories at the University of Minnesota and of equivalent instrumentation (see Figure 2 in the Glossary).

Such developments may renew clearly waning interest in thermometry by gynecologists. In a recent clinically orientated review on secondary amenorrhoea (9), body temperature investigations were not considered. Only lack of familiarity with the availability and the resolving power of methods and the noisiness of conventionally obtained data when they are inspected in time plots can justify why some gynaecologists apparently no longer use the temperature rhythm as a potential measure of menstrual abnormalities.

CONCLUSION

A woman, 39 years of age, recorded 550 voided urine temperatures at different times of day throughout eight consecutive menstrual cycles. A linear least squares spectral analysis revealed statistical significant rhythms with periods of 26 days and 24 hours. Using a 26-day menstrual periodicity, the cycles were superimposed to construct an idealized cycle in order to reveal any menstrual modulation of the circadian component. Some differences were found in the mesor, timing, and amplitude of the circadian rhythm during the first half of the menstrual cycle as compared to the second half.

The concomitant resolution of both circadian and menstrual components in body temperature variation could be further documented by a combined linear-nonlinear analysis of another temperature series from another presumably healthy woman. The analytical tools are now available for a more rigorous scrutiny of body temperature variations *inter alia* in relation to reproductive functions in health and disease.

ACKNOWLEDGMENTS

Financial assistance from the medical Research Council, technical help from Mrs. Hazel McColl, Glasgow, Scotland, and support from U.S.P.H.S. (5-K6-6M-13981 and 1R01-CA-14445-01) and NASA rendered these analyses feasible.

REFERENCES

1. F. Halberg, *Minn. Acad. Sci.*, **28** (1960), 53.
2. F. Halberg, M. Engeli, C. Hamburger and D. Hillman, *Acta Endocr., Suppl.*, **103** (1965), 54 pp.
3. R. W. Johnstone and R. J. Kellar, *A Text Book of Midwifery*, Black, London, 1952, p. 44.
4. R. V. Short, *Brit. J. Hosp. Med.*, **7** (1972), 552.
5. G. H. Bell, J. N. Davidson, and H. Scarborough, *Textbook of Physiology and Biochemistry*, Livingstone, Edinburgh, 1956, pp. 966–967.
6. F. Halberg, and G. S. Katinas, "Chronobiologic Glossary" *International Journal of Chronobiology* **1**: 31–63, 1973.
7. F. Halberg, H. Panofsky, and H. Mantis, *Ann. N. Y. Acad. Sci.*, **117** (1964), 254.
8. J. Rummel, J.-K. Lee, and F. Halberg, in *Biorhythms and Human Reproduction* (M. Ferin, F. Halberg, R. M. Richart, and R. L. Vande Wiele, Eds.), Wiley-Interscience, New York, 1973, Chap. 5.
9. J. Newton, *Brit. J. Hosp. Med.*, **7** (1972), 564.

CHAPTER 37

Circadian Phase-Shifting With and Without Geographic Displacement

HOWARD LEVINE

Department of Medicine, New Britain General Hospital, New Britain, Connecticut; University of Connecticut Health Center, Farmington, Connecticut;

FRANZ HALBERG and ROBERT B. SOTHERN

Chronobiology Laboratory, Department of Laboratory Medicine and Pathology, University of Minnesota, Minneapolis, Minnesota; and

FREDERIC C. BARTTER, WALTER J. MEYER, and CATHERINE DELEA

Endocrinology Branch, National Heart and Lung Institute, Bethesda, Maryland

The phase-shifting of circadian rhythms constitutes a tool for understanding the mechanisms underlying temporal organization along the 24-hour scale. It also promises to shed light upon interactions between circadian rhythms, reproductive cycles, and other variations with yet other frequencies. Unfortunately, only limited work has been done with phase-shifting on the behavior of reproductive cycles and on the synchronization of the latter—topics assigned to others at this symposium. Here we shall focus upon circadian phase-shifting per se. The lessons already learned on this extensively investigated problem may eventually constitute a basis for the study of reproductive cycles, with a view of their interactions with other components in the spectrum of rhythms*.

COMPARATIVE PHYSIOLOGIC BACKGROUND

For the so-called sleep movement of *Mimosa pudica*—the opening and closing of leaves—Augustin Pyramus de Candolle (1) demonstrated not only a persisting circadian rhythm when this "sensitive" plant was kept under a bank of lamps shining continuously but also that (*1*) the continuously illuminated plants had a rhythm with period shorter than 24 hours, and that (2) plants exposed to darkness by day and light by night gradually, though not immediately, adjusted to the new schedule. Extending such observations to *Albizzia julibrissin*, the silk tree, Koukkari et al. (2) have recently demonstrated again not only the rhythm's persistence under conditions of continuous light and its amenability to phase-shifting, but also a polarity to the rhythm's phase-shift (Figure 1). An advance in lighting regimen simulating the adjustment required after the crossing of several time zones in a direction from west to east is faster than the adjustment after a schedule change simulating that required after a flight from east to west. Figures 2 and 3 show the opposite polarity in the phase-shift behavior of the flour beetle, *Tribolium confusum*, and the Minnesota Sprague-Dawley inbred rat, while Figure 4 reveals the phenomenon of polarity in phase-shifting for a group of apparently healthy men. All of the figures as a whole and their summary in Table 1 reveal that the circadian acrophases adjust only gradually to abrupt changes in the synchronizer—the lighting regimen for most forms of life, including socioecologic factors for the case of man. The number of transient cycles varies with the species studied, as is also apparent from the figures and Table 1, in keeping with an earlier postulation (3, 4). Moreover, Figure 5 shows that the speed and extent of the adjustment vary to some extent from function to function in man (10, 12).

* Supported by U.S. Public Health Service (5-K6-Gm-13, 981 and 1R01-CA-14445-01) NASA (NAS 2-2738) and NGR-24-005-006) Auxiliary New Britian General Hospital and Connecticut Regional Medical Program.

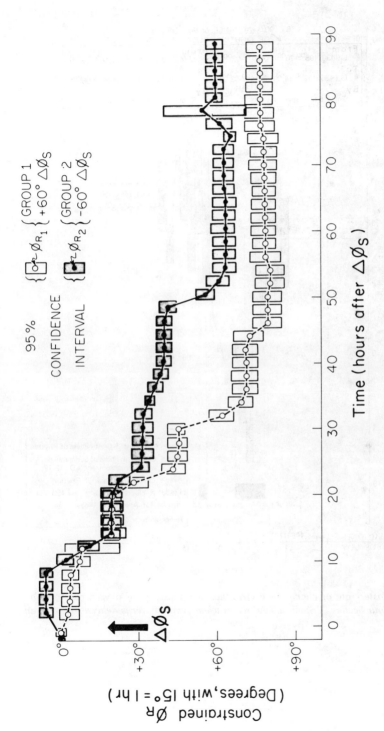

Figure 1 Faster rate of change in circadian acrosphase, ϕ_R, for pinnule movement of the silk tree (Albizzia julibrissin) when the lighting regimen—the synchronizer S—is advanced [+60° (4 hours) $\Delta\phi_S$] rather than delayed. Two groups each of three plants, four pinnae plant. Mean angle or three pinnule pairs from each pinnae.

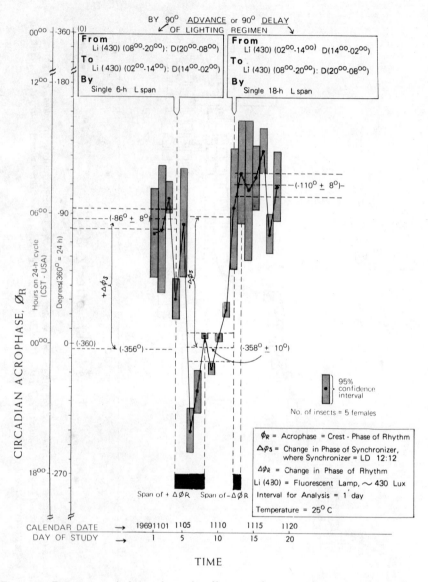

Figure 2   *Faster rate of change in a circadian acrophase, φ, of oxygen consumption rhythm of flour beetles* (Tribolium confusum) *when synchronizer is delayed rather than advanced.*

It is tempting to consider the question of whether the length of the period "self-selected" by human beings in isolation may determine the polarity of the circadian system in phase-shifting, insofar as a longer than 24-hour period might facilitate faster delays than advances, whereas a shorter than 24-hour period would facilitate faster advances than delays. However, the comparative *physiologic* problem posed in Table 1 is complicated by psychologic and sociologic aspects, all of them involved in synchronization. More complete answers are likely to be forthcoming from (*1*) multivariable studies on the same organism, that allow insight into the natural coupling of rhythms within the system; from (*2*) studies of interactions in this connection among various external factors that all may influence, though to a different degree, the synchronization and phase-shifting of different rhythmic body functions; and from (*3*) the further development of techniques as well as models of analysis. Moreover, phase-shift studies on several variables within the given organism gain further in importance when a change in schedule requires multiple repeated shifts, as is the case on a 21-hour day, simulating the need for a 3-hour advance each day rather than requiring a single adjustment to a routine changed but once. Under the conditions of a 21-hour day, it can indeed be demonstrated that different body variables synchronize to a different extent to daily 3-hour phase-shifts. A transient state of desynchronization among rhythms is the consequence of a different extent of adjustment to odd routines or to daily phase-shifts. Thus the question arises whether such single or repeated phase-shifts of human rhythm may be undesirable in that they entail acute performance loss and, if long continued, a chronic deficit—resulting in a shortening of life. Figures 6 and 7 demonstrate a seemingly different kind of acute relative and absolute performance loss in the grip strength and eye-hand skill of a healthy man, following single intercontinental flights. For grip strength we suspect a relative increment, for eye-hand coordination an absolute loss (see later).

AUTORHYTHMOMETRY IN GENERAL

In the study of human rhythms a considerable data base and methods of analyses, illustrated in Figures 6 and 7 (13–16), have been derived from autorhythmometry which involves the self- or automatic measurement and recording of physiological variables as a function of time for analyses of statistical characteristics including rhythms (13,

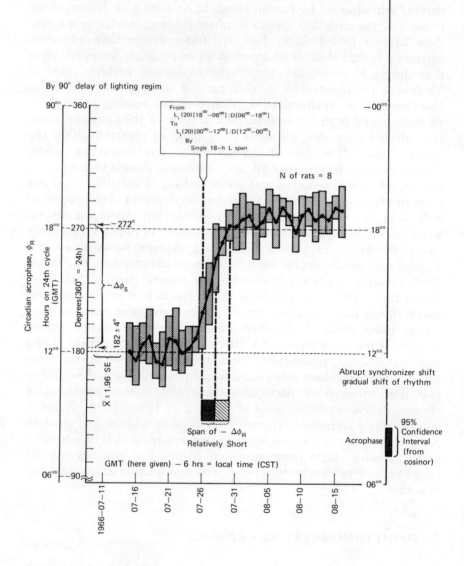

By 90° delay of lighting regim

From
$L_i (20) [18^{00}-06^{00}] : D[06^{00}-18^{00}]$
To
$L_i (20) [00^{00}-12^{00}] : D[12^{00}-00^{00}]$
By
Single 18—h L span

N of rats = 8

Circadian acrophase, ϕ_R

Hours on 24th cycle (GMT)

Degrees(360° = 24h)

$-\Delta\phi_S$

$\bar{X} \pm 1.96$ SE

$182 \pm 4°$

Span of $-\Delta\phi_R$
Relatively Short

GMT (here given) $-$ 6 hrs = local time (CST)

Abrupt synchronizer shift
gradual shift of rhythm

Acrophase

95%
Confidence
Interval
(from
cosinor)

1966—07—11   07—16   07—21   07—26   07—31   08—05   08—10   08—15

By 90° advance of lighting regimen

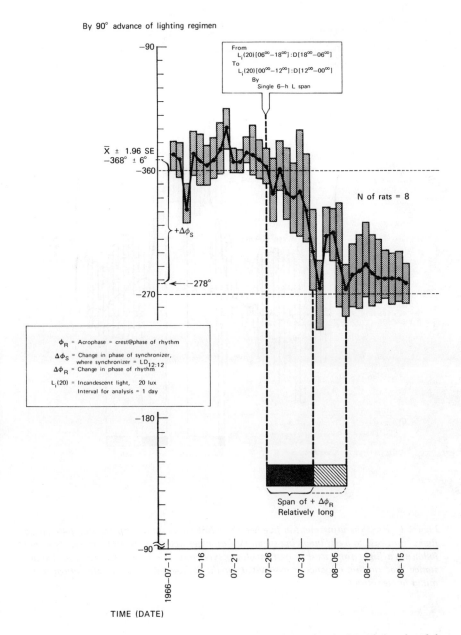

Figure 3 Rhythm adjustment in intraperitoneal temperature in inbred female adult Minnesota Sprague-Dawley rats is faster following a delay than an advance of synchronizer.

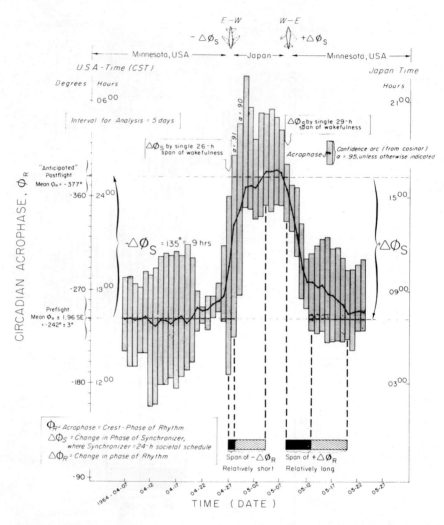

Figure 4   Rhythm adjustment in five healthy adult males (oral temperature), following a flight from east to west, involving social synchronizer delay, seems to be faster than that following a flight from west to east involving synchronizer advance, despite the circumstance that rhythm advance is associated for the individuals studied with return to familiar home setting.

564

Table 1 *Comparative Physiologic Studies on the Phase-Shifting of 24-Hour-Synchronized Circadian Rhythms Reveal Structured Adjustment in the Form of Differences in Rates of Rhythm Advance and Delay*

Organism	Variable	Extent of Shift in Synchronizing Lighting Regime		Hours to Adjust to Regimen			Reference
		Hours (on a 24-hour cycle)	Degrees	Advance	Delay	Difference	
				Faster advance			
Albizzia julibrissin (silk tree)	Pinnule angle	4[a]	60	~24	~52	28	Koukkari et al. (2)
Fringilla coelebs, (chaffinch)	Jumping activity	6	90	~60	~120	60	Aschoff and Wever (5)
				Faster delay			
Tribolium confusum, (flour beetle)	O_2 consumption	6	90	>48	>24	24	Chiba et al., in press (6)
Sprague-Dawley rat	Intraperitoneal temperature	6	90	~312	~120	192	Halberg (7, 8)
Monkey (several species)	Auxillary temperature	6	90	>120	~72	48	Halberg (9)
Man	Oral temperature[b]	6	90	~312	~192	120	Halberg (8, 10)

[a] 4 hours = ½ the dark span in *Albizzia*. 6 hours = ½ the dark span in rat, monkey, and chaffinch.
[b] In social setting. Studies by Aschoff (11) on men isolated in a bunker reveal faster advance than delay.

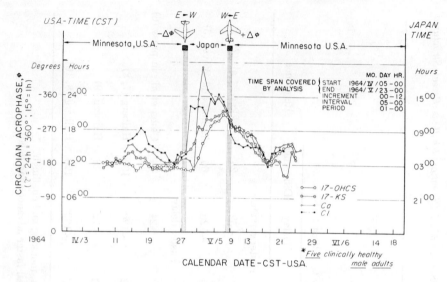

Figure 5 Difference in rate of phase-shift as a function of synchronizer delay (east to west) or advance (west to east) an internal desynchronization during delay in five healthy male adults (urinary excretion rates).

14). The merits of this approach, applicable to the study of reproductive cycles as well, will here be discussed before turning back to the main topic of human phase-shifting.

Autorhythmometry offers at low cost an increased data base, descriptive of the subject in his habitual environment. It further provides both individualized ranges of usual values and estimates of rhythm characteristics not available with single-timepoint sampling. The increased store of individual data will, in the future, aid in defining more precisely individual ranges of normalcy for many physiological variables including those of blood elements being screened by "multiphasic" techniques (13, 14).

The subject thus moves from a passive to an active role and derives potential educational and therapeutic value in the feedback of information. The risk of creating anxiety is small when the subject is properly informed and when the overly reactive individual is either "delabeled" (17) or does not participate in such self-measurement. The concept of timing for diagnosis and treatment is as fundamental in clinical medicine as is the need to introduce a challenge or load for the diabetic who may have a normal fasting blood sugar or the patient with coronary artery disease who has a normal electrocardiogram. With

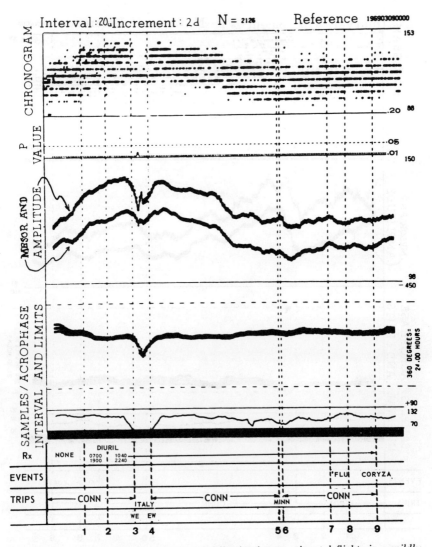

Figure 6 Grip strength in the right hand following intercontinental flights in a mildly hypertensive subject. Subject was off medication for 6 weeks prior to and for the first 6 weeks of the study. Event lines are as follows: (1) chlorothiazide, 0.5 g 07⁰⁰ and 19⁰⁰; (2) chlorothiazide 0.5 G 10⁴⁰ and 22⁴⁰; (3) west to east flight, New York to Italy; (4) east to west flight, Portugal to New York; (5, 6) flights from Connecticut to Minnesota and return; (7, 8) influenza; (9) coryza.

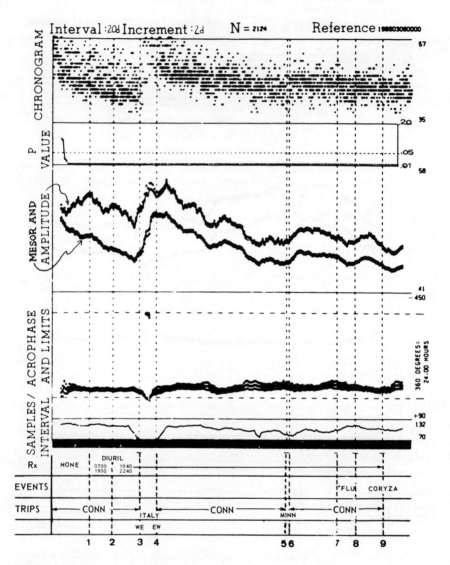

Figure 7   Eye-hand skill (see Figure 6).

regard to susceptibility rhythms as well, Reinberg, Haus, and Halberg have emphasized the need for giving drugs at the time of maximum effectiveness and minimal toxicity (18–20).

SELF-STUDY OF CIRCADIAN PHASE-SHIFTING

The subject of the present study, a 55-year-old physician with mesorhypertension due to primary aldosteronism, performed self-measurement of blood pressure, among other physiological variables including performance tests and estimates of mood and vigor, over a 3 & 1/2-year span repeatedly about every 3 hours during wakefulness from 07^{00} to 23^{00}—an average of six measurements a day. The technique is reported in detail elsewhere (13, 14).

ROUTINE CHANGE WITHOUT DISPLACEMENT ACROSS TIME ZONES

A phase shift of routine by 12 hours took place during a 1-month study at the National Institute of Health (15, 21). The study was divided into three parts: a span of high sodium intake of 240 mEq/day for 10 days and a span of low sodium intake of 9 mEq/day initiated by giving 2 ml of a mercurial diuretic, meralluride, intramuscularly. Urine was collected in 4-hour spans for the entire 30 days and analyzed for sodium, potassium, chloride, calcium, phosphorus, magnesium, 17-hydroxycorticosteroids, and aldosterone (21). On 7 separate days venous blood was collected every 4 hours and analyzed for similar constituents and for plasma renin activity. Stools were collected daily and 4-day spans were analyzed for sodium, potassium, calcium, phosphorus and magnesium. Self-measurements of blood pressure and other physiological as well as physical and mental performance variables were made nine times daily. Saliva was collected five to 11 times daily. Blood pressure was also recorded automatically during the 30-day study every 10 minutes continuously during the day and night during each of four 24- to 48-hour spans totaling 7 days. During the entire study the patient ate a constant metabolic diet (to which sodium chloride tablets were added at mealtime to achieve the high sodium intake), stayed within an air-conditioned environment, and had a definite set routine which he followed each day and maintained after the 12-hour phase-shift.

The strict low-sodium diet lowered blood pressure to values significantly lower than those of high-sodium intake and comparable to those obtained when the subject was on chlorothiazide (13). Both the systolic and diastolic blood pressures continued to fall with the phase-shift. However, the patient was still on an intake low in sodium. Whether the phase-shift contributed to this striking phenomenon and, if so, to what extent remains a problem for further study. There is need for automatic data collection methods that allow frequent sampling and for numerical techniques that resolve the characteristics of phase-shifting. The acrophase for urinary volume, magnesium, calcium, and phosphorus seemed to advance whereas the acrophase for potassium, 17-hydroxycorticosteroids, chlorides and blood pressure apparently retarded.

Plasma renin activity was not detected after the phase shift in the supine position but detectable activity was found in the upright position, a finding suggesting that the secretion of renin quickly adjusted to the phase shift. The blood pressure rhythm—gauged by the peak of a 24-hour cosine function approximating 5-day data intervals displaced in 1-day increments—was roughly in antiphase with the unstimulated salivary sodium potassium ratio or in phase if one takes the inverse of the peak sodium-potassium ratio as indirect index of aldosterone activity. The time relations of rhythms, under different conditions including a phase shift, serve in the differential diagnosis of hypertension (21).

PHASE SHIFTING WITH ACTUAL GEOGRAPHIC DISPLACEMENT

The effect of two round-trip intercontinental flights was also studied by autorhythmometry with some of the results shown in Figures 6 and 7 (13). Unfortunately, endocrine studies were not done on those occasions involving a displacement of 6 hours (rather than 12 hours), with the eastward flight from New York City to Western Europe and the westward flight from Western Europe to New York City.

Future work involving comparable shifts with and without geographic displacement should allow insight into the psychosocial aspects of phase-shifting. To turn back to available methods of analysis, the chronobiologic serial sections in Figures 6 and 7 were prepared by fitting 24-hour cosine functions to consecutive 20-day intervals displaced   by 1-day increments. The 20-day interval corresponds roughly to the total duration of the subject's stay in Europe during the first intercontinental flight (13). In the chronobiologic serial sections one sees after the west to east transmeridian flight a decrement in the circadian rhythm-adjusted mesor of grip strength and eye-hand coordination. The right hand grip strength shows that on the average the values abroad are below those obtained in Connecticut before and after the flight. A gradual adjustment of this rhythm to the 6-hour difference in timing required by social schedules in Italy is apparent in the acrophases, a result confounded by the diluting effect of the long interval analyzed. One can see, however, the relative disadvantage in

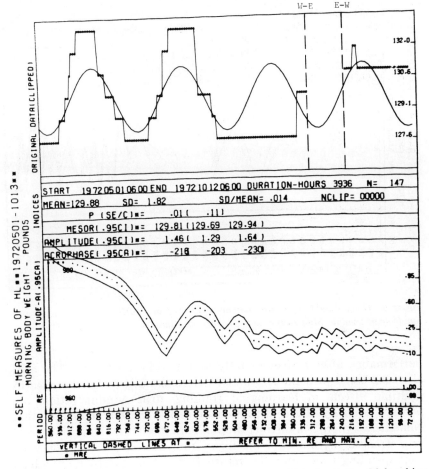

Figure 8   Daily morning body weights analyzed by chronobiologic window. Male subject with primary aldosteronism.

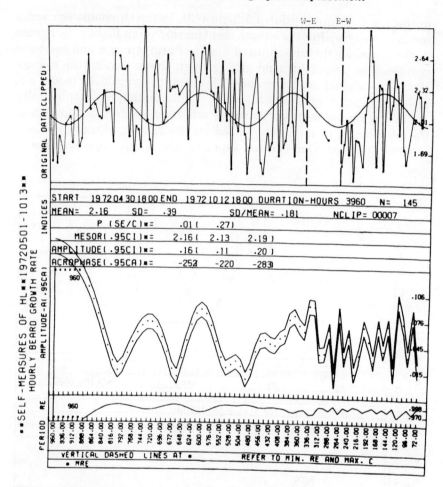

Figure 9 Daily weights of beard shavings expressed as hourly rate. Analyzed by chronobiologic window. Male subject with primary aldosteronism.

performance after arrival in Italy and a while thereafter. The overall rising trend in hand grip mesor before the flight is confounded by multiple effects of learning or training, drug administration, and perhaps even by the contribution of a possible circannual rhythm. Nonetheless there is a distinct drop in both mesor and amplitude of the grip strength rhythm after the flight from west to east but not at the corresponding time after the return flight to Connecticut. Eye-hand skill shows its lowest values at 02^{00} EST (based on data collected only during waking hours). Decrement in coordination is suggested by the increase in number of seconds necessary to complete the eye-hand test following west to east flight, as reflected by the change in mesor. By contrast there is little change immediately after the east to west return flight.

Finally, Figures 8 and 9 introduce limited data on body weight and beard growth deviations with longer than circadian periods. In this case the period lengths are the same. Spans of missing data correspond to the time of an intercontinental flight.

The body weight changes, so regular for a first set of two cycles, obviously are disturbed prior to flight. Indeed such low frequency variations and their possible interaction with other components of the spectrum of rhythms constitute an important problem for future study.

SUMMARY

Changes in the sleep/wakefulness routine and consequent phase-shifting of physiologic functions at different rates have provided an important means for studying circadian rhythms in particular. However, phase-shifting promises to become a biologic tool of even broader scope and may well serve in the future for scrutinizing interactions between circadian rhythms and reproductive cycles, among others.

REFERENCES

1. A. P. de Candolle, *Physiologie Vegetale,* Bechet Jeune, Paris, 1832, p. 859.

2. W. Koukkari, F. Halberg and S. A. Gordon, *Plant Physiol.,* **51,** (1973) 1084–1088.

3. F. Halberg, *Z. Vitamin- Hormon- Fermentforsch.,* **10** (1959), 225.

4. F. Halberg, E. Halberg, C. P. Barnum, and J. J. Bittner, in *Photoperiodism and Related Phenomena in Plants and Animals* (R. B. Withrow, Ed.), AAAS, Washington, D.C., 1959, p. 803.

5. J. Aschoff and R. Wever. *Z. Vergleich. Physiol.,* **46** (1963), 321.

6. Y. Chiba, L. K. Cutkomp, and F. Halberg, in *Chronobiology, Proceedings of the International Society for the Study of Biological Rhythms, Little Rock, Arkansas* (L. E. Scheving, F. Halberg, and J. E. Pauly, Eds.), Igaku Shoin, Tokyo, 1973, pp. 602–606.

7. F. Halberg, W. Nelson, W. J. Runge, O. H. Schmitt, G. C. Pitts, J. Tremor and O. E. Reynolds, *Space Life Sci.,* **2** (1971), 437.

8. F. Halberg, *Ann. Rev. Physiol.* **31** (1969), 675.

9. F. Halberg, in *Circadian Rhythms in Nonhuman Primates* (Bibl. primat. No. 9), (F. H. Rohles, Ed.), Karger, Basel, 1969, p. 106.

10. F. Halberg, E. Halberg, and N. Montalbetti, *Quad. Med. Quant. Sperimentazione Clin. Controllata,* **7** (1969), 5.

11. J. Aschoff, *Aerospace Med.,* **40** (1969), 844.

12. F. Halberg, A. Reinberg, E. Haus, J. Ghata, and M. Siffre *Nat. Speleolog. Soc.,* **32** (1970), 89.

13. H. Levine and F. Halberg, *Circadian Rhythms of the Circulatory System,* SAM-TR-72-3, April 1972.

14. F. Halberg, E. A. Johnson, W. Nelson, W. Runge, and R. Sothern, *Physiol. Teacher,* **1** (1972), 1.

15. F. C. Bartter, W. Meyer, H. Levine, and C. S. Delea, "Circadian Aspects of Hormone and Electrolyte Metabolism in Hypertension," in *Hypertension 72,* Springer, Berlin, Heidelberg, New York, 1972, pp. 220–227.

16. F. Halberg, E. Haus, A. Ahlgren, E. Halberg, H. Strobel, A. Angellar, J. F. W. Kühl, R. Lucas, E. Gedgaudas, and J. Leong, in *Chronobiology, Proceedings of the International Society for the Study of Biological Rhythms, Little Rock, Arkansas* (L. E. Scheving, F. Halberg and J. E. Pauly, Eds.), Igaku Shoin, Tokyo, 1973, pp. 372–378.

17. F. Reeker, in *Chronobiology, Proceedings of the International Society for the Study of Biological Rhythms, Little Rock, Arkansas* (L. E. Scheving, F. Halberg, and J. E. Pauly, Eds.), Igaku Shoin, Tokyo, 1973, pp. 394–398.

18. A. Reinberg and F. Halberg, *Ann. Rev. Pharmacol.,* **2** (1971), 455.

19. E. Haus, F. Halberg, L. E. Scheving, J. E. Pauly, S. Cardoso, J. F. W. Kühl, R. B. Sothern, R. N. Shiotsuka and D. S. Hwang, *Science,* **177** (1972), 80.

20. F. Halberg, E. Haus, S. S. Cardoso, L. E. Scheving, J. Kühl, R. Shiotsuka, G. Rosene, J. E. Paul, W. Runge, and J. Spalding, J. K. Lee and R. A. Good, *Experientia,* **29.** (1973), 909–934.

21. W. F. Meyer, C. S. Delea, H. Levine, F. Halberg, and F. C. Bartter, in *Chronobiology, Proceedings of the International Society for the Study of Biological Rhythms, Little Rock, Arkansas* (L. E. Scheving, F. Halberg, and J. E. Pauly, Eds.), Igaku Shoin, Tokyo, 1973, pp. 100–107.

CHAPTER 38

Rhythms During Hypokinesis

C. M. WINGET, JOAN VERNIKOS-DANELLIS, C. W.
DEROSHIA, S. E. CRONIN, C. LEACH, and P. C.
RAMBAUT

*Biomedical Research Divisions National Aeronautics and
Space Administration Ames Research Center, Moffett
Field, California, and Manned Spacecraft Center Houston,
Texas*

ABSTRACT

Asynchrony of the body temperature (BT) and heart rate (HR) circadian rhythms in man was achieved without changing the photoperiod by complete bed rest without exercise. This asynchrony was characterized by marked phase-shifting and a decrease in amplitude of BT and HR which returned to normal in the post-bed rest period. Bed rest also resulted in hypothermia and a transient bradycardia which were evident within the first 2 days of bed rest. Although the asynchrony of the rhythms was also present at this time, it became most marked on the 23rd day of bed rest. On this day, sudden phase-shifting occurred in all subjects at 0045 hours and was accompanied by a 50% increase in HR. Neither the prolonged inactivity nor the confinement associated with bed rest were responsible for these effects. It is proposed that reduction of stimuli to proprioceptive receptors requires a compensatory increase of other environmental synchronizers in order to maintain rhythm synchrony.

It has generally been agreed in the literature that light acts as the primary influence in the maintenance of synchrony of circadian rhythms (1–6). Research to study the properties and characteristics of rhythms has concentrated on producing rhythm desynchronization by manipulating the photoperiod (7, 8). Even the search for secondary synchronizers such as temperature and magnetic fields has involved analysis of the influence of these variables in the absence of light cues, that is, either continuous light or continuous dark environment (9–11). In a recent study investigating the physiological changes that occur in man in response to prolonged bed rest (12, 13), it was observed that desynchronization of some circadian rhythms occurred in spite of the fact that the subjects were maintained in a highly structured environment including a controlled photoperiod of 14 hours light:10 hours darkness. Since such desynchronization in the presence of a defined light environment has not been described previously in healthy subjects, we decided to use this approach to first confirm this previous finding and, second, to determine the nature of whatever change induced by bed rest was sufficiently powerful to cause rhythm asynchrony in spite of the unchanged photoperiod.

METHODS

Twelve healthy males aged 20 to 26 and weighing 55 $\pm$ 1 kg were selected from 60 applicants based on interviews and psychological tests. All subjects were exposed to a 20-day ambulatory control period. At the end of this period six subjects selected at random (Group I), were subjected to 56 days of absolute bed rest followed by a 20-day post-bed rest recovery period. The other six subjects (Group II) served as ambulatory controls for the entire 96-day period, remaining confined to the metabolic ward in the same controlled envirnoment and following the same daily schedule as all Group I subjects.

An attempt was made to control all known exogenous influences on human circadian rhythms. The subjects were maintained in a controlled environment of 14L:10D (lights on at 0900 hours). Light intensity was 30 foot-candles or greater at eye level, and the rooms were draped to minimize light leakage. Television was permitted only during the lights-on phase. Ambient temperature was maintained at 20 ± 1°C.

Food consisted of a balanced diet of 2.5 kcal/day. Subjects were handfed by dietitians and orderlies. There was no appreciable loss or gain of body weight ($p > .05$) by the end of the study. The daily schedule was as follows: 0900, lights-on, meal; 1300, meal; 1730, meal; 2200, snack; 2300, lights-off.

Heart rate and body temperature were recorded at 2-hour intervals, 24 hours/day. Body temperature data were obtained using ear probes containing a thermistor (Yellow Springs Model 402). Heart rate was measured manually from the pulse rate and by Beckman EKG sensors connected to a cardiotachometer. The data were analyzed by the periodogram and correlogram techniques (14–16) and by the summation dial technique (12, 13, 17). In six men the levels of cortisol and ACTH were also determined. These results are reported elsewhere.

RESULTS

Figure 1 demonstrates the usefulness of the summation dial method to depict concisely a large volume of sequential rhythmic data. Figure 1a shows the actual heart rate (HR) and Figure 1b body temperature (BT) data obtained from one representative ambulatory subject over the entire experimental period. The same data are also presented as summation dials. The main characteristic of the data obtained from the ambulatory subjects was their remarkable consistency. This was particularly true for HR which showed a peak at about 1600 hours in all six subjects and did not deviate throughout the study. The straightness of the summation dial plot pointing toward 1600 hours demonstrates that feature. BT of this subject was somewhat less stable, the peak occurring at about 2200 hours as indicated by the direction of the train of vectors of the summation dial.

Since the summation dial method for analyzing nonstationary data assumes a period of 24 hours, it was important to determine that this was indeed so under the present experimental conditions and that bed rest was not affecting the circadian period of these parameters. Figure 2 compares the individual periodograms of the ambulatory and the bed-rested subjects, indicating the presence of a circadian period in both cases.

In the bed rested group (Figure 3) the HR data can be divided into four segments: a pre-bed rest period of stable rhythms which are similar to those in the ambulatory group; in the first 3 weeks of bed

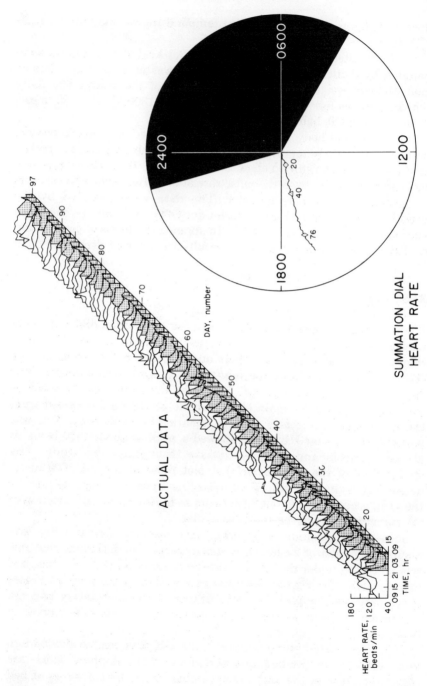

ACTUAL DATA

HEART RATE, beats/min

DAY, number

TIME, hr

SUMMATION DIAL
HEART RATE

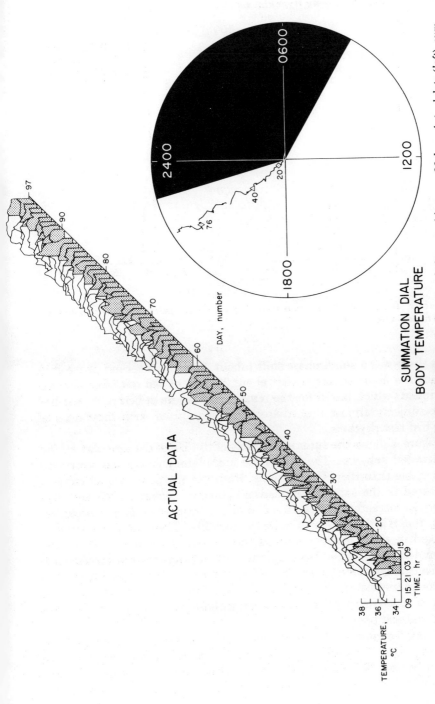

Figure 1 Circadian oscillations in heart rate (a) and body temperature (b) for an ambulatory subject over 96 days. Actual data (left), summation dial (right).

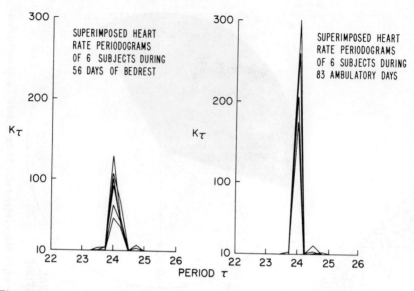

Figure 2   Individual periodograms indicating a circadian period in the ambulatory and bed rested subjects.

rest there was a small phase shift (about 2 hours); a second phase shift of about 4 hours occurred after about 20 days of bed rest (day 40 of experiment) which lasted for the remaining 5 weeks of bed rest; post-bed rest subjects almost immediately resynchronized with their original pre-bed rest rhythms.

Figure 4 shows the summation dials of the BT of the ambulatory and bed-rested subjects. The BT of the ambulatory group was somewhat less stable than their HR rhythms. However, the peaks for all subjects occurred in the same time quadrant, that is, between 2045 and 0245 hours as indicated by the direction of the train of vectors. During bed rest, the BT data were even more variable. Three of these subjects peaked in the same quadrant as the ambulatory and the other three around 0600 hours (60° out of phase). All subjects showed considerable phase changes throughout the bed rest, three of them showing random walks representing rhythm asynchrony. Only two subjects could be said to have resynchronized relative to base at the end of the 3 week post-bed rest period.

Figure 5 compares the daily phase angle, (ϕ), the daily integrated amplitude (area between the daily cycle and its mean), and the daily mean of the HR data, collected from the ambulatory and the bed-rested subjects. It can be seen that the greatest phase angle shift in the

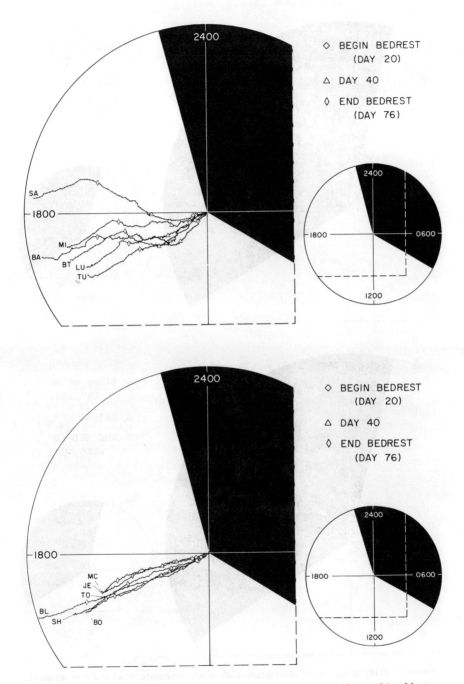

Figure 3   Heart rate summation dials of the bed rested (a) and ambulatory (b) subjects.

581

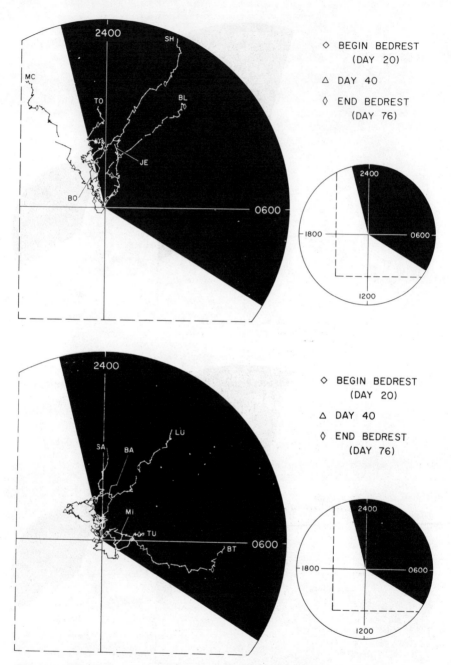

Figure 4   Body temperature summation dials of the ambulatory (a) and bed rested (b) subjects.

582

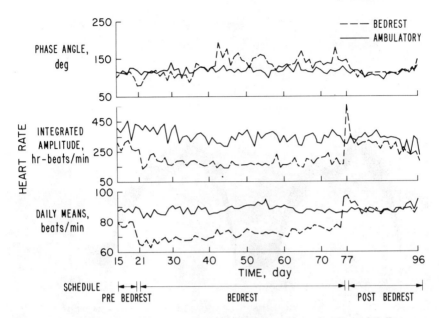

Figure 5 A comparison of daily phase angle (φ), integrated amplitude, and daily means of heart rate rhythm in ambulatory and bed rested subjects.

HR rhythm occurred about the 22nd to 24th day of bed rest (43rd to 45th day of study) and continued for the duration of the bed rest. The amplitude of the rhythm decreased sharply on going to bed and recovered promptly in the post-bed rest period. Similarly, the mean HR dropped initially and then increased sharply when the subjects got out of bed and remained elevated during the 20 days post-bed rest.

Since the sharpest phase shift in the HR rhythm appeared to occur about the 20th day of bed rest, the data were scanned to determine more precisely the actual time and day at which it occurred (Figure 6). It was found that four of the bed-rested subjects showed their sharpest phase shift on the 23rd day of bed rest (day 43 of study) whereas the other two bed-rested subjects showed a similar change on day 24 of bed rest (day 45 of study). On those days the HR of these bed-rested subjects increased sharply to 87 beats/minute by 0100 and did not reach its lowest point (56 beats/minute) until 0700. In contrast, none of the ambulatory subjects showed an increase in HR at these hours but in fact their HR decreased from 74 beats/minute at 2300 hours to 62 beats/minute at 0100 hours.

The BT rhythm did not show a sudden, single phase shift as did the HR. Figure 7 compares the daily phase angle (φ), the daily integrated

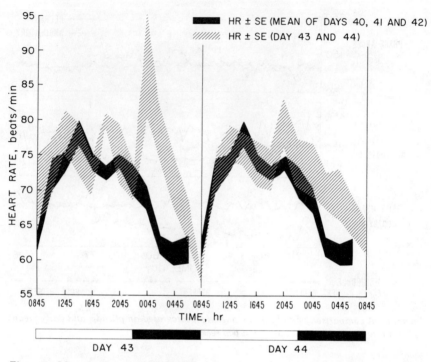

Figure 6   Mean heart rate (+SE) of four subjects on days 23 and 24 during bed rest (Day 43 and 44 of study) compared to that of same subjects during the three previous days.

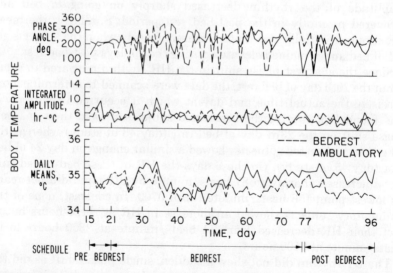

Figure 7   A comparison of daily phase angle (φ), integrated amplitude, and daily means of body temperature in ambulatory and bed rested subjects.

amplitude, and the daily mean of the BT data of both groups of subjects. The results show that numerous phase shifts occurred in the bed-rested subjects as compared to the relatively unchanging phase angle of the ambulatory controls and of these same subjects during their ambulatory periods. There was a progressive decrease in the mean daily BT in spite of an unchanged integrated amplitude, indicating that the BT rhythm fluctuated in bed rest with the same amplitude about a new lower mean level. The decrease in mean daily BT had not recovered by the 21 day post-bed rest period.

DISCUSSION

The results confirm our previous findings that the primary influence of bed rest on BT and HR rhythms is to reduce the amplitude and change their phase relationships (12). The normally entrained rhythms were altered after approximately 20 days of bed rest when they lost their normal relationship to the photoperiod and to each other. In addition, bed rest induced hypothermia and an initial bradycardia. The possibility that this rhythm asynchrony may have been due to the inactivity associated with bed rest was ruled out in a previous study where a moderately heavy exercise regimen did not prevent the observed changes (12). The present study also rules out the possibility that the prolonged confinement associated with this type of experiment may have been causing the observed phase shift since the ambulatory control subjects showed none of the changes seen in the bed-rested individuals.

The dissociation of the HR and BT rhythms from each other and from the light schedule during bed rest and the prompt reassociation of the two rhythms in the post-bed rest ambulatory period suggests that synchrony of these rhythms may be dependent on posture. Hence it seems reasonable that the postural change involved in bed rest or some physiological consequence of that, such as hydrostatic pressure changes or redistribution of body fluids and electrolytes, may be primarily responsible for the rhythm asynchrony.

Various rhythms in man have been reported to be dependent on different cues. For example, the rhythm in aldosterone excretion has been suggested to be primarily posture-dependent (18) while the rhythmicity of other parameters such as plasma cortisol levels appears to be unaffected by bed rest and remains entrained to the light-dark cycle (13). Similarly it has been suggested that like cortisol, the timing of luteinizing hormone (LH) release in the rat is strongly related to the light-dark schedule which synchronizes ovulation by release of LH at a precise time of the cycle (19); it would be of interest to determine if this is true in man and if bed rest disturbs it. On the other hand, HR and to a lesser extent BT appear to require both light and other (postural) cues for maintenance of their rhythm synchrony.

The level of the baseline about which the homeostatic mechanisms operate varies rhythmically, and this rhythmicity is controlled by exogenous and endogenous synchronizers (5). Since the time of Claude Bernard, physiologists have emphasized the study of the mechanisms by which organisms maintain the relative constancy of their internal environment in response to change in the external environment. Such change or stress has generally been considered as an increase in magnitude or duration of inputs to the system. With the exception of behavioral and biologic rhythm research, little attention has been given to the biological consequences of an absence or reduction of input stimuli. The psychological consequences of isolation (20), the disturbance of circadian rhythms in an environment where light intensity was below a certain threshold (2), and the clinical consequences of prolonged bed rest in hospitalized patients have long been recognized. The results of prolonged bed rest studies suggest that neither reduced activity alone nor relative confinement alone, both examples of low input environments, result in HR and BT rhythm asynchrony. On the other hand, it would appear that the reduction of input stimuli to proprioceptive receptors resulting from the postural change alone, or in addition to the confinement and inactivity inherent to prolonged bed rest, was responsible for the observed rhythm asynchrony. The study also gives further support to the hypothesis that maintenance of circadian rhythm synchrony is not dependent on light alone as the environmental synchronizer. It is more likely that the sum of a variety of input stimuli is required to attain a certain threshold before synchrony of these rhythms with the environment and with each other can be maintained.

If this hypothesis were true, then it should be possible to maintain rhythm synchrony in bed rest by changing light intensity or other social and environmental stimuli.

REFERENCES

1. J. Axelrod, R. J. Wurtman, and C. M. Winget, *Nature,* **201,** 4924 (1964), 1134.
2. C. M. Winget and D. H. Card, *Life Sci. Space Res.,* **5** (1967), 148.
3. C. M. Winget, D. F. Rahlman, and N. Pace, in *Circadian Rhythms in Non-Human Primates* (F. Rohles, Ed.), Karger, New York Basel, 1968.
4. C. M. Winget, L. S. Rosenblatt, C. W. DeRoshia, and N. W. Hetherington, *Life Sci. Space Res.,* **8** (1970), 247.

5. E. Bünning, *The Physiological Clock*, rev. 2nd ed., Springer, New York, 1967.

6. J. Aschoff, *Ann. Rev. Physiol.,* **25** (1963), 581.

7. E. B. Flink and R. P. Doe, *Proc. Soc. Exp. Biol. Med.,* **100** (1959), 498.

8. F. Halberg, P. G. Albrecht, and C. P. Barnum, Jr., *Amer. J. Physiol.,* **199** (1960), 400.

9. R. Wever, *Naturwissenschaften, 55,* 1 (1968), 29.

10. F. Wilson and G. J. Snowball, *Aust. J. Zool.,* **7** (1959), 1.

11. J. Aschoff, M. Fatransha, H. Giedke, P. Doen, D. Stamm, and H. Wisser, *Science,* **171** (1971), 213.

12. C. M. Winget, N. W. Hetherington, L. S. Rosenblatt, and P. C. Rambaut. *J. Appl. Physiol.,* In Press.

13. J. Vernikos-Danellis, C. S. Leach, C. M. Winget, P. C. Rambaut, and P. B. Mack, *J. Appl. Physiol.,* In Press.

14. M. G. Kendall, *The Advanced Theory of Statistics*, Vol. 2, Griffin, London, 1948, pp. 363–439.

15. C. I. Bliss, *Conn. Agr. Exp. Sta. Bull.,* **615** (1958), 1.

16. F. Halberg, E. Halberg, C. P. Barnum, Jr., and J. J. Bittner, in *Photoperiodism and Related Phenomena in Plants and Animals* (R. B. Withrow, Ed.), Association for the Advancement of Science, Washington, D.C., 1959, pp. 803–878.

17. N. W. Hetherington, C. M. Winget, L. S. Rosenblatt, and P. B. Mack, *J. Interdisciplinary Cycle Res., 2,* 3 (1971), 365.

18. F. C. Bartter, C. S. Delea, and F. Halberg, *Ann. N. Y. Acad. Sci.,* **98** (1962), 969.

19. N. Schwartz, *Rec. Progr. Homone Res.,* **25** (1969), 1.

20. J. T. Shurley, *Amer. J. Psychiat.,* **117** (1960), 539.

12 Cellular Mechanisms Involved in Biorhythms

LAWRENCE E. SCHEVING, *Moderator*

CHAPTER 39

In Vitro *Adrenal Studies in Relation to Cyclic Reproductive Success*

RICHARD V. ANDREWS

Creighton University, Omaha, Nebraska

RONALD SHIOTSUKA

Chronobiology Laboratory, University of Minnesota, Minneapolis, Minnesota

In vitro studies of adrenal function have confirmed that circadian profiles of secretory products found in the blood and urine of animals result from the separate periodicities in adrenocortical and pituitary function. Transverse (serial sampling) measurements have revealed synchronies of pituitary, adrenal, and target organ function in mice, rats, and men (1–3). Evidence of endogenous adrenal rhythms has resulted from *in vitro* serial incubations (4) and prolonged organotypic culture (5). Pituitary ACTH content peaks in the late afternoon in nocturnal rodents and adrenal steroidogenesis and secretion follow with a peak in the late evening. Even adrenal sensitivity to ACTH follows a daily rhythm (4, 6) in mice and hamsters. Comparative data reveal that *in vitro* rhythms in adrenal secretion, respiration, and ACTH dose responsiveness persist in glands cultured from lemmings, voles, mice, rats, ground squirrels, and weasels (5–9).

Rhythms were apparent in measurements of ^{14}C-acetate incorporation into secreted steroid, as well as in fluorimetrically detected steroid; serial (2 or 3 hour) sampling of the culture medium permits estimations of secretory activity over several days (10). The three measurements (labeled steroid secretion, chemically detectable steroid, and oxygen consumption) revealed daily rhythms with activity maxima which are slightly out of phase, but which show circadian frequencies.

IN VITRO EFFECTS OF ACTH

Variations in acetate incorporation into corticosteroid appear to be based more on synthetic capacity than on the availability of free (labeled) acetate since rapid influx of acetate into whole glands occurs (11). The level of acetate equilibration is time-of-day dependent and can be markedly influenced by ACTH treatment. Measurements of acetate flux into adrenal cells, wherein the extracellular space was accounted for with ^{3}H-manitol, showed that the maximum flux rate coincides with maximum corticosteroid secretory activity. While acetate (and other intermediates) uptake accumulates against a concentration gradient, steroid release into the medium appears to occur with a concentration gradient. Accumulation of labeled steroid within adrenal cells and efflux of the steroid was time-of-day dependent. Flux rates and intracellular steroid concentrations can be increased by ACTH treatment, and the time quadrant during which maximal rates are achieved may be altered by higher doses of ACTH.

COSINAR ANALYSIS OF IN VITRO RHYTHMS

Protracted treatment of adrenal cultures with ACTH increased the amplitude of daily secretary patterns without markedly affecting the phase of the rhythm (10). The least squares fitting of a cosine curve was used to quantify any phase shift resulting from a single pulsed dose of ACTH. This form of analysis yields four primary rhythm parameters, namely period (τ, interval between identifiable points of recurrence), mesor (M, rhythm-determined average), amplitude (A, a measure of extent of rhythmic change), and acrophase (φ, timing of the peak of the fitted curve with respect to some arbitrary reference point in time). The least influence on the phase of the secretory rhythm was

Table 1 *Rhythm Parameters*

Condition	Mesor (95% CI)	Amplitude (95% CI)	Acrophase (95% CA)	p
Control, no ACTH	1.91 (1.81 2.00)	0.70 (0.56 0.84)	−335 (−324 −345)	.01
ACTH pulse (3h) at secretory high	2.13 (2.00 2.27)	0.75 (0.56 0.94)	−349 (−335 −364)	.01
ACTH pulse (3h) at secretory low	2.37 (2.20 2.53)	0.64 (0.41 0.88)	−256 (−233 −273)	.01

CIRCADIAN RHYTHMIC ADRENOCORTICAL ACTIVITY IN VITRO [CIRCADIAN-PHASE-DEPENDENCE OF EXTENT AND LONG-TERM TIMING OF SECRETORY RESPONSE TO AN ACTH-PULSE (▼)]

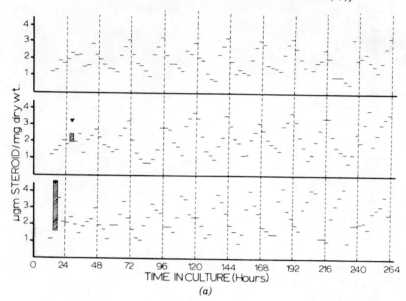

(a)

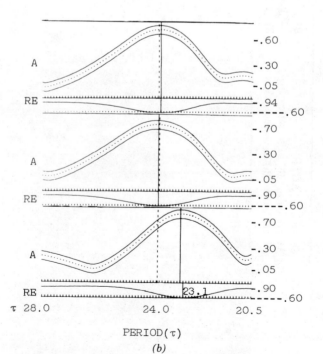

(b)

apparent when the treatment was begun at 24:00 hours during the secretory peak. Pulsed doses of ACTH applied to the cultures during the secretory low were effective in advance-phase-shifting the cycle, while ACTH pulses delivered at the peak of activity were without (phase-shifting) effect.

In Figure 1, three plots on the left show original data (10) as a function of time; amplitude as a function of trial period is displayed on the right. The ordinate scales of each plot are in units of μg steroid/mg dry weight of tissue. The periods fitted range from 28.0 to 20.5 hours, with increments between successive trial periods of a tenth of an hour. The amplitudes for each of the fitted periods form a dotted line bordered by their 95% confidence intervals represented as solid lines. The plot of residual error (RE) is shown below the amplitude diagrams. For a group of 20 paired adrenal cultures incubated without ACTH (control group), the RE reveals a 23.9-hour period (top right). With the data on hand this period is not distinguished from a period of 24 hours. A 24-hour period is also found for the group given a pulsed dose of ACTH at the time of peak secretory activity (middle). However, an average period of 23.1 hours is found when the pulsed dose of ACTH is administered during a period of low secretory activity (lower right). This finding is in keeping with an advance in the peak of steroid secretion (acrophase, φ) resulting in a total acrophase shift of 9 hours in the 10 days of data available from this group of adrenal cultures.

Table 1 shows rhythm parameters obtained from the fit of a 24-hour cosine curve to the same data. A statistically highly significant difference in mesor is detected between the control group and glands treated with ACTH at the secretory trough. Although the glands treated with ACTH at the secretory peak show a higher mesor, the lower limit of its 95% CI is equal to the upper limit of the 95% CI for the control group. Moreover, in the group treated with ACTH during the secretory trough, the acrophase in secretory activity is significantly different from and leads that of the other two groups. This advance in acrophase can be seen macroscopically in a time plot of the data (Figure 1, bottom left) and also is predicted microscopically by the

---

Figure 1   Corticosteroid secretion in vitro analyzed by least squares fit of cosine curves with periods from 28.0 to 20.5 hours, differing by increments of 0.1 hour. Original data as a function of time (a) thus yield amplitude (A) and residual error (RE) diagrams (b). Adrenal cultures in top row received no ACTH; those in second and third row were treated with a pulsed 1.0 IU dose of ACTH at times indicated by the filled triangle (▼). The cross-hatched bars represent the response to such an acute stimulation. In the amplitude diagrams, the vertical dotted lines indicate the 24-hour trial period, and the adjacent solid vertical line indicate the period at which the largest A was found. The solid vertical line in the RE plot indicates the smallest RE, thus the best fitting period.

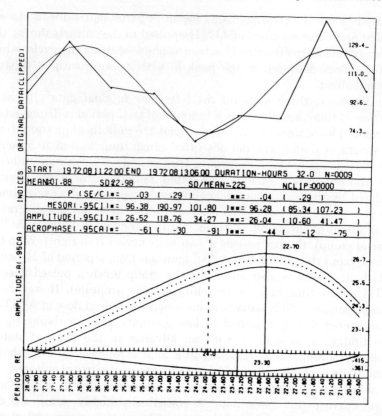

Figure 2   Corticosteroid secretion during a 32-hour incubation fitted by cosine curves with periods ranging from 28.0 to 20.5 hours, with increments of 0.01 hour. A fit significant at the 4% level was obtained for a 24-hour period, but a larger amplitude was found with shorter periods. The best fitting cosine curve had a period of 23.3 hours and accounted for approximately 77% of the variability.

23.1-hour length of the best fitting period shown in the amplitude-residual error diagram (Figure 1, bottom right).

Adrenal incubations using the same technique were carried out at the Chronobiology Laboratory. In a first organ culture series a decreasing trend in steroid production was detected in a data span of about 2 days. A straight line with a negative slope was fitted through all points and the line arbitrarily set equal to 100%. The original data points were then transformed into relative values by expressing each point as a percentage of the corresponding point on the fitted line. A plot of the converted values in Figure 2 shows these values ranging from 70 to 145. A circadian rhythm is described by the fit of a 23.3-

hour cosine curve with a p value of 0.03. The p is 0.04 for the fit of a 24-hour trial period. Thus, in the presence of a decreasing trend in corticosteroid activity, probably resulting for technical reasons, a circadian rhythm continues to be demonstrated in the data here analyzed in Figure 2.

RHYTHMIC ADRENAL CELLULAR MECHANISMS

The mechanism by which ACTH effects phase shifts appears to involve the synthesis of critical enzymes and a feedback of intracellular steroid on the enzyme forming system. For example, when the 11β-hydroxylase inhibitor SU4885 (metapyrapone) is applied with appropriate doses of ACTH, both the phase shifting and the acute steroidogenic responses of ACTH are blocked (10). Puromycin, too, blocks the phase shifting and steroidogenic responses of cultured adrenals to pulse-dosed ACTH (12).

Inhibition of protein synthesis through the action of actinomycin-D produces some paradoxical effects on secretory rhythms (13). Pretreatment of cultures with actinomycin D prior to ACTH administration blocks both acute and phase-shift responses of cultured glands. Treatment during the (presumptive) secretory peak blocks the ACTH effects, obscures the peak, and delays the subsequent peak. Combined ACTH and actinomycin treatment results in the loss of rhythmicity. When both ACTH and actinomycin-D are pulsed at the low point of secretion, the proximate peak is lost but reappears in-phase on the subsequent day.

Accumulation of intracellular steroid during cyclic peaks of adrenal secretion and during ACTH treatment pointed to an effect of steroid on its secretory cycle (7). Estimation of corticosteroid synthesis by precursor incorporation techniques revealed that while exogenous corticosteroids do not affect steroid synthesis *in vitro*, progesterone can inhibit steroidogenesis if administered at other than low points in the cycle (14). A short-loop feedback system of progesterone and ACTH to effect the free cholesterol pool via side-chain-cleaving enzyme has been suggested as a mechanism of action (15). Moreover, when progesterone was administered to adrenal cultures during the rising phase of the secretory cycle, the peak is phase delayed; this phase-shift persisted on subsequent days of culture.

Direct measurements of adrenal synthetic and secretory activity through *in vitro* serial sacrifice and organ culture techniques have enabled us to document seasonal variations in wild rodent populations' endocrine status (9, 16, 17). These studies indicated that *in vitro* secretory responses reflect plasma steroid values in both temporal pattern and steroid constituent profiles (9, 17, 18). A similar series of *in vitro* adrenal incubations and plasma corticosteroid determinations has confirmed the reliability of our methods for another species (19). The seasonal patterns in adrenal function are influenced by both climate and social pressure, and bear interesting relationships to mortality, reproductive success, and gonadal function within wild lemming and rat populations (Figures 3 and 4). Lowered fertility accompanies stress-induced hyperadrenalism in wild populations, particularly when endogenous ACTH stimulation is maximal (7, 9). Moreover, ACTH stimulation alters the rate of corticosteroid precursor utilization, and

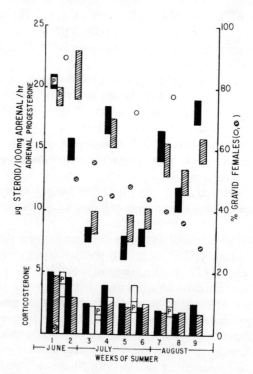

Figure 3 Comparison of in vitro *secretory rates of adrenals from high-density and low-density lemmings and population fertility. The histogram bars represent average corticosterone secretion rates: solid black, high-density males; hatched, high density females; open, low-density females. The bar inserts represent adrenal progesterone secretion rates. The circles represent the percentage of gravid females in the population when the adrenal incubations on cross-sectional samples were made: hatched circles, high-density samples; open circles, low-density samples.*

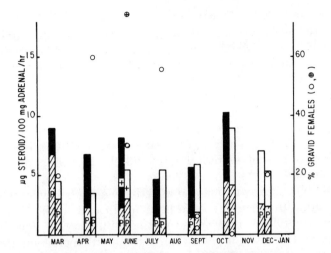

Figure 4   Comparison of in vitro *secretory rates of adrenals from wild rat populations. Shaded bars represent males and open bars represent females; hatched portions of bars represent adrenal progesterone secretion and open bars represent corticosterone secretion rates. Circles represent the pregnancy rate; stars mark comparisons of low-density, expanding population values for an identical time of year.*

increases the amount of adrenal progesterone and corticosterone secreted in a dose-dependent manner (20). The sensitivity of adrenal progesterone secretion has also been shown for laboratory rats (21, 22), and the implications of augmented progesterone sources for reproductive behavior (23) and the timing of LH release (24) have been recently reviewed.

STRESS INTERACTION WITH PRODUCTIVITY

Of particular interest to Christian's hypothesis of adrenal mediated population control are the progesterone data just discussed. Progesterone, besides its major progestational role in the maintenance of pregnancy (25), is now known to exert diverse effects on ovulation and mating behavior when administered alone and together with estrogen (26, 27). Lisk (28) illustrated a biphasic time-dependent response of the female hamster to implanted progesterone. Progesterone implants in the arcuate nucleus region of the brain can alter the onset of heat and ovulation by 24 hour advance or delay, depending on the length of time and the point in the cycle where the hormone was implanted. Adrenalectomy can delay the onset of behavioral estrus in the Sprague-Dawley rat for 2 to 4 hours (29). The critical event in the estrus cycle is an abrupt increase (ovulatory surge) of luteinizing hormone (LH) from 6 to 10 hours prior to ovulation. If the LH surge is blocked, ovulation will not occur. Progesterone injected at physiological doses into rabbits will act as a pregnancy block in inhibiting the ovulatory response of LH-RH infused rabbits (30). Boyarski et al. (31) duplicated the sexual disinterest and infertility of pseudopregnant rabbits by infusing progesterone. Monkeys injected with progesterone also exhibit blocked ovulation, only recovering when given high doses of estradiol immediately following progesterone treatment (32).

Besides blocking ovulation and affecting mating behavior, high levels of circulating progesterone have been implicated in the inhibition of oxytocin secretion in response to mating stimulus (33), facilitation of maternal nest building in mice (34), and inhibition of ovum and sperm transport and implantation by interfering with ovarian hormone secretion (35).

Adrenal progesterone has not been measured through the estrous cycle. Assuming progesterone secretion is ACTH responsive, the rise in adrenal corticosterone on the day of proestrus (36) can be taken as indirect evidence of adrenal sex steroid involvement in the estrous cycle. Further indications of such involvement are seen in the spotted skunk; plasma progesterone titers increase during a period of luteal regression (preimplantation period), indicating a possible adrenal source (37). Evidence of the degree of involvement of adrenal progesterone possible is seen in the maintenance of pregnancy in ovariectomized rats by ACTH supplementation (38).

Further influences of the hyperactive endocrine system of a stressed animal on the reproductive activity of a population have been noted. ACTH injections into juvenile female rats will result in anovulatory adult females (39). This anovulatory state is probably due to increased adrenal androgen secretion since injection of male sex steroids into juvenile female rats has a masculinizing affect (40–42). Thomas and Gerall (42) mimiced the effect of adrenal androgens on behavior by injecting corticosterone into rats. However, the hormones only affected sexual receptivity and lordosis behavior, since the animals still achieved physiological estrous.

SOCIAL EFFECTS ON REPRODUCTION

Social pressure on adult female rats and mice results in decreased fertility in subordinate animals (43), hampered maturation of young females, decreased success of lactation, and aberrant maternal behavior (44). Damping of daily rhythms in corticosterone and luteinizing hormone secretion of stressed laboratory rats (45) may account for lowered fertility among females in a wild rat population (9) if we assume that daily rhythms in adrenal progesterone secretion follow the pattern displayed for corticosterone. Certainly seasonal rhythms in these adrenal steroid secretions are apparent in low density wild rat populations and are attenuated by social disorganization in saturated-density populations (Figure 4).

While daily rhythms in adrenal corticosterone secretion of lemmings do not appear to change in amplitude under severe population stress (17), the level of progesterone secretion remains sensitive to ACTH stimulation longer than corticosterone under chronic stress conditions. While adrenal progesterone secretion has not been followed through the estrus cycle, serial collections of glands from lemmings through a popu-

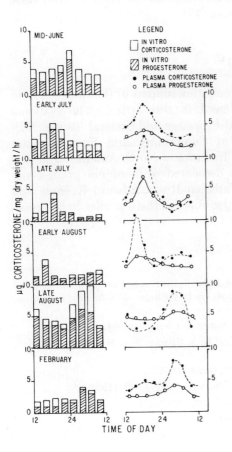

Figure 5 Profiles of adrenal (in vitro) *secretory activity from low-density lemmings at various times during the breeding season. The histogram represents serial sacrifice values: open bars, corticosterone; hatched bars, adrenal progesterone. The right panel represents the time course of adrenal secretion in organ culture: open circle lines, progesterone; filled circle lines, corticosterone.*

lation low reveal that progesterone secretion follows the same time course as corticosterone secretion (Figure 5). Evidence of differential progesterone secretion during a cyclic lemming "high" in 1971 coincided with an early drop in fertility and reproductive potential in the population as compared to earlier low density years (Figure 3). It is interesting to note that both males and females began to show sexual regression at about the same time, and that progestins have produced declines in LH and gonadal function in humans as well (46–48).

In conclusion, *in vitro* adrenal studies have provided clues regarding the potential role of augmented progesterone secretion. Complementary analyses of plasma steroids and of tissue response to this hormone indicate that this role may be critical in population control. The controversies of biphasic progesterone responses regarding ovulation and of behavioral effects of the hormone are unresolved, but careful time analysis of multivariant hormone cycles should aid in answering these questions (23, 24, 28, 34). Precise computer analysis of transverse studies in humans should be of value (49).

ACKNOWLEDGMENTS

I am grateful to the following agencies for support of the research from my laboratory reported or reviewed in this article: The National Science Foundation (Physiological Processes and General Ecology Sections), the U.S.P.H.S. Bureau of Community and Environmental Health Management, The Arctic Institute of North America, and The Office of Naval Reasearch. The dedicated work of my professional collaborators, R. W. Belknap, Kathleen Ryan, Edward Keenan, and my students and technicians has also enabled these studies to progress.

REFERENCES

1. F. Halberg, E. Halberg, C. P. Barnum, and J. J. Bittner, in *Photoperiodism and Related Phenomena in Plants and Animals* (B. B. Withrow, Ed.), AAAS, Washington, D.C., 1959, p. 803.

2. F. Halberg, R. E. Peterson, and R. H. Silber, *Endocrinology,* **64,** (1959), 222.

3. F. Halberg, J. H. Galich, F. Ungar and L. A. Frende, *Proc. Soc. Exp. Biol. Med.,* **118,** (1965), 414.

4. F. Ungar and F. Halberg, *Science,* **137,** (1962), 1058.

5. R. V. Andrews and G. E. Folk, *Comp. Biochem. Physiol.,* **11,** (1964), 393.

6. R. V. Andrews, *Comp. Biochem. Physiol.,* **26,** (1968), 179.

7. R. V. Andrews, *Physiol. Zool.,* **41,** (1968), 86.

8. R. V. Andrews, *Fed. Proc.,* **28,** (1968), 636.

9. R. V. Andrews, R. W. Belknap, J. Southard, M. Lorincz, and S. Hess, *Comp. Biochem. Physiol.*, **41**, (1971), 149.

10. R. V. Andrews, *Comp. Biochem. Physiol.*, **26**, (1968), 179.

11. R. V. Andrews, *Comp. Biochem. Physiol.*, **26**, (1968), 479.

12. R. V. Andrews, *Comp. Biochem. Physiol.*, **30**, (1969), 123.

13. R. V. Andrews, and R. Shiotsuka, *Comp. Physiol. Biochem.*, **36**, (1970), 353.

14. R. V. Andrews, *Proc. XXV Int. Physiol. Congr.*, **9**, (1971), 19.

15. S. Ichii, S. Kobaychi, and M. Matsuba, *Steroids*, **5**, (1965), 663.

16. R. V. Andrews, *Acta Endocr.* (Copenhagen), **65**, (1970), 639.

17. R. V. Andrews, *Acta Endocr. (Copenhagen)*, **65**, (1970), 645.

18. R. V. Andrews and R. Strohbehn, *Comp. Biochem. Physiol.*, **38A**, (1971), 183.

19. L. Adams, *J. Wildlife Dis.*, **8**, (1972), 10.

20. R. V. Andrews, *Endocrinology*, **83**, (1968), 1387.

21. H. H. Feder, J. A. Resko, and R. W. Coy. *J. Endocr.*, **41**, (1968), 563.

22. J. A. Resko, *Science*, **164**, (1969), 70.

23. J. M. Davidson and S. Levine, *Ann. Rev. Physiol.*, **34**. (1972), 1084.

24. D. T. Armstrong and T. G. Kennedy, *Amer. Zool.*, **12**, (1972), 245.

25. C. H. Danforth and S. B. de Aberle, *Amer. J. Anat.*, **41**, (1928), 65.

26. J. B. Powers, *Physiol. Behav.*, **5**, (1970), 831.

27. L. A. Ciaccio and R. D. Lisk, *Amer. J. Physiol.*, **221**, (1971), 936.

28. R. D. Lisk, *Trans. N.Y. Acad. Sci.*, **31**, (1969), 593.

29. L. G. Nequin and N. B. Schwartz, *Endocrinology*, **88**, (1971), 325.

30. J. Hilliard, A. V. Schally, and C. H. Sawyer, *Endocrinology*, **88**, (1971), 730.

31. L. H. Boyarski, H. Baylies, L. E. Casida, and R. K. Meyer, *Endocrinology*, **41**, (1947), 312.

32. H. G. Spies and G. D. Niswender, *Endocrinology*, **90**, (1972), 257.

33. J. S. Roberts and L. Share, *Endocrinology*, **87**, (1970), 812.

34. R. D. Lisk, *Animal Behav.*, **19**, (1971), 606.

35. G. Pincus, *Acta Endocr.* (Copenhagen) *Suppl.*, **28**, (1956), 18.

36. D. Raps, P. L. Barthe, G. Meglioli, and P. L. Desaulles, *Acta Endocr.*(Copenhagen), *Suppl.*, **155**, (1971), 67.

37. R. D. Mead and K. B. Eik-Nes, *J. Reprod. Fert. Suppl.*, **6**, (1969), 397.

38. B. N. Berg and G. M. Carmody, *Endocrinology*, **89**, (1971), 219.

39. J. J. Christian, *Endocrinolgoy*, **74**, (1964), 669.

40. J. Tollman and J. A. King, *J. Animal Behav.*, **4**, (1956), 147.

41. A. W. Schuetz and R. K. Meyer, *Proc. Soc. Exp. Biol. Med.*, **112**, (1963), 875.

42. T. R. Thomas and A. A. Gerall, *Endocrinology*, **85**, (1969), 781.

43. J. B. Calhoun, *Science*, **109**, (1949), 333.

44. J. J. Christian and C. D. Lemunyan, *Endocrinology*, **63**, (1958), 517.

45. J. D. Dunn, A. Arimura and L. E. Scheving, *Endocrinology*, **99**, (1972), 29.

46. M. A. Kirschner and G. Schneider, *Acta Endocr.* (Copenhagen), **69**, (1972), 385.

47. J. J. Strauss and P. E. Pochi, *Rec. Progr. Hormone Res.*, **19**, (1963), 335.

48. R. Swerdloff and W. D. Odell, *Proc. Nat. Acad. Sci., U.S.*, **61**, (1968), 529.

49. F. Halberg, *Ann. Rev. Physiol.*, **39**, (1969), 1036.

13 *Neuroendocrine Mechanisms Involved in Biorhythms*

JOHN C. PORTER, *Moderator*

CHAPTER 40

Neuroendocrine Mechanisms Involved in Biorhythms

JOHN C. PORTER and NIRA BEN-JONATHAN*

Department of Physiology, The University of Texas Southwestern Medical School, Dallas, Texas

The underlying basis for the rhythmic or episodic release of certain hormones from the anterior pituitary remains one of the more poorly understood phenomena in regulatory biology. In some cases the apparent absence of reason for such release makes it difficult to accept the phenomenon as one of central significance. Yet in the case of ovulation, for example, it is evident that a marked increase in the concentration of gonadotropins in plasma of mammals is essential for completion of the ovulatory process. In other cases many periodic hormonal events appear to be mere manifestations of noise in poorly controlled systems having wide ranges of error tolerances.

PERIPHERAL MANIFESTATIONS OF SOME HORMONAL RHYTHMS

One of the first hormonally-related rhythms in mammals to be recognized was the ovulatory cycle. Here, a readily discernible

---

* Postdoctoral fellow of the Population Council, New York, New York.

phenomenon, that is, ovulation, was easily perceived and, in some species, easily demonstrated. With the development of methods for quantitatively determining the concentrations of gonadotropins in the plasma or serum of blood, it was discovered that the ovulatory event was preceded by an accelerated release from the anterior pituitary of gonadotropins, especially luteinizing hormone (LH), into blood. This release of gonadotropins resulted in a many-fold increase in the concentration of LH and to a lesser degree of follicle-stimulating hormone (FSH) in plasma.

In normally cycling mammals the rhythmic release of LH recurs with a period of approximately 4 weeks in humans (1) and in rhesus monkeys (2), 4 to 5 days in the rat (3), and 4 days in the hamster (4). Yet the period of the rhythmic releases of LH is an unstable feature, being greatly affected by such factors as the presence or absence of feedback activities of hormones from the testes, ovaries, possibly the adrenal cortices, and in some species even light.

Dierschke et al. (5) observed that castration of female rhesus monkeys results in the development of a new rhythm which is characterized by a short period of 1 to 2 hours. The rhythmic oscillations appear to be due to a rapid release of LH during a short interval of time followed by a longer interval of a slower release during which the rate of removal of LH from plasma exceeds the rate of release. This regular, pulsatile release of LH was discovered by Dierschke et al. (5) when repetitive determinations of LH levels in plasma were made in the same monkey at short intervals. The rhythmical pattern of LH release was not apparent in those studies in which long sampling intervals were employed or in those studies which relied on data points obtained from many animals. In such instances the data appeared to fluctuate randomly or to approach central values (6–9).

Recently, Gay and Sheth (10) reported results from a skillfully executed study in which they made sequential determinations at 2- to 12-minute intervals in unanesthetized, castrated rats. They found in the rat, as did Dierschke et al. (5) in the monkey, that castration results in a new rhythmic pattern of LH release. The rhythmic alterations in the concentrations of LH in plasma had a period of approximately 30 minutes. In the castrated monkey as well as in the castrated rat, the rhythmic bursts of LH release were superimposed upon an already elevated rate of release characteristic of castrated animals (5, 11). The results of these studies indicated that the duration of the release phase in the monkey was about 10 minutes out of the 1- to 2-hour cycle and in the rat about 5 minutes out of a 30-minute cycle.

In the female rat undergoing ovulation at regular intervals of 4 to 5 days, Neill (12) and Freeman and Neill (13) have observed that prolactin is released during the proestrous phase of the cycle at a time corresponding closely to that seen for the release of LH. Thus, in the cycling rat, the period of the rhythmic release of prolactin is 4 to 5 days. However, in the rat undergoing the state of pseudopregnancy induced by stimulation of the uterine cervix, the rhythmic release of prolactin changes in such a manner that the period becomes 24 hours instead of 4 to 5 days (13). Furthermore, the release of prolactin occurs regularly during the last half of the nocturnal phase of a day consisting of 12 hours of light and 12 hours of darkness.

The anterior pituitary-adrenocortical system has long been recognized to undergo diurnal variation. In persons who sleep at night and are active during daylight hours, the greatest rate of secretion of adrenal corticoids occurs in the early morning hours. Although this general pattern has been confirmed many times, Krieger et al. (14) have shown recently that corticotropin (ACTH) is released in episodic bursts. Although these workers made determinations of the concentration of ACTH in peripheral plasma of normal human subjects, there appeared to be no rhythmic pattern to the changes in the plasma levels of ACTH and consequently no discernible period. One wonders whether a more frequent sampling regimen might not reveal some meaningful rhythm. In the absence of such data, however, we have to assume that fluctuations of ACTH in plasma may result from seemingly random or episodic releases of ACTH from the anterior pituitary.

NEUROENDOCRINE BASIS FOR RHYTHMIC HORMONE PATTERNS

Now, in light of these findings, we can ask: What regulates the rhythmic or episodic release of certain hormones from the anterior pituitary? At this stage of our understanding of anterior pituitary control, we are led to suppose that variation in the output of hormones from this gland is determined to some extent by the rate of delivery of neurohormones to the gland via its portal vasculature. It is generally believed that neurohormones (sometimes called hypophysiotropic substances, releasing factors, or releasing hormones) enter the primary capillary plexus of the hypophysial portal vasculature and are conveyed to the anterior lobe of the pituitary gland by way of the hypophysial portal vessels.

In those mammals which have been examined, the hypophysial portal vasculature has been found to originate in two sets of primary capillary plexuses (15–18). One capillary plexus lies in the infundibular process of the posterior pituitary. These capillaries drain into portal vessels, sometimes called short portal vessels, which empty directly into sinusoids in the anterior pituitary, that is, the secondary capillary plexus of the portal vasculature. Another primary capillary plexus lies in the median eminence of the hypothalamus and pituitary stalk. These capillaries drain into the portal vessels of the pituitary stalk. These vessels are called the long portal vessels, and they too empty into the secondary capillary plexus of the portal vasculature which lies within the anterior pituitary. Although the anterior pituitary receives blood by way of both sets of portal vessels, the long portal vessels are believed to supply 70 to 90% of the blood of the anterior pituitary. The short portal vessels supply the remainder. This arrangement seems to exist in most mammalian species, including man (15, 19–25).

Since many workers agree that there is no direct innervation of the secretory cells of the anterior pituitary, we seem to be left with only two alternatives which may be involved in the control of release of hormones from this gland. These appear to be (*1*) that the cells of the anterior pituitary are innately unstable and spontaneously release hormones in a rhythmic or episodic manner, or (*2*) that hormonally-active substances reach the anterior pituitary via the portal vasculature in a rhythmic or episodic manner and thereby influence the rate of release of anterior pituitary hormones.

Isolated anterior pituitary tissue does release LH spontaneously, but the rate of release is quite constant as shown by Serra and Midgley (26), using glands superfused with a bicarbonate buffered solution. However, when the tissue was superfused with a solution containing an extract of hypothalamic tissue, the release of LH increased quickly but subsided with equal rapidity when the superfusion with hypothalamic extract ceased. We, too, have observed that the anterior pituitaries respond quickly by releasing LH when a solution containing hypothalamic extract was infused into a long hypophysial portal vessel (27). These observations support the view that the rhythmic or episodic release of hormones from the anterior pituitary is a function of the concentration of hypophysiotropic substances in portal blood.

Activities attributed to the following hypophysiotropic substances have been demonstrated in blood from the long portal vessels: corticotropin releasing factor (CRF) (28, 29); luteinizing hormone releasing factor (LRF) (30); follicle-stimulating hormone releasing factor (FRF) (31); prolactin inhibiting factor (PIF) (32); thyrotropin releasing factor (TRF) (33); and growth hormone releasing factor (GRF) (33). However, none of these studies revealed any information about the dynamics of the secretory rates of the hypophysiotropic substances. Recently, we have been able to superfuse anterior pituitary tissue with blood from the pituitary stalk and determine the rate of secretion of LH from the superfused glands (34, 35). We found that the release rate of LH was sometimes erratic but not rhythmic. However, this work was done in rats anesthetized with sodium pentobarbital, and the state of anesthesia may have affected the release of the LRF.

Despite the apparent absence of a rhythmic release of LH from anterior pituitaries superfused with the stalk blood from castrated rats, the release of LH from these pituitaries was 2- to 3-fold greater than that of glands superfused with arterial blood from the same castrated animals. When cycling female rats in diestrus were used as donors of stalk blood, the release of LH from glands superfused with stalk blood was identical with that of glands superfused with arterial blood. These observations indicate that the elevated rate of release of LH in castrated rats may be due to LRF in stalk blood and would suggest that the low rate of LH release in cycling rats during diestrus is due to a low level of LRF in portal blood.

It is not known whether or not hypophysiotropic substances reach the anterior pituitary by way of the short portal vessels. The presence of large quantities of vasopressin and oxytocin and possibly other polypeptides in the infundibular process suggests that hormones from this gland may be involved in the regulation of the anterior pituitary. Goldman and Lindner (36) calculated that the concentration of vasopressin in the effluent blood of the infundibular process can be 20,000 times greater than that seen in well-mixed peripheral blood. This fact, when taken in conjunction with the observation by Porter et al. (37) that blood passes quickly from the infundibular process to the anterior pituitary, leads one to suspect that the short portal vessels may be important in the control of the anterior pituitary.

Most work dealing with the neurohormonal aspects of anterior pituitary regulation has been concerned with the median eminence of the hypothalamus. It is generally believed at this time that hypophysiotropic substances diffuse into the primary capillaries in the median eminence whence they are carried to the anterior pituitary via the long portal vessels. There is much evidence to support this view. The following hypophysiotropic substances have been shown to be present in extracts of hypothalamic tissue: CRF (38), LRF (39), PIF (40), TRF (41), FRF (42, 43), GRF (44), a factor which inhibits the release of growth hormone (45), a factor which inhibits the release of melanocyte stimulating hormone (MIF) (46, 47), MSH releasing factor (48), and possibly a prolactin releasing factor (49). Whether each activity is attributable to a separate hypothalamic substance or whether one substance has two or more activities remains to be established. Recently, Burgus et al. (50) identified a tripeptide in hypothalami which has TRF activity and Matsuo et al. (51) and Amoss et al. (52) identified a decapeptide which has LRF activity.

Although these hypophysiotropic substances can be extracted from hypothalamic tissues, such procedures contribute little to our knowledge concerning their cellular source or sources. Indeed, it is not known whether neuronal, glial, or ependymal elements secrete the hypophysiotropic substances (53):

There has been much work on the content of hypophysiotropic substances in hypothalamic tissue (54, 55), and a diurnal variation in the hypothalamic content of CRF has been observed. David-Nelson and Brodish (56), Takebe et al. (57, 58), Hiroshige and Sakakura (59), and Hiroshige and Sato (60) have shown that the total content of CRF in hypothalamic tissue of the rat varies according to the time of day, being highest during daylight hours and lowest at night. Since the rat is a nocturnal animal, daylight hours correspond to times of least activity. However, it is difficult to interpret the meaning of such results. The justification for such investigations lies in the assumption that the CRF content of the hypothalamus is related to the release rate of CRF and in turn to the rate of release of ACTH. However, there is little basis for supposing that a close correlation between content and release actually exists. It is difficult to see how a rhythmic change in hypothalamic content of CRF which has a period of 24 hours can be related to the rate of release of ACTH which appears to change markedly many times each day (14).

The role of the median eminence of the hypothalamus in the regulation of the concentration of hypophysiotropic substances in hypophysial portal blood is still unknown. And, before an understanding of the rhythmical control of the anterior pituitary can be achieved, several questions have to be resolved. For example: Do some cellular elements of the median eminence—neuronal, glial, or ependymal—synthesize, store, and release hypophysiotropic substances? Or: Is the median eminence a tissue involved in the transport of hypophysiotropic substances from cerebrospinal fluid (CSF) to hypophysial portal blood? Although hypophysiotropic substances are present in the median eminence [cf. McCann and Porter (55)], this fact alone does not resolve either question. It is clear, however, that certain substances such as iodide salts and some polypeptides can pass rapidly from the CSF of the third ventricle into hypophysial portal blood (61).

Furthermore, 10 ng of synthetic LRF, when injected into the CSF of the third ventricle, promptly stimulates the release of LH. When 10 ng of LRF is injected into the CSF at the level of the cisterna magna, LH release is not stimulated. However, 100 ng of LRF injected into the third ventricle or cisterna magna stimulates the release of LH; but the releasing factor, when given via the cisterna magna, did not stimulate LH release until after a delay of 20 to 30 minutes, indicating that the substance probably reached the anterior pituitary via general circulation (62). These findings show that LRF in the CSF of the third ventricle can pass quickly to portal blood and also, when present in sufficiently high concentration in the CSF, LRF can enter general circulation, presumably passing from the subarachnoid space into the subdural sinuses (63), and reach the anterior pituitary via arterial blood. It is of interest that recent evidence by Ben-Jonathan et al. (35) indicates that the LRF levels in the blood of castrated rats is higher than that seen in cycling rats during diestrus. Thus it seems likely that the prime role of the median eminence is to transfer substances in the CSF to portal blood, thereby achieving in portal blood concentrations of hypophysiotropic substances higher than that present in peripheral blood.

Silverman et al. (64, 65), Silverman and Knigge (66), and Knigge and Silverman (67) have demonstrated that median eminence tissue can actively accumulate α-aminoisobutyric acid as well as thyroxin. Synthetic TRF is also accumulated by median eminence tissue (68). Kendall et al. (69) has also found that some substances pass from the CSF to the pituitary.

Much more work is needed before it can be said with assurance that the median eminence is either a secretory structure or a transport structure. However, if the median eminence should be found to be a secretory structure, then the basis for the rhythmic release of certain pituitary hormones may reside in the median eminence. On the other hand, if the median eminence is a transport structure, it is reasonable to suspect that some region of the brain other than the median eminence may regulate the rhythmic release of anterior pituitary hormones.

Certain neurotransmitters such as catecholamines, indoles, and cholinergic substances have been shown to affect the release of such hormones as LH, FSH, prolactin, growth hormone, and ACTH (70–82). However, for the most part the meaning of many of these observations is obscure. These transmitters or inhibitors are often administered by intravenous, intraperitoneal, or intraventricular routes, and, as a result, one rarely knows the site where they may be acting. Therefore, it may be that much refinement of experimental procedures must be made before it can be determined how a particular transmitter is involved in the regulation of the anterior pituitary. It has been difficult to determine whether catecholamines affect the release of pituitary hormones by acting directly on the anterior pituitary or by affecting the release of certain hypophysiotropic substances.

For example, MacLeod (83) and Birge et al. (84) found that catecholamines suppressed the release of prolactin under *in vitro* conditions. Yet Kamberi et al. (72) failed to obtain suppression of prolactin release *in vivo* when catecholamines were infused directly into the anterior pituitary via a cannulated portal vessel. The possible involvement of catecholamines in anterior pituitary regulation by direct action on the anterior pituitary is an attractive view in light of the extensive work of the Swedish workers who have demonstrated so elegantly the presence of catecholamine-containing neurons near the primary capillary plexus of the median eminence (85–89). It is possible that the concentration of dopamine, epinephrine, and norepinephrine in portal blood needs to be rather high to be effective. Kamberi et al. (72) infused solutions having concentrations of catecholamines which ranged from 10^{-8} to 10^{-11} M, whereas MacLeod (83) found concentrations of norepinephrine of 10^{-6} M to be effective. This concentration is not unrealistic since TRF and LRF, when infused into a portal vessel, were effective at concentrations around 10^{-6} M (34). Therefore the possibility that catecholamines may act directly on the anterior pituitary remains viable, and the periodic release of catecholamines into hypophysial portal blood may be important in regulating the rhythmic release of certain hormones from the anterior pituitary.

For this reason it is important to know whether or not various substances which may have a role in neurotransmission also undergo rhythmic changes in concentrations. Unfortunately, the analytical procedures and the design of testing for periodic oscillations are not sufficiently sophisticated to contribute in a meaningful fashion to this problem. Consequently, many pertinent questions remain unanswered. Despite these limitations, diurnal variation of several substances within the brain appears to occur. Reis and Gutnick (90) found that norepinephrine content of the spinal cord of the cat varied throughout a 24-hour period, but the rhythm of the oscillations were different in various parts of the cord. In the C_1 to C_2 segment of the cord, a rhythmic oscillation with a 24-hour period was observed. In the T_9 to T_{13} and in the sacral coccygeal segments, a 12-hour period was observed, whereas in the T_5 to T_8 segment, no rhythmic change in the concentration of norepinephrine was seen. A diurnal variation in the concentration of norepinephrine has been observed in the tuber cinereum, anterior hypothalamus, lateral hypothalamus, and pineal of the cat brain (91), as well as in other parts of the cat brain (92). Minor diurnal changes in the concentration of norepinephrine in the anterior hypothalamus and in the posterior hypothalamus have been observed in the rat (93). Although Schering et al. (94) failed to observe evidence for a diurnal variation in the dopamine or norepinephrine content of the whole brain of the rat, small differences in serotonin content were seen. Friedman and Walker (95) observed a diurnal rhythm for histamine, serotonin, and norepinephrine in the caudate nucleus of the rat. Serotonin levels in various parts of the cat brain varied throughout the day (96). Acetylcholine also varies in a diurnal fashion in the rat brain, being greatest in the presence of light (97, 98). Saito noted that the free as well as bound acetylcholine in the brain of the rat increased during periods of little muscular activity and decreased during periods of increased muscular activity. Thus it seems clear that histamine, norepinephrine, serotonin, as well as acetylcholine levels in various parts of the brain show rhythmical changes during the day.

It is not clear whether these changes are related in a mechanistic sense to the release of hormones from the anterior pituitary. It would seem, at least intuitively, that questions directed toward the total content of a substance in a tissue do not address critical issues. Such measurements do not take into account the fact that the brain is a heterogeneous organ, consisting of many qualitatively different cells. Furthermore, there are probably pools of a given substance within a single cell, and the extent of the involvement of the pools in a particular cellular function is probably dissimilar. There is need for techniques that will enable one to evaluate the dynamic activity, for example, secretory activity, of a discrete set of cells. In addition, the activity needs to be evaluated frequently. It is difficult to relate brain rhythms having 12- to 24-hour periods to anterior pituitary rhythms, for example, LH release in the castrated monkey or rat (5, 10), which have periods of oscillation ranging from a few minutes to 1 to 2 hours. Despite seemingly difficult problems and despite our inability to do more than correlate observations which are themselves often sparsely supported, there is a growing awareness that rhythms are characteristic of neuroendocrine systems.

ACKNOWLEDGMENT

The authors wish to thank Jane C. Gottwald for assistance in the preparation of this manuscript.

This research was supported in part by a grant from the National Institutes of Health, National Institute for Arthritis and Metabolic Diseases (AM01237), and a grant from the Population Council, New York, New York.

REFERENCES

1. C. M. Cargille, G. T. Ross, and T. Yoshimi, *J. Clin. Endocr.*, **29**, (1969), 12.

2. S. E. Monroe, L. E. Atkinson, and E. Knobil, *Endocrinology*, **87**, (1970), 453.

3. B. D. Goldman, I. A. Kamberi, P. K. Siiteri, and J. C. Porter, *Endocrinology*, **85**, (1969), 1137.

4. B. D. Goldman and J. C. Porter *Endocrinology*, **87**, (1970), 676.

5. D. J. Dierschke, A. N. Bhattacharya, L. E. Atkinson, and E. Knobil, *Endocrinology*, **87**, (1970), 850.

6. L. E. Atkinson, A. N. Bhattacharya, S. E. Monroe, D. J. Dierschke, and E. Knobil, *Endocrinology*, **87**, (1970), 847.

7. I. E. Lawton and N. B. Schwartz, *Amer. J. Physiol.*, **214**, (1968), 213.

8. W. J. Digman, A. F. Parlow, and T. A. Daane, *Amer. J. Obstet. Gynecol.*, **105**, (1969), 679.

9. M. Yamamoto, N. D. Diebel, and E. M. Bogdanove, *Endocrinology*, **87**, (1970), 798.

10. V. L. Gay and N. A. Sheth, *Endocrinology*, **90**, (1972), 158.

11. V. L. Gay and A. R. Midgley, Jr., *Endocrinology*, **84**, (1969), 1359.

12. J. D. Neill, *Endocrinology*, **90**, (1972), 568.

13. M. E. Freeman and J. D. Neill, *Endocrinology*, **90**, (1972), 1292.

14. D. T. Krieger, W. Allen, F. Rizzo, and H. P. Krieger, *J. Clin. Endocr.*, **32**, (1971), 266.

15. P. M. Daniel and M. M. L. Prichard, *Quart. J. Exp. Physiol.*, **41**, (1956), 215.

16. H. Duvernoy, "Contribution a l'étude de la vascularisation de l'hypophyse," Thèse, École Natl Méd de Besançon, Paris, 1958.

17. J. M. F. Landsmeer, "Het vaatstelsel van de hypophyse bij de witte rat," Thesis for Doctorate, University of Leiden, 1947.

18. J. M. F. Landsmeer, *Acta Anat.* (Basel), **12**, (1951), 82.

19. J. H. Adams, P. M. Daniel, and M. M. L. Prichard, *Nature*, **198**, (1963), 1205.

20. J. H. Adams, P. M. Daniel, and M. M. L. Prichard, *Acta Endocr.* (Copenhagen) *Suppl.*, **81**, (1963), 3.

21. J. H. Adams, P. M. Daniel, M. M. L. Prichard, and P. H. Schurr, *J. Physiol.* (London), **166**, (1963), 39P.

22. J. H. Adams, P. M. Daniel, and M. M. L. Prichard, *J. Pathol. Bacteriol.*, **87**, (1964), 1.

23. J. H. Adams, P. M. Daniel, and M. M. L. Prichard, *Neuroendocrinology*, **1**, (1965/66), 193.

24. J. H. Adams, P. M. Daniel, and M. M. L. Prichard, *Acta Endocr.* (Copenhagen), **51**, (1966), 377.

25. J. C. Porter, M. F. M. Hines, K. R. Smith, R. L. Repass, and A. J. K. Smith, *Endocrinology,* **80,** (1967), 583.

26. G. B. Serra and A. R. Midgley, Jr., *Proc. Soc. Exp. Biol. Med.,* **133,** (1970), 1370.

27. I. A. Kamberi, R. S. Mical, and J. C. Porter, *Endocrinology,* **88,** (1971), 1294.

28. J. C. Porter and J. C. Jones, *Endocrinology,* **58,** (1956), 62.

29. J. C. Porter and H. W. Rumsfeld, Jr., *Endocrinology,* **58,** (1956), 359.

30. I. A. Kamberi, R. S. Mical, and J. C. Porter, *Science,* **166,** (1969), 388.

31. I. A. Kamberi, R. S. Mical, and J. C. Porter, *Nature,* **277,** (1970), 714.

32. I. A. Kamberi, R. S. Mical, and J. C. Porter, *Experientia,* **26,** (1970), 1150.

33. J. F. Wilber and J. C. Porter, *Endocrinology,* **87,** (1970), 807.

34. J. C. Porter, R. S. Mical, N. Ben-Jonathan, and J. G. Ondo, *Rec. Progr. Hormone Res.,* **29,** (1973), 161.

35. N. Ben-Jonathan, R. S. Mical, and J. C. Porter, Unpublished Observations.

36. H. Goldman and L. Lindner, *Experientia,* **18,** (1962), 279.

37. J. C. Porter, J. G. Ondo, and O. M. Cramer, in *Handbook of Physiology.* (E. B. Astwood and R. O. Greep, Eds.), Williams & Wilkins, Baltimore, In Press.

38. H. W. Rumsfeld, Jr., and J. C. Porter, *Arch. Biochem. Biophys.,* **82,** (1959), 473.

39. S. M. McCann, S. Taleisnik, and H. M. Friedman, *Proc. Soc. Exp. Biol. Med.,* **104,** (1960), 432.

40. J. L. Pasteels, *C. R. Acad. Sci., Paris, Ser. D,* **253,** (1961), 3074.

41. R. Guillemin, E. Yamazaki, M. Jutisz, and E. Sakiz, *C. R. Acad. Sci., Paris, Ser. D,* **255,** (1962), 1018.

42. M. Igarashi and S. M. McCann, *Endocrinology,* **74,** (1964), 446.

43. J. C. Mittler and J. Meites, *Proc. Soc. Exp. Biol. Med.,* **117,** (1964), 309.

44. R. R. Deuben and J. Meites, *Endocrinology,* **74,** (1964), 408.

45. L. Krulich, A. P. S. Dhariwal, and S. M. McCann, *Endocrinology,* **83,** (1968), 783.

46. A. J. Kastin and A. V. Schally, *Gen. Comp. Endocr.,* **7,** (1966), 452.

47. A. V. Schally and A. J. Kastin, *Endocrinology,* **79,** (1966), 768.

48. S. Taleisnik and R. A. Orías, *Amer. J. Physiol.,* **208,** (1965), 293.

49. C. S. Nicoll and R. P. Fiorindo, *Gen. Comp. Endocr. Suppl.,* **2,** (1969), 26.

50. R. Burgus, T. F. Dunn, D. Desiderio, and R. Guillemin, *C. R. Acad. Sci., Paris., Ser. D,* **269,** (1969), 1870.

51. H. Matsuo, Y. Baba, R. M. G. Nair, A. Arimura, and A. V. Schally, *Biochem. Biophys. Res. Commun.,* **43,** (1971), 1334.

52. M. Amoss, R. Burgus, R. Blackwell, W. Vale, R. Fellows, and R. Guillemin, *Biochem. Biophys. Res. Commun.,* **44,** (1971), 205.

53. J. C. Porter, *Progr. Brain Res.,* **39,** In Press.

54. S. M. McCann, A. P. S. Dhariwal, and J. C. Porter, *Ann. Rev. Physiol.,* **30,** (1968), 589.

55. S. M. McCann and J. C. Porter, *Physiol. Rev.,* **49,** (1969), 240.

56. M. A. David-Nelson and A. Brodish, *Endocrinology,* **85,** (1969), 861.

57. K. Takebe, M. Sakakura, Y. Horiuchi, and K. Mashimo, *Endocrinol. Japan* **18,** (1971), 451.

58. K. Takebe, M. Sakakura, and K. Mashimo, *Endocrinology,* **90,** (1972), 1515.

59. T. Hiroshige and M. Sakakura, *Neuroendocrinology,* **7,** (1971), 25.
60. T. Hiroshige and T. Sato, *Endocrinol. Japan* **17,** (1970), 1.
61. J. G. Ondo, R. S. Mical, and J. C. Porter, *Endocrinology,* **91,** (1972), 1239.
62. J. G. Ondo, R. L. Eskay, R. S. Mical, and J. C. Porter, *Endocrinology,* **93,** (1973), 231.
63. H. Davson, *Physiology of the Cerebrospinal Fluid,* Churchill, London, 1967.
64. A.-J. Silverman, K. M. Knigge, and W. A. Peck, *Neuroendocrinology,* **9,** (1972).
65. A.-J. Silverman, K. M. Knigge, J. L. Ribas, and M. N. Sheridan, *Neuroendocrinology,* **11,** (1973b), 107.
66. A.-J. Silverman and K. M. Knigge, *Neuroendocrinology,* **10,** (1972), 71.
67. K. M. Knigge and A.-J. Silverman, In *Brain-Endocrine Interaction Median Eminence: Structure and Function* (K. M. Knigge, D. E. Scott, and A. Weindl, Eds.), Karger, Basel, 1972, p. 350.
68. K. M. Knigge, S. A. Joseph, A.-J. Silverman, and S. Vaala, *Progr. Brain Res.,* In Press.
69. J. W. Kendall, J. J. Jacobs, and R. M. Kramer, In *Brain-Endocrine Interaction. Median Eminence: Structure and Function,* (K. M. Knigge, D. E. Scott, and A. Weindl, Eds.), Karger, Basel, 1972, p. 342.
70. I. A. Kamberi, R. S. Mical, and J. C. Porter, *Endocrinology,* **87,** (1970), 1.
71. I. A. Kamberi, R. S. Mical, and J. C. Porter, *Endocrinology,* **88,** (1971), 1003.
72. I. A. Kamberi, R. S. Mical, and J. C. Porter, *Endocrinology,* **88,** (1971), 1012.
73. I. A. Kamberi, R. S. Mical, J. C. Porter, *Endocrinology,* **88,** (1971), 1288.
74. O.M. Cramer and J. C. Porter, *Progr. Brain Res.,* **39,** In Press.
75. J. C. Porter, R. S. Mical, and O. M. Cramer, *Gynecol. Invest.,* **2,** (1971/72), 13.
76. H. P. G. Schneider and S. M. McCann, *Endocrinology,* **85,** (1969), 121.
77. C. H. Sawyer, *Anat. Rec.,* **112,** (1952), 385.
78. C. Kordon, F. Javoy, G. Vassent, and J. Glowinski, *Eur. J. Pharmacol.,* **4,** (1968), 169.
79. D. T. Krieger and F. Rizzo, *Amer. J. Physiol.,* **217,** (1969), 1703.
80. D. T. Krieger, A. I. Silverberg, F. Rizzo, and H. P. Krieger, *Amer. J. Physiol.,* **215,** (1968), 959.
81. R. Collu, F. Fraschini, P. Visconti, and L. Martini, *Endocrinology,* **90,** (1972), 1231.
82. C. Libertun and S. M. McCann, *Program of the 5th Annual Meeting of the Society for the Study of Reproduction,* 1972, p. 32 (Abstract).
83. R. M. MacLeod, *Endocrinology,* **85,** (1969), 916.
84. C. A. Birge, L. S. Jacobs, C. T. Hammer, and W. H. Daughaday, *Endocrinology,* **86,** (1970), 120.
85. K. Fuxe, *Z. Zellforsch.,* **61,** (1964), 710.
86. B. Falck, N. Å Hillarp, G. Thieme, and A. Torp, *J. Histochem. Cytochem.,* **10,** (1962), 348.
87. N. Å. Hillarp, K. Fuxe, and A. Dahlström, in *Mechanisms of Release of Biogenic Amines* (U. S. von Euler, S. Rosell, and B. Uvnäs, Eds.), Pergamon, New York, 1966, p. 31.
88. K. Fuxe, Hökfelt, and O. Nilsson, *Life Sci.,* **6,** (1967), 2057.

89. K. Fuxe and T. Hökfelt, in *Frontiers in Neuroendocrinology* (W. F. Ganong and L. Martini, Eds.), Oxford Univ. Press, London, 1969, p. 47.

90. D. J. Reis and E. Gutnick, *Amer. J. Physiol.*, **218**, (1970), 1707.

91. D. J. Reis and R. J. Wurtman, *Life Sci.*, **7**, (1968), 91.

92. D. J. Reis, M. Weinbren, and A. Corvelli, *J. Pharmacol. Exp. Therap.*, **164**, (1968), 135.

93. J. Manshardt and R. J. Wurtman, *Nature*, **217**, (1968), 574.

94. L. E. Schering, W. H. Harrison, P. Gordon, and J. E. Pauly, *Amer. J. Physiol.*, **214**, (1968), 166.

95. A. H. Friedman and C. A. Walker, *J. Physiol.* (London), **197**, (1968), 77.

96. D. J. Reis, A. Corvelli, and J. Conners, *J. Pharmacol. Exp. Therap.*, **167**, (1969), 328.

97. I. Hanin, R. Massarelli, and E. Costa, *Science*, **170**, (1970), 341.

98. Y. Saito, *Life Sci.*, **10**, (1971), 735.

CHAPTER 41

Biorhythms in Central Nervous System Disease (Primarily Those of Pituitary-Adrenal Hormones)

DOROTHY T. KRIEGER

Department of Medicine, Division of Endocrinology, Mount Sinai School of Medicine of the City University of New York, New York, New York

Recognition of the periodicity of many biological variables and processes has raised many questions with regard to the origin of such rhythms; for example, whether they are endogenous or exogenous, possible entraining influences for such rhythms, interrelationships between these various rhythms, possible location of "master clocks" and pathways and mediators involved both in translating environmental influences on rhythms, and effect of one biological system on another. With increasing recognition of the role of the central nervous system in the regulation of many endocrine homeostatic mechanisms, and suggestive evidence of circadian nervous system activity as manifest by sleep-wake rhythmicity, circadian variation of the human electroencephalograph (1), and circadian variation in central nervous system neurotransmitter content (2–5), one would expect periodic endocrine rhythms to be affected by disease of areas of the central nervous system involved in endocrine regulation.

This report will not attempt to deal with any possible alteration of the normal rhythmicity of such diverse parameters as body temperature, cardiovascular function, vital capacity, rate of urine flow, sleep, and intraocular tension. These are all areas that require further study in patients with localized central nervous system disease to delineate the role of the central nervous system in their regulation. The major focus of this report will be on the effect of central nervous system disease on endocrine biorhythms. The periodic release of a number of hormones has been established. That of growth hormone (6) is not a truly circadian rhythm, but rather a sleep associated one. There is still controversy with regard to the circadian periodicity of gonadotropin release, with conflicting reports concerning the presence or absence of periodicity of follicle-stimulating hormone release (7–9) and general agreement concerning a lack of circadian periodicity of luteinizing hormone release (9, 10), though smaller cyclic variations may be present (see also Chapters 23 and 29). This latter finding is in keeping with evidence reporting no (11) or minimal (12) circadian variation in human plasma testosterone levels, although a recent report does describe such variation (13). Very recently a circadian periodicity of serum prolactin (14) and thyrotropin (15) levels has been reported in normal subjects. No studies are available as yet as to alterations of these rhythms in central nervous system disease.

The present report will deal mainly with our studies of alterations in pituitary-adrenal (glucocorticoid) rhythms since the periodicity of this endocrine parameter has been most clearly delineated and its underlying neural regulatory pathways have been extensively studied. There is ample evidence, in both the human and many animal species, of a 24-hour periodicity of pituitary-adrenal function (16–20), as well as evidence that ACTH periodicity persists in the absence of the adrenal gland (21), and that corticotropin-releasing factor periodicity persists in hypophysectomized animals (22), indicating a prime role of the central nervous system in the regulation of such periodicity. Hypothalamic lesions (23), anterior hypothalamic deafferentiation (24), fornix section (25), and septal area lesions (26), as well as hippocampal implantation of cortisone (27), or alteration of central nervous system amine levels (28), 29), have been reported to disrupt the circadian periodicity of adrenal corticosteroid levels in many instances without associated disruption of feedback responsiveness or stress-evoked pituitary-adrenal responses. It might therefore be expected that lesions involving the above cited areas in human central nervous disease would be associated with alterations in circadian periodicity of ACTH and corticosteriod levels.

In some of our studies, concomittant blood sampling for growth hormone determination as well as sleep EEG recording were also performed to further determine how specific the conditions studied were for alteration of a given hormone pattern, as well as in an attempt to delineate the association of a given hormonal change with specific neurophysiological events.

The subject material comprising this report consists of:

1. Patients with discrete localized central nervous system disease.
2. Patients with ocular blindness.
3. Normal subjects living under constant light conditions (studied to assess the role of light-dark alternation on periodicity).
4. Patients with Cushing's disease associated with bilateral adrenocortical hyperfunction (in whom physiological central nervous system dysfunction may play an etiological role).
5. Patients with acromegaly.

The studies to be described deal only with circadian periodicity, although other aspects of neuroendocrine function may be altered in these conditions. It is important to note, however, that alterations in circadian periodicity of adrenal corticosteroid levels may occur in patients with central nervous system disease (30), in experimentally produced hypothalamic lesions in animals (23), or following alteration in central nervous system amine levels (28, 29) *without* any alteration in normal hypothalamic-pituitary-adrenal responsiveness to stress. This would imply that there are different pathways and/or transmitters involved in the regulation of these two aspects of hypothalamic-pituitary-adrenal function.

NORMAL PERIODICITY OF PITUITARY-ADRENAL FUNCTION

Before describing alterations in circadian periodicity encountered in disease of the central nervous system, it is important to have an accurate characterization of such periodicity in normal subjects.

The existence of a circadian periodicity of plasma corticosteroid levels in normal human subjects has been well established, (16, 17), and is accompanied by a similar variation in plasma adrenocorticotropic hormone levels (31). These studies, which have been based on sampling every 4 to 6 hours over the 24-hour period, have characterized the circadian pattern as being one in which peak levels

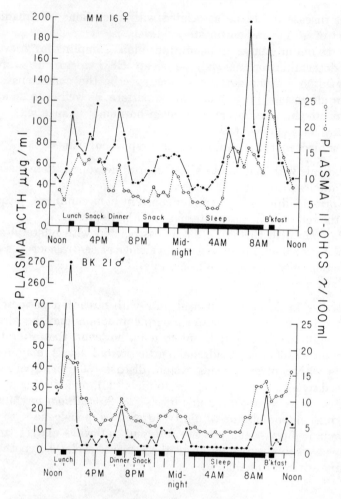

Figure 1 Circadian periodicity of plasma 11-OHCS and ACTH levels over a 24-hour period as determined by half-hourly sampling. Meal times and sleep as indicated (37).

occur in the early morning hours, with a progressive decline over the remainder of the 24-hour period. Reversal of the day-night living schedule results in a phase reversal of the circadian pattern within a period of approximately 8 days (17, 32, 33). As studies progressed it became apparent that further refinements were necessary. In view of the reported half-life of cortisol, which varies between 60 and 90 minutes (34), it was felt that half-hourly sampling would more adequately depict the normal pattern of adrenal secretion. It was felt that criteria derived from such studies could then be used to provide criteria for normal periodicity when sampling was performed less frequently, as well as to define optimal sampling time and frequency. In addition, data in normal individuals as to the effect of age, sex, hospitalization, and pattern reproducibility on successive occasions were necessary before conclusions could be drawn on data obtained from a given patient or experimental subject group.

Studies were therefore performed (35–37) utilizing sampling at half-hourly intervals over a 24- or 48-hour period, the 48-hour studies being either over two sequential days or spaced apart at intervals of a week or several months. In addition, studies were performed comparing the circadian periodicity of ambulatory normal volunteers and similar subjects in a hospital setting, as well of normal subjects of both sexes between 15 and 95 years of age (38).

It became apparent in normal subjects (Figure 1) (37) that synchronous episodic peaks of plasma ACTH and corticosteroid levels occurred throughout the day, with the majority of peaks occurring in the 9-hour period from midnight to 9:00 A.M. and describing a gradual upward course in that time period, with maximum plasma levels occurring 1/2 to 2 hours following awakening. A downward trend then followed, with other peaking activity occurring between 11:30 A.M. and 2:00 P.M. and 4:30 P.M. and 6:00 P.M., with quiescent periods at other intervals. There was close temporal correlation between plasma ACTH and corticosteroid peaks.

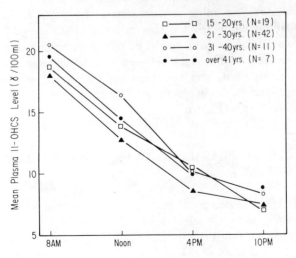

Figure 2   Mean plasma 11-OHCS levels for different age groups (37).

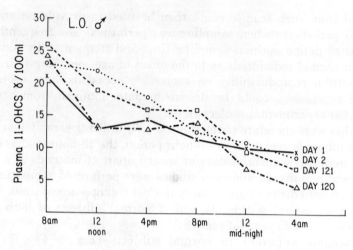

Figure 3   Reproducibility of circadian periodicity in normal subjects over a 120-day period (37).

These half-hourly sampling patterns were utilized in constructing criteria for establishing the presence of normal periodicity under clinical conditions where less frequent sampling is feasible. Normal periodicity of plasma corticosteroid levels was defined as one in which all corticosteroid values after 8:00 A. M. are less than 75% of the 8:00 A.M. level, excluding consideration of noontime levels because of the great variability in the amount and height of peaking at this time. Noon levels 15% or more *greater* than the 8:00 A.M. level are considered abnormal. Utilizing this criteria, a survey of 106 normal subjects gave a false positive rate (i.e., characterization of a given pattern as abnormal) of 2%. Age (between 15 and 95 years), sex (Figure 2), and hospitalization had no effect on circadian periodicity. In a given subject, corticosteroid levels and pattern were reporducible over 1 to 120 day intervals (Figure 3).

PERIODICITY OF PLASMA GROWTH HORMONE LEVELS IN NORMAL SUBJECTS

The occurrence of a sleep-associated rise in plasma growth hormone levels coincident with the electroencephalographic stage of slow-wave sleep is now well established (6). This is not a true circadian periodicity, since the onset of this rise may be delayed if sleep onset is delayed, and a second rise may be seen if the subject is awakened and allowed to resume sleep again. Unlike the stress-provoked release of growth hormone, this nocturnal growth hormone rise is not suppressed by physiological levels of glucose (39).

Figure 4 indicates simultaneous patterns of plasma cortisol, ACTH, and growth hormone levels in a normal male subject; Figure 5 indicates that although male and female subjects both evidence the sleep-associated growth hormone peak, plasma growth hormone levels during the course of the day are more labile in female than in male subjects, similar to the lability of growth hormone levels in response to other stimuli noted in normal females or estrogen-treated males (40,41).

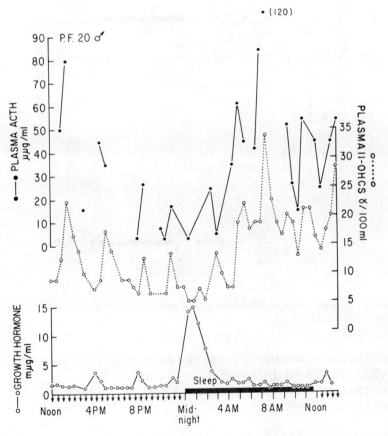

Figure 4 Circadian periodicity of plasma cortisol, ACTH, and growth hormone levels (half-hourly sampling) in a normal male subject. Arrows on time scale indicate feeding times.

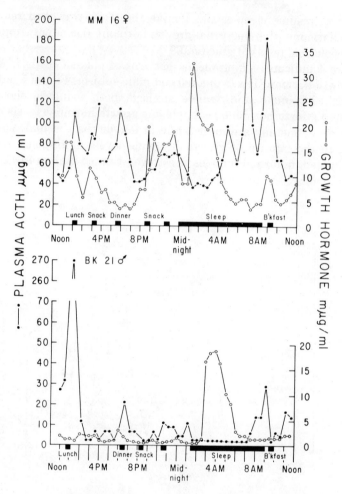

Figure 5   Periodic variation of plasma ACTH and growth hormone levels in a normal male and female subject.

CIRCADIAN PERIODICITY OF PLASMA CORTICOSTEROID LEVELS, GROWTH HORMONE AND URINARY ELECTROLYTES IN LOCALIZED CENTRAL NERVOUS SYSTEM DISEASE

Alterations in the circadian pattern of plasma corticosteriods in patients with chronic, diffuse central nervous system disease accompanied by impairment of consciousness, delirium, or restlessness were first reported by Perkoff et al. (17). In view of the anatomical, physiological, and clinical observations suggesting the major role of the hypothalamus and limbic system in the regulation of pituitary adrenal function, it was felt that alterations in the circadian pattern of plasma corticosteroid levels might occur in patients with central nervous system disease localized to these areas, unaccompanied by any disturbance of consciousness.

To date we have studied 43 such patients utilizing, for the most part, sampling every 4 hours over the 24-hour cycle (30, 42–44). Lesion localization was obtained by clinical and radiographic techniques in all instances. Patients were hospitalized at the time of study; all were afebrile and ambulatory. None exhibited restlessness, delirium, or impairment of consciousness.

Twenty-three of these patients (53%) had abnormal circadian patterns of plasma corticosteroid levels, as considered by the criteria for normalcy presented earlier, as well as by analysis (42) of the slope of linear trend (calculated by the method of least squares) and extent of variation of plasma corticosteroid levels from this fitted line. Representative patterns in one group of these patients are shown in Figure 6. Figure 7 depicts patterns in two patients with hypothalamic tumors who were studied by half-hourly sampling. In one (I.L.), a normal periodicity of plasma corticosteroid levels is seen, in the other (H.G.), an irregularly oscillatory pattern of both ACTH and plasma corticosteroid levels is present. Both of these patients would also have been classified correctly as having normal or abnormal patterns on the basis of their q.4 hourly sampling studies.

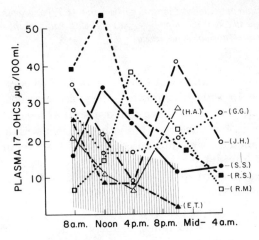

Figure 6 Circadian periodicity of plasma corticosteroid levels in seven patients with disease localized to the pretectal area. Vertical lines indicate two standard deviations (2 SD) about the mean for normal subjects (36).

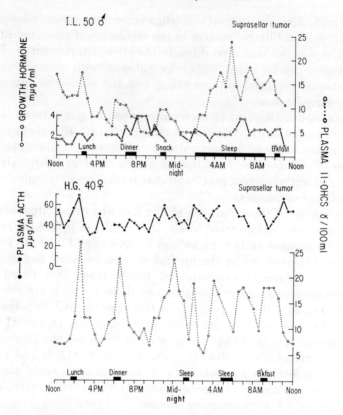

Figure 7   Circadian periodicity of plasma cortisol and ACTH or growth hormone levels in two endocrinologically normal, nonobese patients with radiographically demonstrated hypothalamic tumors (normal sella turcica).

It is of interest that patient I.L. (with a normal plasma corticosteroid pattern) failed to show the normal sleep associated rise in plasma *growth hormone* levels. We have noted a similar absence of the sleep-associated growth hormone rise in another patient with a hypothalamic tumor (not depicted), who additionally manifested abnormal circadian periodicity of plasma corticosteroid levels. No sleep electroencephalograms were performed in these patients; routine electroencephalograms were normal.

Finkelstein et al. (45) cited three cases of children with hypothalamic tumors who failed to display any nocturnal release of growth hormone in the presence of adequate amounts of stage III to IV sleep. These subjects were also reported to manifest absent growth hormone responsiveness to insulin, arginine, or piromen, and had other endocrine deficiencies, as well as being extremely obese. Quabbe (46) has noted absent or reduced nocturnal growth hormone elevations in obese children, and it is known that growth hormone responsiveness to various provocative stimuli may be reduced or even abolished in obese subjects (47). The significance, therefore, of Finkelstein's findings with regard to a correlation between the hypothalamic disease and lack of nocturnal growth hormone elevation awaits further clarification.

There was a paucity of other abnormal endocrine findings in the group of patients with hypothalamic-limbic system disease that we have studied. Five of the 43 had absent or inappropriately low urinary gonadotropin levels; all but one had normal protein-bound iodine levels and RAI uptake; three of 40 patients had levels of urinary 17-OHCS below the lower limit of our normal range (less than 3 mg/24 hours), but only one of these displayed abnormal Metopirone® responsiveness. Utilizing responsiveness to insulin hypoglycemia as a further index of neuroendocrine function, 11 of 11 patients had a normal increase of plasma corticosteroid levels (five of these with abnormal circadian patterns of plasma corticosteroid levels); and 10 of 13 had normal growth

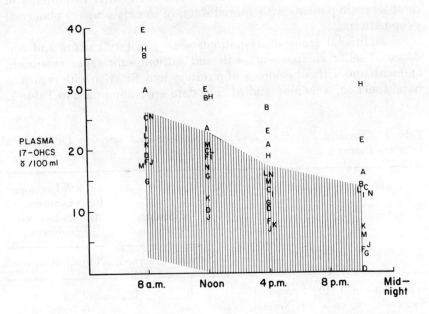

Figure 8   Plasma corticosteroid levels in patients with central nervous system disease radiographically localized to either the cerebral hemisphere, sphenoid ridge, cerebellum, or spinal cord. Letters refer to individual patients, each sampled at all of the times noted.

hormone increments following such hypoglycemia (seven of these had an abnormal plasma corticosteroid circadian pattern.) Only one of the three patients with an abnormal growth hormone response had associated absent gonadotropin levels.

As controls, 21 patients with central nervous system disease not involving the hypothalamus or limbic area were studied. Again, all were ambulatory with no disturbance of consciousness; basal tests of gonadal, adrenal, and thyroid function were normal. Disease was localized to the cerebral hemisphere, sphenoid ridge area, cerebellum, pons, or spinal cord. Only one patient showed an abnormal circadian pattern (patient H, Figure 8, in which patterns of 14 of these patients are depicted). It is of interest that three patients with normal patterns had elevated plasma corticosteroid levels throughout the 24-hour period, although urinary 17-OHCS levels were normal and there were no clinical stigmata of adrenal hyperfunction. It may well be that half-hourly sampling on such patients would show marked variation in plasma corticosteroid levels, so that the 24-hour average would be within the normal range. Perkoff (17) also noted normal circadian variation in patients with elevated 8:00 A.M. levels (disease entity not stated), as well as elevated 8:00 A.M. levels and greater variabililty of these levels in patients with altered states of consciousness or abnormal sleep pattern.

An additional group of 21 patients with pituitary tumors and evidence of sellar enlargement, with and without suprasellar extension, and with and without evidence of pituitary hypofunction with regard to basal function, were also studied. The data are summarized in Table 1.

Table 1   Circadian Periodicity of Plasma Corticosteroid Levels in Patients with Pituitary Tumors

	Basal Hormonal Status	Circadian Periodicity		Growth Hormone Responsiveness to Insulin-Induced Hypoglycemia	
		Normal	Abnormal	Normal	Abnormal
Intrasellar tumor (10)[a]	Normal	6	4	2	8
Extrasellar extension (6)	Normal	5	1	0	6
Intrasellar tumor (5)	Absent FSH	3	2	Not done	

[a] The parenthetical value indicates the number of patients.

In this limited series it would appear that any abnormal periodicity of plasma corticosteroid levels encountered (noted in seven of these 21 patients) was not related to the presence or absence of gonadotropin abnormalities; was, if anything, less frequent in those tumors with extrasellar extension; and was a less frequent finding than abnormalities in growth hormone responsiveness. This is in contrast to the group of patients with hypothalamic disease where abnormal circadian periodicity of plasma corticosteroid levels was more frequently encountered than disturbances in growth hormone responsiveness.

Studies have also been performed on eight patients with acromegaly. Circadian patterns of plasma corticosteroid levels were normal in seven. The one patient with an abnormal pattern had no evidence of extrasellar extension; those with normal patterns had either normal or enlarged sellas, with or without evidence of extrasellar involvement. Figure 9 depicts half-hourly studies on three such patients. In marked contrast to the normal circadian periodicity of plasma corticosteroid levels is the irregular outpouring of growth hormone seen in these patients.

To date we have not investigated circadian periodicity of other pituitary trophic hormones in cases of central nervous system disease. Limited studies have been performed with regard to the *circadian periodicity of urinary electrolyte excretion* in central nervous system disease. Normal subjects show a well-defined pattern of urinary volume and sodium, potassium, and chloride excretion (48), as well as a circadian variation of plasma aldosterone and plasma renin activity independent of position or diet (49). Absence of a circadian pattern in sodium, potassium, and chloride excretion was reported in three patients with disease involving either the upper cervical spinal cord, brain stem, or reticular activating system, and in two comatose patients (50). We have studied (51) eight control subjects, 10 patients with delimited hypothalamic disease, six patients with extrahypothalamic central nervous system disease, and five patients with chromophobe adenomata, all without evidence of clinically significant renal, cardiac, or hepatic disease. All received a calculated diet whose sodium and potassium content was known for 3 days prior to study. Urine volume and electrolytes were analyzed on aliquots of urine obtained at four or six hourly intervals on the study day.

A normal circadian pattern for sodium, potassium, and chloride excretion and urinary volume is considered to be one in which peak values are attained during waking hours, with lowest values being reached during sleep periods. All six patients with extrahypothalamic central nervous system disease exhibited such normal patterns as well

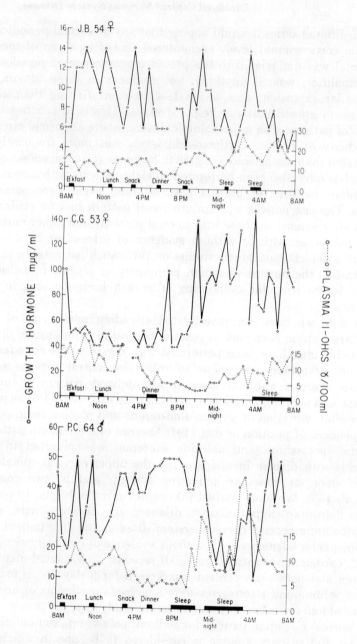

Figure 9 Periodicity of plasma corticosteroid levels in three untreated acromegalic patients with clinically active disease. Sella turcica was roentgenographically normal in patient J.B., and enlarged, but with no evidence of suprasellar extension, in patients C.G. and P.C.

as normal patterns of plasma corticosteroid levels. In contrast, five of the 10 patients with hypothalamic disease had abnormal patterns of urinary potassium excretion, with peak levels noted during the sleep period. Two of these also had reversal of the normal pattern of urinary volume excretion. Urinary sodium and chloride excretion patterns tended to concur with those of potassium excretion but were less consistent. There was no correlation between patients who showed abnormalities in the circadian periodicity of plasma corticosteroid levels and those who showed abnormal electrolyte patterns. The patterns seen in the patients with pituitary disease showed a normal phase of electrolyte excretion but with marked flattening occurring between the peak and trough levels.

SUMMARY OF STUDIES ON PATIENTS WITH DELIMITED CENTRAL NERVOUS SYSTEM DISEASE

It is obvious that even with meticulous radiographic demonstration of the extent of a central nervous system lesion in the human subject, that specific localization to an anatomically discrete area involving a given nucleus or tract is not possible. This may be why only slightly more than half of the patients with hypothalamic-limbic system disease had abnormal circadian patterns of plasma corticosteroid levels. It would appear, however, that such an abnormality is seen only with involvement of this area of the central nervous system, and is also the most sensitive indicator of altered neuroendocrine dysfunction in these patients. On the basis of extremely limited studies, it is suggested that similar findings will pertain to the sleep associated rise in plasma growth hormone levels. Such abnormalities in the circadian periodicity of plasma corticosteroid levels are far less frequent in patients with nonfunctioning pituitary tumors, and are virtually absent in patients with functioning pituitary tumors producing an excess of growth hormone, where erratic growth hormone patterns but normal corticosteroid patterns are seen. Similarly, abnormalities in the pattern of urinary electrolyte excretion are seen in patients with hypothalamic disease but not in those with extrahypothalamic central nervous system disease or nonfunctioning pituitary tumors.

These studies would therefore indicate that the mechanism(s) involved in the regulation of circadian periodicity of a number of variables reside within a relatively localized area of the central nervous system, although they do not define the mechanism of "setting" of such a clock or clocks and their interrelations. It would be expected that similar studies in patients with delimited central nervous system disease of patterns of other physiological or chemical variables known to have a normal circadian variation would yield additional valuable information with regard to these questions.

CIRCADIAN PERIODICITY OF PLASMA CORTICOSTEROID AND GROWTH HORMONE LEVELS IN PATIENTS WITH CUSHING'S DISEASE

Since Cushing's description in 1912 (52) of the syndrome that bears his name, there has been considerable speculation with regard to its etiology in those instances unassociated with adrenal tumor. As early as 1944 the suggestion that the primary site of pathology in this disease was within the central nervous system was put forward by Heinbecker (53) based on observed postmortem pathological changes in the hypothalamic paraventricular nuclei of patients with this syndrome. With growing awareness of the role of the central nervous system in the regulation of ACTH synthesis and release (54) under basal conditions as well as those of steroid feedback, it is possible that a number of the derangements of endocrine function seen in patients with Cushing's syndrome could be secondary to a functional alteration in the neurotransmitter pathways involved in such regulation (55, 56).

Abnormal circadian periodicity of plasma corticosteroid levels in Cushing's disease was first described in a single patient with bilateral adrenal hyperplasia by Lindsay et al. (57), and subsequently confirmed by Doe et al. (58) and Ekman et al. (59) in studies utilizing four hourly sampling. This abnormality consists of a flattening of the circadian pattern, so that virtually no alteration in plasma levels is seen over the course of the 24-hour period. The early morning levels may or may not be elevated; the major abnormality is an absence of the normal subsequent fall of such levels.

With the demonstration, utilizing half-hourly sampling, that the circadian rise of plasma corticosteroid and ACTH levels in normal subjects was attained in a series of abrupt short peaks (35–37), it became of interest to reexamine the circadian periodicity of such levels in greater detail in patients with Cushing's disease (Figures 10 and 11). It is apparent that plasma corticosteroid levels in these patients describe a seemingly random pattern of oscillations; and unlike the pattern seen in normal subjects, there is no tendency for the greatest number of peaks to occur during the midnight to 9:00 A.M. period, and there are fewer periods of quiescent activity.

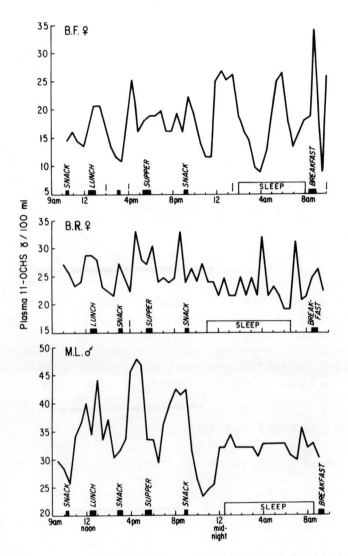

Figure 10 Circadian periodicity of plasma corticosteroid levels in patients with active Cushing's syndrome secondary to bilateral adrenocortical hyperplasia (37).

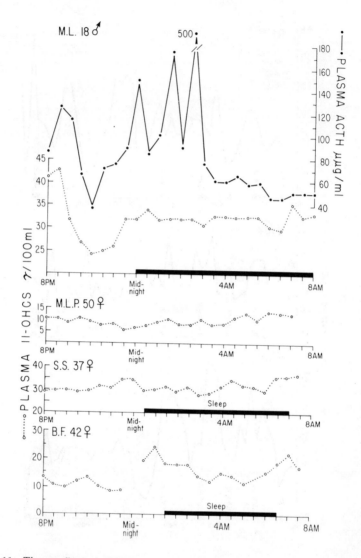

Figure 11 The top figure indicates the irregular nocturnal oscillations (cf. Figure 7) of plasma ACTH and corticosteroid levels in a patient with active Cushing's syndrome. M.L.P. is a patient in clinical remission following unilateral adrenalectomy and pituitary irradiation who still shows a flat corticosteroid pattern; in patient S.S. Cushing's syndrome is associated with a parasellar tumor, patient B.F. is the same as in Figure 10 restudied on another occasion, demonstrating the lack of reproducibility of pattern on separate studies, in contrast to that seen in normal subjects.

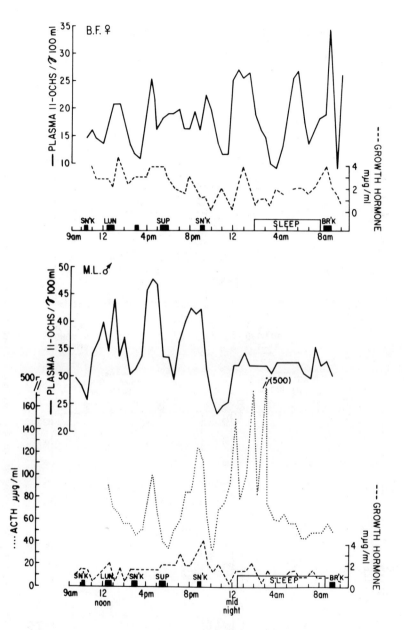

Figure 12 Periodicity of plasma cortisol and growth hormone levels as determined by half-hourly sampling in two patients with active Cushing's syndrome secondary to bilateral adrenal hyperplasia (55).

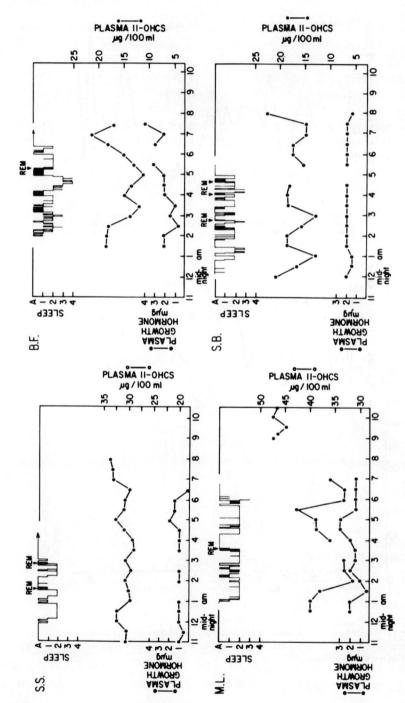

Figure 13 Representative graphs depicting growth hormone and cortisol levels as determined by half-hourly sampling in conjunction with sleep electroencephalographic recording in four patients with clinically active Cushing's syndrome (55).

If our hypothesis is correct that Cushing's syndrome associated with bilateral adrenal cortical hyperplasia is secondary to a functional alteration in neuroendocrine regulatory pathways, then one might expect alterations in the secretion of other pituitary hormones in Cushing's syndrome if similar pathways or transmitters are involved in their regulation. Mention has already been made of the nocturnal rise of plasma *growth hormone* levels in normal subjects which occurs in temporal association with the electroencephalographic stage of slow wave sleep. To date we have studied sleep EEG patterns and the nocturnal periodicity of growth hormone levels in six patients with clinically active Cushing's syndrome associated with bilateral adrenocortical hyperplasia (Figures 12 and 13). None of these patients, in studies performed over three successive nights, exhibited any significant sleep associated rise in plasma growth hormone levels, and all showed no, or markedly reduced, percentages of stage III—IV sleep.

It would appear that these changes in sleep EEG and growth hormone patterns are not secondary to the elevated cortisol levels seen in these patients. We have not (60) been able to suppress either the nocturnal elevation of plasma growth hormone levels or stage III—IV sleep by the acute administration of large doses of hydrocortisone. Preliminary data on two subjects receiving doses of corticosteroid equivalent to 60 mg daily of hydrocortisone over a 2- to 8-year period indicate preservation of stage III—IV sleep in these subjects. In other studies on three patients with Cushing's syndrome in clinical remission, nocturnal growth hormone elevation is still absent, and in two of the three, stage III—IV sleep still remains markedly suppressed. This would indicate that the observed changes in sleep EEG and nocturnal plasma growth hormone levels in the subjects with active disease may be secondary to the disease process and not to the hypercortisolemia per se.

CIRCADIAN PERIODICITY OF PLASMA CORTICOSTEROID AND GROWTH HORMONE LEVELS IN BLINDNESS

The patients comprising this group were specifically chosen for study because of blindness secondary to *intraocular* diseases of varying etiology. Patients with blindness secondary to central nervous system disease were excluded in view of the previously cited alterations in hormone periodicity in such patients. The results of such studies, however, are of relevance in a discussion of biorhythms in central nervous system disease because of the following consideration:

1. Light is known to be a potent factor in the initiation and/or maintenance of many biological rhythms (61).

2. There is evidence that light is involved in the anatomical maturation of central nervous system pathways subserving vision (62) and may perhaps be involved in changes in other pathways. The latter is suggested by the occurrence of gonadal changes in animals exposed to constant light (63), the alteration in pineal enzyme content under different light-dark conditions (64), and the reports of earlier onset of menses in congenitally blind females (65).

We have found (66) in studies of 19 blind subjects that absence of light perception (total or partial), irrespective of etiology of blindness, duration, or age of onset of blindness, is associated with abnormal cir-

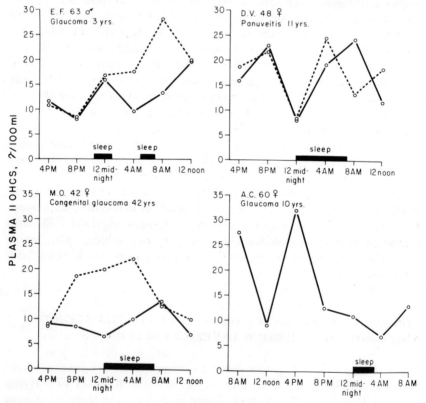

Figure 14   Different circadian patterns of plasma corticosteroid levels in subjects with total absence of light perception. Day 1:——; Day 2:- -. E.F.: abnormal pattern first day, normal second day. M.O.: normal pattern first day, abnormal second day, D.V.: abnormal both days (66).

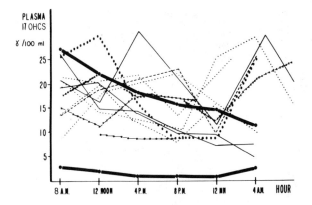

Figure 15 Composite of circadian patterns of plasma corticosteroid levels in subjects with absent or decreased light perception. The upper heavy line depicts 2 SD above the normal mean and the lower heavy line depicts 2 SD below the normal mean (66).

cadian periodicity of plasma corticosteroid levels. This abnormality consists of the occurrence of abnormal secondary peaks during the course of the day or flattening of the circadian pattern, as well as lack of day to day reproducibility of circadian patterns. In all but two of 12 subjects with total lack of light perception, the early morning rise of plasma corticosteroid levels still occurred during the latter part of sleep. Abnormal patterns were encountered more frequently in female subjects (six of seven) than in male subjects (six of 11) (Figures 14 and 15). Although several subjects had abnormal sleep-activity cycles characterized by decreased nocturnal sleep and increased napping during the day, these abnormalities did not recur in regular cycles and were not correlated with the presence or absence of abnormal circadian patterns of plasma corticosteroid levels.

We next attempted to determine whether the absence of a transition from light to dark rather than the absence of light was a factor in the abnormal circadian patterns encountered in the blind subjects. These studies were conducted on normal subjects and were also designed to dissociate the effect of awake-sleep transition from light-dark transition (67). Subjects were studied before and after a 21-day period of constant light exposure on two separate occasions; with subjects having normal day-night activity patterns and then with phase reversal.

Normal circadian patterns of corticosteroid levels with appropriate shifting on phase reversal were obtained during the studies performed at the end of the constant light period. This would indicate that neither

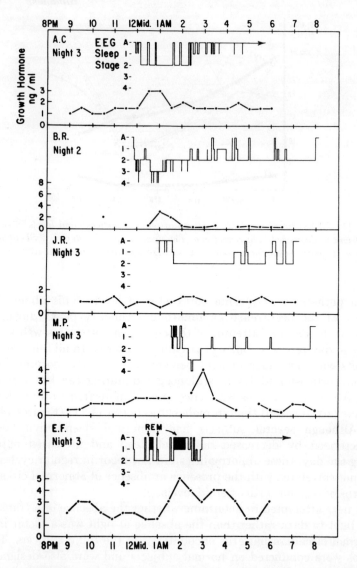

Figure 16 Sleep EEG recording and plasma growth hormone levels in 5 blind subjects. A.C.: female, 60 years, bilateral prostheses, duration of blindness 10 years. B.R.: male, 24 years, bilateral prostheses, duration of blindness 8 years. J.R.: male, 37 years, retinal albinism and choroideremia, partial light perception, duration of blindness 28 years. M.P.: male, 45 years, congenital uveitis, absent light perception, duration of blindness 21 years. E.F.: male, 63 years glaucoma, absent light perception, duration of blindness 3 years (74).

darkness nor a light-dark transition is necessary for the circadian rise in plasma steroid levels to occur. These studies, considered with the observations of a usually normal morning rise in plasma corticosteroid levels in blind subjects, would suggest that the sleep-wake transition is more important than the light-dark transition in the genesis of this aspect of adrenal circadian periodicity.

Orth and Island (68) have also demonstrated abnormal periodicity and lack of reproducibility of circadian patterns of corticosteroid levels in two of three subjects with ocular blindness, whereas Migeon et al. (16) reported normal patterns in six blind patients (etiology unspecified) studied, however, at irregular intervals over one 24-hour period. Animal studies have also reported alteration of circadian periodicity of plasma corticosteroid levels under conditions of either constant light (69, 70), constant dark (71), or blindness (72).

Studies on growth hormone periodicity in blind subjects have been more limited. Such studies were carried out concomittant with sleep EEG recording in five subjects with ocular blindness (73). A reproducible absence of the nocturnal peak of plasma growth hormone levels was observed (Figure 16) as well as a decreased amount of slow wave sleep (stage III—IV), the latter being noted in subjects in their third and fourth decades, so that age associated decreases in stage III—IV sleep (74) cannot be invoked as an etiology for this finding. Insulin hypoglycemia produced a normal rise in plasma growth hormone levels in three subjects so tested. Further studies are indicated to assess the role of defective light perception in the genesis and interrelation of the alterations observed.

Circadian periodicity of other variables has been reported to be altered in blind subjects and with altered lighting regimens. These include alteration of amplitude of urinary volume patterns (16), alteration or absence of the circadian pattern of urinary electrolyte excretion (75), and a delay in the usual rise of both urine flow and sodium and potassium excretion in normal subjects kept in darkness for 3 hours after awakening (76).

SUMMARY

It would appear that the data available to date strongly support the thesis that regulatory "centers" for biorhythms reside within the central nervous system. It is also apparent that in disease of the central nervous system alterations in the periodicity of a given parameter may occur in association with normal periodicity of other parameters. This would suggest the existence of "multiple clocks" and multiple regulatory mechanisms. The need for further study is also apparent. A number of the questions in this field remain unanswered: the role of sleep-wake and light-dark transition; the role of light itself in the genesis and maintenance of periodicity; the anatomical areas involved in the regulation of each parameter known to manifest circadian periodicity (as well as those parameters for which circadian periodicity still remains to be demonstrated); the neurotransmitters involved in these regulatory pathways both with regard to the afferent limb whereby cues are recognized and the efferent limb whereby changes in levels of bodily constituents are effected; the mechanism whereby central nervous system disease disrupts periodicity and the physiological effects of such disruption; as well as the physiological significance of normal periodicity. These are but a few of the areas to be explored.

ACKNOWLEDGMENTS

Supported in part by U.S.P.H.S. Grant NB–02893, Contract DA-1800-106-AMC-147(A) U.S. Army Chemical Center Procurement Agency, Edgewood Arsenal, Maryland; U.S.P.H.S. Research Grant FR–71 from the Division of Research Facilities and Grants; and a grant from the National Society for the Prevention of Blindness.

The excellent technical assistance of Anthony Liotta is gratefully acknowledged.

REFERENCES

1. G. S. Frank, in *Report of the Thirty-ninth Conference on Pediatric Research*, (S. J. Fomon, Ed.), Columbus Ross Laboratories, 1961, p. 48.

2. L. E. Scheving, W. H. Harrison, P. Gordon, and J. E. Pauly, *Amer. J. Physiol.*, **214**, (1968), 166.

3. D. J. Reis, M. Weinbren, and A. Corvelli, *J. Phar. Exp. Ther.*, **164**, (1968), 135.

4. W. B. Quay, *Amer. J. Physiol.*, **215**, (1968), 1448.

5. V. Scapagnini, G. P. Moberg, G. R. VanLoon, J. DeGroot, and W. F. Ganong, *Neuroendocrinology*, **7**, (1971), 90.

6. Y. Takahashi, D. Kipnis, and W. H. Daughaday, *J. Clin. Invest.*, **47**, (1968), 2079.

7. C. Faiman and J. S. Winter, *J. Clin. Endocr.*, **33**, (1971), 186.

8. B. B. Saxena, H. Demura, H. M. Gandy, and R. E. Peterson, *J. Clin. Endocr.*, **28**, (1968), 519.

9. D. T. Krieger, R. Ossowski, M. Fogel, and W. Allen, *J. Clin. Endocr.*, **35**, (1972), 619.

10. D. T. Peterson, Jr., A. R. Midgley, and R. B. Jaffe, *J. Clin. Endocr.*, **28**, (1968), 1473.

11. M. A. Kirshner, M. B. Lipsett, and D. R. Collins, *J. Clin. Invest.*, **44**, (1965), 657.

12. J. A. Resko and K. B. Eik-Nes, *J. Clin. Endocr.*, **26**, (1966), 573.

13. J. L. Evans, A. M. MacLean, A. A. Ismail, and D. Love, *Proc. Royal. Soc. Med.*, **64**, (1971), 841.

14. J. F. Sassin, A. G. Frantz, S. Kapen, and E. D. Weitzman, *Science*, **177**, (1973), 1205.

15. L. Vanhaelst, E. VanCauter, J. P. DeGaute, and J. Goldstein, *J. Clin. Endocr.* **35**, (1972), 479.

16. C. J. Migeon, F. H. Tyler, J. P. Mahoney, A. A. Florentin, H. Castle, E. L. Bliss, and L. T. Samuels, *J. Clin. Endocr.*, **16**, (1956), 622.

17. G. T. Perkoff, K. Eik-Nes, C. A. Nugent, H. L. Fred, R. A. Nimer, L. Rush, L. T. Samuels, and F. H. Tyler, *J. Clin. Endocr.*, **19**, (1959), 432.

18. V. Critchlow, R. A. Liebelt, M. Bar-Sela, W. Mountcastle, and H. Lipscomb, *Amer. J. Physiol.*, **205**, 5 (1963), 807.

19. F. Ungar and F. Halberg, *Experientia*, **19**, (1963), 158.

20. J. W. Mason, G. F. Mangan, D. G. Conrad, and J. V. Brady, *Proc. 39th Meet. Endocr. Soc. N.Y., Trans. N. Y. Acad. Sci.*, 1957, Abstract.

21. C. T. Nichols and F. H. Tyler, *Ann. Rev. Med.*, **18**, (1967), 313.

22. G. Seiden and A. Brodish, *Endocrinology*, **90**, (1972), 1401.

23. M. A. Slusher, *Amer. J. Physiol.*, **206**, (1964), 1161.

24. B. Halasz, M. A. Slusher, and R. A. Gorski, *Neuroendocrinology*, **2**, (1967), 43.

25. G. P. Moberg, V. Scapagnini, J. DeGroot, and W. F. Ganong, *Neuroendocrinology*, **1**, (1971), 11.

26. J. Seggie and G. M. Brown, *Neuroendocrinology*, **8**, (1971), 367.

27. M. A. Slusher, *Exp. Brain Res.*, **1**, (1966), 184.

28. D. T. Krieger, A. I. Silverberg, F. Rizzo, and H. P. Krieger, *Amer. J. Physiol.*, **215**, (1968), 915.

29. D. T. Krieger and F. Rizzo, *Amer. J. Physiol.*, **217**, (1969), 1703.

30. D. T. Krieger, S. Glick, A. Silverberg, and H. P. Krieger, *J. Clin. Endocr.*, **28**, (1968), 1589.

31. R. L. Ney, N. Shimuzu, W. E. Nicholson, D. Island, and G. W. Liddle, *J. Clin. Invest.*, **42**, (1963), 1669.

32. P. J. Martel, P. W. Sharp, S. A. Slovach, and H. J. Vipond, *J. Endocr.*, **24**, (1962), 159.

33. E. B. Flink and R. P. Doe, *Proc. Soc. Exp. Biol. Med.*, **100**, (1962), 498.

34. R. E. Peterson and J. B. Wyngaarden, *Ann. N. Y. Acad. Sci.*, **61**, (1955), 297.

35. S. A. Berson and R. S. Yalow, *J. Clin. Invest.*, **47**, (1969), 2725.

36. D. T. Krieger, *Trans. N. Y. Acad. Sci.*, Series 11, **32**, 3 (1970), 316.

37. D. T. Krieger, W. Allen, F. Rizzo, and H. P. Krieger, *J. Clin. Endocr.*, **32**, (1971), 266.

38. A. Silverberg, F. Rizzo, and D. T. Krieger, *J. Clin. Endocr.,* **28,** (1968), 1661.

39. C. Locke and S. M. Glick, *J. Clin. Endocr.,* **32,** (1971), 729.

40. A. G. Frantz and M. T. Rabkin, *J. Clin. Endocrinol.,* **25,** (1965), 1470.

41. E. T. J. Merimee, J. A. Burgess, and D. Rabinowitz, *J. Clin. Endocrin.,* **26,** (1966), 791.

42. D. T. Krieger, *J. Clin. Endocr.,* **21,** (1961), 695.

43. D. T. Krieger and H. P. Krieger, *Endrocines and the Central Nervous System,* Association for Research in Nervous and Mental Diseases, **43,** (1966), 400.

44. D. T. Krieger and H. P. Krieger, *J. Clin. Endocr.,* **26,** (1966), 929.

45. J. W. Finkelstein, J. Kream, A. Ludan, and H. Hellman, *J. Clin. Endocr.,* **35,** (1972), 13.

46. H. J. Quabbe, H. Helge, and S. Kubicki, *Acta Endocr.,* **67,** (1971), 767.

47. J. Roth, S. M. Glick, R. S. Yalow, and S. A. Berson, *Metabolism,* **12,** (1963), 577.

48. L. G. Wesson, Jr., *Medicine,* **43,** (1964), 547.

49. R. Horton and A. Michelakis, *Clin. Res.,* **17,** (1969), 143, (Abstract).

50. J. B. Dossetor, H. H. Gorman, and J. C. Beck, *Metabolism,* **13,** (1964), 1439.

51. D. T. Krieger and H. P. Krieger, *Metabolism,* **16,** (1967), 815.

52. H. Cushing, *The Pituitary Body and Its Disorders,* Lippincott, Philadelphia, 1912.

53. P. Heinbecker, *Medicine,* **23** (1944), 225.

54. G. Mangili, M. Motta, and L. Martini, in *Neuroendocrinology,* Vol. 1 L. Martini and W. F. Ganong, Eds.), Academic Press, New York, p. 297.

55. D. T. Krieger and S. M. Glick, *Amer. J. Med.,* **52,** (1972), 25.

56. D. T. Krieger, *M. Sinai J. Med.,* **40,** (1973), 302.

57. A. E. Lindsay, C. J. Migeon, C. A. Nugent, and H. Braun, *Amer. J. Med.,* **20,** (1956), 15.

58. R. P. Doe, J. A. Vennes, and E. B. Flick, *J. Clin. Endocr.,* **20,** (1960), 253.

59. H. Ekman, B. Hokansson, J. D. McCarthy, J. Lehman, and B. Sjogren, *J. Clin. Endocr.,* **21,** (1961), 684.

60. D. T. Krieger, J. Albin, S. Paget, and S. M. Glick, *Hormone Metab. Res.,* **4,** (1972), 463.

61. J. Aschoff, *Ann. Rev. Physiol.,* **25,** (1963), 581.

62. T. N. Wiesel and D. H. Hubel, *J. Neurophysiol.,* **26,** (1963), 978.

63. V. M. Fiske, *Endocrinology,* **29,** (1941), 187.

64. R. J. Wurtman, in *Neuroendocrinology,* Vol. 2 (L. Martini and W. F. Ganong, Eds.), Academic Press, New York, 1967, p. 19.

65. L. Zacharias and R. J. Wurtman, *Science,* **144,** (1964), 1154.

66. D. T. Krieger and F. Rizzo, *Neuroendocrinology,* **8,** (1971), 165.

67. D. T. Krieger, J. Kreuzer, and F. Rizzo, *J. Clin. Endocr.,* **29,** (1969), 1634.

68. D. N. Orth and D. P. Island, *J. Clin. Endocr.,* **29,** (1969), 479.

69. V. Critchlow, in *Advances in Neuroendocrinology* (A. Nalbandov, Ed.), Univ. Illinois Press, Urbana, 1963, p. 377.

70. P. Cheifetz, N. Garrud, and J. F. Dingman, *Endocrinology,* **82,** (1968), 1117.

71. L. E. Sheving and J. E. Pauly, *Amer. J. Physiol.,* **210,** (1966), 1112.

72. P. Saba, A. Carnicelli, G. C. Saba, G. Maltinti, and V. Morescotti, *Acta Endocrin.*, **49**, (1965), 289.

73. D. T. Krieger and S. Glick, *J. Clin. Endocr.*, **33**, (1971), 847.

74. J. D. Kales, A. Jacobsen, M. J. Paulson, E. Kollar, and R. D. Walker, *J. Amer. Geriat. Soc.*, **15**, (1967), 405.

References

CHAPTER 42

Cyclic Changes in Integrated Multiple Unit Activity in the Female Rat Brain

JORGE A. COLOMBO, DAVID I. WHITMOYER, and CHARLES H. SAWYER

Department of Anatomy and Brain Research Institute, University of California School of Medicine, Los Angeles, California

Neuroendocrine phenomena have been observed to undergo periodic or cyclic changes of varying frequencies. Among other examples are the approximately 12- and 24-hour fluctuations in circulating prolactin levels under certain physiological conditions (1, 2) and both a circhoral or pulsatile and an approximately 24-hour periodicity in luteinizing hormone values in the ovariectomized rat (3, 4). More complex hormonally-linked events such as motor activity (5) and feeding behavior (6) undergo daily changes as well as periodicities temporally more closely related to the estrous cycle. In considering these relationships it is tempting to speculate about a possible adaptive role of a particular element (e.g., gonadotrophin) or process (e.g., motor activity) undergoing fluctuations at different frequencies. In spontaneously ovulating laboratory animals general attention has been attracted to periodicities occurring in two principal time domains: the 24-hour day and the duration of the sexual cycle. In both circumstances,

651

day length is a crucial factor (e.g., ovulation blocked on proestrus occurs 24 hours late).

Looking at brain-hormonal interactions through the window of electrophysiological output, one is often forced to admit to a somewhat discouraging degree of uncertainty as to possible causal relationships. Several difficulties contribute to the problem of which the intricacies of neuronal networks and hormonal interactions represent only part of the picture. In spite of these difficulties, great advances have been accomplished in this field. The question of interpretation has been treated by other workers in this field, and we would merely like to call attention to one operational factor: the different time constants with which electrical signals and hormonal changes usually do occur. As part of our work at UCLA we have been studying the occurrence of "spontaneous" periodic changes in the integrated multiple unit activity (MUA) from different areas of the brain in cycling rats. Our aims were first to establish the existence of changes appearing at different frequencies and then to define and characterize their periodicities. Second, we proposed to study their possible relationships with hormonal events already described as well as with the effects of applying certain hormonal variables. Part of this work is still in progress. We shall report here the current state of this study which represents a continuation of an investigation started by Terkel et al. (7, 8).

METHODS AND RESULTS

The experimental design consisted of simultaneously recording, from permanently implanted electrodes in unanesthetized freely moving rats, integrated MUA from two different brain locations, together with cortical EEG and movement artifacts. The "chronic" electrodes used for the MUA recording were similar to those described by Johnson et al. (9). The MUA record represents an integration of both amplitude and firing rate of a given population of cells. Adult female rats were submitted to a daily schedule of 14 hours light—10 hours of darkness. One week after surgery the animals were placed in an acoustically isolated recording chamber. Daily vaginal smears were taken in most cases, usually between 9 A.M. and 10 A.M. In some animals smears were omitted to avoid introducing other time cues and disturbances. Recordings were taken continuously for periods of up to 40 to 50 days. Samples of the recorded MUA activity were extracted only during the slow wave sleep stage. Several samples were taken on or near (± 10

minutes) the hour and averaged for input to a spectral analysis and plotting program with a PDP 12 computer.

Although our sampling method did not enable us to look accurately at frequencies higher than about one per 6 hours, it was obvious that the recorded activity across time did have fluctuations of less than a 6-hour period. Filtering was accomplished by looking at the spectrum of a sequence of 10 or more days, erasing the dc component and frequencies higher than one per 12 hours, and resynthesizing the wave form with the remaining frequencies. With this procedure a 24-hour fluctuation was more clearly seen (Figures 1 and 2). The presence of

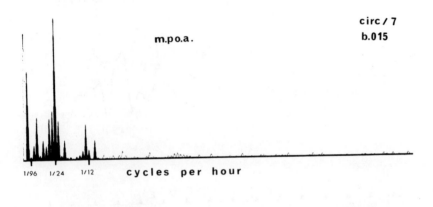

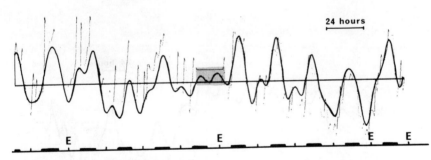

Figure 1 Plot of 10 days of integrated MUA from the medial preoptic area of a cycling female rat (Circ/7) sampled each 60 minutes during slow wave sleep. Bottom record: Time scale showing light-dark (black bars) periods and noon (vertical dash. The E stands for the estrous stage of the cycle. Middle record: Raw data (light plot) and resynthesized waveform (heavy plot). The shaded area indicates lack of data during that period. Top record: Spectral analysis. Heavy plot indicates the components kept to resynthesize the waveform, omitting dc and faster frequencies.

this rhythmicity was further supported by the spectral display in which a 24-hour-like component was consistently present in all regions of the brain so far studied. These include the medial preoptic area, dorsal hippocampus, frontal cortex, and septum. The activity reaches its nadir between 5 and 8 A.M. and slowly rises thereafter to peak between 11 P.M. and 1 A.M. (Figure 3). No significant differences could be seen in this respect when comparing the medial preoptic area, the frontal cortex, and dorsal hippocampus. In addition, the spectrum also disclosed components corresponding to a 3 to 5-day periodicity.

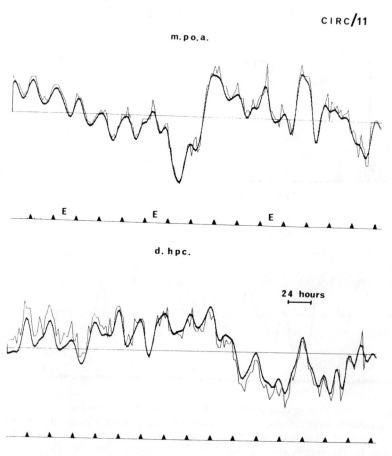

Figure 2   Plot of 16 days of integrated MUA sampled each 60 minutes and averaged for each 2 hours during slow wave sleep in a cycling female rat (Circ/11). Upper record: Raw data (light plot) and resynthetized waveform (heavy plot) from the medial preoptic area. Lower record: Simultaneous recording from the dorsal hippocampus. Time scale: Solid ▲'s indicate midnight. The E stands for the estrous stage of the cycle.

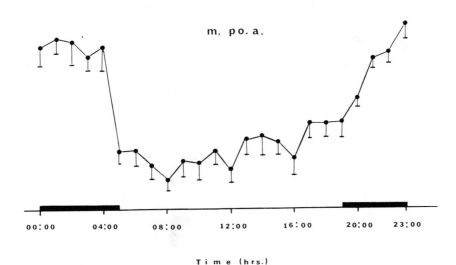

Figure 3 Averaged multiple unit activity levels in the medial preoptic area through-out a composite day representing samples taken during slow wave sleep records every hour for 17 consecutive days. Dark and light periods are included on the time axis.

In summary, a daily rhythmic change in level of integrated MUA from different areas of the brain is described. The results also suggest to 3 to 5-day periodicity in this activity. We are currently investigating the effects of light schedule on these periodic changes as well as their relationships with some known hormonal events.

ACKNOWLEDGMENTS

Supported by grants from NIH (NS 01162) and The Ford Foundation. The PDP 12 computer was used through the courtesy of the Data Processing Laboratory of the Brain Research Institute, supported by NIH grant NS 02501. The technical and secretarial help of Virginia Hoover and Frances Smith is gratefully acknowledged.

REFERENCES

1. M. E. Freeman and J. D. Neill, *Endocrinology,* **90,** (1972), 1292.
2. R. L. Butcher, N. W. Fugo, and W. E. Collins, *Endocrinology,* **90,** (1972), 1125.

3. V. L. Gay and N. A. Sheth, *Endocrinology,* **90,** (1972), 158.

4. I. E. Lawton and S. W. Smith, *Amer. J. Physiol.,* **219,** (1970), 1019.

5. G. B. Colvin and C. H. Sawyer, *Neuroendocrinology,* **4,** (1969), 309.

6. M. F. Tarttelin and R. A. Gorski, *Physiol. Behav.,* **7,** (1971), 847.

7. J. H. Johnson, J. Terkel, D. I. Whitmoyer, and C. H. Sawyer, *Anat. Rec.,* **169,** (1971), 348.

8. J. Terkel, J. H. Johnson, D. I. Whitmoyer, and C. H. Sawyer, *Physiologist,* **14,** (1971), 243.

9. J. H. Johnson, J. A. Clemens, J. Terkel, D. I. Whitmoyer, and C. H. Sawyer, *Neuroendocrinology,* **9,** (1972), 90.

Index